Handbook of Sustainable Materials

Handbook of Sustainable Materials presents recent developments in sustainable materials and how these materials interact with the environment. It highlights the recent advancements involved in proper utilization of sustainable materials, including chemical and biological approaches. With chapters written by global experts, the book offers a guide and insights into sustainable materials from a variety of engineering disciplines. Each chapter provides in-depth technical information on the sustainable materials theory and explores synthesis strategies, green materials, and artificial intelligence. The book considers applications in sectors such as aerospace, automobile, and biomedical for rapid prototyping and customized production without negative environmental impacts. It features research outcomes and case studies of optimization and modeling techniques in practice.

Features:

- Presents recent developments in sustainable materials from various engineering fields and industry applications.
- Emphasizes analytical strategies, computational, and simulation approaches to develop innovative sustainable materials.
- Discusses an artificial intelligence approach, rapid prototyping, and customized production.

This book is designed for researchers and professionals working with sustainable materials, clean manufacturing, and environmental impacts.

Handbook of Sustainable Materials
Modelling, Characterization, and Optimization

Edited by
Ajay, Parveen, Sharif Ahmad,
Jyotsna Sharma, and Victor Gambhir

CRC Press is an imprint of the
Taylor & Francis Group, an **informa** business

Designed cover image: Shutterstock | Rost9.

First edition published 2023
by CRC Press
6000 Broken Sound Parkway NW, Suite 300, Boca Raton, FL 33487-2742

and by CRC Press
4 Park Square, Milton Park, Abingdon, Oxon, OX14 4RN

CRC Press is an imprint of Taylor & Francis Group, LLC

© 2023 selection and editorial matter, Ajay, Parveen, Sharif Ahmad, Jyotsna Sharma, Victor Gambhir; individual chapters, the contributors

Reasonable efforts have been made to publish reliable data and information, but the author and publisher cannot assume responsibility for the validity of all materials or the consequences of their use. The authors and publishers have attempted to trace the copyright holders of all material reproduced in this publication and apologize to copyright holders if permission to publish in this form has not been obtained. If any copyright material has not been acknowledged please write and let us know so we may rectify in any future reprint.

Except as permitted under U.S. Copyright Law, no part of this book may be reprinted, reproduced, transmitted, or utilized in any form by any electronic, mechanical, or other means, now known or hereafter invented, including photocopying, microfilming, and recording, or in any information storage or retrieval system, without written permission from the publishers.

For permission to photocopy or use material electronically from this work, access www.copyright.com or contact the Copyright Clearance Center, Inc. (CCC), 222 Rosewood Drive, Danvers, MA 01923, 978-750-8400. For works that are not available on CCC please contact mpkbookspermissions@tandf.co.uk

Trademark notice: Product or corporate names may be trademarks or registered trademarks and are used only for identification and explanation without intent to infringe.

ISBN: 9781032286327 (hbk)
ISBN: 9781032286334 (pbk)
ISBN: 9781003297772 (ebk)

DOI: 10.1201/9781003297772

Typeset in Times LT Std
by KnowledgeWorks Global Ltd.

Contents

Preface...ix
About the Editors xiii
Contributors xvii

Chapter 1 Strategic Evaluation and Selection for Energy-Effective
Natural Fibers as Alternative Sustainable Insulation
Materials for Green Building Envelope 1

Figen Balo and Lutfu S. Sua

Chapter 2 Sustainable Biodegradable and Bio-based Polymers 19

Bhargav Patel, Bhagwan Toksha, and Prashant Gupta

Chapter 3 Processing of Recycled and Bio-based Plastics 39

Yash Rushiya, Bhagwan Toksha, and Prashant Gupta

Chapter 4 Terpene-based Polymers as Sustainable Materials 55

Eksha Guliani and Christine Jeyaseelan

Chapter 5 Sustainable Biodegradable and Bio-based Materials......................... 71

*Mridul Umesh, Suma Sarojini, Thazeem Basheer,
Sreehari Suresh, Adhithya Sankar Santhosh,
Liya Merin Stanly, Sneha Grigary, and Nilina James*

Chapter 6 Sustainability of Materials: Concepts, History, and Future 93

Rajat Dhawan, Hitendra K. Malik, and Amit Kumar

Chapter 7 Experimental Investigation on Durability Properties of
Concrete Incorporated with Fly Ash... 111

V. G. Meshram, A. M. Pande, and B. P. Nandurkar

Chapter 8 Investigation of Electrode Materials Used in Electrocoagulation
Process for Wastewater Treatment 135

*Mukul Bajpai, Surjit Singh Katoch, Akhilesh Nautiyal,
Rahul Shakya, and Samriddhi Sharma*

v

vi Contents

Chapter 9 Enhancing Mechanical Properties of Adobe by Use of
Vernacular Fibers of Agave Americana and Eulaliopsis Binata
for Sustainable Traditional Mud Houses of Himachal Pradesh 151

Vandna Sharma and Aniket Sharma

Chapter 10 Green Synthesis of Cerium Oxide Nanoparticles and Its Recent
Electrochemical Sensing Applications: A Study............................ 177

Satyajit Das and Partha Pratim Sahu

Chapter 11 Self-lubricating Hybrid Metal Matrix Composite toward
Sustainability.. 193

Sweta Rani Biswal and Seshadev Sahoo

Chapter 12 E-Plastic Waste—A Sustainable Construction Material:
An Indian Perspective 213

Sweta Sinha and Ateeb Hamdan

Chapter 13 Green Synthesis of Sustainable Materials: A Stride toward a
Viable Future.. 233

Mandeep Kaur, Gagandeep Kaur, and Divya Sareen

Chapter 14 Green Materials: Synthesis and Characterization........................... 259

Rajeev Kumar and Jyoti Chawla

Chapter 15 Waste Sewage Sludge Concrete as a Sustainable Building
Construction Material: A Review 273

Jagdeep Singh

Chapter 16 Applications of Sustainable Materials 301

Rajat Dhawan, Hitendra K. Malik, and Davoud Dorranian

Chapter 17 Heusler Alloys: Sustainable Material for Sustainable Magnetic
Refrigeration.. 323

Chhayabrita Maji

Chapter 18 Chemical Synthesis or Phytofabrication of Metal/Metal Oxide
Nanoparticles.. 341

Amulya Giridasappa, Ismail Shareef M., and Gopinath S. M.

Contents

Chapter 19 Antioxidant, Bactericidal, Antihemolytic, and Anticancer Assessment Activities of Al_2O_3, SnO_2, and Green Synthesized Ag and CuO NPs 367

Amulya Giridasappa, Ismail Shareef M., and Gopinath S. M.

Chapter 20 Sustainability of Wind Turbine Blade: Instantaneous Real-Time Prediction of Its Failure using Machine Learning and Solution based on Materials and Design 399

Lohit Malik, Gurtej Singh Saini, Mayand Malik, and Abhishek Tevatia

Chapter 21 A Self-sustained Machine Learning Model to Predict the In-flight Mechanical Properties of a Rocket Nozzle by Inputting Material Properties and Environmental Conditions 431

Lohit Malik, Gurtej Singh Saini, and Abhishek Tevatia

Chapter 22 Heat-Assisted Dieless Sheet Forming Techniques for Hard-to-Form Materials 457

Nikhil Kadian, Rakesh Rathee, and Ajay

Index 469

Preface

Sustainable materials are the materials used throughout our consumer and industrial economy that can be produced in required volumes without depleting nonrenewable resources and without disrupting the established steady-state equilibrium of the environment and key natural resource systems. The detailed description of various sustainable materials like polymers, metals, ceramics, composites, biomaterials, biodegradable materials, smart materials, functionally graded materials and sustainable novel materials etc. is included in this book. The subterranean details like sustainable material characterization techniques, modeling of materials, role of artificial intelligence, Industry 4.0, nature inspired algorithms and optimization possibilities, and various computational and simulation approaches to maintaining the sustainability of materials are covered in this book. In addition to this, various case studies of sustainable materials, especially related to medical, environmental, production/mechanical, and civil engineering, are covered to help the undergraduate and postgraduate students as well as research scholars. This compilation will bring to its readers the latest developments/advancements in sustainable materials, which is currently not available at one place. This book will bridge the gap between R&D in sustainable materials and professionals.

This book is an outcome of the extensive research accomplished by various researchers, academicians, scientists, and industrialists in sustainable materials. The area is under-explored and the outcomes are worth the research effort. Experimentation, modeling, characterization, and simulation techniques are the powerful tools for developing new concepts, approaches, and solutions to devise valuable information on the process, which led to the need of compiling this work. Since the information related to sustainable materials is scattered into patents and research publications and not at one place in a systematic form, editors recognize their ethical responsibility to compile, share, and spread the knowledge accumulated and technology developed, with the students, researchers, and industry people, to draw the benefits of this work in the form of a book and gain technical competence in the frontal area.

The book consists of 22 chapters that describe "Sustainable Materials: Modelling, Characterization, and Optimization" in different aspects. Chapter 1, "Strategic Evaluation and Selection for Energy-Effective Natural Fibers as Alternative Sustainable Insulation Materials for Green Building Envelope," intends to explore natural fiber-based insulation materials in building envelope, their advantages over traditional materials, the challenges and opportunities in relation to their use. Chapter 2, "Sustainable Biodegradable and Bio-based Polymers," focuses on various prospects of sustainable biodegradable plastic concerning social, economic, and environmental stability. Chapter 3, "Processing of Recycled and Bio-based Plastics," focuses on advancements and recent trends in the processing of recyclable polymers and bio-based plastics, challenges and opportunities in the processing, and application of sustainable plastics. Chapter 4, "Terpene-Based Polymers as Sustainable Materials," focuses on the synthesis, characterization, modifications, and applications of

ix

terpene-based sustainable polymers. Chapter 5, "Sustainable Biodegradable and Bio-based Materials," summarizes the important biodegradable and bio-based materials of commercial importance along with their production methods and applications in diverse sectors. Chapter 6, "Sustainability of Materials: Concepts, History and Future," covers the introduction, concept, and history of sustainability along with the applications of sustainability of materials with future prospects and challenges in a detailed manner. Chapter 7, "Experimental Investigation on Durability Properties of Concrete Incorporated with Fly Ash," represents an experimental examination into the sustainability features of fly ash concrete. Chapter 8, "Investigation of Electrode Materials Used in Electrocoagulation Process for Wastewater Treatment," aims at how different anode materials can be used in electrocoagulation units for water and wastewater treatment. Chapter 9, "Enhancing Mechanical Properties of Adobe by Use of Vernacular Fibers of Agave Americana and Eulaliopsis Binata for Sustainable Traditional Mud Houses of Himachal Pradesh," represents experimental investigation of vernacular natural fibers of Agave Americana, and Eulaliopsis Binata in the soil for increasing strength performance. Chapter 10, "Green Synthesis of Cerium Oxide Nanoparticles and Its Recent Electrochemical Sensing Applications: A Study," briefs on recent applications of electrochemical sensing of glucose, hydrogen peroxide, lactone, formalin, dopamine, cholesterol, uric acid, etc. Chapter 11, "Self-lubricating Hybrid Metal Matrix Composite towards Sustainability," introduces self-lubricating composites, which aim to eliminate the usage of external harmful petroleum-based lubricants in sliding contacts and become an energy-efficient sustainable solution for automotive applications. Chapter 12, "E-Plastic Waste—a Sustainable Construction Material: An Indian Perspective," aims to deliver a wholesome context on sustainable construction material by integration and transformation of wastes to value-added products. Chapter 13, "Green Synthesis of Sustainable Materials: A Stride towards a Viable Future," covers an up-to-date overview of the recent approaches and methodologies being adopted for the green synthesis of markedly important sustainable materials, i.e. polymers, metal, and metal oxide nanoparticles and ceramics. Chapter 14, "Green Materials: Synthesis and Characterization," presents different techniques for characterization of materials and available optimization tools in detail. Chapter 15, "Waste Sewage Sludge Concrete as a Sustainable Building Construction Material: A Review," aims at utilization of waste sewage sludge as a viable building material and contributes to waste management and sustainable development. Chapter 16, "Applications of Sustainable Materials," discusses detailed applications of sustainable materials in engineering, architecture, agriculture, aerospace, healthcare, biomedical, and green technology sectors. Chapter 17, "Heusler Alloys: Sustainable Material for Sustainable Magnetic Refrigeration," discusses the Co doped $Ni_2Mn_{1+x}SIn_{1-x}$ Heusler alloys as sustainable material for environment-friendly magnetic refrigeration technique. Chapter 18, "Chemical Synthesis or Phytofabrication of Metal/Metal Oxide Nanoparticles," demonstrates the usage of metal/metal oxide nanoparticles (NPs) synthesized by solution combustion using aluminum oxide (Al_2O_3) and tin dioxide (SnO_2) NPs along with green approach using leaf extract of *Simarouba glauca* (SG) and aerial extract of *Celastrus paniculatus* (CP). Chapter 19, "Antioxidant, Bactericidal, Antihemolytic, and Anticancer Assessment Activities of Al_2O_3, SnO_2, and Green Synthesized Ag and CuO NPs,"

Preface xi

discusses scavenging potential of aluminum oxide (Al_2O_3) and tin dioxide (SnO_2) NPs along with green synthesized silver (Ag) and cupric oxide (CuO) NPs prepared using extracts of *Simarouba glauca* (SG) and *Celastrus paniculatus* (CP) and their antibacterial activity against *E.coli*, *S. pyogenes*, and *P. aeruginosa* microbes, and also their antioxidant capability by ABTS, DPPH, and nitric oxide inhibition assays. Chapter 20, "Sustainability of Wind Turbine Blade: Instantaneous Real-Time Prediction of Its Failure Using Machine Learning and Solution Based on Materials and Design," presents the development of a Machine Learning (ML) model for predicting the failure of a Wind Turbine Blade (WTB) based on the design parameters, properties of material, and environmental conditions inputted. Chapter 21, "A Self-sustained Machine Learning Model to Predict the In-flight Mechanical Properties of a Rocket Nozzle by Inputting Material Properties and Environmental Conditions," describes a self-sustained machine learning model for in-flight prediction of mechanical properties and the gaps generated in a rocket nozzle based on the material properties and environmental conditions comprising bolt preload, gas pressure variation, forces from regeneration channels, and thermal condition inputted. Chapter 22, "Heat-Assisted Dieless Sheet Forming Techniques for Hard-to-Form Materials," aims to provide a review of different methodologies used by different researchers during HA-ISF to heat the work sheet to be deformed in the desired shapes and to study the effect of these HA-ISF techniques on forming forces to accomplish this viable process.

This book is intended for both the academia and the industry. The postgraduate students, Ph.D. students, and researchers in universities and institutions, who are involved in the areas of modeling, characterization, and optimization of sustainable materials, will find this compilation useful.

The editors acknowledge the professional support received from CRC Press and express their gratitude for this opportunity.

Readers' observations, suggestions, and queries are welcome.

Dr. Ajay
Mr. Parveen
Dr. Sharif Ahmad
Dr. Jyotsna Sharma
Dr. Victor Gambhir

About the Editors

Dr. Ajay is currently serving as an Associate Professor in Mechanical Engineering Department, School of Engineering and Technology, JECRC University, Jaipur, Rajasthan, India. He received his Ph.D. in the field of Advanced Manufacturing from Guru Jambheshwar University of Science & Technology, Hisar, India after B.Tech. (Hons.) and M.Tech. (Distinction) from Maharshi Dayanand University, Rohtak, India. His areas of research include Artificial Intelligence, Materials, Incremental Sheet Forming, Additive Manufacturing, Advanced Manufacturing, Industry 4.0, Waste Management, and Optimization Techniques. He has over 60 publications in international journals of repute, including SCOPUS, Web of Science and SCI indexed database, and refereed international conferences. He has also co-authored the textbook *Incremental Sheet Forming Technologies: Principles, Merits, Limitations, and Applications* (CRC Press, Taylor & Francis) and has recently edited a book entitled *Advancements in Additive Manufacturing: Artificial Intelligence, Nature Inspired and Bio-manufacturing* (Elsevier). He has organized various national and international events including an international conference on Mechatronics and Artificial Intelligence (ICMAI-2021) as conference chair. He has more than 15 national and international patents to his credit. He has supervised more than eight M.Tech., Ph.D. scholars, and numerous undergraduate projects/theses and has total 13 years of experience in teaching and research. He is a Guest Editor and Review Editor of reputed journals, including *Frontiers in Sustainability*. He has contributed to many international conferences/symposiums as a session chair, expert speaker, and member of editorial boards. He has won several proficiency awards during the course of his career, including merit awards, best teacher awards, and so on.

He has been adviser of Association of Engineers and Technocrats (AET) and has also authored many in-house course notes, lab manuals, monographs, and invited chapters in books. He has organized a series of faculty development programs, international conferences, workshops, and seminars for researchers, Ph.D., U.G., and P.G. level students. His areas of research include additive manufacturing, dieless sheet forming, and intelligent manufacturing systems. He teaches the following courses at the graduate and post-graduate level: Additive Manufacturing, Manufacturing Technology, Smart Manufacturing, Advanced Manufacturing Processes, Material Science, CAM, Operations Research, Optimization Techniques, Engineering Mechanics, Computer Graphics, Design of Experiments and Research Methodology, and so on. He is associated with many research, academic, and professional societies in various capacities.

Mr. Parveen is currently serving as an Assistant Professor in the Department of Mechanical Engineering, Rawal Institute of Engineering and Technology, Faridabad, Haryana, India. Currently, he is pursuing a Ph.D. from the National Institute of Technology, Kurukshetra, Haryana, India. He completed his B.Tech. (Hons.) from Kurukshetra University, Kurukshetra, India and M.Tech. (Distinction) in Manufacturing and Automation from Maharshi Dayanand University, Rohtak, India.

xiv About the Editors

His areas of research include Materials, Dieless Forming, Additive Manufacturing, CAD/CAM, and Optimization Techniques. He has over 20 publications in international journals of repute, including SCOPUS, Web of Science and SCI indexed database, and refereed international conferences. He has five national and international patents to his credit. He has supervised two M.Tech. scholars and numerous undergraduate projects/theses. He has a total of 12 years of experience in teaching and research. He has organized a series of faculty development programs, workshops, and seminars for researchers and UG level students. He is associated with many research, academic, and professional societies in various capacities.

Dr. Sharif Ahmad is currently working as Pro Vice Chancellor (Research & Consultancy) at SGT University, Gurugram, NCR-Delhi, India and is a Superannuated Professor of Materials Chemistry. He completed his Masters (1976) and Doctoral (1981) in Chemistry at AMU, Aligarh. In recognition of his work, he was nominated as a member to the prestigious American Chemical Society for three years (2015–2018) and has a membership of the Royal Chemical Society, UK. Professor Ahmad has been actively engaged in teaching and research in the frontier areas of Materials Chemistry for the last 40 years. He has made ground breaking contributions, particularly in vegetable oil-based sustainable biodegradable polymers, conducting polymers, hydrogels, hyper branched polymers, polymer-based anti-corrosive coating materials, organic/inorganic nanomaterials, etc. To his credit, Professor Ahmad has more than 238 research papers, four books, and 23 book chapters published in prestigious journals and books with Google Scholar Citations of 8972, h-index 52, and i10-index 167 (till 4 September, 2021). He has guided 31 Ph.D. students through their doctoral work in his laboratory at JMI, New Delhi. Currently, he is guiding four scholars as a co-supervisor. He has contributed his work to 51 national and 52 international conferences. He was a pioneer in designing, developing, and initiating a PG course in Materials Chemistry in 1985 when the said branch was not even formally recognized across the globe. He has been instrumental in starting the two year Unani Pharmacy Course along with the Ph.D. program in Unani Medicine. Further, in order to attract bright faculty (Foreign/National Fellows), a "Multidisciplinary Centre for Advance Research and Studies" was also developed at JMI, New Delhi during the tenure of his deanship.

Professor Ahmad received a research grant worth Rs. 1.32 crore, of which Rs. 68 lakh are from various R&D agencies, including DRDO (ARDB-NRB), Nuclear Science Center, CSIR, and UGC. He has successfully executed and completed these sponsored research projects. Currently, as co-P.I., he is running a DBT project of more than Rs. 64 lakh. Professor Ahmad has been serving as the subject expert in national committees of prestigious academic bodies of various institutions and universities (e.g., IIT, Delhi, NIT Srinagar & Allahabad, BHU, DU, AMU). Besides this, he is also Ph.D. examiner in these universities and acted as subject expert in interviews at CSIR, UPSC, and Uttarakhand PSC. Professor Ahmad has long administrative experience and remains actively engaged in university administration and various national academic committees. He has been the Head, Department of Chemistry, JMI, New Delhi (2009–2012); Dean, Faculty of Natural Sciences (2015–2018), JMI, New Delhi; and acted as Superintendent for UG/PG examinations and entrance tests for different courses at JMI. He has also

About the Editors

performed the job of Deputy Proctor (2002–2005); Assistant Proctor (2001–2002); and Tabulator of UG/PG Examinations (2002, 2004). He has been nominated as a member of the Board of Studies of many departments, faculty committees, academic councils, and executive councils, councils of court and finance committees in JMI and other universities.

His pioneering studies led to the development of volatile organic compounds-free green biodegradable low molecular weight polymers/oligomers from vegetable seeds, which are still underutilized. Vegetable seed oil has wider applications in non-conventional biodiesel and polymer industries. His work on oil-based polyester amides, epoxy, polyurethane, and alkyds has potential scope in industries. Professor Ahmad has also developed various oil-based polyols as well as oleo-polyol techniques for the synthesis of metallic, bimetallic, and trimetallic nano-magnetic ferrites, which have significant applications in the field of anticorrosive coatings, EMI shielding, water treatment, and hydrogels pertaining to drug delivery and the green polymer industry. Furthermore, waterborne alkyds/polyesters are being developed and studied as high performance anti-corrosive coating materials.

Dr. Jyotsna Sharma is currently working as Associate Professor at Amity University Haryana, Gurugram (since July 2018). She obtained her Ph.D. degree in Physics from Jawaharlal Nehru University, New Delhi and completed her Bachelor's and Master's at Kurukshetra University, Kurukshetra. She has been actively engaged in teaching and research for more than 15 years. Her areas of interest include Materials, Experimental and Theoretical Studies of Synthesis and Characterization of Nanomaterials, and Ion Beam Interaction with Plasma/Dusty Plasma. To her credit, she has published more than 30 research papers in international journals of high repute, about 15 research papers in international/national conferences, published two books with Pearson Asia, two book chapters, five patents and one MoU with National Sun-yat Sen University, Taiwan operational from March 2020. She has visited many countries, including the USA, Spain, France, Singapore, and Canada, to present her research work. She has guided two research scholars in their Ph.D. degrees, and currently, she is guiding six students for Ph.D. (four are international students from Taiwan) and many students for UG and PG project work. She has completed (2017–2021) a research project sanctioned from SERB, DST, India of worth Rs. 22 Lakh. She has also organized many international/national conferences/workshops/seminars and is an active member of many professional bodies like IEEE, Physical Society of India, and APCBEES. At AUH, she is a Ph.D. coordinator of ASAS, which includes many departments like Physics, Chemistry, Forensic Science, Biochemistry, and Mathematics. She has many other administrative responsibilities within the department including being a Board of Studies member and program and lab coordinator of B.Sc (H) Physics.

Professor Victor Gambhir is presently working as President/Vice Chancellor and Professor in the Mechanical Engineering Department at JECRC University, Jaipur, Rajasthan (since January 2021). He worked as Vice Chancellor and Professor in the Mechanical Engineering Department at SGT University Gurugram, Haryana from May 2020 to December 2020 and at Maharishi Markandeshwar (deemed to be University), Mullana, Ambala, Haryana from November 2016 to May 2020. He also worked as President/Vice Chancellor and Professor in the Mechanical Engineering Department at JECRC University, Jaipur, Rajasthan from June 2014 to November 2016. Prior to that, he worked as Pro-Vice Chancellor/Director Planning & Coordination and Professor in the Mechanical Engineering Department at Manav Rachna International Institute of Research and Studies from September 2006 to May 2014. He also worked as Workshop Superintendent at YMCA Institute of Engineering (now J. C. Bose University Science & Technology), Faridabad, Haryana from May 2005 to September 2006. Earlier, he worked as Joint Director, Technical Education Department, Government of Haryana from April 1998 to May 2005 and Deputy Director, Technical Education Department, Government of Haryana from April 1991 to June 1994. He has also worked as Head of the Department, Senior Lecturer, Workshop Superintendent and Lecturer in Government Polytechnics under the Department of Technical Education, Government of Haryana from December 1985 to March 1991 and July 1994 to April 1998. Prior to that, he worked in industry from June 1982 to May 1985 in the area of Engineering Projects and Pipelines.

He obtained his Bachelor degree in Mechanical Engineering from the Thapar Institute of Engineering & Technology, Patiala (Punjabi University Patiala) in 1982 and Master degree in Mechanical Engineering with specialization in Rotodynamic Machines from Punjab Engineering College, Chandigarh (Punjab University Chandigarh) in 1998. He obtained his Doctoral Degree (Ph.D.) in Quality in Engineering Education from Manav Rachna International University (now Manav Rachna International Institute of Research and Studies, deemed to be University), Faridabad. He obtained his Diploma in Management from IGNOU, New Delhi in 1989 and attended a Foreign Fellowship Program sponsored by the World Bank at Northern Melbourne Institute of TAFE, Victoria, Australia from January 1998 to April 1998.

He has published five textbooks for polytechnic students in the field of Applied Mechanics, Thermodynamics, Hydraulics and Hydraulic Machines, and Engineering Graphics. He has guided two Ph.D. dissertations. He has published nine research papers in international journals (six SCOPUS indexed) and six research papers in international/national conferences held by IISC Bangalore, IIT Delhi, and YMCA Institute of Engineering, Faridabad. He has been awarded two patents as co-inventor. He is a Member of the American Society for Mechanical Engineers, Life Member of the Indian Society for Technical Education, and Fellow Member of the Institution of Engineers (India).

Contributors

Mukul Bajpai
Department of Civil Engineering
National Institute of Technology Hamirpur
Hamirpur, Himachal Pradesh, India

Figen Balo
Department of METE
Engineering Faculty
Firat University
Elazığ, Turkey

Thazeem Basheer
Waste Management Division
Integrated Rural Technology Centre
Palakkad, Kerala, India

Sweta Rani Biswal
Department of Mechanical Engineering
Institute of Technical Education and
    Research
Siksha O Anusandhan (Deemed to be
    University)
Bhubaneswar, Odisha, India

Jyoti Chawla
Department of Applied Sciences
Manav Rachna International Institute of
    Research and Studies
Faridabad, Haryana, India

Satyajit Das
Department of Electronics and
    Communication Engineering
Tezpur University
Tezpur, Assam, India

Rajat Dhawan
Plasma Science and Technology
    Laboratory
Department of Physics
Indian Institute of Technology Delhi
New Delhi, India

Davoud Dorranian
Plasma Physics Research Center
Science and Research Branch
I. Azad University
Tehran, Iran

Amulya Giridasappa
Department of Biotechnology
Acharya Institute of Technology
Bengaluru, Karnataka, India

Sneha Grigary
Department of Life Sciences
CHRIST (Deemed to be
    University)
Bangalore, Karnataka, India

Eksha Guliani
Amity Institute of Applied
    Sciences
Amity University Uttar Pradesh
Noida, Uttar Pradesh, India

Prashant Gupta
Centre for Advanced Materials
    Research and Technology
Department of Plastic and Polymer
    Engineering
Maharashtra Institute of
    Technology
Aurangabad, India

Ateeb Hamdan
Department of Civil Engineering
Amity University Jharkhand
Ranchi, Jharkhand, India

Nilina James
Department of Life Sciences
CHRIST (Deemed to be University)
Bangalore, Karnataka, India

Christine Jeyaseelan
Amity Institute of Applied Sciences
Amity University Uttar Pradesh
Noida, Uttar Pradesh, India

Nikhil Kadian
University Institute of Engineering &
  Technology (UIET)
Maharshi Dayanand University
Rohtak, Haryana, India

Surjit Singh Katoch
Department of Civil Engineering
National Institute of Technology
  Hamirpur
Hamirpur, Himachal Pradesh, India

Gagandeep Kaur
Department of Chemistry
Sri Guru Teg Bahadur Khalsa College
Sri Anandpur Sahib, Punjab, India

Mandeep Kaur
Department of Chemistry
Sri Guru Teg Bahadur Khalsa College
Sri Anandpur Sahib, Punjab, India

Ajay Kumar
Department of Mechanical
  Engineering
School of Engineering and
  Technology
JECRC University
Jaipur, Rajasthan, India

Amit Kumar
Department of Physics
Amity School of Applied Sciences
Amity University Haryana
Gurugram, Haryana, India

Rajeev Kumar
Department of Applied Sciences
Manav Rachna International Institute of
  Research and Studies
Faridabad, Haryana, India

Gopinath S. M.
Department of studies in
  Biotechnology
Davanagere University
Shivagangotri, Davanagere, India

Ismail Shareef M.
Department of Biotechnology
Acharya Institute of Technology
Bengaluru, Karnataka, India

Chhayabrita Maji
School of Applied Sciences
Centurion University of Technology and
  Management
Bhubaneswar, Odisha, India

Hitendra K. Malik
Plasma Science and Technology
  Laboratory
Department of Physics
Indian Institute of Technology Delhi
New Delhi, India

Lohit Malik
Department of Mechanical and
  Aerospace Engineering
Princeton University
Princeton, New Jersey

Mayand Malik
School of Engineering
Indian Institute of Technology Mandi
Mandi, Himachal Pradesh, India

V. G. Meshram
Department of Civil Engineering
Yeshwantrao Chavan Collage of
  Engineering
Nagpur, Maharashtra, India

B. P. Nandurkar
Deparment of Civil Engineering
Yeshwantrao Chavan Collage of
  Engineering
Nagpur, Maharashtra, India

Contributors

Akhilesh Nautiyal
Department of Civil Engineering
National Institute of Technology
  Hamirpur
Hamirpur, Himachal Pradesh, India

A. M. Pande
Department of Civil Engineering
Yeshwantrao Chavan Collage of
  Engineering
Nagpur, Maharashtra, India

Bhargav Patel
Department of Plastic and Polymer
  Engineering
Maharashtra Institute of
  Technology
Aurangabad, Maharashtra, India

Rakesh Rathee
University Institute of Engineering &
  Technology (UIET)
Maharshi Dayanand University
Rohtak, Haryana, India

Yash Rushiya
Department of Plastic and Polymer
  Engineering
Maharashtra Institute of
  Technology
Aurangabad, Maharashtra, India

Seshadev Sahoo
Department of Mechanical
  Engineering
Institute of Technical Education and
  Research
Siksha O Anusandhan (Deemed to be
  University)
Bhubaneswar, Odisha, India

Partha Pratim Sahu
Department of Electronics and
  Communication Engineering
Tezpur University
Tezpur, Assam, India

Gurtej Singh Saini
Department of Manufacturing
  Processes and Automation
  Engineering
Netaji Subhas University of
  Technology (formerly Netaji
  Subhas Institute of
  Technology)
New Delhi, India

Adhithya Sankar Santhosh
Department of Life Sciences
CHRIST (Deemed to be
  University)
Bangalore, Karnataka, India

Divya Sareen
Department of Chemistry
Faculty of Science
SGT University
Gurugram, Haryana, India

Suma Sarojini
Department of Life Sciences
CHRIST (Deemed to be
  University)
Bangalore, Karnataka, India

Rahul Shakya
Department of Civil Engineering
National Institute of Technology
  Hamirpur
Hamirpur, Himachal Pradesh, India

Aniket Sharma
Department of Architecture
National Institute of Technology
  Hamirpur
Hamirpur, Himachal Pradesh,
  India

Samriddhi Sharma
Department of Civil Engineering
National Institute of Technology
  Hamirpur
Hamirpur, Himachal Pradesh, India

Vandna Sharma
Department of Architecture
National Institute of Technology
  Hamirpur
Hamirpur, Himachal Pradesh, India

Jagdeep Singh
Chitkara University Institute of
  Engineering and Technology
Chitkara University
Rajpura, Punjab, India

Sweta Sinha
Department of Chemistry
Amity University Jharkhand
Ranchi, Jharkhand, India

Liya Merin Stanly
Department of Life Sciences
CHRIST (Deemed to be University)
Bangalore, Karnataka, India

Lutfu S. Sua
Department of Management and
  Marketing
Southern University and A&M College
Baton Rouge, Louisiana,

Sreehari Suresh
Department of Life Sciences
CHRIST (Deemed to be
  University)
Bangalore, Karnataka, India

Abhishek Tevatia
Department of Mechanical
  Engineering
Netaji Subhas University of Technology
  (formerly Netaji Subhas Institute of
  Technology)
New Delhi, India

Bhagwan Toksha
Centre for Advanced Materials
  Research and Technology
Department of Electronics and
  Telecommunication Engineering
Maharashtra Institute of Technology
Aurangabad, Maharashtra, India

Mridul Umesh
Department of Life Sciences
CHRIST (Deemed to be
  University)
Bangalore, Karnataka, India

1 Strategic Evaluation and Selection for Energy-Effective Natural Fibers as Alternative Sustainable Insulation Materials for Green Building Envelope

Figen Balo and Lutfu S. Sua

CONTENTS

1.1 Introduction 1
    1.1.1 Importance of Alternative Sustainable Insulation Materials 2
    1.1.2 Energy-Effective Natural Fibers Used in the Analysis 5
1.2 Selection Criteria and Methodology 5
    1.2.1 Selection Criteria for Natural Fibers Used in Insulation
    Materials 5
    1.2.2 Analytic Hierarchy Process Methodology to Assess
    the Criteria Weights 8
1.3 Discussion and Results 10
1.4 Conclusion 13
References 14

1.1 INTRODUCTION

The need for sustainable insulation materials, eco-friendly nature, and environment is reflected by the research findings indicating that constructions will emit more than 4×10^7 kg of C emissions and exhaust one-third of the worldwide water reserves and annual power demand by the 2030s [1–3]. The building sector supplies employment opportunities indirectly or directly and encourages the improvement of the urbanization process and national economies [4]. Thus, structures have a significant influence both on the economy and ecology. Nevertheless, building structure and material production will give rise to 40 percent of overall emission pollution [5] and roughly one-third of gases in the atmosphere affecting the greenhouse phenomenon [6]. The building sector constitutes a key pressure on energy consumption.

DOI: 10.1201/9781003297772-1

1.1.1 Importance of Alternative Sustainable Insulation Materials

The term "green building" refers to measures taken to maximize the efficiency of material, water, and energy use while reducing the structure's environmental impact. When viewing a structure from outside, the insulation may not be one of the most striking features. The most significant component, however, is the building's insulation, which has the effect of making the building more energy-efficient and comfortable. Inadequate building insulation causes energy loss in the course of cooling and heating processes, resulting in a system that is inefficient. Insulation in green building reduces energy usage while also assisting in the regulation of the habitants' health. Inadequate insulation causes problems that are dangerous for health, like mold. For these reasons, building insulation is a critical issue that must be addressed in a number of ways. The majority of heat losses occur on the most exposed components of residential buildings, like roofs, foundations, floors, and walls. Insulation not only limits heat loss in cold weather and keeps surplus heat out in hot weather, but it also provides comfort inside. To improve comfort and save environment, green building studies need to be taken into consideration. Green building means both the process implementation and a structure that are environmentally responsible and resource-efficient over the course of a structure's life cycle from design to planning, process structure, upkeep demolition, and renovation [7, 8].

Alternative sustainable insulation materials are important to be used. Insulation materials are in a variety of shapes and sizes to suit a variety of purposes [9]. Green building insulation is the healthiest type of insulation for humans and also the most environment-friendly alternative. Though it is not technically a kind of insulation, it is a viewpoint that permits a structure to be more cost-effective, comfortable, and long-lasting from its conception to its everyday use.

If inside walls are insulated, the structure may be shielded from the external elements, heating benefits will be reduced, and more thermal bridges will be existent. On the other side, with insulation on outer walls, the construction may promptly store solar power with low resistance, and heating and cooling impacts will be effective in winter and summer, respectively [10, 11]. Insulation materials operate through preventing the flow of heat, which is gauged through the material's R parameter (the better the insulation, the greater the R parameter). The insolation's R value is determined through its thickness, density, and kind. This research emphasizes the significance of insulation materials with natural fiber in building in the long-term energy performance of green buildings [9]. As a result, local and natural resources with minimal waste are utilized to safeguard sustainability and degrade pollution during and after structure [12–14]. In a nutshell, insulation is a critical component of energy conservation and a green construction's cornerstone. Buildings that are insufficiently or completely uninsulated lose a lot of energy. The term "green insulation material" refers to a type of insulation material that must meet the following criteria: less or no utilization of energy and natural resources; the utilization of clean fabrication technic; a significant quantity of utilization of agricultural, industrial, or municipal solid-waste fabrication with low-pollution characteristics, environmental protection, human health, and recyclability. In this study, the natural fibers within thermal insulation materials are investigated.

Strategic Evaluation and Selection for Energy-Effective Natural Fibers 3

The thermal insulation layer is of critical significance for heat and cold resistance in a building envelope. Thermal insulation can be found everywhere around a structure, and thousands of different materials can be utilized to provide insolation. Natural insulating materials are discussed in depth in this section due to their high potential for long-term sustainability.

Thermal insulation materials are generally made from petroleum-based chemicals in buildings. These materials' manufacturing processes will pollute the environment. In addition, there are numerous issues with the reuse and recovery of industrial products. Developing natural-sourced and environmentally friendly insulation materials promises to be a very important way of constructing maintainable buildings that have a restricted effect on ecology. Several typical materials, such as vegetable-wood, paper, and animal product, are addressed as natural-sourced materials, although the phrase mainly refers to current materials that have undergone numerous in-depth processes. Novel natural-sourced products offer possibilities to utilize them in buildings. Natural-sourced building materials can be quickly renovated through photosynthesis. It has been monitored that this material is an unmatched option not only because it supplies thermal comfort with lesser energy usage for the building's function to change the conventional materials but also helps to the degradation of the energy emissions and carbon. Though some industrial products, such as extruded polystyrene (XPS), glass wool, rock wool, expanded polystyrene (EPS), and other products, have good performance, natural insulating alternatives have a better future. Locally available agricultural waste and other natural fiber composites could be used to replace the building's reinforcement and insulation systems [15]. There are different strategies that can be used for sustainable structures; e.g. natural fiber can be utilized as a raw material for insulation, durability, and abrasion resistance. In literature, coconut pitch, paddy straw, groundnut shell, and maize husk are reported to have the minimum heat conductivity of all natural insulation materials, even lesser than EPS (0.024 $W.m^{-1}.K^{-1}$). Pineapple leaves and fibers have a lesser heat conductivity (0.045 $W.m^{-1}.K^{-1}$) than glass insolation (foamed). The phase change material (0.080 $W.m^{-1}.K^{-1}$) has a higher thermal conductivity than 20 different natural materials. Jute, flax, and hemp have the minimum intensity of lesser than 50 $kg.m^{-3}$. Except for durian, rice hulls, corn peel, and sansevieria fiber, most natural materials have a lesser heat conductivity than ceramic foamed plate (0.1 $W.m^{-1}.K^{-1}$). The density of banana bunches and Sansevieria fiber is greater than 1000 kg/m^3. Instead of being burned, natural bark sheets were shown to have less form-aldehyde emissions and enough heat conductivity as thermal insulation materials. Most natural materials have a lesser density than cement (about 1250 kg/m^3) [16, 17]. Fiberboard made from cotton stalks can be utilized for ceilings and walls [18]. Without the chemical adhesive utilization, bagasse and coconut husk could be turned into less intensity thermally insulating panels [19]. The qualities of date palm wood for building thermal insulation [20], maize cob particle sheet [21], and sugarcane, coconut, and oil palm fiber [22] were found to be appropriate for building insulation constructions. Natural fiber insulation materials, particularly flax, jute, and hemp, can be utilized to construct roofs and external plant walls [23]. Particle panels made from tropical fruit peels could be used for specific purposes like wall and ceiling insulation. Coconut coir and durian peel fibers were used to make low-thermal-conductivity building panels [24]. Pineapple leaves, crushed pecan shells, and rice hulls have all been proven to have high thermal insulation potential [25, 26].

Different researchers have investigated natural-sourced fiber as insulation material and their performances within the sector. Analytic hierarchy process (AHP) methodology was widely employed in diverse material choice scenarios amongst the multi-criteria decision-making procedures (TOPSIS, AHP, and WPIM) [26]. The commercially available "Expert Choice" software can be used to efficiently implement AHP [27–30].

For road transport in Greece, AHP methodology was utilized to determine the optimum fuel option in terms of policy and cost [31]. TOPSIS and AHP methodologies have been used in several research works to determine the most useful ceramic waste to be utilized in place of traditional concrete in terms of environmental impact and compressive strength [32]. Dweiri et al. used AHP methodology to present a decision support model for the purpose of identifying the most effective supplier in the car sector [33]. Furthermore, they used sensitivity analysis to assess their decision. With varying criteria weights, sensitivity analysis verifies the AHP methodology results [34]. For car brake lever implementations, AHP methodology was used to pick the most effective design amongst the creative concepts in design for plastic composites, in addition to material selection [35]. AHP methodology was used to select natural fiber for use in car dashboards [36]. During the selection phase, they used physical and mechanical qualities of natural fiber as selection criteria. AHP methodology was used to choose the plant fiber for the vehicle spall liner [37]. Seven criteria were used, as well as fourteen natural fibers as alternatives. For automotive spall liners, they picked kenaf as a possible natural fiber for hybridization including Kevlar fiber. Salwa used the hierarchy process methodology to choose the most efficient natural fiber for packaging food out of nine options. They developed the model through a sensitivity analysis and found that the preference order determined from the AHP technique was steadfast [38]. Patnaik and his coworkers used the hierarchy process methodology to assess the weights of factors for choosing composites for construction implementations [39].

The fuzzy AHP methodology performed to supply weight values of the criteria for supplier selection, multi-criteria decision-making was used by Chatterjee et al. [40]. Depending on a series of design requirements generated from the novel material design parameters, Mansor et al. were successful in determining the optimal sort of plant-based fiber to be hybridized with fossil-based glass fiber toward the hybrid composites structure for grip brake pry use [27]. The investigation revealed that natural fiber of kenaf is the best suitable product for hybrid materials when employing the AHP technique.

In spite of the expanding requirement to utilize decision-making methodologies and other valued methodologies to help the accomplishment of more sustainability, few studies have been found that consider systematical strategies to suitably assess natural fiber reinforced wall insulation materials and their elements for ecological sustainable items. This study concentrates on different features and is used in area pertaining to materials selection. This research addresses the requirement for materials choice and presents an analysis of strategic evaluation and most effective natural fiber selection in materials selection within sustainable insulation material in building envelope.

Strategic Evaluation and Selection for Energy-Effective Natural Fibers 5

In this study, within the framework of these references, different natural fibers, which have been evaluated in the production of insulation material today have been investigated, whose physical and mechanical properties were tested. Then, among these natural fibers, the material that can provide the most effective thermal insulation to the building envelope in terms of insulating property has been researched. AHP method which enables a hierarchical analysis of multiple factors in the evaluation process was utilized in the research.

1.1.2 Energy-Effective Natural Fibers Used in the Analysis

The natural fiber is derived from natural resources present on our planet. In a latest research, the natural cellulose was the most commonly utilized substance. This is due to ecological regulations causing a paradigm shift in the material sector, as well as a surge in interest in natural fibers with the inclusion of flax, jute, hemp, coir, and sisal. Furthermore, when compared to mineral and animal fiber, cellulose fiber is easier to handle and treat for installation and experimental purposes. The combinations with plant fiber have been used in a range of industries, including food packaging, construction, medicine, and automotive [41–43]. The composites with natural fiber are made up of two different materials: a matrix and natural fiber. The researchers have experimented with many matrices and natural fiber combinations throughout the world. The fibers serve as reinforcement, while thermoplastic, thermoset, polypropylene, and polyester [44–46] are the most common matrix materials. The final material can produce good physical, mechanical, and ecological performance, and can be used to replace steel-based additives, especially in the car industry [47–49]. It also has benefits such as ease of processing, low weight and cost, and good acoustic and thermal insulating qualities [50–52].

There are ways in assembling the matrix and fiber. This would be an issue for planners in determining which products are best for optimizing a design's production process. Planners should choose products attentively since the features of combinations differ from those of metal-based materials, and the qualities of the resulting composites are dependent on elements for which some data is unavailable. The qualities of the combinations could also be customized through the constituents' characteristics. The properties of the most effective natural fibers investigated in this study and used in manufacturing insulation material in buildings are listed in Table 1.1.

1.2 SELECTION CRITERIA AND METHODOLOGY

1.2.1 Selection Criteria for Natural Fibers Used in Insulation Materials

In construction, building materials are the most significant components, and they are immediately exposed to people. Thermal insulation and external-internal wall materials are significant cost sources and the primary elements of the building [4]. Buildings without insulation, especially those in cold regions, have the greatest potential for energy savings because the first layer of insulation can save the most energy. It can also save a significant amount of energy from new buildings in improving nations that are not insulated. The quality and source of the building-insulation

TABLE 1.1
Properties of the Most Effective Natural Fibers Used in Manufacturing of Insulation Material

	Cellulose (%)	Hemicellulose (%)	Lignin (w %)	Microfibrillar Angle (○)	Moisture Content (wt.%)	Pectin (wt.%)	Young's Modulus (G.Pa)	Tensile Strength (M.Pa)	Density (g/cm³)	Therm. Cond. (W/Mk)
Cotton	87.5	5.7	NA	NA	8.17	NA	9.05	NA	1.55	0.234
Oil Palm Mesocarp	60	NA	11	46	NA	NA	0.5	NA	NA	0.179
Palm EFB	65	NA	19	42	NA	NA	3.2	248	NA	0.380
Abaca	59.5	NA	71.5	NA	7.5	1	NA	595	NA	0.298
PALF	76	NA	8.85	14	NA	NA	58.5	1020	NA	0.046
Henequen	77.6	6	13.1	NA	NA	NA	NA	NA	NA	0.134
Nettle	86	NA	NA	NA	14	NA	38	NA	NA	0.026
Ramie	72.4	14.9	0.65	7.5	12.25	1.9	94.7	669	1.50	0.135
Jute	66.25	17	12.5	8	13.1	0.2	19.75	51	1.37	0.111
Hemp	72	20.15	4.7	4.1	9.1	0.9	70	52	1.48	0.081
Flax	71	19.6	2.2	7.5	10	2.3	27.6	850	1.40	0.675
Sisal	69.5	15.35	12	16	16	10	7.9	344	1.42	0.043
Coir	38.73	6.57	40.18	39.5	8	3.5	3.5	83	1.15	0.101
Piassava	43.23	8.34	50	NA	NA	NA	1.82	61	1.57	0.213
Curauá	69	10	13	NA	NA	NA	41.7	620	1.42	0.197
Kenaf	51	21.5	11	NA	NA	4	30.8	1019	1.40	0.055

All the values obtained as maximum and minimum are used as arithmetical average.

material will affect the interior environment and also a building's cost [53]. The green building material's use is an innovative resolution for resource and energy saving throughout the process of construction [54]. The majority of the research focused on green buildings, but little on green building materials. Green building materials are used in all parts of construction. Thermal insulation and external-internal wall materials are significant structure materials in construction, but their alternatives are also researched in detail by scientists.

Selecting the right insulation materials is a vital stage in projecting green building insulation. A properly insulated building will keep the living area cooler in the summer and warmer in the winter, decreasing greenhouse gaseous emissions that contribute to worldwide climate change. Investment in insulation materials for a building is more profitable than an investment in costly calorific systems from the viewpoint of energy efficiency. The elements to be noted are compressive strength, fire resistance, durability, ease of application, water vapor transmission and absorption, thermal conductivity, and cost. Though there are numerous elements included, thermal efficiency is the most significant factor in choosing the insulation materials. Most of the structural elements don't have diverse all-thermal durability according to identical insulation thickness and type. The all-thermal capacitance can be identified without considering where the insulating material is kept; nonetheless, other topics could vary. Properly insulated buildings, on the other side, not only save operating costs and save energy, but also make spaces more healthy and comfortable for people. When construction practices and site conditions are not taken into consideration, comparing R values is one aspect to choose one kind of insulation over another. Insulation is a fundamental structural element that is troublesome to change after it has been installed. Though buildings can be modified with extra insulation, the process is costly and time-consuming. Suggested R values from the United States Department of Energy should be considered as basic minimals rather than maximals [15].

Because density is closely pertaining to the building mass, it is one of the most important parameters in fiber choice. Natural fiber density has a big impact on the strength-to-weight proportion of insulating materials. The insulation material must show development in density to obtain the lightweight in building envelope, that is, natural fiber must be low density. Insulation materials with a higher strength-to-weight rate have better mechanical qualities. The higher density fibers increase the building's overall weight. The natural fiber must also have a high-elongation value at break to ensure toughness. Tensile strength refers to the material's resistance to an externally performed tensile load in terms of field. In general, adding fiber to the matrix can improve the tensile strength of insulation materials since fiber has better stiffness and strength than the matrix [55]. Elastic modulus, often known as Young's modulus, is a measure of a material's stiffness. It is the change in stress induced or divided through a modification in strain in simple terms. The stronger the material, the higher the modulus is [56].

Compared to traditional glass wool, rock wool, etc., cellulose-based insulation offers superior heat resistance, instantly increasing the efficiency of building. Cellulose is also a more durable material and does not collapse, keeps pests away, and is not damaged by mold. It has the disadvantage of a lower thermal insulation value,

8   Handbook of Sustainable Materials

flammability, and high density compared to other insulation systems. Additionally, the natural fiber alternatives must have less humidity content to preserve from the cracks' formation in the insulation material, must have a high percentage of hemicellulose to be able to generate an ecologically beneficial product with good biodegradability, and must have good tensile strength for resilience. Lignin and wax increase durability [57]. The angle made by microfibrils with regard to the fiber axis is known as the microfibrillar angle. The microfibrillar angle is one of the most important factors that influences fiber strength. Higher cellulosic ingredient and a smaller microfibrillar angle are also necessary for good fiber strength [58]. The most important feature of pectin is its ability to form gels. Therefore, the amount of gelled pectin in natural fibers used for insulation material should be low [59].

1.2.2 ANALYTIC HIERARCHY PROCESS METHODOLOGY TO ASSESS THE CRITERIA WEIGHTS

To obtain both environmental-friendly design and consumer satisfaction, selecting the right material for a certain implementation has always been important in material investigation.

AHP methodology is utilized to create rate-scales from paired comparisons, both continuous and discrete. A fundamental scale or true measurements that depict the relative strength of sentiments and preferences can be used to make these comparisons. The AHP methodology is particularly concerned with departures from measurement, consistency, and dependencies within and across groupings of parts. Multi-criteria decision-making, resource allocation and planning, and dispute resolution are just a few of the areas where it has been used extensively [60–66].

In its most basic type, the AHP methodology is a nonlinear system for performing both inductive and deductive thinking without the utilization of the comparing syllogism through simultaneously considering a few elements, permitting feedback and dependence and exchanging numbers to reach a conclusion or synthesis. Generally, AHP is utilized to choose the best option for a specific project or application. The goal-setting phase of the AHP methodology is subsequently through the choice of options and criteria. The relative significance and the weighting of the criteria determine the final sorting of the options. One of the best statistical methodologies for analyzing option preferences utilizing criteria hierarchy is the AHP methodology [67]. Rather than delivering an appropriate decision, this technique assists in providing the best solution. It not only represents and quantifies problem components, but also connects them to larger goals and evaluates various resolutions [68]. The AHP methodology uses a hierarchy modeling to provide an optimal resolution to a multi-criteria decision-making problem, with aims at the top, followed by criteria and sub-criteria, and decision options at last.

Despite the fact that the AHP is one of the most sophisticated methodologies present in the area of operations research and management science, its complexity makes it difficult to employ. The software databases have been developed to automate the math-intensive component of the process. The engineer must follow a simple data gathering process, which is then fed into the program to obtain the desired results.

The process for doing so is as follows:

Stage 1: The AHP method starts with identifying the options to be assessed. These choices could represent many criteria against which solutions must be judged. They could be also the various properties of a material that require to be weighed in order to gain a better understanding of the customer's perspective. A full list of all possible options must be ready at the end of stage 1.

Stage 2: The next stage is to create a modeling of the issue. An issue, according to the AHP method, is a collection of connected subproblems. As a result, the AHP technique divides the problem into smaller problems. Criteria for evaluating solutions come as a result of breaking down to the subproblems. However, similar to root cause analysis, a person can dive deeper and deeper into the problem. It is a matter of subjective judgment when to stop dividing the problem down into smaller subproblems.

Stage 3: To construct a matrix, the AHP approach involves pairwise comparison.

Stage 4: To check, most simulation products that assist with AHP include a consistency stage. Data is inconsistent if they have imparted a weightiness of less than or more than one. Because inconsistent data yields inconsistent results, prevention is preferable to cure.

Stage 5: The simulation will use the data to perform an arithmetical computation and appoint notional weights to the criteria. Once the formula with weight of criteria is complete, one can assess the options to choose the best solution for their requirements.

AHP multi-criteria decision-making methodology is commonly utilized to find the best materials for a certain implementation [68]. Researchers have been pushed to create more reliable and faster techniques for comparing, searching, and choosing materials for diverse implementations owing to the enormous number of available materials with different features. For this reason, there is an increasing demand for diverse approaches. Selecting the right natural fiber composite materials and elements to maximize and improve desired attributes and effectiveness is seen to be a multi-criteria decision-making issue. As a result, dependable and methodical selection methodologies can be expanded to accommodate more sustainable plan alternatives. Convenient methodologies for natural fiber composite materials depend on the AHP methodology and are capable of generating handy decisions that would assist decision makers and designers in making the optimal selection of such composite materials based on their limitations and design criteria. This would aid in the creation of an adequate data for the choice of bio-based fibers and their composite materials to be used as major choice equipment for green composites designers. This study denotes also suitability of the AHP methodology is adequate for choosing bio-sourced materials in a certain working area. In addition, few thorough research works of selecting and assessing plant fiber composite produces and elements by using hierarchy process methodology are described as a reference for the reader for evaluating, choosing, and manufacturing correct plant fiber composite materials for a specific implementation.

The material selection process analyzed significant parameters based on the natural fibers, which is certain characterization properties on the area (for insulation materials of wall) are researched at the outset of the research.

1.3 DISCUSSION AND RESULTS

For the purpose of this research, ten factors are considered when evaluating a set of natural fibers and these criteria are compared based on the provided comparison scale. Table 1.2 provides the resulting decision matrix.

Through pairwise comparisons of the various factors, the following weights are calculated based on expert evaluations. The weights are presented in Figure 1.1. As the figure indicates, the thermal conductivity, which is a mechanical property of the fibers, is evaluated to be the most significant factor in evaluating the alternatives.

The final step of the methodology involves scoring the natural fiber alternatives based on each factor and multiplying the scores with the factor weights. The resulting values are called priority values. These values are presented in Table 1.3. The last row of Table 1.3 displays the total priority figures of each alternative.

Based on the total values, Figure 1.2 exhibits the attractiveness of the selected natural fibers. Figure 1.2 indicates that Cottonseed is the alternative with the highest total score, with a value of 0.048.

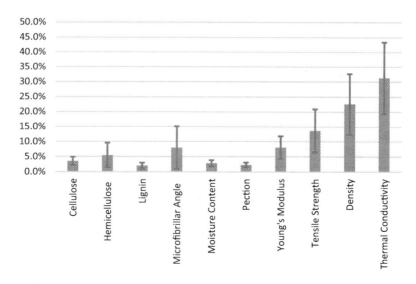

FIGURE 1.1 Factor weights.

TABLE 1.2
Decision Matrix

		Cellulose	Hemicellulose	Lignin	Microfibrillar Angle	Moisture Content	Pection	Young's Modulus	Tensile Strength	Density	Thermal Cond.	Normalized Principal Eigenvector (%)
		1	2	3	4	5	6	7	8	9	10	
Cellulose	1	1	1/4	3	1/5	2	2	1/2	1/3	1/5	1/9	3.54
Hemicellulose	2	4	1	3	1/5	5	3	1/3	1/5	1/7	1/7	5.52
Lignin	3	1/3	1/3	1	1/5	1/2	2	1/5	1/7	1/8	1/9	2.02
Microfibrillar Angle	4	5	5	5	1	2	2	1/2	1/4	1/6	1/7	8.02
Moisture Content	5	1/2	1/5	2	1/2	1	1	1/2	1/4	1/6	1/7	2.85
Pection	6	1/2	1/3	1/2	1/2	1	1	1/4	1/6	1/7	1/8	2.26
Young's Modulus	7	2	3	5	2	2	4	1	1/2	1/4	1/6	8.11
Tensile Strength	8	3	5	7	4	4	6	2	1	1/2	1/3	13.73
Density	9	5	7	8	6	6	7	4	2	1	1/2	22.59
Thermal Conduct.	10	9	7	9	7	7	8	6	3	2	1	31.35

TABLE 1.3
Priority Values

	Cellulose	Hemicellulose	Lignin	Microfibrillar Angle	Moisture Content	Pectin	Young's Modulus	Tensile Strength	Density	Thermal Conduct.	Scores
Castor	−0.028	0.001	0.005	−0.011	−0.006	−0.004	−0.001	−0.002	−0.004	−0.018	**−0.067**
Coconut	−0.019	0.002	0.003	−0.037	−0.004	−0.004	0.000	−0.002	−0.002	−0.014	**−0.077**
Corn	−0.021	0.002	0.005	−0.034	−0.004	−0.004	0.000	−0.002	−0.002	−0.029	**−0.088**
Cottonseed	−0.019	0.002	0.020	−0.011	−0.005	−0.002	−0.003	−0.004	−0.002	−0.023	**−0.048**
Crambe	−0.024	0.002	0.002	−0.011	−0.004	−0.004	−0.006	−0.007	−0.002	−0.004	**−0.057**
DF No.2	−0.024	0.001	0.004	−0.011	−0.004	−0.004	−0.003	−0.002	−0.002	−0.010	**−0.057**
Lard	−0.027	0.002	0.000	−0.011	−0.010	−0.004	−0.004	−0.002	−0.002	−0.002	**−0.060**
Linseed	−0.023	0.003	0.000	−0.006	−0.009	−0.004	−0.009	−0.004	−0.004	−0.010	**−0.066**
Olive	−0.021	0.003	0.003	−0.006	−0.009	0.000	−0.002	0.000	−0.003	−0.009	**−0.045**
Palm Oil	−0.023	0.004	0.001	−0.003	−0.007	−0.002	−0.007	0.000	−0.004	−0.006	**−0.046**
Peanut	−0.022	0.004	0.001	−0.006	−0.007	−0.004	−0.003	−0.006	−0.003	−0.052	**−0.099**
Rapeseed	−0.022	0.003	0.003	−0.013	−0.012	−0.019	−0.001	−0.002	−0.003	−0.003	**−0.069**
Safflower	−0.012	0.001	0.011	−0.032	−0.006	−0.004	0.000	−0.001	−0.003	−0.008	**−0.053**
Safflower (high–oleic)	−0.014	0.002	0.014	−0.011	−0.004	−0.004	0.000	0.000	−0.004	−0.016	**−0.038**
Sesame	−0.022	0.002	0.004	−0.011	−0.004	−0.004	−0.004	−0.004	−0.003	−0.015	**−0.062**
Soybean	−0.016	0.004	0.003	−0.011	−0.004	−0.008	−0.003	−0.007	−0.003	−0.004	**−0.050**

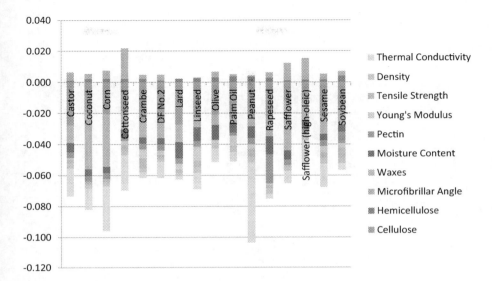

FIGURE 1.2 Factor priorities.

1.4 CONCLUSION

The right material choice plays a significant part in the manufacturing processes, acceptance, and the functionality of a novel product. Various needs and difficulties should be considered when selecting a material for a certain product. These topics include the product plan, safety, the production process, recyclability, functionality, energy needed for manufacturing, toxicity, human ergonomic, and the price or an evaluation of the product's life cycle at the conclusion of its useful life. Thus, natural fibers as designed functional insulation materials have different topics to be noted throughout their selection process.

Generation of green building elements in the insulation material sector could be used as the natural resources like plant and animal fibers. Natural fiber material selection for fabricating operation of insulation materials is a significant step and should be applied in parallel with the selection of insulation material fabricating operation. Recently, a number of multi-criteria decision-making methods are proposed to select the best materials for specific implementations. This research work deals with right natural fiber selection to be used as insulation materials for external walls.

For the purpose of this study, a number of criteria are included in the investigation by assigning them different weights based on their relative importance. Consequently, a set of alternative natural fibers are evaluated depending on the criteria's same class. The alternative with the highest score is presented as the optimum material to be used as an insulation material based on the set of criteria used here. Including new and diverse criteria as well as adjusting the criteria weights are the two venues for future research in this field.

REFERENCES

1. WBCSD. Energy efficiency inbuildings, business realities and opportunities. Geneva, Switzerland: The World Business Council for Sustainable Development (WBCSD), 2007.
2. United Nation Environment Program. Environment for development. Nairobi, Kenya: United Nation Environment Program, 1 July 2016.
3. Zhang Y., Wang J., Hu F. and Wang Y. Comparison of evaluation standards for green building in China, Britain, United States. Renew. Sustain. Energy Rev. 2017; 68: 262–71.
4. Zuo J. and Zhao Z. Y. Green building research—current status and future agenda: A review. Renew. Sustain. Energy Rev. 2014; 30: 271–81.
5. Ofori G., Briffett C., Gang G. and Ranasinghe M. Impact of ISO 14000 on construction enterprises in Singapore. Constr. Manag. Econ. 2000; 18: 935–47.
6. USEIA. International Energy Outlook 2010; U.S. Department of Energy, Ed. Washington, DC, USA: U.S. Energy Information Administration, 2010; Volume DC20585.
7. Hoffman A. J. and Henn R. Overcoming the social and psychological barriers to green building. Org. Environ. 2008; 21: 390–419.
8. The US Environmental Protection Agency. Green Buildings. Available online: https://archive.epa.gov/greenbuilding/web/html/about.html (accessed on 1 July 2021).
9. Özgür B. Insulation materials, Energy Generation and Efficiency Technologies for Green Residential Buildings, 2019.
10. Al-Homoud M. S. Performance characteristics and practical applications of common building thermal insulation materials. Build. Environ. 2005; 40: 357.
11. Kruger A. and Seville C. Green building: Principles and practices in residential construction, International edition. USA: Cengage Learning, 2013.
12. Building Design and Construction, "White Paper on Sustainability", p. 4, November 2006.
13. Office of the Federal Environment Executive, "The Federal Commitment to Green Building: Experiences and Expectations," 18 September 2003.
14. Johnston D. and Gibson S. Green from the ground up, sustainable, healthy and energy efficient home construction. Newtown, CT: Tounton Press Inc, 2008.
15. Roldan L. V., Perez L. G., Amores L. F. and Ibarra A. Potential use of vegetal biomass as insulation in extreme climates of Ecuador. Enfoque UTE. 2015; 6: 23–41.
16. Knapic S., Oliveira V., Machado J. S. and Pereira H. Cork as a building material: A review. Eur. J. Wood Prod. 2016; 74: 775–91.
17. Pasztory Z., Mohacsine I. R. and Borcsok Z. Investigation of thermal insulation panels made of black locust tree bark. Constr. Build. Mater. 2017; 147: 733–35.
18. Zhou X. Y., Zheng F., Li H. G. and Lu C. L. An environment-friendly thermal insulation material from cotton stalk fibers. Energy Build. 2010; 42: 1070–74.
19. Palomo A., Grutzeck M. W. and Blanco M. T. Alkali-activated fly ashes—A cement for the future. Cem. Concr. Res. 1999; 29: 1323–29.
20. Agoudjil B., Benchabane A., Boudenne A., Ibos L. and Fois M. Renewable materials to reduce building heat loss: Characterization of date palm wood. Energy Build. 2011; 43: 491–97.
21. Paiva A., Pereira S., Sá A., Cruz D., Varum H. and Pinto J. A contribution to the thermal insulation performance characterization of corn cob particleboards. Energy Build. 2012; 45: 274–79.
22. Manohar K. Experimental investigation of building thermal insulation from agricultural by-products. Br. J. Appl. Sci. Technol. 2012; 2: 227–39.

23. Korjenic A., Zach J. and Hroudová J. The use of insulating materials based on natural fibers in combination with plant facades in building constructions. Energy Build. 2016; 116: 45–58.
24. Khedari J., Charoenvai S. and Hirunlabh J. New insulating particleboards from durian peel and coconut coir. Build. Environ. 2003; 38: 435–41.
25. Yarbrough D. W., Wilkes K. E., Olivier P. A., Graves R. S. and Vohra A. Apparent thermal conductivity data and related information for rice hulls and crushed pecan shells. Therm. Conduct. 2005; 27: 222–30.
26. Tangjuank S. Thermal insulation and physical properties of particleboards from pineapple leaves. Int. J. Phys. Sci. 2011; 6: 4528–32.
27. Mansor M. R., Sapuan S., Zainudin E. S., Nuraini A. and Hambali A. Hybrid natural and glass fibers reinforced polymer composites material selection using analytical hierarchy process for automotive brake lever design. Mater. Des. 2013; 51: 484–92.
28. Erdogan S. A., Šaparauskas J. and Turskis, Z. Decision making in construction management: AHP and expert choice approach. Proc. Eng. 2017; 172: 270–76.
29. Hani A., Roslan A., Mariatti J. and Maziah M. Body armor technology: A review of materials, construction techniques and enhancement of ballistic energy absorption. Adv. Mat. Res. 2012; 806–12.
30. Dalalah D., Al-Oqla F. and Hayajneh M. Application of the analytic hierarchy process (AHP) in multi-criteria analysis of the selection of cranes. Jordan J. Mech. Indust. Eng. 2010; 4: 567–78.
31. Tsita K. G. and Pilavachi P. A. Evaluation of alternative fuels for the Greek road transport sector using the analytic hierarchy process. Ener. Policy. 2012; 48: 677–86.
32. Rashid K., Razzaq A., Ahmad M., Rashid T. and Tariq S. Experimental and analytical selection of sustainable recycled concrete with ceramic waste aggregate. Constr. Build Mater. 2017; 154: 829–40.
33. Dweiri F., Kumar S., Khan S. A. and Jain V. Designing an integrated AHP based decision support system for supplier selection in automotive industry. Expert Syst. Appl. 2016; 62: 273–83.
34. Chang C. W., Wu C. R., Lin C. T. and Chen H. C. An application of AHP and sensitivity analysis for selecting the best slicing machine. Comput. Ind. Eng. 2007; 52: 296–307.
35. Mansor M. R., Sapuan S., Zainudin E. S., Nuraini A. and Hambali A. conceptual design of kenaf fiber polymer composite automotive parking brake lever using integrated trizmorphological chart-analytic hierarchy process method. Mater. Des. 2014; 54: 473–82.
36. Sapuan S., Kho J., Zainudin E., Leman Z., Ali B. and Hambali A. Materials selection for natural fiber reinforced polymer composites using analytical hierarchy process. Ind. J. Eng. Mater. Sci. 2011; 18: 255–67.
37. Yahaya R., Sapuan S., Leman Z. and Zainudin E. Selection of natural fibre for hybrid laminated composites vehicle spall liners using analytical hierarchy process (AHP). Appl. Mech. Mater. 2014; 564: 400–5.
38. Salwa H. N., Sapuan S. M., Mastura M. T. and Zuhri M. Y. M. Analytic hierarchy process (AHP)-based materials selection system for natural fiber as reinforcement in biopolymer composites for food packaging. BioResources. 2019; 14 (4): 10014–46.
39. Patnaik P. K., Swain P. T. R., Mishra S. K., Purohit A. and Biswas S. Composite material selection for structural applications based on AHP-MOORA approach. Mater. Today Proc. 2020; 33: 5659–63.
40. Chatterjee P. and Stevi ć Ž. A two-phase fuzzy AHP-fuzzy TOPSIS model for supplier evaluation in manufacturing environment. Oper. Res. Eng. Sci. Theory Appl. 2019; 2(1): 72–90.
41. Chithra S., Kumar S. R. R. S., Chinnaraju K. and Alfin Ashmita F. A comparative study on the compressive strength prediction models for high performance concrete containing nano silica and copper slag using regression analysis and artificial neural networks. Constr. Build Mater. 2016; 114: 528–35.

42. Sharabi M., Benayahu D., Benayahu Y., Isaacs J. and Haj-Ali R. Laminated collagen-fiber bio-composites for soft-tissue bio-mimetics. Compos. Sci. Technol. 2015; 117: 268–76.
43. Macuvele D. L. P., Nones J., Matsinhe J., Lima M., Soares C., Fiori M. and Riella H. Advances in ultra-high molecular weight polyethylene/hydroxyapatite composites for biomedical applications: A brief review. Mater. Sci. Eng. C. 2017.
44. AL-Oqla F. M., Sapuan S. M., Ishak M. R. and Nuraini A. A. A. Model for evaluating and determining the most appropriate polymer matrix type for natural fiber composites. Int. J. Polym. Anal. Charact./J. Mech. Eng. Sci. 2018; 12(1).
45. Jumaidin R., Sapuan S. M., Jawaid M., Ishak M. R. and Sahari J. Effect of seaweed on physical properties of thermoplastic sugar palm starch/agar composites. J. Mech. Eng. Sci. 2016; 10(3): 2214–25.
46. Lee C. H., Sapuan S. M., Lee J. H. and Hassan M. R. Mechanical properties of kenaf fibre reinforced floreon biocomposites with magnesium hydroxide filler. J. Mech. Eng. Sci. 2016; 10(3): 2234–48.
47. Fairuz A. M., Sapuan S. M., Zainudin E. S. and Jaafar C. A. Effect of filler loading on mechanical properties of pultruded kenaf fibre reinforced vinyl ester composites. J. Mech. Eng. Sci. 2016; 10(1): 1931–42.
48. Yahaya R., Sapuan S. M., Jawaid M., Leman Z. and Zainudin E. S. Effect of layering sequence and chemical treatment on the mechanical properties of woven kenaf-aramid hybrid laminated composites. Mater. Des. 2015; 67: 173–79.
49. Bartosz T. W., Fan M. and Hui D. Compressive behaviour of natural fibre composite. Composites Part B. 2014; 67: 183–91.
50. Al-Oqla F. M. and Salit M. S. Natural fiber composites in materials selection for natural fiber composites, 1st ed., Cambridge, USA: Woodhead Publishing, Elsevier. 2017, 23–45.
51. Mustafa A., Abdollah M. F., Shuhimi F. F., Ismail N., Amiruddin H. and Umehara N. Selection and verification of kenaf fibres as an alternative friction material using Weighted Decision Matrix method. Mater. Des. 2015; 67: 577–82.
52. Al-Oqla F. M. and Sapuan S. M. Natural fiber reinforced polymer composites in industrial applications: Feasibility of date palm fibers for sustainable automotive industry. J. Clean. Prod. 2014; 66: 347–54.
53. Kuo C. F. J., Lin C. H., Hsu M. W. and Li M. H. Evaluation of intelligent green building policies in Taiwan—using fuzzy analytic hierarchical process and fuzzy transformation matrix. Energy Build. 2017; 139: 146–59.
54. Yue W., Cai Y., Xu L., Tan Q. and Yin X. A. Adaptation strategies for mitigating agricultural GHG emissions under dual-level uncertainties with the consideration of global warming impacts. Stoch. Environ. Res. Risk Assess. 2017; 31: 961–79.
55. Naveen J. Jawaid M., Zainudin E. S., Sultan Mohamed T. H. and Yahaya Ridwan B. Selection of natural fiber for hybrid kevlar/natural fiber reinforced polymer composites for personal body armor by using analytical hierarchy process. Front. Mater. 2018.
56. Mourad C., Boudjemaa A. and Fatiha M. A possible correlation investigation between Young's modulus and thermal properties of green composites. Mater. Sci. Forum. 2017; 895: 52–55.
57. https://tr.zhonyingli.com/zellulosed-mmung-vor-und-nachteile-preis-beispiele
58. Mohanty A. K., Misra M., Drzal L. T., et al. Natural fibers, biopolymer and biocomposites. Boca Raton, FL: CRC Press, 2005, 20–21.
59. Textile Technology-Natural Fibers. Ankara: Ministry of National Education Publications, 2014.
60. Da silva neves A. J. and Camanho R. The use of AHP for it project prioritization—a case study for oil & gas company. Proc. Comput. Sci. 2015; 55: 1097–105.

61. Saaty T. L. Absolute and relative measurement with the AHP. The most livable cities in the United States. Socio-econ. Plum Sci. 1986; 20(6): 3277331.
62. Saaty T. L. and Alexander J. A new logic for conflict resolution. In preparation.
63. Saaty T. L. and Kearns K. P. Analytical planning. Oxford: Pergamon Press, 1985.
64. Saaty T. L. Decision making for leaders. Belmont, CA: Wadsworth, 1982.
65. Saaty T. L. and Vargas L. G. The logic of priorities, applications in business, energy, health, transportation. The Hague: KluwerNijhoff, 1981.
66. Saaty T. L. The analytic hierarchy process. New York, NY: McGraw-Hill, 1980.
67. Mitra A., Majumdar A., Ghosh A., Majumdar P. K. and Bannerjee D. Selection of handloom fabrics for summer clothing using multi-criteria decision making techniques. J. Nat. Fibers. 2015; 12(1): 61–71.
68. Al-Oqla F. M., Sapuan S., Ishak M. and Nuraini A. Decision making model for optimal reinforcement condition of natural fiber composites. Fiber Polym. 2015b; 16: 153–63.

2 Sustainable Biodegradable and Bio-based Polymers

Bhargav Patel, Bhagwan Toksha, and Prashant Gupta

CONTENTS

2.1 Introduction 19
2.2 Biopolymers 22
    2.2.1 Bio-based/Biodegradable Polymers 22
        2.2.1.1 Polyesters 22
        2.2.1.2 Polylactic Acid (PLA) 24
        2.2.1.3 Starch 24
        2.2.1.4 Cellulose 25
2.3 Sustainability of Biopolymers 25
    2.3.1 Sustainable Biopolymers from the Nature 26
    2.3.2 Opportunities for the Development of Biopolymers from Food/Vegetative Waste 27
        2.3.2.1 Bioplastics Produced by Cyanobacteria through Photosynthesis 28
        2.3.2.2 Seaweed Polysaccharide Bioplastics 29
        2.3.2.3 Bioplastics from Crab Shells and Tree Discards 30
2.4 Life Cycle Assessment and Analysis 30
2.5 Future Applications with Respect to Increase in Demand for Such Sustainable Materials 31
2.6 Limitations and Challenges 32
2.7 Conclusion 33
References 33

2.1 INTRODUCTION

India is one of the fastest growing and developing economic nations, which aims to be a leading superpower by the year 2030 in various sectors, like automobile, automation, defense, space, agriculture, etc., requiring vast human needs [1]. The advancements in such breakthrough technologies, post their use, result in the generation of a lot of solid waste. Plastics have found their use in almost all of the abovementioned technologies because of the virtue of their versatility, cost, high strength to weight ratios, etc. Today, the world is concerned about global industrial development. Similar concerns have increased multifold over the last two decades which resonate with the same idea toward sustainability and the environment [2]. The global consumption of plastic has increased drastically in recent years, because of its versatile characteristics. Approximately 370 MMTA of plastic are being manufactured annually by various related industries across the globe, of which the majority is single-use plastic [3]. Out of total plastic produced, discarded plastic waste may take hundreds of years to degrade, out of which only 8–9% is recycled, 8% is incinerated, and the rest is landfilled [4]. This criterion shift has influenced many developing countries, including India, to pay more attention toward the life cycle assessment of synthetic plastics. Looking toward the environmental consequences, economic characteristics, and social use of plastics, both industry and consumers have felt a strong pull toward the use of biodegradable and bio-based plastics [5]. Increased interest among researchers is observed in finding an alternative for these petroleum-based products that are renewable as well as biodegradable along with a minimal negative impact on the environment. The current bio-based/biodegradable materials, which are far costly, have found their use in various applications in the automotive industry, packaging, agriculture, and consumer electronics [6]. A comparative analysis of petro-based versus bio-based plastics is given in Figure 2.1.

Are plastics nondegradable? They have chemical structures that are unable to be easily disintegrated by microorganisms [7]. The major factor that typically reduces the degradability of plastic is its high molecular weight, high degree of crystallinity, and insolubility in water. Some organic materials, such as paper, animal carcasses, and wood, are disintegrated into basic molecules and the process thus carried out is called degradation [8]. To provide a long-lasting effect, the plastic quality is improved drastically by the use of functional additives that allow higher heat resistance and durability, so they cannot be broken down into simple molecules and cannot be decomposed. With the addition of such additives, plastics are considered

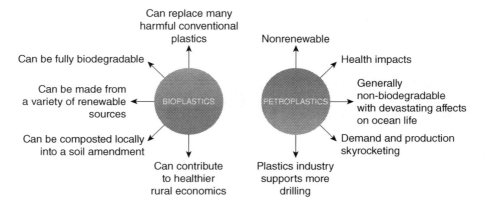

FIGURE 2.1 A representation of comparison between bioplastics and conventional petroleum-based counterparts [10].

Sustainable Biodegradable and Bio-based Polymers

nondegradable. However, there is an alternative methodology by which plastic can be changed into biodegradable with the addition of certain chemicals that can break down polymer chain and structure, thus offering another alternative for the use of "biodegradable" and "bio-based plastic" [9].

We are presenting our view through this work, which is going to convey the missing knowledge evident by the gaps visible in it. There might be a series of queries addressing the following topics:

1. How are biodegradable plastics going to greatly revolutionize the industrial sector in terms of high commercial use?
2. Which type of bioplastic can contribute to higher commercial use?
3. What raw materials are used for synthesis and what are their availability?
4. How is the performance of what is currently used versus bioplastics?
5. Are these bioplastics a boon or curse for the environment?
6. What will be the impact of such sustainable materials on human life and the environment?

Looking at global industry demand, biodegradable plastic is primarily made from petrochemicals, and the chemistry of the material allows it to break down very quickly. However, it needs specific environmental conditions to break down. In general, degradation starts with a chemical process, which brings about thermal/photocatalytic/hydrolytic-induced oxidation, followed by a biological process, which evolves CO_2 and methane as the degradation products [11]. The terms bio-based plastic and biodegradable plastic are sometimes used instead of each other but are technically not the same. Bio-based plastic is obtained via nonfossil or biological (plant/animal) resources. On the contrary, biodegradable plastic is degraded via contact with the environmental microorganisms. The idiom "bioplastic" narrates an evolving and increasing futuristic family of plastics. Bio-based plastic usually originates from renewable biomass, i.e., plants [12]. Some common plants, which are a source of their extraction, include sugarcane, corn, etc. Biopolymers like starch and cellulose are mainly employed as a precursor for the synthesis of complex biodegradable polymer matrices such as thermoplastic starch and cellulose acetate (CA) [13].

In this chapter, we are going to focus on the prospects of sustainable biodegradable plastic concerning social, economic, and environmental stability and discuss the latest advancements in enzyme-based biodegradable and bio-based plastic in bringing down the negativity surrounding the use of plastics. As a result, it will have a positive impact on plastic waste challenges by bringing them under control at a much faster rate and decreasing their percentage [14]. In addition, we are going to discuss the prospect of deriving bio-based plastics from vegetative waste, which is a sustainable pathway toward mass production. We are further going to deliver a multi-disciplinary approach to bring this sustainable material to 80% of its total use in all areas by abandoning the use of single-use plastic [15]. The chapter discusses the critical aspects of biomaterial application and provides insights into production, types, challenges, fermentation process, sustainability, process development, market, and rising demand.

2.2 BIOPOLYMERS

Biopolymers can be classified into many different categories. From a degradability perspective, biopolymers can be classified into two groups: biodegradable and nonbiodegradable. They can alternatively be divided into bio-based and synthetic biopolymers as well. Also, looking at the chemistry of their polymeric backbone, the biopolymers can be classified into polyester, polysaccharides, polycarbonate, polyamides, etc. The nature of the repeating unit helps in the classification of biopolymers as follows:

   I. Polysaccharides made of sugar (cellulose found in plastic)
   II. Protein made of amino acids (myoglobin found in muscle tissue)
   III. Nucleic acid of nucleotides (DNA)

They can be further classified, on the basis of their application, into bioadhesives, biosurfactants, bioflocculants, and so on [16]. The popular bio-based/biodegradable materials, as shown in Figure 2.2, which are also currently on the commercial market are discussed in this section.

2.2.1 Bio-based/Biodegradable Polymers

2.2.1.1 Polyesters

2.2.1.1.1 Polyhydroxyalkanoates (PHA)

Among all the bio-based materials, PHA, from a group of polyoxoesters, has received massive interest due to its thermoplastic properties [17]. Bacteria have the capability to synthesize PHA from a range of carbon-based materials such as complex waste, plant oils, inexpensive fatty acids, alkanes, simple carbohydrates, etc. [18]. PHA is being used with many bacterial cultures. The microorganisms are *Cupriavidus necator*, *Pseudomonas*, *Aeromonads*, *hydrophila*, and *Rhodopseudomonas palustris* [19].

Choi and Lee reported the commercial route for the synthesis of PHAs by recombinant *Escherichia coli* bacteria [20]. However, they can also be made using different microorganisms such as *Pseudomonas*, *Aeromonas*, *Azotobacter*,

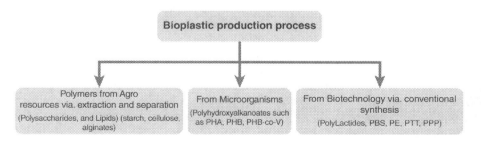

FIGURE 2.2 Different methodologies of synthesis of bio-based polymers. Adapted from ref. [10].

Sustainable Biodegradable and Bio-based Polymers 23

Ralstonia, Cupriavidus, Clostridium, Methylobacterium, Syntrophomonas, etc. [21]. *Pseudomonas* is the best species for making PHAs with longer carbon chains [22]. However, with the addition of certain nutrients and considering the energy cost, the cost of production rises by 75% of the product cost. The manufacturing process makes PHA's products relatively costly as compared to commercially used petroleum-based plastics having similar characteristics. For the production of bacterial PHAs, numerous nonrenewable sources of carbon feed, such as organic waste, peanut, methane, soybean, palmitate oil, glycerol, etc., are used [23]. Despite the advantages of using these renewable materials, the cost associated with the production of PHAs makes the decision to use this, a dicey one. However, to bring down the overall cost of production in order to make its manufacturing a sustainable one, PHAs are incorporated with petroleum-based biodegradable plastics and the functionality can be achieved.

Media selection is one of the important conditions for high production so that the final product is cost-effective. The media used are starch, molasses, wheat, and rice bran. Some of the salient features that PHA offers are degradation at a very high rate, by microorganisms breaking it down into CO_2 and water, produced from renewable sources, and hence it is eco-friendly in nature. Zhila and Shishatskaya discussed the similarity in properties of conventional petroleum-based polypropylene (PP) with that of PHA [24].

2.2.1.1.2 Polyhydroxybutyrate (PHB)

PHB belongs to the broader family of PHA biopolymers. These classes resemble those of synthetic plastics, polyester in particular, but their attractive features include manufacturing and especially degradation that is contradictory to artificial plastics. When cells are subjected to nutrient pressure through the usage of excess carbon with a poor nutrient, synthesis of PHB takes place in cells. In addition, a few other chemical reactants like P, Mg, Fe, and O are regulating elements for making PHB. Other factors such as pH and temperature have an effect on PHB manufacturing [25]. A host of bacterial species are well reported to yield PHB. Some of them are *Pseudomonas putida, Alcaligenes eutrophus, Methylobacterium rhodesianum, Bacillus megaterium, Methylorubrum extorquens*, and *Sphaerotilus natans* [26]. The varieties of bacteria have their own combinations for the use of media sources. Some mixtures of carbon and nitrogen include carbon source as methanol along with malate, acetic acid, n-alkanoic acid, ammonium chloride, ammonium sulfate as a nitrogen source. The media along with the reaction conditions are immensely important for the manufacturing of polymers. All the essential elements like carbon, nitrogen, pH, temperature, pressure, media transfer/movement rate, and oxygen supply are chosen primarily on the basis of aerobic and anaerobic processes. Some of the sources with lower cost include molasses, corn steep liquor, wheat bran, and starch. The selection of a reasonable priced supply of media ensures an economy of scale in terms of manufacturing. For example, activated sludge and co-culturing structures are recognized to be worthwhile, and primarily profitable when used [27]. Some of the salient features of PHB include, eco-friendly, easy decomposition into water molecules, and microorganisms that can eliminate CO_2, etc.

2.2.1.1.3 Poly (Butylene Adipate-Co-terephthalate) (PBAT)

PBAT is employed as an alternative to high density polyethylene, mainly for packaging items such as garden waste sacks, meal bins, or movie wraps, among others [28]. PBAT is synthesized via the reaction of 1,4-butanediol with adipic and terephthalic acid or butylene adipate [28] along with the use of rare earth compounds/zinc acetate as a catalyst. However, these catalysts have been scaled up one level for use in synthesizing PBAT at pilot plant level [29]. The preparation of PBAT necessitates lengthy reaction times, excessive vacuum, and temperatures of up to 190°C or more [30]. There are three primary methods by which PBAT composites can be fabricated, i.e., in-situ polymerization, solvent casting, and melt mixing. Each approach has benefits and disadvantages [31]. In-situ polymerization has shown promise in manufacturing composites since the dispersion of natural fillers in the uncured polymeric solution yields homogeneous mixture, leading to the proper load transfer and offers good tensile loading characteristics. Also, this approach has the capability of being used in industrial-scale processes [32].

Apart from its salient features like biodegradability, PBAT shows interesting physical properties that include good strength, flexibility, tear resistance, and recommended for food contact applications [33]. In addition to the above, quick molding cycles and remarkable sealing performance, to go along with simplicity in mass production at excessive extrusion speeds, make it a suitable material of choice [34].

2.2.1.2 Polylactic Acid (PLA)

PLA is a thermoplastic that is synthesized via the condensation reaction of lactic acid or ring-opening polymerization of lactide. The use of bioplastic PLA has been elevated because of its excessive mechanical strength, clean process ability, and good thermal properties in comparison to different commercially used polymers based on fossil resources [35]. PLA composites have been blended with other natural compounds to improve their utility in different fields. There have been reports of blending PLA with wood and other natural fibers to make PLA composites [36]. Unlike the biomaterial itself, such composites don't create many issues during processing through extrusion and compression molding [37]. It is a flexible polymer that finds its use in sectors like textiles and medicine. It is also used in the food sector, especially in the packaging of meals and single-use products. PLA packaging, used for yogurt, butter, margarine, and cheese, has found excellent mechanical stability and good permeability characteristics for moisture, fats, and gases [38].

2.2.1.3 Starch

The starch, available as white powder, doesn't possess any characteristic taste or odor. Starch is biologically absorbable, nontoxic, and semi-permeable to CO_2 and possesses amylose along with branched amylopectin [39]. The amylose-content material can also range from 20 to 25%, even as the amylopectin-content material varies from 75 to 80% of weight, depending on the type of plant. If heated, starch might become water soluble as the granules swell and burst, dissociating and liberating glucose molecules in water, thereby increasing the viscosity of the mixture. This method is referred to as starch gelatinization, as reported for cornstarch materials [40, 41].

Sultan and Johari specifically experimented with banana peel (BP) and cornstarch. The banana skin was engineered with cornstarch in between 1 and 5%. The samples were tested and analyzed with numerous durability checks and characterization methods. Based on the outcomes received, the BP films with 4% cornstarch exhibited the best tensile strength of 3.72 N/m^2. The water absorption of BP testing showed that with 3% cornstarch, the product had been immune to water uptake with the aid of soaking up water as much as 60.65%. In terms of characterization, the Fourier transform infrared (FTIR) spectra for BP control film and BP film with 4% cornstarch were similar, with the maximum of the peaks present. The combination of starches from diverse sources can be assessed as an opportunity in generating bioplastics [42]. The manufacturing of starch-based bioplastics is simple, and they are broadly used for packaging applications. The tensile strength of starch is appropriate for the manufacturing of packing materials [43] with the addition of glycerol as a plasticizer.

2.2.1.4 Cellulose

Cellulose bioplastics are produced using cellulose or its derivatives. Cellulose plastics can be synthesized using softwood bushes as raw material. The barks of the trees are separated and may be used as a source of fuel while manufacturing. To segregate cellulose fiber from the tree, the tree is cooked or heated in a digester. The pulp produced from the process comprises hemicelluloses and alpha-cellulose. The pulp is then handled with bleaching chemical compounds to dispose of any residual resins and lignin and to reduce the hemicellulose content of the pulp. The processed pulp consists of water that is eliminated from the pulp before processing it with excessive alpha-cellulose content. Cellulose esters are produced by reacting processed pulp with acids and anhydrides at various concentrations and temperatures, depending on the end-user application [44]. Cellulose has the potential for use in the food packing industry due to characteristics like the lowest water permeability and highest biodegradability. It can be one of the best stabilizers for the food industry and one of the best bioplastic raw materials for packaging. Cellulose can be employed to make more durable films [45].

2.3 SUSTAINABILITY OF BIOPOLYMERS

The sustainable features that biopolymers offer have given the recycling concept a very new orientation with respect to biological recycling. The technicality of the term "biodegradable" needs to be very well disseminated within the community and the market at large with respect to the sustainable use of such material in the near future [46]. This would ensure more stringent adoption from both the government administration and consumers' perspectives. From a sustainability point of view, this material is very useful in agriculture, commercial retail, food packaging, etc. These materials have come into the limelight by increasing the efficiency of fossil fuels, reducing CO_2 emissions, and curbing the generation of unmanaged plastic waste [47].

They are eco-friendly and compatible with the environment throughout their life cycle. The development of new sustainable bioplastics also helps to resolve the issues surrounding the use of fossil-based plastics. Han et al. reported a sustainable bioplastic made from DNA and ionomers derived from biomass. The sustainability analysis involves all the aspects of this material, right from production, its use, and the end-life of DNA plastics [48]. The raw material used was developed from bio-renewable resources, and the process employed was eco-friendly. The water-based process did not employ as much energy as conventional processes do and refrained from the use of organic solvents or the production of by-products. Also, the end-of-life cycle phase is ensured via a couple of green methodologies, wherein the recycling of waste plastics can be done with an enzyme-catalyzed degradation that is controlled under milder environments.

2.3.1 Sustainable Biopolymers from the Nature

Bioplastics are typically synthesized from bio-based polymeric materials, which contribute greatly to sustainable and commercial plastic life cycles. They are produced into commercial products from natural or renewable sources [49]. In this process, virgin polymers are synthesized from renewable or recycled raw materials. Further, such materials are produced with the help of a carbon-neutral energy process, and, at the end of their lives, they are reused or recycled, depending on the life of the material and the chemistry involved. Bioplastics from nature are sustainable, largely biodegradable, and biocompatible [50]. Still, in the initial developing stages of the industrial community with regard to its widespread commercial use, bioplastic materials have become essential for food packaging, agriculture, horticulture, composting bags, and hygiene from a sustainability perspective [51]. Bioplastic has also found its deep roots in sectors like biomedical, structural, electrical and other consumer products. Preservation of nonrenewable fossil fuel is only possible by large-scale production and utilization of bioplastics in industrial as well as in day-to-day life, which will help us to better solve the environmental problems concerning single-use plastic waste [52].

Since bioplastics are mainly plant-based products, the amount of petroleum raw material consumed for the production of plastic is predicted to be decreased by 20–25% by 2025 [53]. The largest share of bioplastics will come from Asia and Europe, out of which 34% of the share will come from Asia alone, followed by other countries [54]. In fact, it is becoming the next big thing. Whenever scientists are developing a new biodegradable polymer, the important factors they consider are biodegradable behavior and antimicrobial activity of the material. Varieties of processes are being used for the preparation of biodegradable products with various types of material. A prominent biomaterial that is being used for polymeric applications that can biodegrade is cellulose. It is a natural, bio-based polymer that is obtained from almost all trees and plants [55]. Cellulose can be used either individually or in a blended form with other polymers to create a sustainable biodegradable material with desirable tensile strength and elongation properties. Ebnesajjad reported some formulations which showed the improvement of thermal and mechanical properties [11]. Another commercial synthetic bioplastic, PLA, is a biodegradable polyester material that is made from lactic acid through the process of fermentation from renewable crops such as sugar beets and corn [56]. Polycaprolactone (PCL) is also a biodegradable

Sustainable Biodegradable and Bio-based Polymers 27

polyester and is synthesized through the polymerization of ε-caprolactone. It possesses very good thermal process ability, a low melting point, and low viscosity [57]. Meereboer et al. reported the blending of PCL with other polymers, such as PHB, that resulted in increased ductility, impact strength, and heat stability [58]. The other materials, which are also used for developing biodegradable polymers, are PHA and PHB, which are commonly used in food packaging [25].

Polybutylene succinate is a pliable and more economical form of polyester. It can be made more sustainable. It is another common biodegradable polyester polymer made from a monomer of succinic acid [59]. The commercial polymer available on the market is based on fossil-fuel-derived succinic acid, which is expensive. Researchers are trying to develop a renewable alternative to this monomer, which can make it more affordable and sustainable. Xu and Guo have reviewed and reported the biological starting materials for succinic acid, such as corn, starch, sugarcane, whey, wood, and soybean, which can allow low-cost production of the polymer [60].

Also, starch-based durable bioplastics can be produced on an industrial scale for use in different areas such as flexible films, agriculture mulch film, food packaging, and plant pots [61]. The process for making this bioplastic involves blending plant-derived starches with a plasticizer such as glycol or glycerol [62]. Such ingredients aid in improving the flexibility along with improvements in processing due to processing aids. The advantages of such products include, but are not limited to, being inexpensive and biodegradable. In addition, they are entirely made up of renewable sources. However, its demerits include its tendency to absorb water, leading to hydrolysis and rapid degradation in highly humid environments. In the event of a temperature increase, degradation effects are amplified [63].

2.3.2 Opportunities for the Development of Biopolymers from Food/Vegetative Waste

Food and vegetable waste can be a foundation source of raw materials for the development of biorefinery concepts. The processes involved in the sustainability of this material cycle are the concentration of valuable components, regeneration of biopolymers, conversion of waste by anaerobic digestion, and producing biomaterials by the process of fermentation [64]. Perotto et al. reported the development of sustainable bioplastic from vegetative waste using a water-based process. The bioplastic films made from vegetative waste were reported to have mechanical properties similar to those of commercially used PP and those of starch-based bioplastics. The developed films based on bioplastics were degradable. However, a disadvantage of the use of such a material is that it loses its characteristic properties and mass when it comes into contact with water [65]. Orenia et al. further worked on such vegetative waste-based bioplastic films and developed one made using carrot vegetative waste with characteristic hydrophobic properties. Furthermore, nuclear magnetic resonance (NMR) and associated structural analysis revealed that cellulose crystals bonded the film together. They were further blended by adding a soluble component like pectin to provide additional binding, along with sugar acting as a plasticizer [66]. Luca Ceseracciu et al. reported the blending of polyvinyl alcohol (PVA) and carrot-based bioplastic, which exhibited very low oxygen permeability as compared

28   Handbook of Sustainable Materials

to other commercially used synthetic polymers [65]. Another study that used PVA/chitosan film to create intelligent packaging nanofilms with improved mechanical, thermal, and antibacterial properties using clay [67]. Blending can be done in many variations and formations for making bioplastic, considering it one of the smart materials, making its availability for a wide range of applications in cosmetics, disposable objects [68], and biodegradable electronics [69].

Furthermore, valuable biomaterials, such as proteins, polysaccharides, and lipids, can be extracted from vegetative waste and used to create bio-based sustainable packaging materials alternatives to the same single-use concept [64]. A host of vegetative source-based wastes have an abundant presence of film-forming biopolymeric materials as listed above. Common protein sources include soy, wheat gluten, corn, lentils, peas, mung beans, triticale, etc. In addition, polysaccharides like starch can be extracted from rice, wheat, potato, and cellulose from cotton, wood, hemicellulose, etc. They have lower water vapor permeability characteristics due to hydrophilicity. Lipids, such as beeswax, carnauba wax, free fatty acids, and gum, are hydrophobic and can be used with the above materials to improve water vapor permeability.

The increase in plastic pollution has been due to improper management of plastic waste and a lack of awareness in social communities. However, such issues can be tackled by introducing a renewable source of material for the production of bioplastic from plants, bacteria, algae, seaweed, tree discards, etc. We will look at specific cases of the development of biodegradable plastics, which can produce from waste materials in nature.

2.3.2.1   Bioplastics Produced by Cyanobacteria through Photosynthesis

Cyanobacteria are from the well-known class of phylum of bacteria that gain energy through the process of photosynthesis. They are the only kinds of prokaryotes that can produce oxygen from photosynthesis. The idiom "cyanobacteria" is generally derived from the Greek word meaning "blue," which indicates the main color of cyanobacteria. In other words, they are also called "blue-green algae." It has been discovered that cyanobacteria can be synthesized to produce PHA or PHB. The synthesis process for producing PHA by these routes makes short chain monomeric alkanoates that conjointly lead to medium-chain PHA; but, the starting materials of the processing route are straightforward carbon compounds like sucrose, aldohexose, and ketohexose, thereby, making it a cost-effective method [70]. This pathway leads to the presence of each sugar and lipid, such as a glycolic precursor and the use of carboxylic acid synthesis intermediates. The *Entner-Doudoroff* pathway is reportedly employed by pseudomonas with the contact action of the sugar supply from aldohexose to acid. This is made durable by the action of PHA synthase, which catalyzes the synthesis of PHA from fatty acids and sugar [71]. The PHB acquired from cyanobacteria is positioned to be higher than the PHB from heterotrophic bacteria. Such a polymer from Spirulina platensis has a melting temperature of 192.9°C, accompanied by *Nostoc muscorum* (176°C), which is higher than bacterial PHB (174.2°C). The crystallinity of the polymer acquired from *Aulosira fertilissima* (60.7%) and *Nostoc muscorum* (62.4%) is better than bacterial PHB (40.7%). PHB from *fertilissima* confirmed higher tensile strength (37.6 MPa) than bacterial PHB

Sustainable Biodegradable and Bio-based Polymers 29

(32.0 MPa) [72, 73]. They have the potential to eliminate CO_2 and can greatly contribute toward the betterment of the environment as they are one of the most sustainable materials for the production of bioplastics [74].

2.3.2.2 Seaweed Polysaccharide Bioplastics

Seaweeds such as Chlorophyta, Phaeophyta, and Rhodophyta are one of the most sustainable bioplastic materials because of their biodegradability, edibility, and eco-friendly characteristics [75]. The synthesis of sustainable bioplastics can also be done using seaweed, a living organism rich in polysaccharides and positioned at the bottom of the food chain. Due to its versatility, it is widely used in areas like food technology, microbiology, biotechnology, and even in the medical domain. Due to its abundant availability, it has very high demand in the food industry. Seaweed is divided into three major groups based on the form in which it is found [76]:

1. Brown Seaweeds (*Laminaria pallida*, *Fucus* and *Zonaria* species-polysaccharide-alginate, fucoidan): They vary greatly when it comes to their size, species of origination, and morphology. However, they are brown in color due to the presence of photosynthetic pigment. They possess characteristics such as antitumor, antiviral, anticoagulant, anti-inflammatory, etc. Therefore, they are essentially used in the biomedical field.
2. Green Seaweeds (*Monostroma* species, polysaccharide Ulvan): Green seaweeds are commonly found in salt and freshwater, and possess a green color due to the presence of chlorophyll A and B. Due to its huge potential, it has many uses in agriculture, pharmaceutical, and food applications.
3. Red Seaweeds (*Porphyra capensis*, *Notogenia striata*-polysaccharide-agar, carrageenan): Red seaweeds are different from brown and green seaweeds due to the presence of blue and red pigments in them, as phycocyanin and phycoerythrin, as well as chlorophyll A, are present in their composition. These red seaweeds contain sulfated galectins, such as agar, which are broadly employed for the manufacturing of biopolymers for food and industrial applications.

Commonly, seaweed derivatives, specifically alginate, carrageenan, and agar, are used in a host of applications as they exhibit excellent film-forming abilities and feature negligible lignin content. However, seaweed is obtained from water bodies or extracted from the by-products as given above. However, they are expensive in comparison to conventional plastic [77]. Recent research reports the incorporation of *Kappaphycus alvarezii*, a seaweed, as an alternative to its derivatives, making it a less expensive option, fit for human consumption with no use of chemical compounds that may cause any harmful effects [78]. Even though they have got lower water vapor barrier properties and mechanical strength [79], exotic seaweed species may be combined or mixed collectively with different substances to enhance their performance. Seaweeds are best utilized in the food industry because they come directly into contact with food materials as an active agent and are composed of polysaccharides. Products manufactured from seaweed can improve sustainability, functionality, and sensory properties. For its further use in Future Commission

30 Handbook of Sustainable Materials

Merchant, seaweed material needs to be combined with other materials like essential oils or plant extracts to improve its thermal, barrier, mechanical, antioxidant, and antimicrobial properties. Seaweed can be actively combined with other polymers to actively manufacture packing films [80].

2.3.2.3 Bioplastics from Crab Shells and Tree Discards

Atiwesh et al. developed a sustainable bioplastic wrap using crab shell and wood pulp. Plant-based cellulose is one of the most commonly used natural biopolymers in terms of its usage for producing biofilms. Chitin, a recently developed bioplastic, is mostly found in shellfish, insects, and fungi. It has featured in limited discussions despite having potential industrial value. They developed flexible packaging film featuring similar characteristics to that of a packaged plastic product by spraying multiple layers of chitin extracted from a crab shell and cellulose from a tree. This sustainable material and its process are highly versatile as they enable the formulation scientist to achieve different properties by adjusting the ratio and proportions of ingredients. The stiffness, flexibility, and optical clarity of the material, along with its thickness, can be tailored in order to allow its use in packaging materials. They reportedly compared the developed film with PET transparent plastic packaging film and the results revealed that the new bio-based food packaging film exhibited reduced oxygen permeability (73%) against that of its counterpart. It enables the food packaging material developed from the given system to allow the food to stay fresh a lot longer as compared to commercial petroleum-based PET packing films [81].

2.4 LIFE CYCLE ASSESSMENT AND ANALYSIS

Looking at the life cycle assessment (LCA) analysis, we have to differentiate between sustainable bioplastics and other commercially used petroleum-based plastics. This analysis is a crucial part of evaluating the environmental impact of bioplastics right from their synthesis, production, and utilization up until their final treatment and disposal. The most convenient tool of analysis to evaluate the impact on the environment of bioplastics or commercially used plastics is LCA, which helps us to track down the overall impact of bioplastics on nature at each stage of its life cycle, followed by the same for petroleum-based plastics [82]. The analysis signifies chemistry, structure, behavior, and end use during the whole life of the product, assessing its right from raw material or its form of origin to the environment, processing or synthesis at various stages, manufacturing, distribution, use of the product, and its disposal. According to the data, we can assess the overall impact that can be individually measured with global warming, human toxicity, abiotic depletion, eutrophication, acidification, etc. After all, the final assessment and test results provided by LCA can be taken into major consideration, probably for the creation of different policy laws. One of the factors, for example, global warming potential, can guide the system to calculate the greenhouse gas emissions, i.e., CO_2 emitted. LCA analysis will help by identifying the major areas that are required to be worked upon in order to possibly reduce greenhouse emissions [81]. It can be done by substituting the use of petroleum-based plastics, which are responsible for adverse environmental effects with more sustainable natural waste/vegetative waste-based bioplastic materials [83].

Sustainable Biodegradable and Bio-based Polymers 31

A certain amount of energy is needed for the process to convert renewable biomass into fully functionable bioplastics through the biological process of fermentation. The major negative impact of using petroleum-based plastics is that they need energy for manufacturing, which comes from nonrenewable sources such as coal-fired thermal power plant-based power. Hence, the negative impact and ecological profile can be greatly improved by using renewable feedstock. The modification of agro-waste-based polymers, such as starch or protein, may require less energy as compared to some other fermentation processes based on bioplastic technologies [84].

Choosing the best option for bioplastic waste management and disposal, LCA gives a very clear ideology. Waste management for bioplastics cannot be carried out by incineration or landfilling, as they are not better options due to the wet nature of waste and the value that the waste can offer for making sustainable alternative materials. However, looking from the future perspective, smart management of waste plastic is only possible by reducing greenhouse gas emissions as much as possible to zero land use change emissions.

2.5 FUTURE APPLICATIONS WITH RESPECT TO INCREASE IN DEMAND FOR SUCH SUSTAINABLE MATERIALS

Global consumption and production of petroleum-based plastics have increased by 36% in the year 2021, as compared to the last decade. Nonbiodegradable plastics are continuously finding their way to the open landfills and being dumped into the oceans, which has led to a very serious hazardous impact on the environment and the living beings that live in the ecosystem. To overcome the issue, and by finding an alternative, humans have shown a keen interest in bioplastics to meet the global plastic demand and requirements. Bioplastic materials have distinct features from the other conventionally available plastics on the market such as biodegradability, low carbon footprint, energy efficiency, versatility, and comparable mechanical and thermal stability. Due to the great potential of bioplastic, it is going to substitute petroleum-based plastic in many sectors, right from automobiles to biomedical fields. In addition, because of its exceptional use in the food packing sector, it is the first preference to be employed in food packaging films [85]. Because of the versatile benefits on offer in terms of horticulture and agriculture, it has been a go-to material for use in mulching films. One of the examples of using such films is to cover banana bushes from environmental influences like insects, dust, external effects, etc. [86].

Bioplastic has also been used in the field of electronics such as making casings, circuit boards, computer casings, loudspeakers, laptop mouse, etc. [69]. To a partial extent, the automotive industry has shown great interest in bioplastics due to their sustainability for making dashboard components as well as solid interior and exterior products in cars [87]. When it comes to the food packing industry, keeping food fresh for a long time by reducing oxygen permeability characteristics and prolonging its shelf life, the performance of these sustainable bioplastic films and containers has shown comparable performance to other conventional petroleum-based plastic films and containers [88]. Also, these sustainable materials offer features which enable their use in wound management, tissue engineering, drug delivery, production of

2.6 LIMITATIONS AND CHALLENGES

Biodegradable and bio-based plastics are, as such, designed and manufactured in a way that they should decompose or degrade after observing changes in the surrounding atmosphere such as microorganisms, moisture, humidity, water, temperature, sunlight, etc. However, problems arise if these bioplastics are not properly handled and well managed, and when they are discarded inappropriately, this material goes directly into the water stream and then enters deep into the ocean, polluting the ocean on a large scale. Also, the absence of sunlight inside the ocean makes it difficult for bioplastic material to initiate degradation for decomposition to happen. As we know, these bioplastics need sunlight to accelerate their process of decomposition. Material entering into the ocean is a problem, and material cannot be decomposed. The same problem arises when this material is dumped or landfilled [90]. As referred to in the relevant biodegradable American Society for Testing and Materials (ASTM) standards, all biodegradable materials require a special and adequate compostable environment for the material to decompose within the stipulated time period of 90 days. The material may degrade later, but won't fit into the category of biodegradable materials in definitive terms.

According to the European Plastics Recyclers Association, they recommend that, while making petroleum-based plastic bags and films, it requires a lot of energy and essential oils. As a result, why waste them by creating a material that will eventually self-destruct in nature? The future approach we should be implementing right from the present day is to consider sustainable bioplastic as an incredibly valuable material and bring it into widespread use where the requirements are very high and according to its desirable application. Otherwise, we have to follow the rule of "recycle, reuse, and reduce," which the industry will not be very fond of, as it would certainly reduce the quantity of material manufactured and sold in the market.

Also, some bioplastics offer a shorter lifespan as compared with other conventional oil-based plastics because of their weaker mechanical properties. One of the limiting reasons for using some bioplastics is that they have greater water permeability than conventionally used plastics. According to the research work carried out, these bioplastic films are sometimes very soft and can be torn apart like tissue paper, or sometimes very brittle and stiff, which makes them break easily. One of the best examples is algae-based bioplastic that will breakdown in just a few hours when coming in contact with excess water, making it biodegradable but also delicate [91]. Sustainable bioplastic material being labeled as compostable and biodegradable gives it a very distinct feature from commercial plastic, but every bioplastic should follow a standard disposable procedure and should be very well managed. The local administration and community should also promote the use of such materials and follow proper handling of this bioplastic waste and subsequent management for use as humus. Precautions and awareness need to be regulated among human societies for proper use and disposal of bioplastics.

2.7 CONCLUSION

At different times, crude oil and petroleum-based raw materials have skyrocketed. More of such volatility in pricing is expected in the coming century, with the fossil nature of the material. The need to find an alternative material to such fossil-based plastic products is the benchmark for sustainability in such terms. The renewable characteristics of nature, the superiority of biodegradability, and the sustainability of bioplastic materials make it an exclusive option that scientists have been able to conjure in the last 4–5 decades of work put in and substitute the use of petroleum-based products in numerous high-end applications. However, research has revealed that when producing a bio-based sustainable product, the production cost is expensive for certain products, but then again, these technologies are still in the initial stages of their development. There is an ideology to bring down the production cost by using raw materials with zero value along with other methods such as recombinant microbial strains, mixed cultures, efficient fermentation, purification, and recovery. Studies have certainly displayed that the environmental damage caused by using these bioplastic products is actually a smaller amount as compared to our commercially used plastic products and not even severe as compared to conventional plastic. Research and development have been modernized in academics, as well as research has been conducted in the industrial sector, to mitigate the current drawbacks upcoming in sustainable bioplastic manufacturing, which can be addressed effectively. The future of bioplastics is only possible if all efforts and responsibility are put forth in meeting its requirements, as well as considering the desirable price and performance. However, because of their wide-ranging capability to have special characteristics and wide application use in biotechnology, bioplastics have an enormously promising future.

REFERENCES

1. Ashter, S. A. *Introduction to Bioplastics Engineering*; William Andrew: Oxford, UK, 2016.
2. Marjadi, D.; Dharaiya, N.; Ngo, A. D. Bioplastic: A Better Alternative for Sustainable Future. *Everyman's Sci*, 2010, *15* (2), 90–92.
3. Tieso, I. *Global Plastic Production 1950–2020*; 2021.
4. Lebreton, L.; Andrady, A. Future Scenarios of Global Plastic Waste Generation and Disposal. *Palgrave Commun.*, 2019, *5* (1), 1–11.
5. Kalia, V. C.; Raizada, N.; Sonakya, V. Bioplastics. *J. Ind. Res.*, 2000, *59*, 433–445.
6. Pilla, S. *Handbook of Bioplastics and Biocomposites Engineering Applications*; Scrivener Publishing LLC: Salem, MA, 2011; Vol. 81.
7. Taghavi, N.; Udugama, I. A.; Zhuang, W.-Q.; Baroutian, S. Challenges in Biodegradation of Non-Degradable Thermoplastic Waste: From Environmental Impact to Operational Readiness. *Biotechnol. Adv.*, 2021, *49*, 107731.
8. Chamas, A.; Moon, H.; Zheng, J.; Qiu, Y.; Tabassum, T.; Jang, J. H.; Abu-Omar, M.; Scott, S. L.; Suh, S. Degradation Rates of Plastics in the Environment. *ACS Sustain. Chem. Eng.*, 2020, *8* (9), 3494–3511.
9. Ramesh Kumar, S.; Shaiju, P.; O'Connor, K. E. Bio-Based and Biodegradable Polymers-State-of-the-Art, Challenges and Emerging Trends. *Curr. Opin. Green Sustain. Chem.*, 2020, *21*, 75–81.

10. Z. Naser, A.; Deiab, I.; M. Darras, B. Poly(Lactic Acid) (PLA) and Polyhydroxyalkanoates (PHAs), Green Alternatives to Petroleum-Based Plastics: A Review. *RSC Adv.*, 2021, *11* (28), 17151–17196. https://doi.org/10.1039/D1RA02390J.

11. Ebnesajjad, S. *Handbook of Biopolymers and Biodegradable Plastics: Properties, Processing and Applications*; William Andrew: Oxford, UK, 2012.

12. Wu, F.; Misra, M.; Mohanty, A. K. Sustainable Green Composites from Biodegradable Plastics Blend and Natural Fibre with Balanced Performance: Synergy of Nano-Structured Blend and Reactive Extrusion. *Compos. Sci. Technol.*, 2020, *200*, 108369.

13. Muthusamy, M. S.; Pramasivam, S. Bioplastics—An Eco-Friendly Alternative to Petrochemical Plastics. *Curr. World Environ.*, 2019, *14* (1), 49.

14. Boucher, J.; Billard, G. The Challenges of Measuring Plastic Pollution. *Field Actions Sci. Rep. J. Field Actions*, 2019, No. Special Issue 19, 68–75.

15. Van, L.; Hamid, N. A.; Ahmad, F.; Ahmad, A. N. A.; Ruslan, R.; Tamyez, P. F. M. Factors of Single Use Plastic Reduction Behavioral Intention. *Emerg. Sci. J.*, 2021, *5* (3), 269–278.

16. Hassan, M. E. S.; Bai, J.; Dou, D.-Q. Biopolymers; Definition, Classification and Applications. *Egypt. J. Chem.*, 2019, *62* (9), 1725–1737. https://doi.org/10.21608/ejchem.2019.6967.1580.

17. Tan, G.-Y. A.; Chen, C.-L.; Li, L.; Ge, L.; Wang, L.; Razaad, I. M. N.; Li, Y.; Zhao, L.; Mo, Y.; Wang, J.-Y. Start a Research on Biopolymer Polyhydroxyalkanoate (PHA): A Review. *Polymers*, 2014, *6* (3), 706–754. https://doi.org/10.3390/polym6030706.

18. Lee, C. H.; Sapuan, S. M.; Ilyas, R. A.; Lee, S. H.; Khalina, A. Chapter 5— Development and Processing of PLA, PHA, and Other Biopolymers. In *Advanced Processing, Properties, and Applications of Starch and Other Bio-Based Polymers*; Al-Oqla, F. M., Sapuan, S. M., Eds.; Elsevier, 2020; pp. 47–63. https://doi.org/10.1016/B978-0-12-819661-8.00005-6.

19. Sruamsiri, D.; Thayanukul, P.; Suwannasilp, B. B. In Situ Identification of Polyhydroxyalkanoate (PHA)-Accumulating Microorganisms in Mixed Microbial Cultures under Feast/Famine Conditions. *Sci. Rep.*, 2020, *10* (1), 3752. https://doi.org/10.1038/s41598-020-60727-7.

20. Choi, J.; Lee, S. Y. Efficient and Economical Recovery of Poly (3-Hydroxybutyrate) from Recombinant *Escherichia coli* by Simple Digestion with Chemicals. *Biotechnol. Bioeng.*, 1999, *62* (5), 546–553.

21. Rebah, F. B.; Prévost, D.; Tyagi, R. D.; Belbahri, L. Poly-β-Hydroxybutyrate Production by Fast-Growing Rhizobia Cultivated in Sludge and in Industrial Wastewater. *Appl Biochem Biotech*, 2009, *158* (1), 155–163.

22. Ramsay, B. A.; Saracovan, I.; Ramsay, J. A.; Marchessault, R. H. A Method for the Isolation of Microorganisms Producing Extracellular Long-Side-Chain Poly (β-Hydroxyalkanoate) Depolymerase. *J. Environ. Polym. Degrad.*, 1994, *2* (1), 1–7.

23. Fedorov, M. B.; Vikhoreva, G. A.; Kil'deeva, N. R.; Mokhova, O. N.; Bonartseva, G. A.; Gal'braikh, L. S. Antimicrobial Activity of Core-Sheath Surgical Sutures Modified with Poly-3-Hydroxybutyrate. *Appl. Biochem. Microbiol.*, 2007, *43* (6), 611–615.

24. Zhila, N.; Shishatskaya, E. Properties of PHA Bi-, Ter-, and Quarter-Polymers Containing 4-Hydroxybutyrate Monomer Units. *Int. J. Biol. Macromol.*, 2018, *111*, 1019–1026. https://doi.org/10.1016/j.ijbiomac.2018.01.130.

25. Bharti, S.; Swetha, G. Need for Bioplastics and Role of Biopolymer PHB: A Short Review. *J. Pet. Environ. Biotechnol.*, 2016, *07*. https://doi.org/10.4172/2157-7463.1000272.

26. Sarma, A.; Das, M. K. Improving the Sustainable Performance of Biopolymers Using Nanotechnology. *Polym.-Plast. Technol. Mater.*, 2021, *60* (18), 1935–1965.

27. Rosero-Chasoy, G.; Rodríguez-Jasso, R. M.; Aguilar, C. N.; Buitrón, G.; Chairez, I.; Ruiz, H. A. Microbial Co-Culturing Strategies for the Production High Value Compounds, a Reliable Framework towards Sustainable Biorefinery Implementation— An Overview. *Bioresour. Technol.*, 2021, *321*, 124458.

28. Vroman, I.; Tighzert, L. Biodegradable Polymers. *Materials*, 2009, *2* (2), 307–344.
29. Zhu, K.; Zhu, W.-P.; Gu, Y.-B.; Shen, Z.-Q.; Chen, W.; Zhu, G.-X. Synthesis and Characterization of Poly (Butylene Adipate-Co-Terephthalate) Catalyzed by Rare Earth Stearates. *Chin. J. Chem.*, 2007, *25* (10), 1581–1583.
30. Sousa, A. F.; Vilela, C.; Fonseca, A. C.; Matos, M.; Freire, C. S.; Gruter, G.-J. M.; Coelho, J. F.; Silvestre, A. J. Biobased Polyesters and Other Polymers from 2, 5-Furandicarboxylic Acid: A Tribute to Furan Excellency. *Polym. Chem.*, 2015, *6* (33), 5961–5983.
31. Siqueira, G.; Bras, J.; Dufresne, A. Cellulosic Bionanocomposites: A Review of Preparation, Properties and Applications. *Polymers*, 2010, *2* (4), 728–765.
32. Mittal, G.; Dhand, V.; Rhee, K. Y.; Park, S.-J.; Lee, W. R. A Review on Carbon Nanotubes and Graphene as Fillers in Reinforced Polymer Nanocomposites. *J. Ind. Eng. Chem.*, 2015, *21*, 11–25.
33. Al-Tayyar, N. A.; Youssef, A. M.; Al-Hindi, R. Antimicrobial Food Packaging Based on Sustainable Bio-Based Materials for Reducing Foodborne Pathogens: A Review. *Food Chem.*, 2020, *310*, 125915.
34. Dammak, M.; Fourati, Y.; Tarrés, Q.; Delgado-Aguilar, M.; Mutjé, P.; Boufi, S. Blends of PBAT with Plasticized Starch for Packaging Applications: Mechanical Properties, Rheological Behaviour and Biodegradability. *Ind. Crops Prod.*, 2020, *144*, 112061. https://doi.org/10.1016/j.indcrop.2019.112061.
35. Fortunati, E.; Armentano, I.; Iannoni, A.; Kenny, J. M. Development and Thermal Behaviour of Ternary PLA Matrix Composites. *Polym. Degrad. Stab.*, 2010, *95* (11), 2200–2206.
36. Lu, T.; Liu, S.; Jiang, M.; Xu, X.; Wang, Y.; Wang, Z.; Gou, J.; Hui, D.; Zhou, Z. Effects of Modifications of Bamboo Cellulose Fibers on the Improved Mechanical Properties of Cellulose Reinforced Poly (Lactic Acid) Composites. *Compos. Part B Eng.*, 2014, *62*, 191–197.
37. Shih, Y.-F.; Huang, C.-C. Polylactic Acid (PLA)/Banana Fiber (BF) Biodegradable Green Composites. *J. Polym. Res.*, 2011, *18* (6), 2335–2340.
38. National University of Costa Rica; J, V.-B. Polylactic Acid (PLA) As a Bioplastic and Its Possible Applications in the Food Industry. *Food Sci. Nutr.*, 2019, *5* (2), 1–6. https://doi.org/10.24966/FSN-1076/100048.
39. Sanyang, M. L.; Ilyas, R. A.; Sapuan, S. M.; Jumaidin, R. Sugar Palm Starch-Based Composites for Packaging Applications. In *Bionanocomposites for Packaging Applications*; Jawaid, M., Swain, S. K., Eds.; Springer International Publishing: Cham, 2018; pp. 125–147. https://doi.org/10.1007/978-3-319-67319-6_7.
40. Ratnayake, W. S.; Jackson, D. S. Gelatinization and Solubility of Corn Starch during Heating in Excess Water: New Insights. *J. Agric. Food Chem.*, 2006, *54* (10), 3712–3716.
41. Wang, H.; Sun, X.; Seib, P. Mechanical Properties of Poly (Lactic Acid) and Wheat Starch Blends with Methylenediphenyl Diisocyanate. *J. Appl. Polym. Sci.*, 2002, *84* (6), 1257–1262.
42. Sultan, N. F. K.; Johari, W. L. W. The Development of Banana Peel/Corn Starch Bioplastic Film: A Preliminary Study. *Bioremediation Sci. Technol. Res.*, 2017, *5* (1), 12–17. https://doi.org/10.54987/bstr.v5i1.352.
43. Sasan, J. M. Bioplastic Made from Starch as a Better Alternative to Commercially Available Plastic. *Sci. Educ.*, 2021, *2* (11), 257–277.
44. Yu, Z.; Ji, Y.; Bourg, V.; Bilgen, M.; Meredith, J. C. Chitin-and Cellulose-Based Sustainable Barrier Materials: A Review. *Emergent Mater.*, 2020, *3* (6), 919–936.
45. Yaradoddi, J. S.; Banapurmath, N. R.; Ganachari, S. V.; Soudagar, M. E. M.; Mubarak, N. M.; Hallad, S.; Hugar, S.; Fayaz, H. Biodegradable Carboxymethyl Cellulose Based Material for Sustainable Packaging Application. *Sci. Rep.*, 2020, *10* (1), 21960. https://doi.org/10.1038/s41598-020-78912-z.

46. Thakur, S.; Chaudhary, J.; Sharma, B.; Verma, A.; Tamulevicius, S.; Thakur, V. K. Sustainability of Bioplastics: Opportunities and Challenges. *Curr. Opin. Green Sustain. Chem.*, 2018, *13*, 68–75. https://doi.org/10.1016/j.cogsc.2018.04.013.
47. Arikan, E. B.; Ozsoy, H. D. A Review: Investigation of Bioplastics. *J. Civ. Eng. Archit.*, 2015, *9* (2). https://doi.org/10.17265/1934-7359/2015.02.007.
48. Han, J.; Guo, Y.; Wang, H.; Zhang, K.; Yang, D. Sustainable Bioplastic Made from Biomass DNA and Ionomers. *J. Am. Chem. Soc.*, 2021, *143* (46), 19486–19497. https://doi.org/10.1021/jacs.1c08888.
49. Cosma, I. G. Bio-Plastic–between Current Practices and the Challenges of a Sustainable Future. *Manager*, 2018, 27(1), 51–63.
50. Dash, S. S.; Kanungo, A. P.; Behera, R. K. Bio-Plastic for Sustainable Future. *Vigyan Varta*, 2022, *3* (1), 23–25.
51. Reddy, R. L.; Reddy, V. S.; Gupta, G. A. Study of Bio-Plastics as Green and Sustainable Alternative to Plastics. *Int. J. Emerg. Technol. Adv. Eng.*, 2013, *3* (5), 76–81.
52. Momani, B. Assessment of the Impacts of Bioplastics: Energy Usage, Fossil Fuel Usage, Pollution, Health Effects, Effects on the Food Supply, and Economic Effects Compared to Petroleum Based Plastics. *Interact. Qualif. Proj. Rep. Worcest. Polytech. Inst.*, 2009.
53. Gironi, F.; Piemonte, V. Bioplastics and Petroleum-Based Plastics: Strengths and Weaknesses. *Energy Sources Part Recovery Util. Environ. Eff.*, 2011, *33* (21), 1949–1959.
54. Meeks, D.; Hottle, T.; Bilec, M. M.; Landis, A. E. Compostable Biopolymer Use in the Real World: Stakeholder Interviews to Better Understand the Motivations and Realities of Use and Disposal in the US. *Resour. Conserv. Recycl.*, 2015, *105*, 134–142.
55. Bhagwat, G.; Gray, K.; Wilson, S. P.; Muniyasamy, S.; Vincent, S. G. T.; Bush, R.; Palanisami, T. Benchmarking Bioplastics: A Natural Step towards a Sustainable Future. *J. Polym. Environ.*, 2020, *28* (12), 3055–3075.
56. Benn, N.; Zitomer, D. Pretreatment and Anaerobic Co-Digestion of Selected PHB and PLA Bioplastics. *Front. Environ. Sci.*, 2018, *5*, 93.
57. Al Hosni, A. S. K. *Biodegradation of Polycaprolactone Bioplastic in Comparision with Other Bioplastics and Its Impact on Biota*; The University of Manchester: United Kingdom, 2019.
58. Meereboer, K. W.; Misra, M.; Mohanty, A. K. Review of Recent Advances in the Biodegradability of Polyhydroxyalkanoate (PHA) Bioplastics and Their Composites. *Green Chem.*, 2020, *22* (17), 5519–5558.
59. Xu, Y.; Xu, J.; Liu, D.; Guo, B.; Xie, X. Synthesis and Characterization of Biodegradable Poly (Butylene Succinate-Co-Propylene Succinate) s. *J. Appl. Polym. Sci.*, 2008, *109* (3), 1881–1889.
60. Xu, J.; Guo, B.-H. Poly (Butylene Succinate) and Its Copolymers: Research, Development and Industrialization. *Biotechnol. J.*, 2010, *5* (11), 1149–1163.
61. Gadhave, R. V.; Das, A.; Mahanwar, P. A.; Gadekar, P. T. *Starch Based Bio-Plastics: The Future of Sustainable Packaging*, 2018.
62. Gozan, M.; Noviasari, C. The Effect of Glycerol Addition as Plasticizer in Spirulina Platensis Based Bioplastic. In *E3S Web of Conferences*; EDP Sciences, 2018; *67*, 03048.
63. Oluwasina, O. O.; Akinyele, B. P.; Olusegun, S. J.; Oluwasina, O. O.; Mohallem, N. D. Evaluation of the Effects of Additives on the Properties of Starch-Based Bioplastic Film. *SN Appl. Sci.*, 2021, *3* (4), 1–12.
64. Gupta, P.; Toksha, B.; Rahaman, M. A Review on Biodegradable Packaging Films from Vegetative and Food Waste. *Chem. Rec.*, 2022, *22* (7), e202100326. https://doi.org/10.1002/tcr.202100326.

Sustainable Biodegradable and Bio-based Polymers

65. Perotto, G.; Ceseracciu, L.; Simonutti, R.; Paul, U. C.; Guzman-Puyol, S.; Tran, T.-N.; Bayer, I. S.; Athanassiou, A. Bioplastics from Vegetable Waste via an Eco-Friendly Water-Based Process. *Green Chem.*, 2018, *20* (4), 894–902. https://doi.org/10.1039/C7GC03368K.

66. Orenia, R. M.; Iii, A. C.; Magno, M. G.; Cancino, L. T. Fruit and Vegetable Wastes as Potential Component of Biodegradable Plastic. *Asian J. Multidiscip. Stud.*, 2018, *1* (1), 61–77.

67. Koosha, M.; Hamedi, S. Intelligent Chitosan/PVA Nanocomposite Films Containing Black Carrot Anthocyanin and Bentonite Nanoclays with Improved Mechanical, Thermal and Antibacterial Properties. *Prog. Org. Coat.*, 2019, *127*, 338–347. https://doi.org/10.1016/j.porgcoat.2018.11.028.

68. Cinelli, P.; Coltelli, M. B.; Signori, F.; Morganti, P.; Lazzeri, A. Cosmetic Packaging to Save the Environment: Future Perspectives. *Cosmetics*, 2019, *6* (2), 26. https://doi.org/10.3390/cosmetics6020026.

69. Vasquez, E. S. L.; Vega, K. From Plastic to Biomaterials: Prototyping DIY Electronics with Mycelium. In *Adjunct Proceedings of the 2019 ACM International Joint Conference on Pervasive and Ubiquitous Computing and Proceedings of the 2019 ACM International Symposium on Wearable Computers*; UbiComp/ISWC '19 Adjunct; Association for Computing Machinery: New York, NY, USA, 2019; pp. 308–311. https://doi.org/10.1145/3341162.3343808.

70. Doi, Y. Microbial Synthesis and Properties of Polyhydroxy-Alkanoates. *MRS Bull.*, 1992, *17* (11), 39–42.

71. Huijberts, G. N.; Eggink, G.; De Waard, P.; Huisman, G. W.; Witholt, B. *Pseudomonas putida* KT2442 Cultivated on Glucose Accumulates Poly (3-Hydroxyalkanoates) Consisting of Saturated and Unsaturated Monomers. *Appl. Environ. Microbiol.*, 1992, *58* (2), 536–544.

72. Kundu, P. P.; Nandy, A.; Mukherjee, A.; Pramanik, N. Polyhydroxyalkanoates: Microbial Synthesis and Applications. *Encycl. Biomed. Polym. Polym. Biomater.* CRC Press: Boca Raton, FL, 2015; *11*, 10444.

73. Stiles, W. A.; Styles, D.; Chapman, S. P.; Esteves, S.; Bywater, A.; Melville, L.; Silkina, A.; Lupatsch, I.; Grünewald, C. F.; Lovitt, R. Using Microalgae in the Circular Economy to Valorise Anaerobic Digestate: Challenges and Opportunities. *Bioresour. Technol.*, 2018, *267*, 732–742.

74. Price, S.; Kuzhiumparambil, U.; Pernice, M.; Ralph, P. J. Cyanobacterial Polyhydroxybutyrate for Sustainable Bioplastic Production: Critical Review and Perspectives. *J. Environ. Chem. Eng.*, 2020, *8* (4), 104007. https://doi.org/10.1016/j.jece.2020.104007.

75. Lim, C.; Yusoff, S.; Ng, C. G.; Lim, P. E.; Ching, Y. C. Bioplastic Made from Seaweed Polysaccharides with Green Production Methods. *J. Environ. Chem. Eng.*, 2021, *9* (5), 105895. https://doi.org/10.1016/j.jece.2021.105895.

76. Carina, D.; Sharma, S.; Jaiswal, A. K.; Jaiswal, S. Seaweeds Polysaccharides in Active Food Packaging: A Review of Recent Progress. *Trends Food Sci. Technol.*, 2021, *110*, 559–572. https://doi.org/10.1016/j.tifs.2021.02.022.

77. Khalil, H. P. S.; Tye, Y. Y.; Saurabh, C. K.; Leh, C. P.; Lai, T. K.; Chong, E. W. N.; Fazita, M. R.; Hafiidz, J. M.; Banerjee, A.; Syakir, M. I. Biodegradable Polymer Films from Seaweed Polysaccharides: A Review on Cellulose as a Reinforcement Material. *Express Polym. Lett.*, 2017, *11* (4).

78. Baskar, S. *Food Grade Bioplastics From Red Seaweed*, 2020.

79. Cazón, P.; Velazquez, G.; Ramírez, J. A.; Vázquez, M. Polysaccharide-Based Films and Coatings for Food Packaging: A Review. *Food Hydrocoll.*, 2017, *68*, 136–148.

80. Rajendran, N.; Puppala, S.; Sneha, R.; Angeeleena, R. *Seaweeds Can Be a New Source for Bioplastics*, 2012.

81. Atiwesh, G.; Mikhael, A.; Parrish, C. C.; Banoub, J.; Le, T.-A. T. Environmental Impact of Bioplastic Use: A Review. *Heliyon*, 2021, *7* (9), e07918. https://doi.org/10.1016/j.heliyon.2021.e07918.

82. Yates, M. R.; Barlow, C. Y. Life Cycle Assessments of Biodegradable, Commercial Biopolymers—A Critical Review. *Resour. Conserv. Recycl.*, 2013, *78*, 54–66. https://doi.org/10.1016/j.resconrec.2013.06.010.

83. Chalermthai, B.; Giwa, A.; Schmidt, J. E.; Taher, H. Life Cycle Assessment of Bioplastic Production from Whey Protein Obtained from Dairy Residues. *Bioresour. Technol. Rep.*, 2021, *15*, 100695.

84. Kendall, A. A Life Cycle Assessment of Biopolymer Production from Material Recovery Facility Residuals. *Resour. Conserv. Recycl.*, 2012, *61*, 69–74. https://doi.org/10.1016/j.resconrec.2012.01.008.

85. Grujic, R.; Vukic, M.; Gojkovic, V. Application of Biopolymers in the Food Industry. In *Advances in Applications of Industrial Biomaterials*; Pellicer, E., Nikolic, D., Sort, J., Baró, M., Zivic, F., Grujovic, N., Grujic, R., Pelemis, S., Eds.; Springer International Publishing: Cham, 2017; pp. 103–119. https://doi.org/10.1007/978-3-319-62767-0_6.

86. Şengör, S. S. Review of Current Applications of Microbial Biopolymers in Soil and Future Perspectives. In *Introduction to Biofilm Engineering*; ACS Symposium Series; American Chemical Society, 2019; Vol. 1323, pp. 275–299. https://doi.org/10.1021/bk-2019-1323.ch013.

87. Mallick, P. K. Advanced Materials for Automotive Applications: An Overview. In *Advanced Materials in Automotive Engineering*; Rowe, J., Ed.; Woodhead Publishing, 2012; pp. 5–27. https://doi.org/10.1533/9780857095466.5.

88. Jariyasakoolroj, P.; Leelaphiwat, P.; Harnkarnsujarit, N. Advances in Research and Development of Bioplastic for Food Packaging. *J. Sci. Food Agric.*, 2020, *100* (14), 5032–5045.

89. Kulma, A.; Skórkowska-Telichowska, K.; Kostyn, K.; Szatkowski, M.; Skała, J.; Drulis-Kawa, Z.; Preisner, M.; Żuk, M.; Szperlik, J.; Wang, Y. F.; et al. New Flax Producing Bioplastic Fibers for Medical Purposes. *Ind. Crops Prod.*, 2015, *68*, 80–89. https://doi.org/10.1016/j.indcrop.2014.09.013.

90. Porta, R. Plastic Pollution and the Challenge of Bioplastics. *J. Appl. Biotechnol. Bioeng.*, 2017, *2* (3). https://doi.org/10.15406/jabb.2017.02.00033.

91. Abe, M. M.; Martins, J. R.; Sanvezzo, P. B.; Macedo, J. V.; Branciforti, M. C.; Halley, P.; Botaro, V. R.; Brienzo, M. Advantages and Disadvantages of Bioplastics Production from Starch and Lignocellulosic Components. *Polymers*, 2021, *13* (15), 2484. https://doi.org/10.3390/polym13152484.

3 Processing of Recycled and Bio-based Plastics

Yash Rushiya, Bhagwan Toksha, and Prashant Gupta

CONTENTS

3.1 Introduction 39
3.2 Classification of Plastics Derived from Renewable
and Nonrenewable Resources 40
3.3 Life Cycle Assessment...42
3.4 Impact of Recycling on Polymer Properties...43
3.5 Bio-Based Plastics and Mechanism of Biodegradation....................................43
3.6 Processing..44
    3.6.1 Additives..44
    3.6.2 Extrusion..45
    3.6.3 Compression Molding..46
    3.6.4 Injection Molding 47
3.7 Applications of Recycled and Bio-based Materials...47
3.8 Challenges and Limitations 48
3.9 Conclusion 49
References..49

3.1 INTRODUCTION

The huge technical advancements made in the last few decades have led to plastics becoming an essential part of human life, and one cannot live without them. In comparison to other materials, plastics have the advantages of being inexpensive, lightweight, strong, and readily available. Plastics are used in a wide range of applications, including packaging, food, consumer goods, and medical devices. For example, fuel consumption in automobiles is reduced due to various components used in an automobile being made from plastic [1]. Plastics are made from a variety of sources, including petroleum and natural resources such as plants and animals. Looking ahead to gasoline and diesel crisis in a few years, it's possible that plastics made from fossil fuels will be phased out. Plastic has become increasingly harmful to the environment in recent years, causing global warming, littering, and the depletion of natural resources [2–4]. It is estimated that the global use of plastic will reach 236–417 million tons per year by 2030 [5]. From the overall consumption of plastic, only 16% of plastic is recycled, while 40% of it ended up in landfills, and 25% is incinerated. As per the report in the UK, plastics collected in 2015–2016 were 22% polyethylene, 10.2% polypropylene, 40% PET, PVC, and PS as 2% [5].

DOI: 10.1201/9781003297772-3

Plastics derived from petroleum are referred to as nonbiodegradable plastics, whereas those derived from natural resources are referred to as biodegradable plastics, which imply that they are derived from a biological source such as a plant or animal. Biodegradation is the breakdown of a polymer by microorganisms, humidity, temperature, and the presence of oxygen, which disintegrate the polymer into CO_2, methane, and water [6]. The processing of bio-based and fossil-based plastic is similar to that of conventional plastic and is processed by the same processing equipment [7]. Biodegradable and bio-based materials are used for applications such as shopping bags, flexible and rigid packaging, nonwovens, and agricultural mulching films. Biodegradable and bio-based materials have poorer properties than fossil-based materials, including mechanical, water vapor barrier, heat stability, and are difficult to process [8]. Widespread use of biopolymers is limited due to their short shelf life, poor mechanical and barrier properties, and, therefore, conventional plastics are not replaced by bio-based materials due to their versatile properties and long shelf life. It has, thus, been a challenge to the researchers working in biomaterials domain. [9]. Few researchers have worked on it by modifying bio-based materials, blending, and using nanocomposite materials [10, 11].

Sustainable materials have a lower environmental impact, and their primary goal is public safety. Bio-plastics can be easily made from agricultural resources such as proteins, starch, cellulose, and soybeans, and contain 100% carbon sourced from forests and agriculture [12]. Thorpe and Vander developed a plastics pyramid that helps in material selection by displaying the life cycle hazards of different plastic materials. The development of the pyramid shows the recyclability of different plastics with their life cycle hazards. This pyramid spectrum suggests the use of easily recyclable plastics along with the ones that do not emit hazardous chemicals [13]. Most European countries have imposed taxes on landfilling, which has increased the recycling rate. Out of 27 EU countries, 24 of them have imposed taxes on landfilling [14]. The importance of plastic in health care (face shields, gloves, masks, PPE kit, etc.) has been demonstrated by the COVID-19 situation, and during the lockdown, several norms such as social distancing and public gathering have been restricted, as a result of which most people have done their shopping online, which has resulted in increased use of plastic in the packaging sector [15]. In the context of the aforesaid discussion, the objective of this chapter is set to touch upon the sustainability of bio-based and recycled materials, challenges while processing, improvement in processability by addition of additives, blending and use of nanocomposite, recent advancements in bio-based and recycled materials. The life cycle of plastics as identified by Filiciotto and Rothenberg is depicted in Figure 3.1 [16].

3.2 CLASSIFICATION OF PLASTICS DERIVED FROM RENEWABLE AND NONRENEWABLE RESOURCES

Plastics are classified broadly based on renewable and nonrenewable resources as shown in Figure 3.2. Bio-based and biodegradable plastics are those that are made from renewable resources. These plastics are safer to use and will not pollute the environment if disposed of inland. Nonrenewable plastics are those that are derived from nonrenewable sources, such as fossils, and are limited in use. Because fossils

Processing of Recycled and Bio-based Plastics

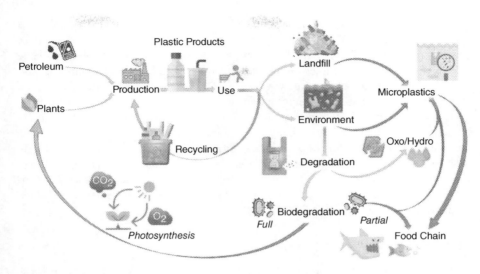

FIGURE 3.1 The evolution, usage and fate of fossil- and bio-based plastics [16].

are formed from the remains of dead animals, their formation is difficult and takes millions of years. Due to the demand versus supply scenario for bio-based plastics, most fossil-based plastics are made biodegradable by adding either additives or bio-based materials. These materials enhance properties such as biodegradability along with other functional aspects and potentially degrade them after use due to microorganism attack, heat, or environmental conditions [17].

Nonbiodegradable plastic is classified as plastic obtained from a nonrenewable resource like a fossil. Figure 3.2 represents plastics that are obtained from nonrenewable resources and biodegradable. They are Polyethylene (PE), Polypropylene (PP),

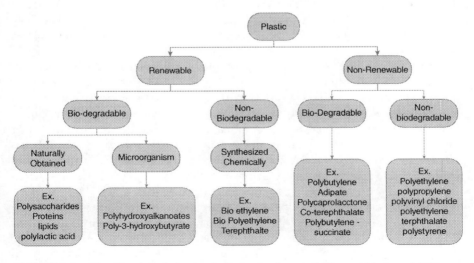

FIGURE 3.2 Classification of plastics on the basis of renewability and biodegradability.

Polyvinylchloride (PVC), and Polyethylene terephthalate (PET). All fossil-based plastics are not nonbiodegradable. A few of them, i.e., polybutylene adipate (PBA), polycaprolactone, poly-butylene succinate, and co-terephthalate, are biodegradable plastics [18]. Plastics obtained from renewable resources are biodegradable as well as nonbiodegradable, as illustrated in Figure 3.2. These types of plastics are derived from sources, such as plants, microorganisms, and chemicals, and are chemically synthesized; some are obtained naturally. Polyhydroxyalkanoates (PHA) and poly-3-hydroxybutyrate (PHB) are biodegradable plastics obtained from microorganisms or plants. Naturally obtained plastics are polysaccharides (starch, cellulose, lignin, and chitin), proteins (gelatin, casein, wheat gluten, silk, and wool), and lipids (plant oils and animal fats) [18]. Bio-PE and Bio-PET are plastics that are prepared chemically from biomaterials obtained from a renewable resource, such as Bio-PE, which is synthesized from bioethanol [19].

3.3 LIFE CYCLE ASSESSMENT

Life cycle assessment is a process that expresses the effect of any product or process on the environment that can be either positive or negative. Cradle-to-gate and cradle-to-grave are two types of commonly used studies [20]. In a cradle-to-gate methodology, the upstream impact is studied, which includes the impact from raw material till the point of product is dispatched. Similarly, in the cradle-to-grave study, the downstream impact is studied, which includes impact starting from raw material to its end of life. The study of the impact of a process and product on the environment requires the absolute finest details of an upstream and downstream process to be studied. One impact category involves the study of amount of greenhouse gas emitted by the polymer. The life cycle assessment involves environmental management system standards such as ISO 14040 and 14044. The limitation of these standards is that they do not describe the proper method of assessment or impact category and data quality. The European Union Product Environmental Footprint (EU PEF) and organizational impact assessment (OEF) were developed over the standard method. This method overcame the limitations of the standard method, which involves an impact category and a requirement for data quality.

Sperling et al. reviewed bio-based plastic for the bio-plastic life cycle from all aspects, which included economic (life cycle costing, LCC), environmental (LCA), and social perspectives [21]. The results of life cycle analyses of fossil-based and bio-based plastics were categorized as energy use, acidification, ecotoxicity, eutrophication, climate change, particulate matter formation, and ozone depletion. The result of the life cycle assessment was carried out by comparing 39 fossil-based polymers and 50 bio-based polymers. Various studies on life cycle assessment show that bio-based plastics achieve good results in energy use and climate change categories. Bio-based PET and PBS show poor result in acidification, while bio-based PVC, PHB, and PET are reported as significantly worse. Bio-based PLA and many other polymers have values twice those of fossil and bio-based alternatives. The ecotoxicity of bio-based PET is twice that of fossil and bio-based alternatives [20]. The negative result in the bio-based polymer is due to CO_2 getting absorbed by biomass during its growth to the polymer product.

3.4 IMPACT OF RECYCLING ON POLYMER PROPERTIES

Recycling is the process of treating polymers so that they can be used again. Recycling of polymers may be either mechanical or chemical recycling. Mechanical recycling of waste plastic involves various steps, such as separating, grinding, washing, drying, regranulating, and compounding, which lead to the deterioration of polymer properties. Similarly, in chemical recycling, the polymer is broken into its components, which lead to a change in the structure of the polymer [18]. Virgin PET has good mechanical, barrier, and processability properties. When it is mechanically recycled, it is found that the value of elongation at break is reduced by a factor of 4 [22]. Various carbon-based molecules, such as ketones, alcohols, aldehydes, acids, short-chain hydrocarbons, and esters, are formed as degradation side products during mechanical recycling [23]. Oblak et al. performed 100 consecutive trials of HDPE on extrusion and a significant change in its structure was reported. They reported the dominance of chain branching during the first 30 cycles, with chances of chain scission prominent between 30–60 cycles and again crosslinking above 60 cycles, thereby increasing the viscosity by 5 times [24]. The study done on lowering the screw speed concluded that it will reduce the crosslinking while chain scission can be enhanced by varying screw speed. Peroxides are used to minimize chain scission, induce crosslinking, and maintain mechanical properties. The use of trimethylolpropane triacrylate (TMPTA) to minimize chain scission and disproportionate reaction with a reduction in MFI value by more than a factor of 10 were among the primary work [25].

Virgin PP has good optical properties, a high volume-to-weight ratio, a glossy appearance, and high-tensile strength [26]. Tertiary carbon atoms present in the backbone increase flammability but are more prone to shear degradation [27]. Recycling of PP, on multiple occasions, reduces molecular weight, which will cause thermomechanical and thermo-oxidative degradation. An increase in the degree of crystallinity increases Young's modulus and reduces elongation at break. PP is stable to process at a lower temperature up to five extrusion cycles, while, at a high temperature, chain scission will take place. Virgin LDPE has a low density due to reduced packing, which will create space between polymers due to branching. Rheology, FTIR, and GPC tests of LDPE show an increase in molecular weight during recycling. PVC is light in weight, flexible, and has good oxygen and water barrier properties and is used in the food and construction industries. PVC is difficult to process due to the presence of chlorine atoms, which produce toxic HCl gas when released into the atmosphere. Mechanical recycling of polystyrene decreases the viscosity, elongation at break, and molecular weight. In their work on the processing of polystyrene, Remili et al. found that during ten cycles, the molecular weight of polystyrene decreased by 50% [28]. High-impact polystyrene when processed at a high temperature will lead to more chain scission than crosslinking and a decrease in elongation at the break by 38%.

3.5 BIO-BASED PLASTICS AND MECHANISM OF BIODEGRADATION

Bio-based plastics are made from biological resources, meaning they use fewer fossil fuels and are therefore more sustainable than fossil fuels. Bio-based plastics are the only type of plastic that does not harm the environment, as fossil-based plastics do not degrade and render the land unfit for agriculture. These plastics have a short shelf life and degrade in a usage cycle. In the presence of sunlight or heat, the fossil-based polymer may degrade. There are many fossil-based plastics that are biodegradable. Biodegradability can be improved by adding additives such as scavengers. The mechanism of biodegradation involves the attack of microorganisms, such as fungi or bacteria, on plastic after its use. Biodegradable polymers get easily degraded by microorganisms if no preservative is added, making it difficult to use for a long time, and their use in the packaging sector will degrade the food material inside them. These polymers can be degraded either physically or biologically. Physical biodegradation of polymers involves breaking down the polymer chains and making them inactive so that fresh monomers cannot attack them. The physical mechanism is either photodegradation, in which the polymer is degraded by sunlight, or hydrolysis, in which the polymer is broken down into its constituents using water. When all of the monomers present inside the polymer become inactive, i.e., when they have no charge or free radicals, then only polymer can be completely degraded. Natural polymers are easily degradable, whereas synthetic polymer chains must be broken by interventions from an external source.

Degradation of a polymer depends on properties such as the structure of the polymer, solubility, the presence of a functional group, and environmental factors such as temperature, pH, moisture, and microorganisms [6, 29]. Polymer biodegradation by microorganism attack eventually converts it to CO_2, methane, water, and biomass [6, 30]. Some polymers are degraded by the presence of oxygen, called aerobic biodegradation, and others by the absence of oxygen, which is called anaerobic biodegradation. These come under the category of biological degradation.

Various steps that involved in biodegradation [31] are given as under,

I. Fragmentation of biodegradable materials into small, tiny fragments by the combined action of decomposers, an abiotic factor which includes sunlight, temperature, etc., and microorganisms that will lead to biodeterioration.
II. Dimers, monomers, and oligomers are formed by the homolytic cleavage of polymer molecules by catalytic agents secreted by microorganisms. This step is known as depolymerization.
III. The stage of mineralization converts them into simple molecules, such as methane, CO_2, water, nitrogen, and other salts, are oxidized and released to the environment from intracellular metabolites.
IV. Primary and secondary metabolites, new biomass, storage vesicles, and energy are produced from the transported molecules which integrate the microbial metabolism into the cytoplasm.

3.6 PROCESSING

3.6.1 ADDITIVES

The addition of additives before processing of recycled material is a must to form a product with improved properties. The role of using an additive is to achieve

Processing of Recycled and Bio-based Plastics

good dispersion, complete solubility, and stability for the polymer during processing. During recycling of PET, metal-based stabilizers, such as lead phthalate or tin mercaptide, are incorporated to prevent thermal oxidation, organic phosphates are used as radical scavengers to prevent radical attack on the chain, ((1-X oxide,6,7-trioxane-1-phosphabicyclo-octan-4-yl) methoxy) dibenzo-oxaphosphinine 6-oxide (DP) used as a flame retarder [32]. PE, including HDPE and LDPE, requires stabilization before reprocessing to prevent it from thermo-oxidative degradation and to maintain the quality of the recyclate. Ketones and peracids are formed as a degradation product at different UV rates, which will impart a combination of thermal stabilizers, antioxidants, and UV stabilizers [33]. A commonly used antioxidant is Irganox 1010, which is a phenol that functions as an alkyl peroxy radical trap and hydrogen bonding stabilizer [34]. Another system used for stabilization includes the use of hindered amine stabilizer and hindered phenol, which forms nitroxyl radical upon photooxidation. Further, it reacts with phenol, making them responsible for affecting the redox system [34]. To maintain mechanical properties and to minimize chain scission in HDPE, peroxides are used to form crosslinking in chains [25]. TMPTA was used to reduce chain scission with increased branching, even though the addition of TMPTA in HDPE reduced MFI by a factor of 10 [35]. Chains are linked to other polymers by using diamines, epoxides, phosphates, acrylates, and silane [36].

Concerns about public health and the environment are becoming more prominent. The majority of research is focused on preventing antioxidant migration that causes cancer in humans [37]. Antioxidants are grafted onto nano-silica to prevent loss due to migration or volatilization. The main issue with recycling PVC is that if it is not properly stabilized, HCl gas is released. Some stabilizers can replace the lost chlorine atom in the polymer backbone. Because the stabilizer added during the processing of virgin polymer is consumed completely in its first life, more stabilizer is added to process recycled PVC. Sulfur organotin, metal soap, lead, and calcium zinc are among the metal-based stabilizers. The use of metal-based stabilizers will result in dark spots on the product, as well as the release of toxic gases that are harmful to humans [38]. Instead of a metal-based stabilizer, $CaCO_3$ is used to prevent the release of HCl and improve the mechanical properties such as elongation at break, elastic modulus, and impact strength by 20% [38, 39]. Polystyrene is a brittle material at room temperature. The incorporation of flexible material, such as a plasticizer, that makes it easily processable. Poly (styrene-*co*-2-ethylhexyl acrylate) is used as a compatibilizing copolymer to plasticize polymer [40]. The viscosity of HIPS can be increased by the addition of montmorillonite clay [41].

3.6.2 Extrusion

Extrusion is a process of making granules from virgin, recycled, or virgin and recycled materials that are fed into a hopper and, due to shearing of material inside the barrel above which heaters are fixed, in between rotating screws, makes the material homogeneously mixed. Shearing and thermal conduction polymer within the extruder will result in thermo-oxidation degradation, breaking of chains, or crosslinking between chains [42]. Extrusion is the primary process used in processing any plastic through different processing techniques such as injection, blow, and rotational molding. Products formed from extrusion are generally 2-D rod shaped called extrudates. The properties of the extrudate form depend on the processing parameters such as screw rotation speed, temperature, and pressure. Processing recycled material is challenging because of certain properties lost during its use and the various steps involved before processing such as collection, sorting, etc. Proper collection and sorting make recycling efficient.

Degradation kinetics are controlled by the polymer chain length while the thermo-oxidative degradation process is due to structure and oxygen [43]. Therefore, the susceptibility of a polymer to shear-induced degradation is directly proportional to the degree of chain branching and chain length for PET [44]. During the mechanical recycling and extrusion processes, the mechanical properties such as elongation at break, impact strength, and elongation at break decrease for recycled PP, HDPE, and LLDPE [42, 45]. Other polymers present in the recycled polymer will create processing difficulties due to variation in their melting point, screw speed rotation, etc. Proper sorting of material to be recycled will eliminate this problem of degradation of polymer which ultimately affects the quality of recyclate [46]. The presence of pigment which gives color to plastic has a potential to cause thermal degradation of polymer within the extruder. The presence of volatile ink component in final recyclate is due to label and printing ink present in the mix [47].

Innovation in the extrusion process has improved the reprocessing of polymers. Polymer melt quality is improved due to advancements in technology, which include the addition of various sections such as degassing, drying, softening, and filter extrudate [48]. The degassing section allows the volatile material within the polymer to escape from an open vent or by vacuum. Removal of volatile matter prevents the material from acidolysis, hydrolysis, and improves the melt odor of the polymer [49]. Nonvolatile and larger contaminants, such as oil or dust, are removed from the polymer melt, which improves the mechanical, optical, and homogeneity of the blend [50]. Melt filters are used to remove contamination and can include screen changers such as filter cartridges, woven screens, or slide plates [51].

3.6.3 Compression Molding

Compression molding is a process in which plastic materials, either virgin or recycled, are charged into the mold cavity, melted by the heaters, after which they are suppressed by applying pressure to the core half of the mold to get the shape of the mold. Complex-shaped products are not possible in compression. Compression molding products are primarily used in the electrical, packaging, and transportation industries. Recycling of electrical plastic is difficult due to the presence of elastomers and thermosets that show different melting temperatures during reprocessing. Therefore, proper sorting of material will eliminate this problem. Compression molding is a less expensive process than other processes, and therefore has its importance in less developed countries. Modification in mold design could help the polymer to be processed in compression molding by escaping the volatile gases through the vents provided in the mold, which improves the moldability and strength by minimizing the voids.

Processing of Recycled and Bio-based Plastics

Various studies show that variation in processing parameters, such as molding temperature, molding pressure, and molding time, reduces the challenges that occur during molding. The main problem during processing is the proper design of vents in the mold. Polymer when processed forms a gas, i.e., an exothermic reaction. Proper removal of these will help to achieve a good quality product and avoid breaking of mold due to excess heat generated, which causes high-pressure buildup in the mold [52].

3.6.4 INJECTION MOLDING

The injection molding process is used universally and, nowadays, a huge advancement in injection molding, from manual to fully automatic, has made the molding of complex parts within a second. The raw material used in injection molding is either virgin or recycled. In injection molding, material is melted inside the barrel by the heaters mounted on top of it. Material melted inside the barrel is plasticized and pushed into the mold where it takes the shape of a complex part.

Different research has made it easy to process recycled plastic material in injection molding machines by varying the processing parameters such as temperature, pressure, and cycle time. Recycled material before processing is made capable of processing by the addition of additives to make them stable during processing. Taguchi developed a method to optimize the processing parameters, such as melt temperature, mold temperature, injection speed, and packing pressure, to obtain end products with improved mechanical properties. In his work, he has used various ratios of virgin and recycled PP at different mold temperatures, melt temperatures, packing pressures, and injection pressures and studied the variation in properties. With the addition of recycled PP, both density and MFI have been increased. For better impact and flexural strength, it is recommended from the result that injection speed should be increased. Blends processed at low temperatures have good mechanical properties, as low temperatures prevent the material from degradation. Higher packing pressure results in good impact and flexural properties, while lower packing pressure will result in improved tensile strength [53].

3.7 APPLICATIONS OF RECYCLED AND BIO-BASED MATERIALS

The use of bio-based plastics all over the world has increased in the past few years. Bio-based plastics are used in the packaging sector because of their importance, which results in biodegradation, preventing pollution, and making them sustainable to use. Bio-based materials are finding new packaging and textile applications. The end of life and recycling are possible with industrial and home composting, mechanical and chemical/catalytic recycling, enzymatic depolymerization, and biogas installations. PLA, being commercially available, has a market of around 700 million USD in 2019 and is expected to increase by 2500 million USD by 2025. PLA is used as an opaque and rigid plastic for packaging, durable goods, disposable goods, and bottles, as well as for films and fibers [54, 55]. PHA is used for disposable food containers and utensils, yogurt containers, cold and hot cups, trays, tubs, and single use for food packaging [56].

Polybutylene succinate adipate and Polybutylene succinate are used for making semirigid bowls and films [57]. The global capacity of PBS is around 140,000 tons per annum which was 12.8% of overall biodegradable plastic produced, as reported in 2014 [58]. Other bio-based materials, such as bio-based PET and 1,3-propanediol, are obtained from plants and used for applications, such as film and carpeting, while B-PET is for clear plastic bottles. Starch-based bio-plastic is used for mulch films, clips, and carrier bags. Natural biopolymer films have good gas barrier properties, but poor mechanical properties, such as being brittle and mechanically weak, are the main challenges in their applications. Their shelf life is limited, and they are susceptible to microbial attack and oxidation. Under the influence of government policies and internal initiatives, beverage companies are using partially bio-based PET bottles for packaging and practicing recyclability [59]. About 30% of the PET is used as bio-based PET and 72% of recycled PET is used for the application of fibers.

3.8 CHALLENGES AND LIMITATIONS

Plastic recycling is important nowadays for plastic obtained from petroleum, such as HDPE, PP, and PET, which is used more and more all over the world, which creates the problem of pollution, ultimately affecting the living organisms on earth. Plastic is misused due to a lack of awareness in humans, which is creating different types of pollution. This includes water pollution due to plastics being thrown in oceans after use, which is affecting living things present in water. Burning waste plastics leads to air pollution which causes humans and animals to breathe hazardous gases in the air. Disposing plastic inside the Earth makes the soil unable to use, which has a great impact on agriculture. Countries, such as the UK and France, use incineration and landfilling as common practices. Identifying proper techniques and their effective implementation regarding the collection, sorting, and recycling of plastic after its use will really help us to minimize environmental damage. Waste plastic produced in a year is huge in quantity and goes without recycling [60]. A major challenge in the recycling sector is the separation of mixed plastics. Flexible plastic recycling is difficult during the collection and sorting of flexible bags and films due to the low weight to volume ratio. This requires high-performance sorting, which sorts the plastic to high purity [61]. A major challenge in the recycling sector is the separation of mixed plastics. Flexible plastic recycling is difficult during the collection and sorting of flexible bags and films due to their low weight-to-volume ratio. This requires high-performance sorting, which sorts the plastic to high purity.

Processing of recycled material is difficult without the addition of additives. Due to various steps involved in recycling, certain properties, such as mechanical, optical, and melting points, are reduced, which make the material burn inside a barrel of injection molding without the addition of heat stabilizer. The mechanical properties of recycled material are induced by the addition of filler in different proportions according to end-use application. The melting point of plastic is increased by the additional plasticizer, which makes the polymer chains slide past each other freely. Challenges offered by traditional plastics are replaced by bio-based and biodegradable plastics. Bio-based materials have yet to gain their significance, unlike traditional plastics, due to the degradation of plastic taking place only under controlled

Processing of Recycled and Bio-based Plastics

conditions. The degradation of bio-based plastics in the sea is imposing survival threats on aquatic vertebrates. Researchers are also faced with the challenge of developing approaches to protect the sea ecosystem [61].

3.9 CONCLUSION

Bio-based plastics are the future of plastics to end the dependency on fossil-based plastics. There are a few challenges for the use of bio-based plastic that make it uncomfortable to use over fossil-based plastic. The complex structure and difficulty of their processing without additives is a research gap yet to be overcome. When plastic undergoes recycling, it passes through various steps such as collection, sorting, shredding, and processing in an extruder to form granules. The addition of additives in plastic and blending the recycled plastic with virgin plastic reduce the challenges that occur during processing and regenerate most of the functional properties that virgin polymer has lost during its use and recycling step. Polymers, either bio-based or fossil-based, are reprocessed by varying the processing parameters such as injection pressure, clamping pressure, melting temperature, and mold temperature. The focus is on the development of bio-based plastics, including their recycling without environmental degradation, achieving mechanical properties similar to those of conventional plastics, and easy processability on injection, blow, and compression.

REFERENCES

1. Szeteiová, K. Automotive Materials Plastics in Automotive Markets Today. **2010**, 7.
2. Mangaraj, S.; Yadav, A.; Bal, L. M.; Dash, S. K.; Mahanti, N. K. Application of Biodegradable Polymers in Food Packaging Industry: A Comprehensive Review. *J. Packag. Technol. Res.*, **2019**, *3* (1), 77–96. https://doi.org/10.1007/s41783-018-0049-y.
3. Schmidt Rivera, X. C.; Leadley, C.; Potter, L.; Azapagic, A. Aiding the Design of Innovative and Sustainable Food Packaging: Integrating Techno-Environmental and Circular Economy Criteria. *Energy Procedia*, **2019**, *161*, 190–197. https://doi.org/10.1016/j.egypro.2019.02.081.
4. Wohner, B.; Pauer, E.; Heinrich, V.; Tacker, M. Packaging-Related Food Losses and Waste: An Overview of Drivers and Issues. *Sustainability*, **2019**, *11* (1), 264. https://doi.org/10.3390/su11010264.
5. Recycling and the future of the plastics industry | McKinsey https://www.mckinsey.com/industries/chemicals/our-insights/how-plastics-waste-recycling-could-transform-the-chemical-industry (accessed Mar 13, **2022**).
6. Guzman, A.; Gnutek, N.; Janik, H. Biodegradable Polymers for Food Packaging—Factors Influencing Their Degradation and Certification Types—A Comprehensive Review. **2011**, 8.
7. Rujnić-Sokele, M.; Pilipović, A. Challenges and Opportunities of Biodegradable Plastics: A Mini Review. *Waste Manag. Res. J. Sustain. Circ. Econ.*, **2017**, *35* (2), 132–140. https://doi.org/10.1177/0734242X16683272.
8. Nilsen-Nygaard, J.; Fernández, E. N.; Radusin, T.; Rotabakk, B. T.; Sarfraz, J.; Sharmin, N.; Sivertsvik, M.; Sone, I.; Pettersen, M. K. Current Status of Biobased and Biodegradable Food Packaging Materials: Impact on Food Quality and Effect of Innovative Processing Technologies. *Compr. Rev. Food Sci. Food Saf.*, **2021**, *20* (2), 1333–1380. https://doi.org/10.1111/1541-4337.12715.

9. Pettersen, M. K.; Bardet, S.; Nilsen, J.; Fredriksen, S. B. Evaluation and Suitability of Biomaterials for Modified Atmosphere Packaging of Fresh Salmon Fillets. *Packag. Technol. Sci.*, **2011**, *24* (4), 237–248. https://doi.org/10.1002/pts.931.

10. Imre, B.; Pukánszky, B. Compatibilization in Bio-Based and Biodegradable Polymer Blends. *Eur. Polym. J.*, **2013**, *49* (6), 1215–1233. https://doi.org/10.1016/j.eurpolymj.2013.01.019.

11. Basavegowda, N.; Baek, K.-H. Advances in Functional Biopolymer-Based Nanocomposites for Active Food Packaging Applications. *Polymers*, **2021**, *13* (23), 4198. https://doi.org/10.3390/polym13234198.

12. Mohanty, A. K.; Misra, M.; Drzal, L. T. Sustainable Bio-Composites from Renewable Resources: Opportunities and Challenges in the Green Materials World. *J. Polym. Environ.*, **2002**, *10* (1), 19–26. https://doi.org/10.1023/A:1021013921916.

13. Alvarez-Chavez, C. R.; Edwards, S.; Moure-Eraso, R.; Geiser, K. Sustainability of Bio-Based Plastics: General Comparative Analysis and Recommendations for Improvement. *J. Clean. Prod.*, **2011**, *23*, 47–56. https://doi.org/10.1016/j.jclepro.2011.10.003.

14. CEWEP—The Confederation of European Waste-to-Energy Plants https://www.cewep.eu/landfill-taxes-and-bans/ (accessed Mar 13, **2022**).

15. COVID-19 Pandemic Repercussions on the Use and Management of Plastics | Environmental Science & Technology. https://pubs.acs.org/doi/abs/10.1021/acs.est.0c02178 (accessed Mar 13, **2022**).

16. Filiciotto, L.; Rothenberg, G. Biodegradable Plastics: Standards, Policies, and Impacts. *Chemsuschem*, **2021**, *14* (1), 56–72. https://doi.org/10.1002/cssc.202002044.

17. Rydz, J.; Musioł, M.; Zawidlak-Węgrzyńska, B.; Sikorska, W. Present and Future of Biodegradable Polymers for Food Packaging Applications. *Biopolym. Food Des.*, **2018**, 431–467.

18. Wojnowska-Baryła, I.; Kulikowska, D.; Bernat, K. Effect of Bio-Based Products on Waste Management. *Sustainability*, **2020**, *12* (5), 2088. https://doi.org/10.3390/su12052088.

19. Hottle, T. A.; Bilec, M. M.; Landis, A. E. Sustainability Assessments of Bio-Based Polymers. *Polym. Degrad. Stab.*, **2013**, *98* (9), 1898–1907. https://doi.org/10.1016/j.polymdegradstab.2013.06.016.

20. Walker, S.; Rothman, R. Life Cycle Assessment of Bio-Based and Fossil-Based Plastic: A Review. *J. Clean. Prod.*, **2020**, *261*, 121158. https://doi.org/10.1016/j.jclepro.2020.121158.

21. Spierling, S.; Knüpffer, E.; Behnsen, H.; Mudersbach, M.; Krieg, H.; Springer, S.; Albrecht, S.; Herrmann, C.; Endres, H.-J. Bio-Based Plastics—A Review of Environmental, Social and Economic Impact Assessments. *J. Clean. Prod.*, **2018**, *185*, 476–491. https://doi.org/10.1016/j.jclepro.2018.03.014.

22. Hong, M.; Chen, E. Y.-X. Chemically Recyclable Polymers: A Circular Economy Approach to Sustainability. *Green Chem.*, **2017**, *19* (16), 3692–3706. https://doi.org/10.1039/C7GC01496A.

23. Bikiaris, D. N.; Karayannidis, G. P. Effect of Carboxylic End Groups on Thermooxidative Stability of PET and PBT. *Polym. Degrad. Stab.*, **1999**, *63* (2), 213–218. https://doi.org/10.1016/S0141-3910(98)00094-9.

24. Oblak, P.; Gonzalez-Gutierrez, J.; Zupančič, B.; Aulova, A.; Emri, I. Processability and Mechanical Properties of Extensively Recycled High Density Polyethylene. *Polym. Degrad. Stab.*, **2015**, *114*, 133–145. https://doi.org/10.1016/j.polymdegradstab.2015.01.012.

25. Kim, K. J.; Kim, B. K. Crosslinking of HDPE during Reactive Extrusion: Rheology, Thermal, and Mechanical Properties. *J. Appl. Polym. Sci.*, **1993**, *48* (6), 981–986.

26. Gu, F.; Hall, P.; Miles, N. J. Performance Evaluation for Composites Based on Recycled Polypropylene Using Principal Component Analysis and Cluster Analysis. *J. Clean. Prod.*, **2016**, *115*, 343–353. https://doi.org/10.1016/j.jclepro.2015.12.062.
27. Boldizar, A.; Jansson, A.; Gevert, T.; Möller, K. Simulated Recycling of Post-Consumer High Density Polyethylene Material. *Polym. Degrad. Stab.*, **2000**, *68* (3), 317–319. https://doi.org/10.1016/S0141-3910(00)00012-4.
28. Remili, C.; Kaci, M.; Benhamida, A.; Bruzaud, S.; Grohens, Y. The Effects of Reprocessing Cycles on the Structure and Properties of Polystyrene/Cloisite15A Nanocomposites. *Polym. Degrad. Stab.*, **2011**, *96* (8), 1489–1496. https://doi.org/10.1016/j.polymdegradstab.2011.05.005.
29. Mohan, S. K.; Srivastava, T. Microbial Deterioration and Degradation of Polymeric Materials. *J. Biochem. Tech.*, **2010**, *2* (4), 210–215.
30. Mohanty, A. K.; Wibowo, A.; Misra, M.; Drzal, L. T. Development of Renewable Resource-Based Cellulose Acetate Bioplastic: Effect of Process Engineering on the Performance of Cellulosic Plastics. *Polym. Eng. Sci.*, **2003**, *43* (5), 1151–1161. https://doi.org/10.1002/pen.10097.
31. Walsh, J. H. Ecological Considerations of Biodeterioration. *Int. Biodeterior. Biodegrad.*, **2001**, *48* (1), 16–25. https://doi.org/10.1016/S0964-8305(01)00063-4.
32. Gooneie, A.; Simonetti, P.; Salmeia, K. A.; Gaan, S.; Hufenus, R.; Heuberger, M. P. Enhanced PET Processing with Organophosphorus Additive: Flame Retardant Products with Added-Value for Recycling. *Polym. Degrad. Stab.*, **2019**, *160*, 218–228. https://doi.org/10.1016/j.polymdegradstab.2018.12.028.
33. Rosales-Jasso, A.; Allen, N. S.; Sasaki, M. Evaluation of Novel 4,4-Dimethyloxazolidine Derivatives as Thermal and UV Stabilisers in Linear Low Density Polyethylene (LLDPE) Film. *Polym. Degrad. Stab.*, **1999**, *64* (2), 277–287. https://doi.org/10.1016/S0141-3910(98)00203-1.
34. Schwetlick, K.; Habicher, W. D. Antioxidant Action Mechanisms of Hindered Amine Stabilisers. *Polym. Degrad. Stab.*, **2002**, *78* (1), 35–40. https://doi.org/10.1016/S0141-3910(02)00116-7.
35. Su, F.-H.; Huang, H.-X. Rheology and Thermal Properties of Polypropylene Modified by Reactive Extrusion with Dicumyl Peroxide and Trimethylol Propane Triacrylate. *Adv. Polym. Technol.*, **2009**, *28* (1), 16–25. https://doi.org/10.1002/adv.20146.
36. Hamielec, A. E.; Gloor, P. E.; Zhu, S. Kinetics of, Free Radical Modification of Polyolefins in Extruders—Chain Scission, Crosslinking and Grafting. *Can. J. Chem. Eng.*, **1991**, *69* (3), 611–618. https://doi.org/10.1002/cjce.5450690302.
37. Dintcheva, N. Tz.; D'Anna, F. Anti-/Pro-Oxidant Behavior of Naturally Occurring Molecules in Polymers and Biopolymers: A Brief Review. *ACS Sustain. Chem. Eng.*, **2019**, *7* (15), 12656–12670. https://doi.org/10.1021/acssuschemeng.9b02127.
38. Everard, M. Twenty Years of the Polyvinyl Chloride Sustainability Challenges. *J. Vinyl Addit. Technol.*, **2020**, *26* (3), 390–402. https://doi.org/10.1002/vnl.21754.
39. Braun, D. Recycling of PVC. *Prog. Polym. Sci.*, **2002**, *27* (10), 2171–2195. https://doi.org/10.1016/S0079-6700(02)00036-9.
40. Nagalakshmaiah, M.; Nechyporchuk, O.; El Kissi, N.; Dufresne, A. Melt Extrusion of Polystyrene Reinforced with Cellulose Nanocrystals Modified Using Poly[(Styrene)-Co-(2-Ethylhexyl Acrylate)] Latex Particles. *Eur. Polym. J.*, **2017**, *91*, 297–306. https://doi.org/10.1016/j.eurpolymj.2017.04.020.
41. Nunes, M. A. B. S.; Galvão, L. S.; Ferreira, T. P. M.; Luiz, E. J. F. T.; Bastos, Y. L. M.; Santos, A. S. F. Reprocessability of High Impact Polystyrene/Clay Nanocomposites in Extrusion. *Polym. Degrad. Stab.*, **2016**, *125*, 87–96. https://doi.org/10.1016/j.polymdegradstab.2015.12.013.

42. Gryn'ova, G.; Hodgson, J. L.; Coote, M. L. Revising the Mechanism of Polymer Autooxidation. *Org. Biomol. Chem.*, **2010**, *9* (2), 480–490. https://doi.org/10.1039/C0OB00596G.
43. Luzuriaga, S.; Kovářová, J.; Fortelný, I. Degradation of Pre-Aged Polymers Exposed to Simulated Recycling: Properties and Thermal Stability. *Polym. Degrad. Stab.*, **2006**, *91* (6), 1226–1232. https://doi.org/10.1016/j.polymdegradstab.2005.09.004.
44. La Mantia, F. P.; Vinci, M. Recycling Poly(Ethyleneterephthalate). *Polym. Degrad. Stab.*, **1994**, *45* (1), 121–125. https://doi.org/10.1016/0141-3910(94)90187-2.
45. Eriksen, M. K.; Christiansen, J. D.; Daugaard, A. E.; Astrup, T. F. Closing the Loop for PET, PE and PP Waste from Households: Influence of Material Properties and Product Design for Plastic Recycling. *Waste Manag.*, **2019**, *96*, 75–85. https://doi.org/10.1016/j.wasman.2019.07.005.
46. Burgess, M.; Holmes, H.; Sharmina, M.; Shaver, M. P. The Future of UK Plastics Recycling: One Bin to Rule Them All. *Resour. Conserv. Recycl.*, **2021**, *164*, 105191. https://doi.org/10.1016/j.resconrec.2020.105191.
47. Faraca, G.; Astrup, T. Plastic Waste from Recycling Centres: Characterisation and Evaluation of Plastic Recyclability. *Waste Manag.*, **2019**, *95*, 388–398. https://doi.org/10.1016/j.wasman.2019.06.038.
48. Alshahrani, S. M.; Morott, J. T.; Alshetaili, A. S.; Tiwari, R. V.; Majumdar, S.; Repka, M. A. Influence of Degassing on Hot-Melt Extrusion Process. *Eur. J. Pharm. Sci.*, **2015**, *80*, 43–52. https://doi.org/10.1016/j.ejps.2015.08.008.
49. Marschik, C.; Löw-Baselli, B.; Miethlinger, J. Modeling Devolatilization in Single- and Multi-Screw Extruders. *AIP Conf. Proc.*, **2017**, *1914* (1), 080006. https://doi.org/10.1063/1.5016746.
50. Pachner, S.; Aigner, M.; Miethlinger, J. Modeling and Optimization of Melt Filtration Systems in Polymer Recycling. *AIP Conf. Proc.*, **2017**, *1914* (1), 080004. https://doi.org/10.1063/1.5016744.
51. Pachner, S.; Roland, W.; Aigner, M.; Marschik, C.; Stritzinger, U.; Miethlinger, J. Using Symbolic Regression Models to Predict the Pressure Loss of Non-Newtonian Polymer-Melt Flows through Melt-Filtration Systems with Woven Screens. *Int. Polym. Process.*, **2021**, *36* (4), 435–450. https://doi.org/10.1515/ipp-2020-4019.
52. Mills, W.; Tatara, R. Potential for Reuse of E-Plastics through Processing by Compression Molding. *Challenges*, **2016**, *7* (1), 13. https://doi.org/10.3390/challe7010013.
53. Gu, F.; Hall, P.; Miles, N. J.; Ding, Q.; Wu, T. Improvement of Mechanical Properties of Recycled Plastic Blends via Optimizing Processing Parameters Using the Taguchi Method and Principal Component Analysis. *Mater. Des. 1980–2015*, **2014**, *62*, 189–198. https://doi.org/10.1016/j.matdes.2014.05.013.
54. Vink, E. T. H.; Rábago, Karl. R.; Glassner, D. A.; Springs, B.; O'Connor, R. P.; Kolstad, J.; Gruber, P. R. The Sustainability of NatureWorks™ Polylactide Polymers and Ingeo™ Polylactide Fibers: An Update of the Future. *Macromol. Biosci.*, **2004**, *4* (6), 551–564. https://doi.org/10.1002/mabi.200400023.
55. Madhavan Nampoothiri, K.; Nair, N. R.; John, R. P. An Overview of the Recent Developments in Polylactide (PLA) Research. *Bioresour. Technol.*, **2010**, *101* (22), 8493–8501. https://doi.org/10.1016/j.biortech.2010.05.092.
56. Philip, S.; Keshavarz, T.; Roy, I. Polyhydroxyalkanoates: Biodegradable Polymers with a Range of Applications. *J. Chem. Technol. Biotechnol.*, **2007**, *82* (3), 233–247. https://doi.org/10.1002/jctb.1667.
57. Vytejčková, S.; Vápenka, L.; Hradecký, J.; Dobiáš, J.; Hajšlová, J.; Loriot, C.; Vannini, L.; Poustka, J. Testing of Polybutylene Succinate Based Films for Poultry Meat Packaging. *Polym. Test.*, **2017**, *60*, 357–364. https://doi.org/10.1016/j.polymertesting.2017.04.018.

Processing of Recycled and Bio-based Plastics

58. Hu, X.; Li, Y.; Gao, Y.; Wang, R.; Wang, Z.; Kang, H.; Zhang, L. Renewable and Super-Toughened Poly (Butylene Succinate) with Bio-Based Elastomers: Preparation, Compatibility and Performances. *Eur. Polym. J.*, **2019**, *116*, 438–444. https://doi.org/10.1016/j.eurpolymj.2019.03.057.

59. James Quincey. The Coca-Cola Company: World without Waste 2019. The Coca-Cola Company. **2019**.

60. Gebre, S.H.; Sendeku, M.G.; Bahri, M. Recent Trends in the Pyrolysis of Non-Degradable Waste Plastics. *ChemistryOpen*, **2001**, *10* (12), 1202–1226.

61. Hopewell, J.; Dvorak, R.; Kosior, E. Plastics Recycling: Challenges and Opportunities. *Philos. Trans. R. Soc. B Biol. Sci.*, **2009**, *364* (1526), 2115–2126. https://doi.org/10.1098/rstb.2008.0311.

4 Terpene-based Polymers as Sustainable Materials

Eksha Guliani and Christine Jeyaseelan

CONTENTS

4.1 Introduction 55
    4.1.1 Terpenes and Terpenoids 56
4.2 Green Chemistry and Sustainable Approach 57
4.3 Synthesis of Terpene-Based Polymers.. 58
    4.3.1 Synthesis of Poly-Alloocimene.. 58
    4.3.2 Synthesis of Allyl-Terpene Maleate Polymer 58
    4.3.3 Synthesis of Polymyrcene... 59
    4.3.4 Synthesis of Poly(Stearyl Methacrylate).. 59
    4.3.5 Synthesis of Tetrahydrogeraniol Acrylate (THGA)............................ 60
4.4 Characterization of Synthesized Polymers.. 60
    4.4.1 Characterization of Poly-Alloocimene 60
    4.4.2 Characterization of Allyl-Terpene Maleate Polymer.......................... 61
    4.4.3 Characterization of Polymyrcene 61
    4.4.4 Characterization of Poly(Stearyl Methacrylate) 61
    4.4.5 Characterization of THGA 62
4.5 Mechanism of Polymer Formation 62
    4.5.1 Chain-Growth Polymerization 62
    4.5.2 Step-Growth Polymerization 63
4.6 Modifications in Terpene-Based Polymers.. 64
    4.6.1 Lignin-Polymer Blends and Oligomers 64
    4.6.2 Grafting of Terpenes... 65
4.7 Applications 65
4.8 General Discussion 66
4.9 Conclusion and Future Prospects 66
References.. 67

4.1 INTRODUCTION

Plastics are one of the greatest inventions of humans that have made our life very much easier. These are versatile in nature and have amazing material properties like great strength, water and shock resistance, readily available, and cheap. These are widely used in almost every industry and one of the most common applications of plastic is its use as polythene. These are also used to make cars, houses, clothing, electronics, etc. Plastics are mostly formed from petroleum that is depleting.

DOI: 10.1201/9781003297772-4

These are the nonrenewable resources as these are limited in nature and their rapid consumption is ultimately causing global warming. Research is mainly focused to replace the polymers derived from petroleum products with renewable alternatives so that the materials synthesized are biodegradable and easily recyclable.

The renewable natural resources could be used as precursors for synthesizing eco-friendly polymers that are known as sustainable polymers and these could be fatty acids, lignin, cellulose, and plant oils. Greenhouse gases are hampering the environment a lot and one such gas is carbon dioxide. This gas can be used as a reagent to produce some valuable and useful polymers that has become a major interest for researchers. About 30–50% of the mass of polymer is obtained from carbon dioxide that is basically the remainder obtained from petroleum products, and hence it gives both kinds of benefits, i.e., environmental and economic [1].

Over the past many years, synthetic polymers have increased their demand among the people as they have tremendous uses and applications like food packaging, water purification, etc. It is because of the amazing properties they exhibit like lightweight, easy transportation, eco-friendly, biocompatibility, and many more. All these applications lead them to attain sustainability, and these add to the fact of bringing additional benefits for their expansion in various applications. In today's scenario, about 8% of fossil fuels are required to produce fuels and this estimation would reach 20% by the year 2050 [2–4].

The sustainable polymers can be categorized into two divisions namely, natural polymers and synthetic polymers. The natural polymers basically include lignin, protein, starch, cellulose, etc., and are used for bioplastics production and its composites. The synthetic polymers are basically obtained from various biomass like fatty acids, terpenes, amino acids, and plant oils. In some cases, it is observed that these sustainable polymers are nonbiodegradable, and their application can reduce the dependence on petroleum resources and even a reduction in the carbon emissions could be recorded [5,6].

4.1.1 TERPENES AND TERPENOIDS

The terpenes and terpenoids are the group of molecules which are nonpolar and contain an isoprene unit which is the elementary portion of the compound. A very common terpene comprises two isoprene units which is regarded as monoterpenes that are produced from various plants and trees biosynthetically. Some common terpenes are limonene, α-terpinene, α-pinene and β-pinene, 3-carene, and myrcene. Terpenes are abundantly present in the environment and are inexpensive, which makes them the best small molecules to act as the building blocks for various applications. These comprise one or more double bonds between carbon-carbon having a general formula $C_{10}H_{16}$. The composition of such a molecule depends on the species and the age of the tree. It can also be easily interconverted to obtain the basic monoterpene skeletons that are present in it from various kinds of precursors. The monoterpenes are diverse in structure because of the presence of various stereogenic centers and the oxygenated compounds that can be easily obtained from the basic skeletons as shown in Figure 4.1. These are often studied for the applications of polymerization because of their low cost, easy availability, high abundance, and easy isolation.

Terpene-based Polymers as Sustainable Materials

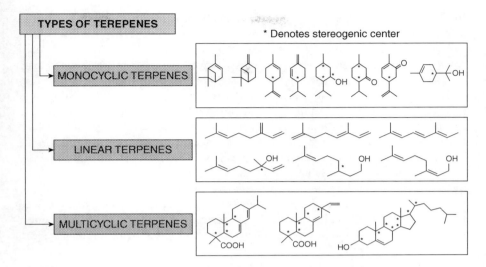

FIGURE 4.1 Types of terpene molecules.

Therefore, the sustainable polymers obtained from terpenes and terpenoids are the focus of this chapter [7].

4.2 GREEN CHEMISTRY AND SUSTAINABLE APPROACH

A lot of importance is being gained by the sustainable polymers which are obtained from alternative sources like renewable resources. This is because the people basically require materials which have a smaller ecological footprint and are durable in future to lead to sustainability. The production of sustainable polymers follows all principles of green chemistry which utilizes a basic set of principles eventually eliminating or reducing the production of some harmful materials or substances, application, and also manufacturing of the products. But the challenge lies to identify the platform of chemicals that are obtained from abundantly available feedstock which does not compete for the resources required for feedstock or disturb the ecosystem. Modifications in the agricultural method for harvesting and crop production increase the economic impact [8].

The petrochemical industry today is basically centralized and highly integrated. The process of cracking naphtha leads to the production of alkenes like ethylene, and mixtures of C4 and C5 molecules. Also, the catalytic reforming is required for the manufacturing of the aromatic compounds like xylene, benzene, toluene, and naphthalene. These hydrocarbons act as important precursors which are essential for the synthesis of all those products required for the use of monomers [9].

The bio-derived feedstocks can help in neutralizing the carbon footprint. These can create a balance between the utilization and production of carbon dioxide. The sequestration of the carbon dioxide gas is very much slower in comparison to its release from various petroleum feedstocks. Therefore, the use of natural polymers,

58 Handbook of Sustainable Materials

like lignin, hemicellulose, and cellulose, should be promoted as these are the renewable materials acting as a source of carbon.

The polycarbonates and polyesters add to the examples of tenable feedstocks which are rich in oxygen. A large number of renewable monomers are obtained from starch or sugar-rich crops like sugar cane and corn. The conversion involved basically works in two directions: the first one is the replacement of the existing ones that are derived from petroleum with their counterparts which are renewable and the second is to produce new monomers which are bio-derived and the polymers which have new structures and properties that plastic does not possess. This results in the utilization of those monomers that can act as precursors for this synthesis [10].

The production of recyclable sustainable polymers is very much desirable as these can be completely converted back into the monomers by the process of depolymerization. Also, the monomers obtained are highly selective, pure, and obtained in good yield with the ability to be re-polymerized. This process can be infinitely repeated, and these kinds of polymers have a great potential to prevent wastage [11].

4.3 SYNTHESIS OF TERPENE-BASED POLYMERS

Recently, research is going on to find such renewable monomers that could be consumed for the synthesis of sustainable polymers. It consists of finding those kinds of new materials which are derived completely from feedstocks, are renewable, and blend with petroleum and renewable monomers to achieve sustainable alternatives in the chemical industry [12]. Below are some of the methods to synthesize terpene-based polymers along with their characterization.

4.3.1 SYNTHESIS OF POLY-ALLOOCIMENE

Pranabesh Sahu et al. have synthesized poly-alloocimene which is a terpene-based polymer by the process of redox-initiated emulsion polymerization as shown in Figure 4.2. Initially, deionized water, potassium chloride, potassium oleate, and potassium phosphate tribasic buffer were poured into a round bottom flask. The stirring was done at 280 rpm for about 20 min. Then the nitrogen was added to create an inert atmosphere so that a stable emulsion is formed upon the addition of Allo. $Na_2S_2O_5$, $FeSO_4.7H_2O$, and Na-EDTA were added and then tert-butyl hydroperoxide solution was poured into the flask. The washings were done, and the product was then dried in a vacuum desiccator to obtain a pale-yellow polymer with 20–25% yield. This indicated the formation of a polymer (PAllo) by the redox emulsion polymerization process as shown in Figure 4.2 [12].

4.3.2 SYNTHESIS OF ALLYL-TERPENE MALEATE POLYMER

The terpene maleate adduct was prepared using 49 g maleic anhydride that was added in a 250 mL four-necked flask and then refluxed along with an electric agitator. After continuous heating was done up to 145°C, p-toluene sulfonic acid was introduced as a catalyst along with 96 g of other reactant moieties, like α-pinene, β-pinene, dipentene, and terpinolene, in 30 mins and was then stirred for another 2 hours. Then, maleic adduct was removed with distilled water and treated with

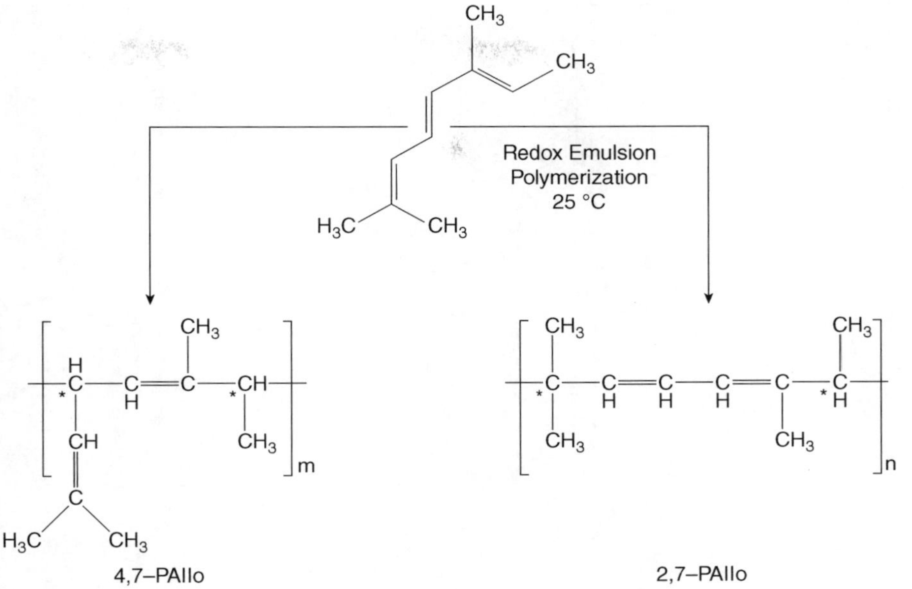

Terpene-based Polymers as Sustainable Materials

FIGURE 4.2 Schematic synthesis of poly-alloocimene.

sodium hydroxide for 2 hours at 60°C followed by drying to produce the terpene maleate sodium salt. 2.96 g of this salt was taken in a three-necked flask along with 4.59 g allyl chloride, 0.66 g KI and then 3.03 g of trimethylamine and was refluxed for 16 hours at 60°C. Then the extraction was done with n-hexane. This led to the synthesis of the desired polymer as shown in Figure 4.3 [13].

4.3.3 Synthesis of Polymyrcene

Myrcene is also known as β-myrcene and is an alkene which is a natural hydrocarbon. It is basically a monoterpene and is an intermediate during the formation of various fragrances. It can be extracted from various plants and has a pleasant odor. Polymyrcene is broadly available in nature and is versatile in the polymerization. The monomer β-myrcene was distilled and an inert atmosphere was created with argon. $Nd(Oi-Pr)_3$ was added as a catalyst along with cyclohexane into the reactor with 100 rpm and stirring. The catalyst was injected into the reactor system with the help of a syringe. It was then kept for 1 hour at room temperature and it was deactivated by the addition of methanol and then dried at 25°C in a vacuum. Before using cyclohexane, it was made sure that it was distilled from sodium. The polymer obtained was hence stabilized by Irganox 1076 [14].

4.3.4 Synthesis of Poly(Stearyl Methacrylate)

The poly(stearyl methacrylate) was synthesized by the emulsion polymerization method. The deionized water along with a buffer was collected in a round bottom

60   Handbook of Sustainable Materials

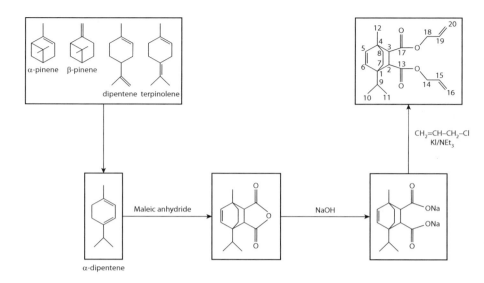

FIGURE 4.3 Synthetic route adopted to prepare terpene-diallyl maleate adduct.

flask and mixed at 350 rpm for about 20 mins. The monomer was added slowly for 10 mins, and it was left for 20 mins more to obtain a stable emulsion. Then an inert atmosphere was created with nitrogen and the temperature was set to 70°C and then the initiator was injected dropwise. For about 20 hours, the polymerization continued and then ethanol was poured with vigorous stirring to enable coagulation. The polymer was washed and dried for 24 hours in a vacuum at 50°C [15].

4.3.5 Synthesis of Tetrahydrogeraniol Acrylate (THGA)

A three-necked round bottom flask of 500 mL was taken on a magnetic stirrer. Then, tetrahydrogeraniol, trimethylamine, and dichloromethane were poured into the flask and left for stirring for about 2 hours in an ice bath. Then acryloyl was added dropwise and left to stir for 24 hours in an ice bath. After this filtration was carried out and the mixture produced, it was given several washings with deionized water and brine solution. Then it was passed through alumina, and by a rotatory evaporator, the volatiles were removed. Toluene was then poured into it and the mixture was deoxygenated by enabling the passage of nitrogen for 30 mins and then the polymerization was initiated. It was kept for 7 hours at 80°C [16]. The final product obtained was THGA.

4.4 CHARACTERIZATION OF SYNTHESIZED POLYMERS

4.4.1 Characterization of Poly-Alloocimene

The synthesized poly-alloocimene can be characterized by FT-IR and NMR techniques. Various peaks could be recorded in the FT-IR spectra like the stretching for =C-H, stretching frequency for $-CH_3$, $-CH_2$, and asymmetric stretching for $-CH$

Terpene-based Polymers as Sustainable Materials 61

group. Also, there exists the stretching frequency of C-C and the bending vibrations of −CH and −CH$_3$ groups. A decrease in the intensities was observed as there is an unsaturation in the monomer. When considering the proton NMR technique for characterization, it could be observed that the olefinic hydrogens of the corresponding carbon would appear in the downfield region and the protons present in the methyl groups would be observed in the upfield region. But in the case of ^{13}C-NMR, the olefin carbon atoms would occur at a larger chemical shift that were appearing in the downfield region, and the methyl groups attached to the olefinic carbon would be detected in the upfield region [12].

4.4.2 Characterization of Allyl-Terpene Maleate Polymer

The synthesized allyl-terpene maleate polymer can be characterized by FT-IR and NMR techniques. Various peaks could be recorded for the terpene-maleate adduct, terpene-maleate sodium salt, and terpene-diallyl maleate adduct in the FT-IR spectra. A characteristic peak will be observed for C=O coupling vibration that is particularly because of the anhydride. Also, the low wave number peak would be slightly stronger because of the typical characteristic of the cyclic anhydride. The stretching vibration absorption peaks of methyl and methylene groups will be observed. The C-H bending vibration absorption peaks along with the stretching vibration absorption peak of C-O will also be observed. Additionally, there will be a stretching vibration absorption peak of C-H of the double bond in the ring and the peak for the C=C will be observed, respectively. When considering the proton NMR technique for recording the data of terpene-diallyl maleate adduct, a total of 13 signals would be recorded as there are 13 protons having chemically different environments [13].

4.4.3 Characterization of Polymyrcene

The characterization of polymyrcene can be done by size exclusion chromatography by using PLGel mixed column by carrying out the process of calibration with THF and polystyrene. The microstructures so formed can be characterized by proton NMR and ^{13}C-NMR by using CDCl$_3$ as the solvent at 25°C. The characteristic peak will be observed for the olefinic carbon group. Also, other techniques, like differential scanning calorimetry thermograms, can be used to illustrate the elimination of thermal history of the sample [14].

4.4.4 Characterization of Poly(Stearyl Methacrylate)

The molecular weight of the polymer synthesized can be obtained using PLGel column using THF as the eluent. The percentage of gel could be evaluated by calculating the ratio of the polymer (dried) to its initial value. The FT-IR spectra could be recorded for the polymer synthesized in the range of about 4000–650 cm^{-1}. Also, the proton NMR spectra and ^{13}C-NMR spectra could be recorded for the polymer synthesized by using tetramethylsilane as the internal standard. The XRD analysis could also be carried out for this polymer along with thermogravimetric analysis with a heating rate of about 10°C/min [15].

4.4.5 Characterization of THGA

There is the presence of hydroxyl group that can accommodate acrylic group. It could be characterized by proton NMR and ^{13}C-NMR that would show the presence of vinyl bonds which are very much reactive. It can exhibit the presence of the macromonomer formation by proton NMR and the presence of quaternary carbon atoms through ^{13}C-NMR [16].

4.5 MECHANISM OF POLYMER FORMATION

The polymers are formed by connecting various structural units which are also known as monomers in different ways. The monomers are basically linked in such a way that they have covalent connectivity between them, and these add together to result in a long-chain product having different properties.

4.5.1 Chain-Growth Polymerization

Terpenes behave as a raw material for the production of polyterpene hydrocarbons as they do not consist of a different functionality. 1,4-Polyisoprene, latex, has been used before the time of synthetic polymers and is a natural terpene-based polymer. There exist two monomers of pinene that can be easily isomerized into one another.

The isomer given in Figure 4.4 is more reactive as there is the presence of exocyclic double bond. The cationic polymerization of this compound existed long

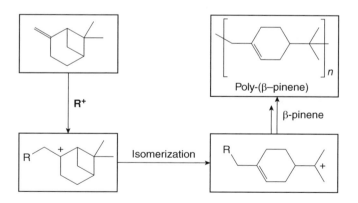

FIGURE 4.4 Cationic polymerization of pinene.

before 1937 using AlCl$_3$ as the catalyst. A lot of work has been done on this study and in 1997, it got confirmed that the polymerization of the given isomer of pinene compound is much easier.

The cationic polymerization of pinene was performed and the molecular weight distribution was 1.3 along with the use of H$_2$O/AlCl$_3$/OPh$_2$ as the co-initiator catalyst performed at room temperature which eventually gave high molecular weight polymer having 82–87°C as its glass transition temperature. The most important factor governing this scheme is the basicity of diphenyl ether that acts as a donating group, leading to the β-hydrogen abstraction resulting in polymers having low molecular weight [17].

4.5.2 Step-Growth Polymerization

There exists one more promising option which involves the step-growth polymerization of the terpene monomers. Meier and coworkers were successfully able to explain radical initiator-free addition along with the best suitable solvent. The addition was carried out using thiols to pinene and limonene that resulted in the formation of monomers having ester and alcohol as the functional groups. Here, Triazabicyclodecene was used as the catalyst that enabled to study the polycondensation behavior of the polymer formed. This process resulted in the production of short-chain diols that ultimately gave oligomers having semi-crystalline nature of the polyesters obtained with glass transition temperature of about 45°C. Also, the melting point reported for these polymers was in the range of -15 to 50°C as shown in Figure 4.5 [18].

The bicyclic anhydride monomers were also derived from nature as they occurred in monoterpene form of phellandrene (BCA1 and BCA2). This monoterpene was easily obtained by melt polymerization having tetra-functional diglycerol to yield the polyester that has a great sustainability. Phellandrene has undergone the

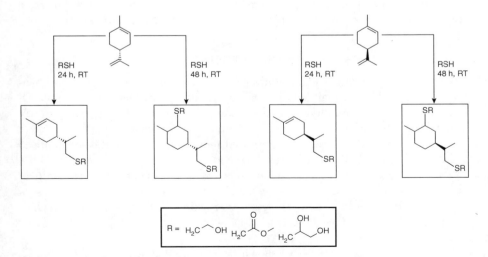

FIGURE 4.5 Step-growth polymerization of pinene.

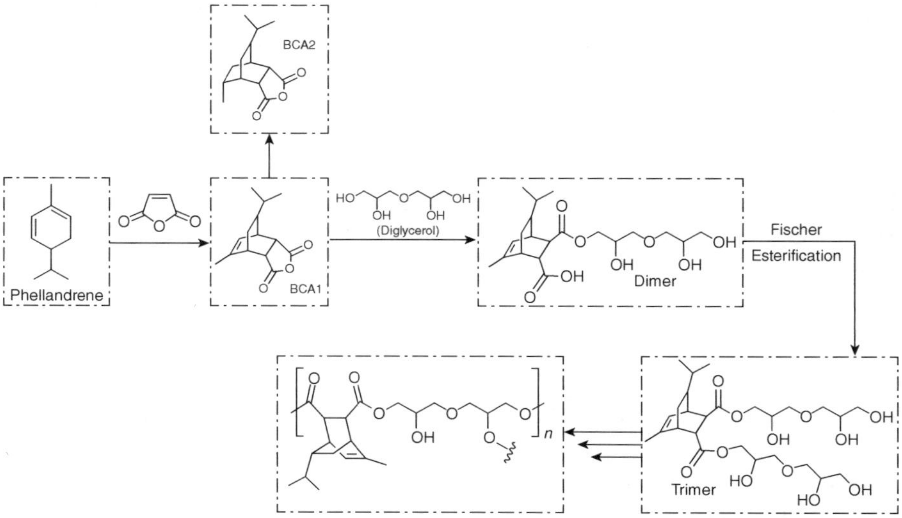

FIGURE 4.6 Phellandrene undergoes the transformation with maleic anhydride leading to Diels–Alder reaction with no solvent to produce hydrophobic anhydrides.

transformation with maleic anhydride leading to Diels–Alder reaction with no solvent to produce hydrophobic anhydrides. The compounds obtained till this step were eventually analyzed and the carboxylic acid groups present in the compounds have undergone the Fisher esterification with diglycerol as shown in Figure 4.6 [18].

4.6 MODIFICATIONS IN TERPENE-BASED POLYMERS

There have been specific modifications made in the terpene-based polymers to improve certain characteristics, and these are described in this section [19].

4.6.1 LIGNIN-POLYMER BLENDS AND OLIGOMERS

These are produced at high temperatures via the process of melt mixing with mechanical blending led by thermal extrusion. The observation is gradually recorded since it depends on the compatibility that exists between the lignin and the polymer matrix blends. The observation is also done if there exists any phase separation system. The formation of blends basically improved the interfacial properties of lignin to increase its compatibility with the synthetic polymers. The cationic polymerization of terpenes leads to the formation of oligomers which can undergo further modification to prepare epoxy resins and polyols. As shown in Figure 4.7, the epoxy resins can be transformed by using ring opening polymerization (ROP) along with secondary amines. These then ultimately react with polyisocyanates to yield polyurethanes that are cross-linked [20–22].

The compounds obtained have shown flexibility, good strength, thermal resistance, and chemical resistance. Therefore, these properties are a result of the

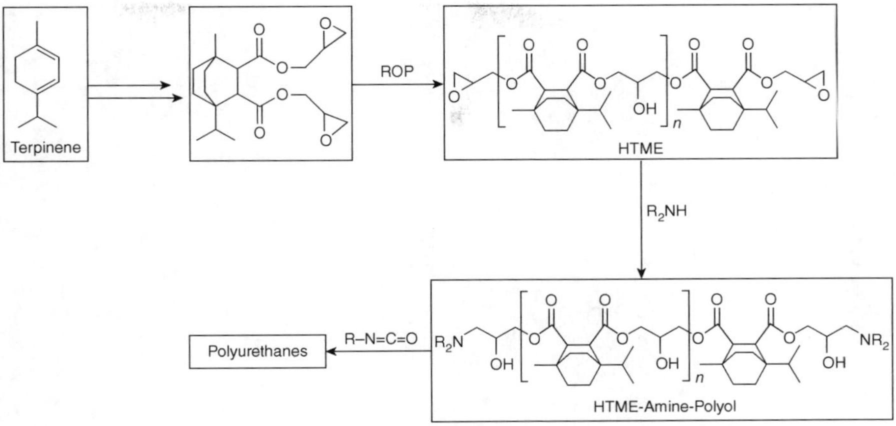

Terpene-based Polymers as Sustainable Materials

FIGURE 4.7 Polyol synthesis from epoxy resins.

combination of polyurethanes and epoxy resins. The glass transition temperature lies in the range of 5–37°C and some thermostability tests performed have shown about 5% loss in weight. They also have the ability to form some transparent films that have a great strength of impact of more than 50 cm and 0.5 mm of flexibility with good adhesion. The presence of large number of OH groups resulted in a higher value of glass transition temperature and hardness [23–25].

4.6.2 GRAFTING OF TERPENES

The grafting of the terpenes is one of the best methods of modifications. The attachment of menthol or cholesterin to polysiloxanes through hydrosilation is described. There is an introduction of the double bonds at the hydroxyl (OH) groups in terpenes that is governed by monomer grafting that have been modified. It is done on polysiloxanes by the action of a catalyst as shown in Figure 4.8. Chitosan–abietic acid conjugates were synthesized and investigated for their drug-release behavior [2,26–28].

4.7 APPLICATIONS

The terpenes are used as one of the important components in many applications since ancient times. The modifications can be done through the double bonds and even by using pure polymer. These have an increasing application in the polymer industry in the future. Some of the important applications of them are listed below.

- These are used as solvents in many chemical industries and are a vital component of essential oils.
- These are used as additives for food and even as the agents required to produce fragrances in the perfume industry.

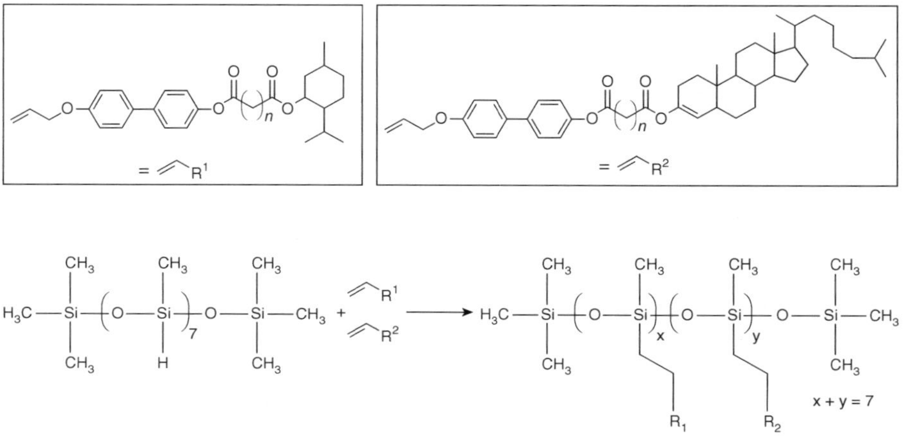

FIGURE 4.8 Grafting of terpenes.

- They also have antifungal, antianxiety, and antidepressant properties. Therefore, it can be used as an antiseptic, antibacterial, and anti-inflammatory component for medical benefits.
- Many polyterpenes are used as additives or tackifiers in various hot-melt and pressure adhesives as these are very much compatible with many elastomers that mainly include natural rubber, acrylics, and polyolefins. The importance of polymeric myrcene is basically its bio-based origin because of its possible application potential as well.
- These days terpene-based polymers are used for many engineering applications in various fields mainly including biodegradable packaging, coatings, fibers, antimicrobial films, material reinforcement, and in medicine and pharmaceuticals [29–32].

4.8 GENERAL DISCUSSION

There are basically three types of monoterpenes. These are acyclic, monocyclic, and bicyclic. There are various examples of all three categories and those examples have been discussed in Table 4.1. Also, a small description is there to depict the occurrence and its use in daily life industries [33–39].

4.9 CONCLUSION AND FUTURE PROSPECTS

The sustainable polymers could be defined by considering all the factors such as the resources from which they are produced, utilized, recovered, and stored, and how they are relating to the environment. The most significant challenge that is observed is that plastic is an amazing material having low cost. Therefore, the sustainable polymers formalized should be economically competitive in order to be competitive in the market. The production of sustainable polymers has a higher potential to protect natural resources and present flexible solutions in the field of synthetic polymers.

TABLE 4.1
Explanation of Types of Monoterpenes and Their Examples with a Short Description

S. No.	Type of Monoterpene	Example	Description
1	ACYCLIC	α-Ocimene	It has a lime flavor and is used in perfume industry
2		Citral A	It is found in Australian lemon and is used in cosmetics
3		Geraniol	It is used as a mosquito repellent
4		Citronellol	It is found in lemon grass and is used as a mosquito repellent and attracts mites
5		Linalool	It is found in those compounds that are produced by mint family
6	MONOCYCLIC	s-Limonene	It has a pleasant smell of pine trees and is used in perfume industry
7		Phellandrene	It smells like koala bears and has an aroma of Eucalyptus
8		Menthol	It has a strong and cooling mint-like aroma and is used as a mild anesthetic
9		Safranal	It has a saffron-like aroma and is used as an antidepressant and antioxidant
10		Thymol	It has a thyme-like aroma and has the ability to kill fungi
11	BICYCLIC	Sabinene	It contributes to the spiciness of black pepper
12		Thujone	It has a menthol-like smell that is found in the alcoholic drinks
13		Pinene	It acts as an anti-inflammatory and an antibiotic
14		Ascaridole	It is a poisonous and an explosive compound
15		Verbenone	It has a fragrance of L'Occitane and an aroma like Spanish verbena

Research is going on in order to contribute toward the sustainable polymers' development from various new perspectives.

Terpenes are made and produced on the scale of 300 kilotons per annum and have a great potential to be produced on a larger scale mainly from waste streams if there lies a great demand. These are made by the pathways that are derived from renewable feedstocks. The terpene-based monomers and their polymerization techniques have a vital role in addition to other natural plant resources and these will be extremely important in the future.

REFERENCES

1. Zhu, Y., Romain, C., & Williams, C. K. 2016. Sustainable polymers from renewable resources. Nature. 540(7633), 354–362.
2. Schneiderman, D. K., & Hillmyer, M. A. 2017. 50th Anniversary perspective: There is a great future in sustainable polymers. Macromolecules. 50(10), 3733–3749.

3. Hillmyer, M. A. 2017. The promise of plastics from plants. Science. 358, 868–870.
4. Hong, M., & Chen, E. Y. X. 2017. Chemically recyclable polymers: A circular economy approach to sustainability. Green Chem. 19, 3692–3706.
5. Kristufek, S. L., Wacker, K. T., Tsao, Y. Y. T., Su, L., & Wooley, K. L. 2017. Monomer design strategies to create natural product-based polymer materials. Nat. Prod. Rep. 34, 433–459.
6. Wang, Z., Ganewatta, M. S., & Tang, C. 2020. Sustainable polymers from biomass: Bridging chemistry with materials and processing. Prog. Polym. Sci. 101, 101197.
7. Zhang, X., Fevre, M., Jones, G. O., & Waymouth, R. M. 2018. Catalysis as an enabling science for sustainable polymers. Chem. Rev. 118(2), 839–885.
8. Hong, M., & Chen, E. Y. X. 2019. Future directions for sustainable polymers. Trends Chem. 1(2), 148–151.
9. Tang, X.-Y. & Chen, E. Y.-X. 2019. Toward infinitely recyclable plastics derived from renewable cyclic esters. Chem. 5, 284–312.
10. Hong, M. & Chen, E. Y.-X. 2016. Completely recyclable biopolymers with linear and cyclic topologies via ring-opening polymerization of G-butyrolactone. Nat. Chem. 8, 42–49.
11. Zhu, J.-B. 2018. A synthetic polymer system with repeatable chemical recyclability. Science. 360, 398–403.
12. Sahu, P., Sarkar, P., & Bhowmick, A. K. 2017. Synthesis and characterization of a terpene-based sustainable polymer: Poly-alloocimene. ACS Sustain Chem. Eng. 5(9), 7659–7669.
13. Gu, Y., Hummel, M., & Muthukumarappan, K. 2019. Synthesis and characterization of allyl terpene maleate monomer. Sci. Rep. 9, 19149.
14. Sarkar, P., & Bhowmick, A. K. 2017. Terpene Based Sustainable Methacrylate Copolymer Series by Emulsion Polymerization: Synthesis and Structure-Property Relationship. 1–11.
15. Noppalit, S., Simula, A., Ballard, N., Callies, X., Asua, J. M., & Billon, L. 2019. Renewable terpene derivative as a biosourced elastomeric building block in the design of functional acrylic copolymers. Biomacromolecules. 20(6), 2241–2251.
16. Frolov, A. N., Kostjuk, S. V., Vasilenko, I. V., & Kaputsky, F. N. 2010. Controlled cationic polymerization of styrene using AlCl3OBu2 as a coinitiator: Toward high molecular weight polystyrenes at elevated temperatures. J Polym. Sci. Part A Polym. Chem. 48, 3736–3743.
17. Satoh, K., Nakahara, A., Mukunoki, K., Sugiyama, H., Saito, & H., Kamigaito, M. 2014. Sustainable cycloolefin polymer from pine tree oil for optoelectronics material: living cationic polymerization of β-pinene and catalytic hydrogenation for high-molecular-weight hydrogenated poly(β-pinene). Polym. Chem. 5, 3222–3230.
18. Ye, Dz., Jiang, L., Ma, C., Zhang, M. H., & Zhang, X. 2014. The graft polymers from different species of lignin and acrylic acid: Synthesis and mechanism study. Int. J Biol. Macromol. 63, 43–48.
19. Liu, H., & Chung, H. 2017. Lignin-based polymers via graft copolymerization. J. Polym. Sci. Part A Polym. Chem. 55, 3515–3528.
20. Yu, J., Wang, J., Wang, C., Liu, Y., Xu, Y., & Tang, C. 2015. UV-absorbent lignin-based multi-arm star thermoplastic elastomers. Macromol. Rapid. Commun. 36, 398–404.
21. Leavell, M. D., McPhee, D.J., & Paddon, C.J. 2016. Developing fermentative terpenoid production for commercial usage. Curr. Opin. Biotechnol. 37, 114–119.
22. Hazan, Z., & Amselem, S. Compositions of Polymeric Myrcene. U.S. Patent WO2010100651A2, 10 September 2010.
23. Wypych, G. 2018. Handbook of Surface Improvement and Modification; ChemTec Publishing: Toronto, ON, Canada.

24. Rubulotta, G., & Quadrelli, E. A. 2019. Terpenes: A valuable family of compounds for the production of fine chemicals. In Studies in Surface Science and Catalysis; Elsevier: Amsterdam, The Netherlands. pp. 215–229.
25. United States Department of Agriculture. 2016. USDA BioPreferred Program Guidelines How to Display and Promote the USDA Biobased Product Label.
26. Papageorgiou, G. Z. 2018. Thinking green: Sustainable polymers from renewable resources. Polymers. 10, 952.
27. Herbert, K. M., Schrettl, S., Rowan, S. J., & Weder, C. 2017 50th anniversary perspective: Solid-state multistimuli, multiresponsive polymeric materials. Macromolecules. 50, 8845–8870.
28. Hazan, Z., Adamsky, K., & Lucassen, A. C. B. 2016. Use of Isolated Fractions of Mastic Gum for Treating OpticNeuropathy. U.S. Patent WO2016142936A1.
29. Billiet, L., Hillewaere, X. K. D., Du Prez, F. E. 2012. Highly functionalized, aliphatic polyamides via CuAAC and thiolyne chemistries. Eur. Polym. J. 48, 2085–2096.
30. Kolb, N., Winkler, M., Syldatk, C., & Meier, M. A. 2014. Long-chain polyesters and polyamides from biochemically derived fatty acids. Eur. Polym. J. 51, 159–166.
31. Sahu, P., Bhowmick, A. K., & Kali, G. 2020. Terpene based elastomers: Synthesis, properties, and applications. Processes. 8, 1–21.
32. Breitmaier, E. 2006. Terpenes: Flavors, Fragrances, Pharmaca, Pheromones; Wiley-VCH Verlag GmbH: Weinheim, Germany.
33. Wilbon, P. A. Chu, F., & Tang, C. 2013. Progress in renewable polymers from natural terpenes, terpenoids, and rosin. Macromol. Rapid. Commun. 34, 8–37.
34. Gandini, A., & Lacerda, T. M. 2015. From monomers to polymers from renewable resources: Recent advances. Prog. Polym. Sci. 48, 1–39.
35. Zhao, J., & Schlaad, H. 2013. Synthesis of terpene-based polymers. Adv. Polym. Sci. 253, 151–190.
36. Schoenberg, E., Marsh, H. A., Walters, S. J., & Saltman, W. M. 1979. Polyisoprene. Rubber Chem. Technol. 52, 526–604.
37. Mooibroek, H., & Cornish, K. 2000. Alternative sources of natural rubber. Appl. Microbiol. Biotechnol. 53, 355–365.
38. Ouardad, S., Deffieux, A., & Peruch, F. 2012. Polyisoprene synthesized via cationic polymerization: State of the art. Pure Appl. Chem. 84, 2065–2080.
39. Moad, G. 2017. RAFT (co) polymerization of the conjugated diene monomers: Butadiene, isoprene and chloroprene. Polym. Int. 66, 26–41.

5 Sustainable Biodegradable and Bio-based Materials

*Mridul Umesh, Suma Sarojini, Thazeem Basheer,
Sreehari Suresh, Adhithya Sankar Santhosh,
Liya Merin Stanly, Sneha Grigary, and Nilina James*

CONTENTS

5.1 Introduction 72
    5.1.1 Problem Associated with Nonbiodegradable Materials 72
    5.1.2 Need for Biodegradable Materials... 72
    5.1.3 Sustainable Production: Waste Valorization as a Key Tool
        for Production 73
5.2 Types Of Biodegradable And Bio-Based Materials 73
    5.2.1 Plant-based Biodegradable Materials 73
        5.2.1.1 Cellulose and Cellulose Derivatives – Production
               Methodology and Applications... 73
        5.2.1.2 Starch and Starch Derivatives – Production
               Methodology and Applications... 75
    5.2.2 Animal-based Biodegradable Materials... 76
        5.2.2.1 Chitin and Chitosan – Production Methodology
               and Applications 76
        5.2.2.2 Keratin and Keratin Derivatives – Production
               Methodology and Applications... 78
        5.2.2.3 Collagen and Elastin – Production Methodology
               and Applications 78
        5.2.2.4 Hydroxyapatite – Production Methodology and
               Applications 80
    5.2.3 Microbe-based Biodegradable Materials... 81
        5.2.3.1 Polyhydroxyalkanoates – Production Methodology
               and Applications 81
        5.2.3.2 Levan – Production Methodology and Applications 82
        5.2.3.3 Polylactides – Production Methodology and Applications....... 83
5.3 Challenges in Commercialization and Future Prospects.................................. 84
5.4 Conclusion 85
References... 85

DOI: 10.1201/9781003297772-5

5.1 INTRODUCTION

5.1.1 PROBLEM ASSOCIATED WITH NONBIODEGRADABLE MATERIALS

Humans have always been on the quest to maximize comforts by making use of products which can be produced swiftly, easily, and cheaply. There has not been much thought on the sustainability of the products or the processes. This has been continuing incessantly for many thousands of years. With the expanding population, the need for such cheap products grew manifold. Polymeric substances have multiple uses in many domains. The industrial revolution accelerated the production of such polymers thereby leading to extremely cheap products which found uses in our daily life – like the plastics used in wrapping materials (waterproof nature, long shelf life, and resistance to corrosion being some of the positive traits). But the same positive trait of stability and shelf life, which gained a lot of customers for these materials, has turned into a bane – we cannot get rid of this material from our planet even after thousands of years. Since these materials will not degrade if left in the environment, people try to burn them off. This results in more severe and serious consequences. Burning of plastic releases huge quantities of noxious organic pollutants, including polycyclic aromatic hydrocarbons and dioxins, primarily due to incomplete combustion processes. A recent study has shown that the burning of plastic foam emitted aerosols with the highest emission factors – close to 35 g/kg. These figures are multiple folds of those emitted from the burning of other types of wastes [1]. Since most of the gasses cannot be seen, their grave danger goes undetected in many countries, especially the developing ones, where there are no strict waste-burning regulations.

All the plastic debris ultimately ends up in oceans and other water bodies. There have been systematic studies projecting more than 50 million tons of plastic ending up in water bodies every year by 2030, despite many countries desperately taking necessary actions to reduce the reliance on plastics [2]. Yet another grave threat associated with increased amounts of plastics on the Earth is that of microplastics – particles less than 5 mm – which can directly (by blocking gastrointestinal tract or by giving a feeling of satiety) or indirectly (by the accumulation of pollutant chemicals on its surface) affect our health. Plastics make up more than three-quarters of all marine debris. Millions of marine animals and seabirds perish annually due to this plastic pollution. Microplastics in water bodies undergo horizontal transport, vertical settlement, and aggregation, which are ending up in the bodies of aquatic animals. The dynamics of this process is decided by a multitude of factors as evidenced by a recent study concluding that bent-body fish takes more time to excrete polyethylene microbeads when compared to the straight body fish [3]. These materials undergo biomagnification and finally end up in the bodies of humans, higher up in the food chain.

5.1.2 NEED FOR BIODEGRADABLE MATERIALS

Realizing the dangers of such nonbiodegradable materials which have crept into each and every aspect of our life needs, there has been a realization of going back to nature by the application of the concept of sustainability. The biggest advantage of biodegradable substances is their ability to be broken down into simpler compounds

Sustainable Biodegradable and Bio-based Materials 73

by microbes in soil and water and hence preventing recalcitrance. Various biodegradable substances including polylactides, polyhydroxybutyric acid, chitin, chitosan, levan, strach, keratin, and collagen have been demonstrated to be useful in food, agricultural, and pharmaceutical industries. Some of these are easily biodegradable whereas some require specific physicochemical conditions to degrade. While advocating the large-scale production and use of biodegradable materials, this aspect of readiness for degradation should also be taken into consideration. Complete biodegradation encompasses different steps, finally mineralizing into gaseous end products depending on factors like temperature, moisture, and microbial flora of the environment. Biodegradable plastic production on a global scale was 1.2 Mt/year in 2020, which is expected to have a steep growth.

5.1.3 SUSTAINABLE PRODUCTION: WASTE VALORIZATION AS A KEY TOOL FOR PRODUCTION

The concept of zero waste assumes a lot of importance nowadays. Earlier, environmentalists were more concerned about recycling products, but of late there has been a shift in the thought process – the concept of zero waste has been given a lot of thrust. This idea goes far beyond the concept of just reusing or recycling materials. Agroindustrial and marine product wastes are two such prominent materials which can be effectively utilized for production of biodegradable materials. More than one-third of the global harvest is either lost or wasted globally in aquaculture and fisheries [4] which can be channelized for biopolymer production. Recently, an effort has been made to study 39 agricultural waste and by-product valorization projects spread across Asian and European countries to have a sustainable circular economy-based strategy by integrating waste valorization for product development [5]. The design of the product, raw material used, processing mechanisms, and reagents used for production, all these should promote the sustainable angle. One has to envision this based on the concept of a circular economy which involves reusing, repairing, refurbishing, and recycling materials in a sustainable and cost-effective manner.

5.2 TYPES OF BIODEGRADABLE AND BIO-BASED MATERIALS

5.2.1 PLANT-BASED BIODEGRADABLE MATERIALS

5.2.1.1 Cellulose and Cellulose Derivatives – Production Methodology and Applications

Cellulose is the most abundant biopolymer found in nature accounting for about 33% of plant polysaccharides. Its production is approximately 50 billion tons per year [6]. The cellulose polymer is composed of thousands of repetitive D-glucose molecules with β-1,4 glycosidic linkages [7]. Cellulose is being used in a wide variety of fields like food packaging, paper production, biomaterial production, animal food additives, and pharmaceutical industries because of its structure, biodegradability, renewability, and harmless effect on the environment [8, 9]. There are different derivatives of cellulose with varying properties and applications, these derivatives are produced as a result of different chemical modifications (by

74 Handbook of Sustainable Materials

functional groups) and mechanical or physical complexations using different polymers. Some of the cellulose derivatives with commercial importance are discussed in brief below.

5.2.1.1.1 Cellulose Acetate (CA)

CA was first produced by treating wood pulp with acetic anhydride. It can also be synthesized by treating cellulose-rich substances like cotton with acetic acid (acetylation) followed by dehydration and treatment with acetic anhydride. This process will yield cellulose triacetate which can be converted to cellulose diacetate and to CA by acid hydrolysis. CA has the same mechanical properties as cellulose, they are water insoluble and have stable hydrolytic properties, and they are decomposable, biocompatible, and nontoxic. CA are used mainly in biomedical fields as membranes, their applications involve catalysis, separation, biosensing, drug delivery, and tissue regeneration [10].

5.2.1.1.2 Cellulose Sulfate (CS)

When cellulose undergoes any of the sulfation procedures (homogeneous or heterogeneous), it results in an ester which is called as CS. This cellulose derivative is also having antimicrobial properties, and at higher concentrations it is soluble in water. The applications of CS are based on biological activities, such as its effect on blood pressure, anticoagulant properties, anti-AIDS virus activity, and periodontitis treatment [11].

5.2.1.1.3 Cellulose Nitrate (CN)

Commonly called "gun cotton", CN breaks down explosively and is widely used in smokeless gunpowder. Nitration of cellulose by treating wood pulp or cotton with strong nitrating agents like nitric acid will yield CN. The reaction is achieved by the substitution of hydroxyl groups of cellulose with nitrate esters. CN is used in hemodialysis and biosensors. Other than these applications CN is used in osmosis and ultrafiltration techniques [12].

5.2.1.1.4 Carboxymethyl Cellulose (CMC)

Cellulose is treated first with sodium hydroxide solution and then extracted as alkali cellulose. This alkali cellulose is further reacted with monochloroacetic acid or sodium monochloroacetate solution in alcohol to obtain the cellulose ether known as CMC. They are classified as a technical grade, semi-purified and purified, based on the purity of CMC. Cellulose gum, which is mainly used as a food additive, is one of the highly purified CMC [13].

5.2.1.1.5 Ethyl Cellulose (EC)

EC is obtained when alkali cellulose is treated with ethyl chloride. EC derivatives have industrial applications and are insoluble in water, but a wide variety of organic solvents, like chloroform, ethanol, methanol, toluene, and so on, are able to dissolve the compound [10, 14]. EC combined with compounds like hydroxytyrosol are used in the preparation of functional food products and pharmaceuticals/nutraceuticals with improved antioxidant activity [15].

5.2.1.1.6 Nanocellulose (NC)

NC is a nano-sized natural nanomaterial produced in plant cell walls. It is biodegradable and the presence of many hydroxyl groups in the structure makes them suitable for modification according to the intended application. Based on the size, function, source, and synthesis method, NC is classified as nano-fibrillated cellulose, cellulose nanocrystals, and bacterial NC, these classifications all have the same chemical composition but differ in their morphology [16].

5.2.1.1.7 Microbial Cellulose (MC)

MC is the cellulose derivatives naturally produced by certain microbes like the genera of *Agrobacterium*, *Sarcina ventriculi*, and *Acetobacter*. Over the last few decades, bacterial cellulose gained more attention because of its idiosyncratic structural and functional properties when weighed up against the plant cellulose. These cellulose are 100 times smaller than that of the plant cellulose, but the presence of intermolecular and intramolecular bonds between the cellulose fibrils aids in its structural integrity. They are ideal for the use in material industries because of their high mechanical and tensile strength. MC possesses excellent crystallinity and good water-holding capacity [17].

5.2.1.2 Starch and Starch Derivatives – Production Methodology and Applications

Extraction of starch and its derivatives exists from the remote future. The standard method followed in industries has the following steps (Figure 5.1). The raw materials for starch extraction are washed and peeled if tuberous, or threshed if grains [18]. The preprocessed materials are then soaked in water to soften the cells. Tubers are grated or grounded. Degermination is a process to remove the kernels from grains. The raw materials are then centrifuged to separate starch and proteins. The separated starch is further processed to separate fibers from it. The cleaned and concentrated starch is then converted to starch milk. The starch milk is dried and used [19]. Starch and its derivatives have abundant utility in industries like textile, paper, food, medicine, agriculture, casting, animal feed, construction, oil drilling, and

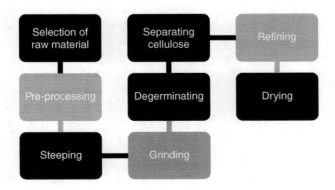

FIGURE 5.1 Standard method for starch extraction from raw materials.

76 Handbook of Sustainable Materials

chemical [20]. Starch-based sweeteners, like glucose syrup and maltose syrups, are used in the food industries for the production of milk powder, solid drinks, confectionery, sorbitol caramel-coloring, antifreeze, soft-drinks, injection solution, etc. [21]. Hydrogenated sweetener sorbitol is used for the production of vitamin C, detergents, cosmetics, toothpaste, medical injection solution, food gum, etc. [22]. Mannitol is used as raw material in the production of polyester and polyether; it is used in medicine for diuretic and dehydration problems [23]. Maltol is a dietary supplement used to prevent dental decay and obesity due its low calories, used as a thickener due to its high viscosity, used for moisture adjustment, and used as food for diabetic patients as it shows very minimum decomposition by insulin [24].

5.2.2 ANIMAL-BASED BIODEGRADABLE MATERIALS

5.2.2.1 Chitin and Chitosan – Production Methodology and Applications

Chitin and chitosan are bio-based polymers, which have gained massive attention in the last decade because of their promising biological and physical properties. Properties, like biocompatibility, nontoxicity, and eco-friendly nature, increased the commercial interest in chitin and its derivative chitosan. Crustacean shells and insect cuticles are the most common source of chitin, and other major sources that include fungi, some mushroom envelopes, green algae, etc. Chemical and biological methods are the two types of extraction processes for extracting chitin and chitosan. The chemical methods have short-processing time and are commonly used at the industrial scale, where strong acids and bases are used to dissolve calcium carbonates and proteins [25]. Biological extraction on the other hand has a long-processing time, but provides high-quality final product and is an eco-friendly process through which acidic and alkali treatments can be avoided. Both chemical and biological extraction processes of chitin and chitosan have three major steps; demineralization, deproteinization, and deacetylation along with a facultative step of decolorization process, which is generally used to eliminate common pigments such as Astaxanthin and β-carotene [25]. The steps of extraction of chitin and derivation of chitosan are shown in Figure 5.2.

In the chemical extraction of chitosan, the primary step is demineralization which is performed using dilute organic acid to eradicate the calcium carbonate and calcium chloride [26]. These minerals form the major inorganic part of the crustacean exoskeleton. Chitin is formed after the step of deproteinization which is performed using alkaline treatments to remove adherent proteins [27]. Chitin is converted to chitosan by the process of deacetylation where the acetyl groups are eliminated using concentrated alkali and high temperature [28]. In biological extraction, demineralization is performed using lactic acid-producing bacteria, deproteinization using proteases from bacteria and the final step of deacetylation is carried out using enzymatic methods. Various extraction procedures for chitin and chitosan along with their sources are given in Table 5.1.

Remarkable properties of chitin and chitosan, such as antitumor, immunoenhancing, antibacterial, wound healing, antioxidant, nontoxicity, biodegradability, and biocompatibility, make chitosan a major applicant in various industries. The low cost of chitosan production is also an attractive factor for industries to assimilate chitosan

Sustainable Biodegradable and Bio-based Materials

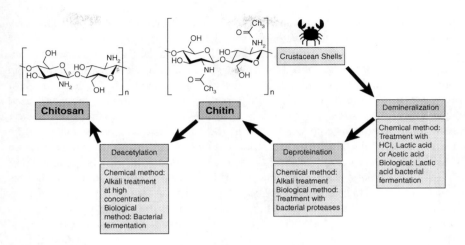

FIGURE 5.2 Steps in chitin extraction and chitosan preparation.

TABLE 5.1
Various Extraction Procedures for Chitin and Chitosan from Different Source Materials

Source Material	Sequential Procedure	Reference
Fish scale (*Oreochromis niloticus*)	• 1 M HCl solution at 75°C for 2 hours • 1 M NaOH solution at 80°C for 2 hours • 50% NaOH solution at 100°C for 2 hours	[29]
Shrimp shells (*Penaeus monodon*)	• 1 M HCl solution at 25°C for 75 min. • 3 M NaOH at 25°C for 75 min. • 50% NaOH at 90°C for 50 min.	[30]
Squid gladius (*Loligo vulgaris*)	• 1.5 M HCl at 50°C for 8 hours • Alcalase enzyme 10 U/mg at pH 8 at 50°C for 3 hours • 50% NaOH at 120°C for 4 hours	[31]
Squid pen (*Illex argentinus*)	• 6% HCl for 2 hours • 10% NaOH at 100°C for 1 hour • 50% NaOH at 60°C for 4 hours	[32]
Beetle cuticle	• 1 M HCl at optimum temperature • 1 M NaOH at 100°C for 8 hours • 50% NaOH at 100°C for 8 hours	[33]
Honey bee waste	• 1 M HCl, ratio 1:15, at ambient temperature • 1 M NaOH, ratio 1:15, at 100°C for 8 hours • 50% NaOH, ratio 1:15, at 100°C for 8 hours	[33]
Crab shell (*Crangon crangon*)	• 3% NaOH at 80°C for 30 min. • 3% HCl at room temperature for 30 min. • 40% KOH at 90°C for 6 hours	[34]
Catharsius molossus	• 1.3 M HCl at 80°C for 30 min. • 4 M NaOH at 90°C for 6 hours • 18 M NaOH at room temperature for 24 hours and heated at 90°C for 7 hours	[35]

for economic benefits. In the pharmaceutical sector, the biocompatibility property of chitosan is exploited for developing drug delivery carriers, gene delivery carriers, and as a scaffold in tissue engineering [36–38]. In recent years, an extensive development has occurred using chitosan in bioimaging, making contact lenses and protein binding, widening the plausible applications of chitosan in the pharmaceutical sector furthermore [39]. In future, chitosan can replace various chemical treatments used in the textile industries, as they are eco-friendly in nature and can provide better outcomes. The use of chitosan as an adsorbent for wastewater treatment from the textile industry, the development of antimicrobial fabricant, and the immobilization of enzymes for improving the enzyme characteristics, in turn reducing the final cost of enzymatic reactions are also some major highlights of chitin and chitosan application in the textile industry [40]. The feeding trials demonstrated on domestic animals show the biological safety of chitosan and the possibility of indulging it in the food and packaging industry [41]. The antimicrobial property of chitosan against a vast spectrum of foodborne fungi, yeast, and bacteria makes chitosan a potential food preservative of natural origin [42]. Formation of bioplastic, food packaging, cosmetics, nutritional enhancement and wastewater treatment are other major sectors for chitosan application.

5.2.2.2 Keratin and Keratin Derivatives – Production Methodology and Applications

Animals are the natural source of some of the most important biopolymers like keratin. Keratin is an essential biopolymer in animals. The main sources of keratin include nails, hairs, hooves, horns, wool, feathers and scales [43]. Keratin follicles called keratinocytes present inside the epithelial cells of animals divide and differentiate to form external appendages containing keratin protein [44]. Keratin protein extracted from animal-based sources find application in various fields like agriculture, aquaculture, biomedical application, bioremediation and cosmetic industry [45]. Keratin molecules in their native form find very little application on the industrial and commercial scale as they are generally insoluble due to the strong disulfide bonds present in their polypeptide chains [46]. Several methods are available for modifying native keratin structure including reduction, oxidation, sulfitolysis, enzymatic hydrolysis, using ionic solutions, quaternization, succinylation and coupling [47]. The common site of modification in a keratin protein is its basic building blocks which are the amino acids. The amino terminal region (-NH$_2$) and the disulfide bond region (-SH) are the main sites where modifications in the forms of additions or deletions can be done. The list of modification processes and application of keratin derivatives are summarized in the following Table 5.2.

5.2.2.3 Collagen and Elastin – Production Methodology and Applications

Collagen, the most significant fibrous structural protein in mammals (30% of the total proteins) is abundantly found in bones, skin, cornea, ligaments, gut and blood vessels. Collagen fibers (strong bundles of collagen) form the major constituent of the extracellular matrix which in turn reinforces many tissues and helps cells to develop structure. Till date, about 28 types of collagen have been found amongst which collagen types I, II and III are more prominent and escort cell functions [60]. Collagen

Sustainable Biodegradable and Bio-based Materials

TABLE 5.2
Keratin Derivatives and its Application

Source of keratin protein	Reaction	Form	Application	Reference
Human hair	Reduction/Oxidation	Hydrogels/Sponges/Films	Biomedical Applications	[48]
	Etherification/Carboxymethylation	fibers	Cosmetics industry	[49]
Wool	Succinylation	Fibers	Textile industry	[50]
Chicken feather	Coupling	Films/Fibers	Adsorbent	[51]
	Etherification/Carboxymethylation	Films	Biodegradable packing materials	[52]
	Hydrolysis (Acid/alkali/enzymatic)	Soluble proteins	Biofertilizer, Feed formulation	[53, 54]
	Acetylation	Fibers/Films/Extrudates	Bioplastic	[55]
	cyanoethylation	Films		[56]
	Graft Polymerization	Films		[57]
	Quaternization	Powder/solution	Textile industry	[58]
Pigeon feather	sulfitolysis	sponges	Oil adsorption	[59]

for industries is chiefly gained from fish, bovine and porcine origin. Collagen has appreciable applications in food, pharmaceutical and cosmetic (skincare products) industries. It serves as an ideal biomaterial with significant properties of interest in industrial application. It is processed into different forms for wound dressing systems, drug delivery (mini pellets and tablets) and tissue engineering constructs. In the food industry, collagen supplements nutrition and improves the functional property of the food, ensuring enhanced health benefits. Consumption of collagen-supplemented food (as functional foods, drinks, dietary supplements, desserts and confectioneries) is essential, as the synthesis of collagen in our body decreases with the increase in age. Hide contains about 96.5% fibrous proteins (98% collagen, 1% keratin and 1% elastin). Bovine hides of bull, calf, cow, ox and face pieces were successfully subjected to collagen extraction via three different methods like acid-solubilization, acid-enzyme solubilization and modified acid-enzyme solubilization [61]. Significantly higher content of collagen was obtained from the third method (75.13% from cow hides) followed by bull hides (74.45%). By-products of fish processing plants (skin, scales, bones and fins) are considered as the safest source of collagen and recently gained attention as promising substitutes for mammalian collagens. Researchers were successful in isolating and characterizing acid-soluble and pepsin-soluble type I collagen from the skin, scales and fins of *C. catla* and *C. mrigala*, with alanine as the prominent amino acid, followed by glycine [62]. Extraction of collagen from seafood waste involves its pretreatment (to remove the impurities) and collagen extraction. Various extraction methods such as acid solubilization, salt solubilization, pepsin solubilization, ultra-sonication assisted extraction and physical-aided extraction are available for the recovery of collagen from wastes. Collagen proves as a valid resource for biomaterials and bioplastics. Fish biomass and by-catch marine organisms have been considered as promising sources of collagen; their usage as food additives and in packaging stays unique [63]. Collagen-based edible films and coatings have been postulated to serve as an effective barrier and safeguard, maintain and prolong the shelf life of various food products. They are well-known for their ability to control microbial growth, oxygen diffusion, fat oxidation and discoloration and sensory qualities.

Elastin, a hydrophobic protein present in the connective tissues (aorta, dermis, lung, ligament, skin, tendon, vascular wall and blood vessels) along with collagen provides elasticity to organs.

Next to poultry, tanneries serve as a hub of underutilized proteinaceous solid wastes rich in collagen and elastin [64]. Recovery of proteins and various active biomolecules from tannery solid waste, as raw materials for other industries, has been accomplished by many researchers through solid waste management techniques and microbial interventions [65-67]. Antihypertensive elastin peptides were extracted from poultry skin waste [68] and were water-soluble. Elastin was successfully isolated from the broiler and spent hen skin using four different solubilizing liquids (sodium chloride, sodium hydroxide, acetone, and oxalic acid) followed by freeze-drying. The ultra-filtered elastin fractions present in broiler and spent hen skin hydrolysate [≤3KDa] exhibited the highest ACE inhibitory activity. Elastin-derived peptides (EDPs) also have roles in breast tumor progression; their biological activities towards cancer cells and stroma were experimentally studied [69]. They were found to trigger the proliferation and migration of monocytes and advocate elastase release. EDPs were reported to be the strong cues, as they helped to resist the process of apoptosis.

5.2.2.4 Hydroxyapatite – Production Methodology and Applications

Hydroxyapatite(HAp) is a naturally occurring mineral, which has many applications as a biomaterial. It is an aptite family biopolymer of calcium phosphate, with the chemical formula $Ca_{10}(PO_4)_6(OH)_2$, They have a similar composition and morphology to that of human hard tissue, and this material makes up about 50% of the volume and 70% of the weight in human bone [70]. Other than in bones they are present in teeth and hard tissues of pineal gland. HAp grafts are used to replace damaged and to assist bone in-growth coating in orthopaedic, dental and maxillofacial applications [71]. HAp has low water solubility, ion-exchange capability and high adsorption capacity. The high adsorption capacity makes it an ideal candidate for the use as an adsorbent to remove highly hazardous pollutants from wastewater [72]. Their high affinities to bone tissues are employed in developing site-specific drug delivery systems to treat bone-related diseases [73]. HAp can be derived from many natural sources rich in calcium, phosphate, and carbonate; in fact in the field of waste management, HAp production is important as this biopolymer can be obtained from animal bones, seafood shells and eggshells [74]. These waste materials cause major problems as these hard materials take a very long time to disintegrate completely. The most used methods of extraction of HAp are the alkaline hydrolysis method and the wet chemical precipitation method. In the alkaline hydrolysis method, the

Sustainable Biodegradable and Bio-based Materials 81

substrates like fish bones or scales are first washed to remove contaminants like salt and dirt. This is followed by deproteinization step, in which the substrate is treated with acid (usually HCl), then with an alkali (NaOH), followed by heat treatment prolonging for hours. For the alkali treatment, other alkalies like KOH or K_2CO_3 can also be used [75]. The resultant product is a white precipitate which is again subjected to alkaline heat treatment to obtain HAp, which is then washed and dried at high temperature [76]. The deproteinization acts as the primary step in wet chemical precipitation method as well, the deproteinized eggshell powder is mixed in distilled water followed by treatment with different concentrations of ammonium dihydrogen phosphate. Before subjecting to precipitation, an acid treatment is done to remove the membranes and a nitric acid treatment can bring up all the impurities into froth [71]. Later the precipitation is attained by alkali treatment using NaOH for several hours. The resulting HAp is washed several times and dried. Other than these two, other methods, like enzymatic hydrolysis, thermal calcination, microwave irradiation, combustion, and hydrothermal, are used for the extraction of HAp.

5.2.3 MICROBE-BASED BIODEGRADABLE MATERIALS

5.2.3.1 Polyhydroxyalkanoates – Production Methodology and Applications

Polyhydroxyalkanoates (PHAs) are polyesters of hydroxyalkanoates (HAs), synthesized majorly from bacteria under nutrient-limited conditions. They are the only class of bioplastic regarded as 100% truly biodegradable and are widely preferred as they do not require any additives for their degradation [77]. They can be degraded both aerobically and anaerobically without the production of any toxic end products. Although PHA biosynthesis follows various pathways, the pathway utilizing sugar molecules is well studied in *Ralstonia eutropha*. The most common PHA biosynthetic pathway involves the conversion of glucose to pyruvate through glycolysis. This is then converted into acetyl-CoA in the Krebs cycle. In PHA accumulating organisms, these acetyl-CoA will not enter into further steps of Krebs cycle but are channelized for PHA biosynthesis. Two acetyl-CoA units condense to form acetoacetyl-CoA aided by enzyme β-ketoacyl-CoA thiolase. Acetoacetyl-CoA is then reduced to (R)-3-hydroxybutyrate monomer by NADPH-dependent acetoacetyl-CoA dehydrogenase. In the last step, (R)-3-hydroxybutyrate is polymerized to PHA by the PHA synthase enzyme [78]. There are several other biosynthesis pathways established in other bacteria and other substrates for PHA biosynthesis. Being an intracellular PHA, extraction of PHA from the bacterial cells is done using solvent extraction, alkaline hydrolysis, acid treatment, and enzymatic method [79]. The cost associated with PHA fermentation and downstream processing is the major factors that restrict their commercialization. Production of PHA from waste valorization and integrated strategies were focused on as alternatives to reduce production costs [80, 81]. PHA is widely applied in diverse sectors because of its excellent material properties. Biocompatibility of PHAs makes them ideally suitable for biomedical applications and drug delivery [82]. PHA is used for manufacture of food packaging and the controlled release of pesticides in agriculture. Other applications of PHA include its incorporation as a feed additive due to its immunomodulatory activities [83]. It is also used in the cosmetic

82   Handbook of Sustainable Materials

industry [84], biofuel production and wound healing applications by incorporating herbal extracts and other compounds [85].

5.2.3.2 Levan – Production Methodology and Applications

Levan is a homopolysaccharide containing β-(2,6)-linked fructose backbone, with numerous properties like biocompatibility, biodegradability, flexibility, and nonhazardous nature [86]. Levan is a heat-stable biopolymer with peculiar thermal properties that makes it a suitable candidate for industries to produce film-forming agents and stable polymers [87]. Levan, in nature, is produced from sucrose by a broad range of microorganisms, as exopolysaccharides for the development of the exopolysaccharide matrix, which in turn aids in the formation of microbial biofilm [88]. Levan is a homopolymer which is produced by the addition of a fructose unit to the substrate biomolecule. The biosynthesis of levan by levansucrase is a gradual process that initiates with the acceptance of a hexosyl group by a growing acceptor molecule transport from a donor molecule. The enzyme, levansucrase, needed for the synthesis of levan is produced inside the cell as a premature enzyme. The enzyme is then transported to the periplasmic space where it attains its active conformation. After maturation, the enzyme is secreted out of the cell either through signal peptide cleavage followed by protein folding or through a signal peptide pathway itself. The surrounding outside of a cell is a pool of various molecules like glucose and fructose. Levansucrase acts on its substrate outside the cell and adds fructose subunits to its substrate molecule [89]. To the C6 hydroxyl of the non-reducing fructose terminal unit of a slowly developing levan polysaccharide chain, single fructofuranosyl units get attached and, as the process goes on, levan polysaccharide grows. This reaction is called the transfructosylation reaction. Levansucrase is the enzyme which initiates and terminates the process of transfructosylation [90]. Biosynthesis of levan from sucrose is presented in Figure 5.3.

The application of levan is extended to the fields of medical, chemical, pharmaceutical, cosmetic, and food industries. The factors, which help levan to be applicable in the industrial sector, are its biodegradability, biocompatibility and film-forming nature [91]. One of the finest applications of levan can be seen in the food and feed industry. In the food industry, leaven could be used as a prebiotic. Feed formulated with levan increased the prebiotic colonization of the pig gut [92]. Levan-supplemented feed of mice increased the colonization of useful bacteria in the gut and elevated the amount of short-chain fatty acids which was seen in the feces of mice [93]. In the aquaculture field also, levan can be used as a prebiotic and immunostimulant [94]. Not only as prebiotics has levan also found applications in the beverage and dairy industry. In the beverage and dairy industry, levan can act as stabilizers, emulsifiers, and flavor enhancers. Levan is one of the main constituents of ultra-high fructose syrup which is served as a commercial product of the food industry [95]. The film- and hydrogel-forming potential of levan makes it applicable in the biomedical field for the preparation of adhesives, films, and gels. Sulfated levan, along with alginate and chitosan, can form free-standing membranes which can be used as adhesive and for tissue engineering. Such sulfated levan membranes are found to be mechanically stable and more stress tolerant than any of the other used bioadhesive membranes [96]. One of the major factors that helps levan to be

Sustainable Biodegradable and Bio-based Materials

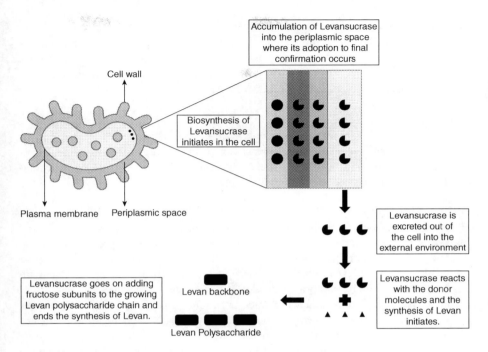

FIGURE 5.3 Biosynthetic pathway for levan from sucrose.

used in the medical field is its antioxidant property. The reactive oxygen scavenging activity of levan in HepG2 cells was reported [97]. Also, the quantity and availability of two major enzymes, catalase and superoxide dismutase, involved in oxidative stress response was increased by levan which in turn shows that levan is an excellent antitumor agent [98]. Levan biopolymer of molecular weight between 400–800 kDa showed to inhibit tumor cells in hepatic and gastric cancers [99, 100]. Hydrogels made of biopolymers are of great use in the biomedical industry. One of the major problems associated with such hydrogels is their low stability. By employing levan in the preparation of hydrogels, hydrogels with more stability, less toxicity, and with self-healing properties can be made [101]. In major and minor, levan plays a role in drug delivery, nanoparticle synthesis, cell proliferation, anti-inflammatory activity, and film-forming activity [102]. Levan is one of the biopolymers which could be molded and used for various industrial applications. But still, a large-scale production methodology and structural modification studies need to be conducted based on levan so that levan can become a more industrially valuable raw material in the near future.

5.2.3.3 Polylactides – Production Methodology and Applications

Polylactides are commendable biodegradable materials obtained from fermentation of lactic acid by microbes. Lactic acid (LA) is naturally obtained by the fermentation of sugar and starch products by microbes [103]. The most efficient and widely used method in production of polylactides involves three main steps.

5.2.3.3.1 Lactic Acid Production by Microbial Fermentation

The existing biomass containing sugars and starch will be pretreated. The pretreatment of biomass includes enzyme hydrolysis, after which the fermentable carbohydrates will be subjected to fermentation by LA bacteria, the facultative anaerobic bacteria undergoes glycolysis and produces LA.

5.2.3.3.2 Lactic Acid Purification

LA is purified and converted into lactide by the two processes. At first, LA is converted into an oligomer by polycondensation reaction. Secondly, the oligo LA is depolymerized thermally to form the cyclic lactide through a mechanism called unzipping mechanism or back-biting reaction [104].

5.2.3.3.3 Ring-opening Polymerization (ROP)

The ring of lactides opens and forms a high-molecular weight Polylactides in the presence of a catalyst which is a transition heavy metal. The presence of trace amounts of catalyst in the PLA product is a concern making it unfit to be used in the medical and food industry. Therefore, the replacement of heavy metal catalysts with environmental-friendly catalysts is needed in this method of production [104].

PLA is a biodegradable polyester made from renewable natural resources such as corn or sugarcane that is harvested once a year [105]. The mechanical properties of PLA have contributed much to the polymer world. PLA is recommended as a base resin for packaging applications. This becomes more flexible when it is plasticized with its own monomers enabling it to prepare products that can imitate PVS, LDPE, PP, LLDPE, and PS. As these PLAs are biodegradable, it can also be used for preparing compost bags, food packaging disposable tableware, and many more [106]. PLA has been extensively used because of their potential material properties. PLAs also have various applications in the biomedical field, which includes their use in bone fixation material, drug delivery microspheres as well as tissue engineering. In the horticulture production, PLAs are used in order to decrease the environmental problems caused by the use of plastics and also used as a controlled liberation of herbicides. Research data on application of lactide and PLA revealed that they accelerated soybean pod quantity, bean quantity, leaf area, and dry weight. This indicates that the use of PLA as an encapsulation matrix for herbicides may want to offer decreased environmental effect and progressed weed manipulate and on the identical time growing yield of soybeans via the launch of plant increase stimulants with inside the form of oligomeric or monomeric LA [107]. Oligomers of PDLA are used for the formation of stereo-complex crystals with L-lactate with inside the human frame to set off lactate deficiency in most cancer cells. This proves its application in anticancer therapy [108].

5.3 CHALLENGES IN COMMERCIALIZATION AND FUTURE PROSPECTS

Ever since the development of bio-based polymers, the major constraints that restricted their commercialization are attributed to the cost associated with their production and extraction from various sources. Although the development in science

Sustainable Biodegradable and Bio-based Materials

and technology in the last few decades has provided novel and cost-effective strategies for sustainable production of biopolymers, their competence to replace the synthetic polymers is a very long way ahead. The issues associated with scaling up of pilot scale strategy to fermenters and the technical challenges are the major issues associated with large-scale production process. The development of green methods for polymer extraction despite being eco-compatible cannot fully replace the efficiency of solvent extraction protocols when applied on a commercial scale. The integration of genetic modification for polymer production, especially in the case of microbial polymers levan and PHA, is still under development and requires more research focus. Understanding the market demands and following the circular economics concept can be the better options in the way ahead for sustainable development and commercialization of bio-based materials.

5.4 CONCLUSION

The global issue associated with the unscientific use of plastic and the ever-increasing stigma of use and overuse of plastics can be tackled only through the sustainable development of biopolymers and bio-based materials. Although the potential of commercially important biopolymers, like PHA, chitosan, collagen, elastin, HAp, levan, and PLA, were widely studied in the scientific literature, their sustainable production still faces many challenges. The concept of circular bioeconomy by turning the waste materials as biorefineries for deriving commercially important biomaterials can reach its sustainable tag only with the participation of government agencies, private stakeholders, and general public. A proper framework and guidelines for the collection, segregation, and strategy for developing bio-based materials in tune with waste valorization can be the best sustainable alternative for pollution caused by plastics. More research with special emphasis on integrated strategy for the development of multiple products should be developed for a sustainable tomorrow.

REFERENCES

1. Abdel-Fattah, A. M., Gamal-Eldeen, A. M., Helmy, W. A. and Esawy, M. A. (2012) Antitumor and antioxidant activities of levan and its derivative from the isolate *Bacillus subtilis* NRC1aza. Carbohydrate Polymers 89, no. 2: 314–322. doi:10.1016/j.carbpol.2012.02.041.
2. Abdelmalek, B. E., Sila, A., Haddar, A., Bougatef, A. and Ayadi, M. A. (2017) β-Chitin and chitosan from squid gladius: Biological activities of chitosan and its application as clarifying agent for apple juice. International Journal of Biological Macromolecules 104, no. A: 953–962. doi:10.1016/j.ijbiomac.2017.06.107.
3. Abdou, E. S., Nagy, K. S. A. and Elsabee, M. Z. (2008) Extraction and characterization of chitin and chitosan from local sources. Bioresource Technology 99, no. 5: 1359–1367. doi:10.1016/j.biortech.2007.01.051.
4. Adamberg, K., Tomson, K., Talve, T., et al. (2015) Levan enhances associated growth of bacteroides, *Escherichia, Streptococcus* and *Faecalibacterium* in fecal microbiota. PLOS ONE. doi:10.1371/journal.pone.0144042.
5. Akram, M., Ahmed, R., Shakir, I., Ibrahim, W. A. W. and Hussain, R. (2014) Extracting hydroxyapatite and its precursors from natural resources. Journal of Materials Science 49, no. 4: 1461–1475. doi:10.1007/s10853-013-7864-x.

6. Anbu, P., Hilda, A., Sur, H.-W., Hur, B.-K. and Jayanthi, S. (2008) Extracellular keratinase from *Trichophyton sp.* HA-2 isolated from feather dumping soil. International Biodeterioration & Biodegradation. doi:10.1016/j.ibiod.2007.07.017.

7. Avigad, G., Feingold, D. S. and Hestrin, S. (1956) The mechanism of polysaccharide production from sucrose. 3. Donor-acceptor specificity of levansucrase from *Aerobacter levanicum*. Biochemical Journal 64, no. 2: 340–351. doi:10.1042/bj0640340.

8. Azmana, M., Mahmood, S., Hilles, A. R., et al. (2021) A review on chitosan and chitosan-based bionanocomposites: Promising material for combatting global issues and its applications. International Journal of Biological Macromolecules 185: 832–848. doi:10.1016/j.ijbiomac.2021.07.023.

9. Bardhan, R., Mahata, S. and Mondal, B. (2011) Processing of natural resourced hydroxyapatite from eggshell waste by wet precipitation method. Advances in Applied Ceramics 110, no. 2: 80–86. doi:10.1179/1743676110Y.0000000003.

10. Basheer, T. and Umesh, M. (2018) Valorization of Tannery Solid Waste Materials Using Microbial Techniques: Microbes in Tannery Solid Waste Management. In Handbook of Research on Microbial Tools for Environmental Waste Management, 127–145, IGI Global. doi:10.4018/978-1-5225-3540-9.ch007.

11. Bee, S.-L. and Hamid, Z. A. A. (2020) Hydroxyapatite derived from food industry bio-wastes: Syntheses, properties and its potential multifunctional applications. Ceramics International 46, no. 11, Part A: 17149–17175. doi:10.1016/j.ceramint.2020.04.103.

12. Bello, F. D., Walter, J., Hertel, C. and Hammes, W. P. (2001) In vitro study of prebiotic properties of levan-type exopolysaccharides from *Lactobacilli* and non-digestible carbohydrates using denaturing gradient gel electrophoresis. Systematic and Applied Microbiology 24, no. 2: 232–237. doi:10.1078/0723-2020-00033.

13. Bhagwat, P. K., Bhise, K. K., Bhuimbar, M. V. and Dandge, P. B. (2018) Use of statistical experimental methods for optimization of collagenolytic protease production by *Bacillus cereus* strain SUK grown on fish scales. Environmental Science and Pollution Research International 25, no. 28: 28226–28236. doi:10.1007/s11356-018-2859-4.

14. Borrelle, S. B., Ringma, J., Law, K. L., Monnahan, et al. (2020) Predicted growth in plastic waste exceeds efforts to mitigate plastic pollution. Science 369, no. 6510: 1515–1518. doi:10.1126/science.aba3656.

15. Cadano, J. R., Jose, M., Lubi, A. G., et al. (2021) A comparative study on the raw chitin and chitosan yields of common bio-waste from Philippine seafood. Environmental Science and Pollution Research International 28, no. 10: 11954–11961. doi:10.1007/s11356-020-08380-5.

16. Calazans GMT, Lima, R. C., de França, F. P. and Lopes, C. E. (2000) Molecular weight and antitumour activity of *Zymomonas mobilis levans*. International Journal of Biological Macromolecules 27, no. 4: 245–247. doi:10.1016/s0141-8130(00)00125-2.

17. Cerning, J. (1990) Exocellular polysaccharides produced by lactic acid bacteria. FEMS Microbiology Reviews 7, no. 1-2: 113–130. doi:10.1111/j.1574-6968.1990.tb04883.x.

18. Cheng, R., Cheng, L., Zhao, Y., Wang, L., Wang, S. and Zhang, J. (2021) Biosynthesis and prebiotic activity of a linear levan from a new *Paenibacillus* isolate. Applied Microbiology and Biotechnology 105, no. 2: 769–787. doi:10.1007/s00253-020-11088-8.

19. Coppola, D., Oliviero, M., Vitale, G. A., Lauritano, C., D'Ambra, I., Iannace, S. and de Pascale, D. (2020) Marine collagen from alternative and sustainable sources: Extraction, processing and applications. Marine Drugs. doi:10.3390/md18040214.

20. Dahech, I., Harrabi, B., Hamden, K., Feki, A., et al. (2013) Antioxidant effect of nondigestible levan and its impact on cardiovascular disease and atherosclerosis. International Journal of Biological Macromolecules 58: 281–286. doi:10.1016/j.ijbiomac.2013.04.058.

21. Donner, M., Gohier, R. and de Vries, H. (2020) A new circular business model typology for creating value from agro-waste. The Science of the Total Environment 716: 137065. doi:10.1016/j.scitotenv.2020.137065.

Sustainable Biodegradable and Bio-based Materials

22. Dürig, T. and Karan, K. (2019) Binders in Wet Granulation. In A. S. Narang & S. I. F. Badawy (Eds.), Handbook of Pharmaceutical Wet Granulation, 317–349, Academic Press. doi:10.1016/B978-0-12-810460-6.00010-5.
23. Umesh, M., Choudhury, D. D., Shanmugam, S., Ganesan, S., Alsehli, M., Elfasakhany, A., Pugazhendhi, A. Eggshells biowaste for hydroxyapatite green synthesis using extract piper betel leaf - Evaluation of antibacterial and antibiofilm activity. (2021) Environmental Research 200: 111493. doi:10.1016/j.envres.2021.111493.
24. Eggum, B. O. (1970) Evaluation of protein quality of feather meal under different treatments. Acta Agriculturae Scandinavica. 20 no. 4: 230–234 doi:10.1080/00015127009433412.
25. Eliasson, A.-C. (2004) Starch in Food: Structure. Function and Applications, CRC Press.
26. El Knidri, H., Belaabed, R., Addaou, A., Laajeb, A. and Lahsini, A. (2018) Extraction, chemical modification and characterization of chitin and chitosan. International Journal of Biological Macromolecules 120, no. Pt A: 1181–1189. doi:10.1016/j.ijbiomac.2018.08.139.
27. FAO (2020) The State of World Fisheries and Aquaculture 2020: Sustainability in Action. Food and Agriculture Organization of the United Nations.
28. Fathima, N. N., Nishad Fathima, N., Raghava Rao, J. and Nair, B. U. (2014) Effective utilization of solid waste from leather industry. The Role of Colloidal Systems in Environmental Protection. doi:10.1016/b978-0-444-63283-8.00023-5.
29. Gao, D., Chen, J., Qian, W., He, Y., Song, P. and Wang, R. (2020) Improving wettability of feather fiber by surface modification. Waste and Biomass Valorization. doi:10.1007/s12649-019-00885-6.
30. Gherasim, O., Grumezescu, A. M., Grumezescu, V., Iordache, F., Vasile, B. S. and Holban, A. M. (2020) Bioactive surfaces of polylactide and silver nanoparticles for the prevention of microbial contamination. Materials 13, no. 3: 768. doi:10.3390/ma13030768.
31. Gomes, T. D., Caridade, S. G., Sousa, M. P., et al. (2018) Adhesive free-standing multilayer films containing sulfated levan for biomedical applications. Acta Biomaterialia 69: 183–195. doi:10.1016/j.actbio.2018.01.027.
32. Guktur, R. E., Nep, E. I., Asala, O., et al. (2021) Carboxymethylated and acetylated xerogel derivatives of *Plectranthus esculentus* starch protect Newcastle disease vaccines against cold chain failure. Vaccine 39, no. 34: 4871–4884. doi:10.1016/j.vaccine.2021.06.062.
33. Gupta, S. K., Pal, A. K., Sahu, N. P., Dalvi, R., Kumar, V. and Mukherjee, S. C. (2008) Microbial levan in the diet of *Labeo rohita* Hamilton juveniles: Effect on non-specific immunity and histopathological changes after challenge with *Aeromonas hydrophila*. Journal of Fish Diseases. doi:10.1111/j.1365-2761.2008.00939.x.
34. Haroon, M., Yu, H., Wang, L., Ullah, R. S., Haq, F. and Teng, L. (2019) Synthesis and characterization of carboxymethyl starch-g-polyacrylic acids and their properties as adsorbents for ammonia and phenol. International Journal of Biological Macromolecules 138: 349–358. doi:10.1016/j.ijbiomac.2019.07.046.
35. Heinze, T. and Liebert, T. (2012) 10.05 - Celluloses and Polyoses/Hemicelluloses. In K. Matyjaszewski & M. Möller (Eds.), Polymer Science: A Comprehensive Reference, 83–152, Amsterdam: Elsevier. doi:10.1016/B978-0-444-53349-4.00255-7.
36. Hindi, S. S. Z. (2017) Microcrystalline cellulose: The inexhaustible treasure for pharmaceutical industry. Nanoscience and Nanotechnology Research 4, no. 1: 17–24. doi:10.12691/nnr-4-1-3.
37. Hirano, S., Itakura, C., Seino, H., et al. (1990) Chitosan as an ingredient for domestic animal feeds. Journal of Agricultural and Food Chemistry 38, no. 5: 1214–1217. doi:10.1021/jf00095a012.

38. Hj Latip, D. N., Samsudin, H., Utra, U. and Alias, A. K. (2021) Modification methods toward the production of porous starch: A review. Critical Reviews in Food Science and Nutrition 61, no. 17: 2841–2862. doi:10.1080/10408398.2020.1789064.
39. Hoang, T. C. and Felix-Kim, M. (2020) Microplastic consumption and excretion by fathead minnows (*Pimephales promelas*): Influence of particles size and body shape of fish. The Science of the Total Environment 704: 135433. doi:10.1016/j.scitotenv.2019.135433.
40. Huang, Y.-L. and Tsai, Y.-H. (2020) Extraction of chitosan from squid pen waste by high hydrostatic pressure: Effects on physicochemical properties and antioxidant activities of chitosan. International Journal of Biological Macromolecules 160: 677–687. doi:10.1016/j.ijbiomac.2020.05.252.
41. Hu, C., Reddy, N., Yan, K. and Yang, Y. (2011) Acetylation of chicken feathers for thermoplastic applications. Journal of Agricultural and Food Chemistry. doi:10.1021/jf2023676.
42. Hu, Y., Daoud, W. A., Cheuk, K. K. L. and Lin, C. S. K. (2016) Newly developed techniques on polycondensation, ring-opening polymerization and polymer modification: Focus on poly(lactic acid). Materials 9, no. 3. doi:10.3390/ma9030133.
43. Ikada, Y., Cheng-Xiu, X., Kubo, K. and Kk., S. P. (1984) Method of Production of High Molecular Weight Polylactides.
44. Im, S. H., Im, D. H., Park, S. J., Chung, J. J., Jung, Y. and Kim, S. H. (2021) Stereocomplex polylactide for drug delivery and biomedical applications: A review. Molecules 26, no. 10: 2846. doi:10.3390/molecules26102846.
45. Iriarte-Velasco, U., Sierra, I., Zudaire, L. and Ayastuy, J. L. (2015) Conversion of waste animal bones into porous hydroxyapatite by alkaline treatment: Effect of the impregnation ratio and investigation of the activation mechanism. Journal of Materials Science 50, no. 23: 7568–7582. doi:10.1007/s10853-015-9312-6.
46. Jeencham, R., Sutheerawattananonda, M., Rungchang, S. and Tiyaboonchai, W. (2020) Novel daily disposable therapeutic contact lenses based on chitosan and regenerated silk fibroin for the ophthalmic delivery of diclofenac sodium. Drug Delivery 27, no. 1: 782–790. doi:10.1080/10717544.2020.1765432.
47. Jin, E., Reddy, N., Zhu, Z. and Yang, Y. (2011) Graft polymerization of native chicken feathers for thermoplastic applications. Journal of Agricultural and Food Chemistry. doi:10.1021/jf1039519.
48. Jinks, I. R. (2014) The Viscoelastic Properties of Chemically Modified A-keratins in Human Hair.
49. Klemm, D., Heublein, B., Fink, H.-P. and Bohn, A. (2005) Cellulose: Fascinating biopolymer and sustainable raw material. Angewandte Chemie 44, no. 22: 3358–3393. doi:10.1002/anie.200460587.
50. Kohara, N., Kanei, M. and Nakajima, T. (2001) Succinylation of chemically modified wool keratin - The effect on hygroscopicity and water absorption. Recent Advances in Environmentally Compatible Polymers: 91–96. doi:10.1533/9781845693749.2.91.
51. Kumari, S., Kumar Annamareddy, S. H., Abanti, S. and Kumar Rath, P. (2017) Physicochemical properties and characterization of chitosan synthesized from fish scales, crab and shrimp shells. International Journal of Biological Macromolecules 104, Pt B: 1697–1705. doi:10.1016/j.ijbiomac.2017.04.119.
52. Kumari, S., Rath, P., Sri Hari Kumar, A. and Tiwari, T. N. (2015) Extraction and characterization of chitin and chitosan from fishery waste by chemical method. Environmental Technology & Innovation 3: 77–85. doi:10.1016/j.eti.2015.01.002.
53. Liu, K., Du, H., Zheng, T., Liu, H., Zhang, M., Zhang, R., ... Si, C. (2021) Recent advances in cellulose and its derivatives for oilfield applications. Carbohydrate Polymers 259: 117740. doi:10.1016/j.carbpol.2021.117740.

Sustainable Biodegradable and Bio-based Materials   89

54. Liu, Y., Ahmed, S., Sameen, D. E., Wang, Y., Lu, R., Dai, J., ... Qin, W. (2021) A review of cellulose and its derivatives in biopolymer-based for food packaging application. Trends in Food Science & Technology 112: 532–546. doi:10.1016/j.tifs.2021.04.016.

55. Li, Z., Tan, B. H., Lin, T. and He, C. (2016) Recent advances in stereocomplexation of enantiomeric PLA-based copolymers and applications. Progress in Polymer Science. doi:10.1016/j.progpolymsci.2016.05.003.

56. Madhavan Nampoothiri, K., Nair, N. R. and John, R. P. (2010) An overview of the recent developments in polylactide (PLA) research. Bioresource Technology 101, no. 22: 8493–8501. doi:10.1016/j.biortech.2010.05.092.

57. Mahboob, S. (2015) Isolation and characterization of collagen from fish waste material - Skin, scales and fins of *Catla catla* and *Cirrhinus mrigala*. Journal of Food Science and Technology 52, no. 7: 4296–4305. doi:10.1007/s13197-014-1520-6.

58. Maier, S. S., Rajabinejad, H. and Buciscanu, I.-I. (2019) Current approaches for raw wool waste management and unconventional valorization: A review. Environmental Engineering and Management Journal. doi:10.30638/eemj.2019.136.

59. Ma, J., Xin, C. and Tan, C. (2015) Preparation, physicochemical and pharmaceutical characterization of chitosan from *Catharsius molossus* residue. International Journal of Biological Macromolecules 80: 547–556. doi:10.1016/j.ijbiomac.2015.07.027.

60. Marei, N. H., El-Samie, E. A., Salah, T., Saad, G. R. and Elwahy, A. H. M. (2016) Isolation and characterization of chitosan from different local insects in Egypt. International Journal of Biological Macromolecules 82: 871–877. doi:10.1016/j.ijbiomac.2015.10.024.

61. Meng, G. and Fütterer, K. (2003) Structural framework of fructosyl transfer in *Bacillus subtilis* levansucrase. Nature Structural Biology 10, no. 11: 935–941. doi:10.1038/nsb974.

62. Mescher, A. L. (2013) Junqueira's Basic Histology: Text and Atlas, 12, 13th ed., New York, NY: McGraw-Hill Medical.

63. Nair, R. and Roy Choudhury, A. (2020) Synthesis and rheological characterization of a novel shear thinning levan gellan hydrogel. International Journal of Biological Macromolecules 159: 922–930. doi:10.1016/j.ijbiomac.2020.05.119.

64. Nandgude, T. and Pagar, R. (2021) Plausible role of chitosan in drug and gene delivery against resistant breast cancer cells. Carbohydrate Research 506: 108357. doi:10.1016/j.carres.2021.108357.

65. Noorzai, S. (2020) Extraction and Characterization of Collagen from Bovine Hides for Preparation of Biodegradable Films.

66. Oprea, M. and Voicu, S. I. (2020) Recent advances in composites based on cellulose derivatives for biomedical applications. Carbohydrate Polymers 247: 116683. doi:10.1016/j.carbpol.2020.116683.

67. Panda, H. (2004) The Complete Technology Book on Starch and Its Derivatives, Asia Pacific Business Press Inc.

68. Paulo, F. and Santos, L. (2018) Inclusion of hydroxytyrosol in ethyl cellulose microparticles: In vitro release studies under digestion conditions. Food Hydrocolloids 84: 104–116. doi:10.1016/j.foodhyd.2018.06.009.

69. Pavon, C., Aldas, M., López-Martínez, J., Hernández-Fernández, J. and Arrieta, M. P. (2021) Films based on thermoplastic starch blended with pine resin derivatives for food packaging. Foods (Basel, Switzerland) 10, no. 6. doi:10.3390/foods10061171.

70. Perța-Crișan, S., Ursachi, C. Ștefan, Gavrilaș, S., Oancea, F. and Munteanu, F.-D. (2021) Closing the loop with keratin-rich fibrous materials. Polymers 13, no. 11. doi:10.3390/polym13111896.

71. Phanthong, P., Reubroycharoen, P., Hao, X., Xu, G., Abudula, A. and Guan, G. (2018) Nanocellulose: Extraction and application. Carbon Resources Conversion 1, no. 1: 32–43. doi:10.1016/j.crcon.2018.05.004.

72. Preethi, K. and Vineetha, U. M. (2015) Water hyacinth: A potential substrate for bioplastic (PHA) production using *Pseudomonas aeruginosa*. International Journal of Applied Research in Veterinary Medicine 1, no. 11: 349–354.
73. Priyanka, K., Umesh, M., Thazeem, B. and Preethi, K. (2020) Polyhydroxyalkanoate biosynthesis and characterization from optimized medium utilizing distillery effluent using *Bacillus endophyticus* MTCC 9021: A statistical approach. Biocatalysis and Biotransformation.
74. Radley, J. A. (2012) Examination and Analysis of Starch and Starch Products, Springer Science & Business Media.
75. Raghavendran, V., Asare, E. and Roy, I. (2020) Chapter Three - Bacterial Cellulose: Biosynthesis, production, and applications. In R. K. Poole (Ed.), Advances in Microbial Physiology, 77: 89–138, Academic Press. doi:10.1016/bs.ampbs.2020.07.002.
76. Rajabi, M., Ali, A., McConnell, M. and Cabral, J. (2020) Keratinous materials: Structures and functions in biomedical applications. Materials Science & Engineering. C, Materials for Biological Applications 110: 110612. doi:10.1016/j.msec.2019.110612.
77. Reddy, C. C., Khilji, I. A., Gupta, A., Bhuyar, P., Mahmood, S., AL-Japairai, K. A. S. and Chua, G. K. (2021) Valorization of keratin waste biomass and its potential applications. Journal of Water Process Engineering. doi:10.1016/j.jwpe.2020.101707.
78. Reddy, N., Hu, C., Yan, K. and Yang, Y. (2011) Thermoplastic films from cyanoethylated chicken feathers. Materials Science and Engineering: C. doi:10.1016/j.msec.2011.07.022.
79. Sagoo, S., Ron Board and Roller, S. (2002) Chitosan inhibits growth of spoilage micro-organisms in chilled pork products. Food Microbiology 19, no. 2-3: 175–182. doi:10.1006/fmic.2001.0474.
80. Salesse, S., Odoul, L., Chazée, L., et al. (2018) Elastin molecular aging promotes MDA-MB-231 breast cancer cell invasiveness. FEBS Open Bio 8, no. 9: 1395–1404. doi:10.1002/2211-5463.12455.
81. Satpathy, A. A., Dash, S., Das, S. K., Shyamasuta, S. and Pradhan, S. (2021) Functional and bioactive properties of chitosan from Indian major carp scale. Aquaculture International. doi:10.1007/s10499-020-00622-0.
82. Schrooyen, P. M., Dijkstra, P. J., Oberthür, R. C., Bantjes, A. and Feijen, J. (2000) Partially carboxymethylated feather keratins. 1. Properties in aqueous systems. Journal of Agricultural and Food Chemistry 48, no. 9: 4326–4334. doi:10.1021/jf9913155.
83. Sharma, S. and Kumar, A. (2018) Keratin as a Protein Biopolymer: Extraction from Waste Biomass and Applications, Springer.
84. Siró, I. and Plackett, D. (2010) Microfibrillated cellulose and new nanocomposite materials: A review. Cellulose 17, no. 3: 459–494. doi:10.1007/s10570-010-9405-y.
85. Song, Y., Joo, K. I. and Seo, J. H. (2021) Evaluation of mechanical and thermal properties of hydroxyapatite-levan composite bone graft. Biotechnology and Bioprocess Engineering: BBE. doi:10.1007/s12257-020-0094-6.
86. Srikanth, R., Reddy, C. H. S. S. S., Siddartha, G., Ramaiah, M. J. and Uppuluri, K. B. (2015) Review on production, characterization and applications of microbial levan. Carbohydrate Polymers 120: 102–114. doi:10.1016/j.carbpol.2014.12.003.
87. Srinivasan, H., Kanayairam, V. and Ravichandran, R. (2018) Chitin and chitosan preparation from shrimp shells *Penaeus monodon* and its human ovarian cancer cell line, PA-1. International Journal of Biological Macromolecules 107, Pt A: 662–667. doi:10.1016/j.ijbiomac.2017.09.035.
88. Srivastava, B., Singh, H., Khatri, M., Singh, G. and Arya, S. K. (2020) Immobilization of keratinase on chitosan grafted-β-cyclodextrin for the improvement of the enzyme properties and application of free keratinase in the textile industry. International Journal of Biological Macromolecules 165, no. Pt A: 1099–1110. doi:10.1016/j.ijbiomac.2020.10.009.

Sustainable Biodegradable and Bio-based Materials

89. Thazeem, B., Preethi, K., Umesh, M. and Radhakrishnan, S. (2018) Nutritive characterization of delimed bovine tannery fleshings for their possible use as a proteinaceous aqua feed ingredient. Waste and Biomass Valorization 9, no. 8: 1289–1301. doi:10.1007/s12649-017-9922-0.

90. Thazeem, B., Umesh, M., Mani, V. M., Beryl, G. P. and Preethi, K. (2020) Biotransformation of bovine tannery fleshing into utilizable product with multifunctionalities. Biocatalysis and Biotransformation: 1–19. doi:10.1080/10242422.2020.1786071.

91. Trache, D., Hussin, M. H., Hui Chuin, C. T., et al. (2016) Microcrystalline cellulose: Isolation, characterization and bio-composites application - A review. International Journal of Biological Macromolecules 93, no. Pt A: 789–804. doi:10.1016/j.ijbiomac.2016.09.056.

92. Umesh, M. and Basheer, T. (2021) Microbe Mediated Bioconversion of Fruit Waste into Value Added Products. In Research Anthology on Food Waste Reduction and Alternative Diets for Food and Nutrition Security, 604–625, IGI Global. doi:10.4018/978-1-7998-5354-1.ch031.

93. Umesh, M. and Mamatha, C. (2018) Comparitive study on efficacy of alkaline and acid extraction for PHA recovery from *Bacillus subtilis* NCDC0671. International Journal of Recent Scientific Research 9: 29467–29471.

94. Umesh, M., Mani, V. M., Thazeem, B. and Preethi, K. (2018) Statistical optimization of process parameters for bioplastic (PHA) production by *Bacillus subtilis* NCDC0671 using orange peel-based medium. Iranian Journal of Science and Technology, Transactions A: Science 42, no. 4: 1947–1955. doi:10.1007/s40995-017-0457-9.

95. Umesh, M. and Preethi, K. (2017) Fabrication of antibacterial bioplastic sheet using orange peel medium and its antagonistic effect against common clinical pathogens. Research Journal of Biotechnology 12, no. 7: 67–74.

96. Umesh, M., Priyanka, K., Thazeem, B. and Preethi, K. (2017) Production of single cell protein and polyhydroxyalkanoate from *Carica papaya* waste. Arabian Journal for Science and Engineering 42, no. 6: 2361–2369. doi:10.1007/s13369-017-2519-x.

97. Umesh, M., Sankar, S. A. and Thazeem, B. (2021) Fruit Waste as Sustainable Resources for Polyhydroxyalkanoate (PHA) Production. In M. Kuddus & Roohi (Eds.), Bioplastics for Sustainable Development, 205–229, Singapore: Springer Singapore. doi:10.1007/978-981-16-1823-9_7.

98. Umesh, M. and Santhosh, A. S. (n.d.) 'A Strategic Review on Use of Polyhydroxyalkanoates as an Immunostimulant in Aquaculture. Applied Food Biotechnology Retrieved from https://iranjournals.nlai.ir/handle/123456789/777364.

99. Venkatesan, J., Lowe, B., Manivasagan, P., et al. (2015) Isolation and characterization of nano-hydroxyapatite from salmon fish bone. Materials 8, no. 8: 5426–5439. doi:10.3390/ma8085253.

100. Wang, X.-Y. and Liu, Y. (2014) The dyeing dynamics and structure of modified cotton fabric with cationic chicken feather keratin agent. Textile Research Journal. doi:10.1177/0040517513495947.

101. Wells, C. M., Coleman, E. C., Yeasmin, R., et al. (2021) Synthesis and characterization of 2-decenoic acid modified chitosan for infection prevention and tissue engineering. Marine Drugs 19, no. 10. doi:10.3390/md19100556.

102. Wu, D., Li, Q., Shang, X., et al. (2021) Commodity plastic burning as a source of inhaled toxic aerosols. Journal of Hazardous Materials 416: 125820. doi:10.1016/j.jhazmat.2021.125820.

103. Xu, W., Zhang, W. and Mu, W. (2021) Characteristics of Levansucrase and Its Application for the Preparation of Levan and Levan-Type Oligosaccharides. In W. Mu, W. Zhang, & Q. Chen (Eds.), Novel Enzymes for Functional Carbohydrates Production: From Scientific Research to Application in Health Food Industry, 175–198, Singapore: Springer Singapore. doi:10.1007/978-981-33-6021-1_9.

104. Yang, Y., Galle, S., Le, M. H. A., Zijlstra, R. T. and Gänzle, M. G. (2015) Feed fermentation with reuteran- and levan-producing Lactobacillus reuteri reduces colonization of weanling pigs by enterotoxigenic *Escherichia coli*. Applied and Environmental Microbiology 81, no. 17: 5743–5752. doi:10.1128/AEM.01525-15.
105. Yoo, S.-H., Yoon, E. J., Cha, J. and Lee, H. G. (2004) Antitumor activity of levan polysaccharides from selected microorganisms. International Journal of Biological Macromolecules 34, no. 1-2: 37–41. doi:10.1016/j.ijbiomac.2004.01.002.
106. Yu, M., Yue, Z., Wu, P., et al. (2004) The biology of feather follicles. The International Journal of Developmental Biology 48, no. 2-3: 181–191. doi:10.1387/ijdb.031776my.
107. Yusop, S. M., Nadalian, M., Babji, A. S, et al. (2016) Production of antihypertensive elastin peptides from waste poultry skin. ETP International Journal of Food Engineering. doi:10.18178/ijfe.2.1.21-25.
108. Zhou, L.-T., Yang, G., Yang, X.-X., Cao, Z.-J. and Zhou, M.-H. (2014) Preparation of regenerated keratin sponge from waste feathers by a simple method and its potential use for oil adsorption. Environmental Science and Pollution Research International 21, no. 8: 5730–5736. doi:10.1007/s11356-014-2513-8.

6 Sustainability of Materials
Concepts, History, and Future

Rajat Dhawan, Hitendra K. Malik, and Amit Kumar

CONTENTS

6.1 Introduction to Sustainability 93
6.2 Concepts of Sustainability...96
6.3 History of Sustainability...99
6.4 Metrics For Sustainbility 101
6.5 Sustainable Logistics: A Factor of Modern Economics 103
6.6 Brief Applications of Sustainability of Materials...105
6.7 Future Prospects and Challenges...107
References..108

6.1 INTRODUCTION TO SUSTAINABILITY

On the nonrenewable resources' exploitation, there is a large dependence of supply of goods and services. It includes gas, oil, coal, metallic, and nonmetallic minerals. Several imperative questions have been raised by this. For example, in the near future, can we continue to satisfy indefinitely the global material needs? In the atmosphere without enhancing the CO_2 levels, can we meet with the future energy needs? In order to attain the needs of energy and material for society, the production and the consumption have been enhanced. Consequently, can the enhanced waste resulted from the production and the consumption be coped by the environment? It becomes necessary to address these questions. In order to address these questions, this chapter has been divided into several sections where we have discussed a detailed introduction to sustainability and have described the concept of sustainability along with history of sustainability in a deep manner by describing its environmental perspective also.

Throughout history for several hundreds of years, there have been many persisting preindustrial civilizations and indigenous cultures. Some of them have been collapsed now. This is because of the spread of infectious diseases or as a result of invasion. The collapse of societies has been historically contributed by eight factors, as identified, summarized, and explained by Diamond [1]. These are overpopulation; overfishing; overhunting; on native species, the impacts of introduced species; water management problems; soil problems; habitat destruction and deforestation; on the environment, enhanced per capita effect of people. The world is facing these issues today which are environmentally related and requires a sustainable solution.

DOI: 10.1201/9781003297772-6

Nowadays, there is an increasing decline in the environment health key indicators. The United Nations Environment Programme [2] has coordinated a study on the state of the environment. In their report, more than two thousand reviewers and authors have contributed. The key findings of this report can be summarized as follows. In their study, several ecosystem services have been analyzed and a degradation in around 3/5th of the ecosystem services has been observed. It composed of water and air purification, fisheries, freshwater, pests, natural hazards, and local and regional climate regulation. These services show a substantial growth in the full costs of degradation and loss. In order to enhance the other services supply, many actions have been taken time-to-time. As a result of these actions, a degradation in several ecosystem services has been recorded. From one group of people to another, the cost of degradations is being shifted due to these trade-offs and the future generations due to this may also be deferred. In ecosystems, the likelihood of nonlinear changes is increasingly occurring due to the changes taking place in the ecosystems. The changes in regional climate, the collapse of fisheries, the dead zones creation in coastal waters, unexpected changes in water quality, and emergence of diseases are included in such nonlinear alterations. Across the groups of people, disparities and inequities are contributing to grow as a consequence of these degradation of ecosystem services. Social conflict and poverty are being caused by these principal factors.

Since World War II, several benefits as well as several environmental degradations have been caused by the green revolution in agriculture. Adopting monocultures (single commodities), intensive cultivation, and clearing of the land have been performed. For this, artificial fertilizers and pesticides are being adopted increasingly which result in the damage of ecosystem. Energy- and water-intensive are the main characteristics of the modern agriculture. It requires mechanized farming practices, transport of fertilizers, and manufacture of artificial fertilizers. The food production quantity is greatly increasing by these modern agricultural techniques, which are also contributing to the ecosystems general degradation.

Social sustainability, economy sustainability, and environmental sustainability are the three pillars of sustainability (Figure 6.1). Including both exhaustible and renewable inputs, physical inputs are provided for economic production by a particular portion of natural resources. This particular portion is focused in economic

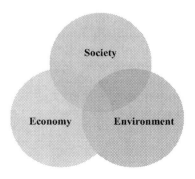

FIGURE 6.1 Three pillars of sustainability.

Sustainability of Materials

TABLE 6.1
Superiorly Known Sustainable Frameworks

S. No.	Framework	Basis
1	Green Engineering	Physical Sciences
2	Green Chemistry	Physical Sciences
3	Eco-efficiency	Science & Engineering
4	Natural Capitalism	Social & Economics
5	Triple Bottom Line	Social & Economics
6	Five Capitals Model	Social & Economics

sustainability. Life-support systems are greatly emphasized in the environmental sustainability. For human life to occur or for economic production, such systems must be maintained. Soil and atmosphere are the examples of life-support systems. The human impacts of economic system are focused in social sustainability. The efforts to eradicate hunger and poverty are included in the category of social sustainability.

In order to overcome these difficulties, several sustainable frameworks have been proposed time-to-time. There are many approaches through which we can lead to transition to sustainability. Table 6.1 has shown a few of superiorly-known sustainable frameworks. Sustainability frameworks or sustainable developments constitute series of principles, rules, and guidelines. At a general audience, these have been promoted by not-for-profit organizations and through books. For a variety of purposes, these sustainability frameworks have been proposed. In a specific economic sector, groups of institutions or companies and also individual institutions or companies may be intended to use these frameworks. At national, regional, and local levels, governments and communities may also be intended to use these frameworks.

Green engineering and green chemistry are two technically based frameworks. The term sustainable chemistry is also used to describe green chemistry framework. The generation and adoption of hazardous substances can be eliminated or reduced by the designs of chemical processes and products which fall under the category of green chemistry. Widely accepted principles of green chemistry have been developed by Anastas and Warner [3]. In comparison with inorganic substances, the manufacturing of organic substances is more applicable by these principles. Therefore, metal production and minerals are not directly relevant by these principles. Nowadays, one of the concepts which requires a well development is the green engineering. The adoption, commercialization, and design of products and processes which are economical and feasible with eliminating the risk to environment and human health and minimizing the generation of pollution at the source are coming under the concept of green engineering. Nine principles of green engineering have been proposed by the United States Environmental Protection Agency, [4] which are looking adequately generic for all the industrial sectors to obey.

Enhancing service intensity, extending product durability, maximizing use of renewables, enhancing recyclability, reducing toxic substances dispersion, reducing energy intensity, and reducing material intensity are the general principles of eco-efficiency. The toxic substances dispersion, waste disposal, water discharges, and air emissions must be minimized, and usage of renewable resources must be fostered to diminish the impact on nature. Through increasing product durability and recyclability and reducing the usage of land, water, materials, and energy, we can diminish the resources consumption. Hawkins et al. [5] in their book titled *Natural Capitalism* have emphasized that four central strategies must be adopted in order to have next industrial revolution. These are: sustaining and restoring natural resources, or investing in natural capital; from quantity to quality: a change in the social values; recycling and reuse of materials; more effective manufacturing processes for the conservation of resources.

The social–environmental–financial accountability is the triple bottom line (TBL) concept and for sustainable development, it was an early attempt [6]. People–planet–prosperity is the other form of TBL concept. A simple method for several companies to demonstrate their social credentials and environment, during the 1990s, was the TBL. In the business community, an imperative role has been played by TBL concept to raise consciousness of sustainable development. For sustainable development framework, TBL has been adopted by South African Government in 2006. Financial, manufactured, social, human, and nature are the five forms of capital on which the five capitals model is based. A sustainable society must have the following features. The environmental capacity to recycle, absorb, and disperse must be exceeded than the substances taken from the Earth. Otherwise, environment and humans must be protected by neutralizing their harmful effects. An enhanced or a protected capacity of the environment is requisite to support biological productivity and ecological system integrity. There should be maximum use of human skills and innovations and minimum use of natural resources in all processes, technologies, and infrastructures. The manufactured, social, human, and natural capital are accurately represented by the financial capital.

6.2 CONCEPTS OF SUSTAINABILITY

Without degrading the future generation's abilities, achieving our own needs is defined as sustainability. Sustainable development broadly depends on this concept. In order to attain present and future requirements, concept of limits and concept of needs to the ability of the environment are the two concepts contained in this definition. The phrase 'sustainable development' is often preferred by the term sustainability. Implying growth may be perceived by the word 'development'. Thus, continuing economic growth, ameliorating the problems caused by, is the meaning of the sustainable development.

There are several different ways in which we use the term 'sustainability'. When we speak about sustainability, then economic sustainability and ecological sustainability, respectively, are widely used by many business people and environmentalists. Economic, social, and ecological aspects are the three aspects of sustainability. A specific level of economic, social, or ecological sustainability cannot be achieved independently until and unless basic levels of all three forms are achieved simultaneously. Within an unsustainable global system, any subsystem cannot be sustainable. Especially in the lithosphere, biosphere and hydrosphere, a property of the Earth

Sustainability of Materials

system is sustainability. In terms of the global system, a community's sustainability can only be described. If the society is unsustainable at large, then an organization or a firm cannot be sustainable.

In ecological or economic terms, generally sustainability is defined more rigorously. According to economic definitions, if over time there is no declination in production or consumption then development is sustainable or if in order to supply non-declining per capita utility, there is no diminishment in the capacity of a system then development is sustainable. According to ecological definitions, if over time there is a maintenance of ecosystem resilience and stability's minimum conditions, then development is sustainable; or if ecosystem services sustainable yield is maintained by managing resources, then development is sustainable; or if over time there is no declination in the stock of natural capital, then development is sustainable.

In order to attain material consumption or production indefinitely in terms of the economy's ability, the sustainability is defined in economic definitions. However, without the usage of environmental resources, this is not possible. Therefore, some degree of environmental sustainability must be existed as per the implications of economic interpretations. In terms of the environment, sustainability is defined directly by the ecological interpretations. Saying the preservation of capital assets stocks indefinitely is the same as saying the maintenance of level of production or consumption or the productive capacity indefinitely (as per the economic definition). For consumption or output's non-declining flow, sufficient productive capacity will be ensured by this stock. Several types are composed in the capital assets stock. Up to what amount several varieties of capital can be substituted for one another is an imperative question in the theory of sustainability. For environmental resources declining stocks, the substituting possibilities of productive capital's other forms is also an important issue. Financial, manufactured, social, human, and nature are the five forms of capital which are described as follows.

In the form of currency, an organization's assets are existed in the financial capital. Banknotes, bonds, and shares are included here which can be traded or owned. All other forms of capitals are represented by the values of financial capital. The infrastructure and material goods are consisted in manufactured capital which are controlled, leased, or owned by an organization. The provision or production of services is contributed by this organization. In social capital, the development and maintenance of human capital of individuals in partnership with others are enabled by the relationships, institutions, and structures. In comparison with doing work in isolation, more productive output is resulted when working together. It comprises educational and health bodies, legal and political systems, voluntary organizations, schools, trade unions, businesses, communities, families, communication channels, and networks. For individual's spiritual and emotional capacity, and for productive work, the required motivation, skills, knowledge, and health come under human capital. Intellectual property (intellectual capital) is understood to be included by human capital in this context. Over time, the accumulation of useful knowledge societies stocks derives the intellectual capital. The processes which provide superior services and goods, and stock of natural resources (matter and energy) are coming under the natural capital. Ecosystem services and renewable and nonrenewable resources are consisted in this capital. The sum of financial, social, human, and manufactured capital assets is the human-made capital. Hence, the stock of productive assets is the sum of natural capital and human-made capital. It is important to have knowledge of distinction among natural capital and human-made capital. For natural capital, the degree up to which human-made capital can be substituted decides whether we can maintain the productive capacity indefinitely or not. The weak and strong sustainability concepts are leading by this behavior.

If for both directly as a provider of ecosystem services and as an input for services for consumption and production of goods, natural capital can be substituted by human-made capital, then the assumption leads to the concept of weak sustainability. As far as we have built-up sufficient human-made capital, it is allowed to degrade the natural capital. In other words, it is allowed to decline a few portions of total stock of assets. For declining natural capital, if other varieties of capital are substituting adequately, then such declines are not imperative. An optimistic view of resources can be taken by the model of weak sustainability. In four propositions, this view can be summarized. For natural capital, the substitution of human-made capital mostly occurs in economic modeling. For natural capital, if the elasticity of substitution of human-made capital is equal to or higher than 1, then it is possible to have perpetual economic growth of consumption, as shown in the mathematical works of Stiglitz [7] and Solow [8]. In practice, magnitude of at least 1 is required for the elasticity of substitution. No theoretical justifications are there to describe the significant substation of human-made capital for the natural capital. However, a possibility of little substitution is indicated by the empirical evidence. Clearly it is possible to have significant substitutions among the several varieties of human-made capital. Because of distinguish characteristics of natural capital from human-made capital, it becomes problematic with human-made capital to have substitutions of natural capital. Basic life-support functions have been provided by several varieties of natural capital and it cannot be provided by any other capital variety. On Earth, human life is possible only because of such ecosystem services. In addition to this, there are several varieties of natural capital with unique characteristics in such a way that once destroyed they cannot be rebuilt. For human-made capital, the situation is not same in general. In principle, it is plausible to have slow or expensive reconstruction for human-made capital. For most varieties of natural capital, up to a greater extent, human-made capital can be substituted.

If for either directly as a provider of ecosystem services or as an input for services for consumption and production of goods, natural capital can be substituted by human-made capital, then the assumption leads to the concept of strong sustainability. Strong sustainability has two interpretations. In first interpretation, it is necessary to preserve total value of the natural capital. In order to maintain total value of natural capital at a constant level, nonrenewable resource extraction's scarcity rents can be invested in alternative developments such as by the development of renewable energy sources, the royalties from coal mining can be invested. Non-substitutable varieties of natural capital must be preserved, as per the second interpretation. Such natural capitals are designated by critical natural capital. Among distinctive varieties of critical natural capital, there should be no substitution. According to second interpretation, one should use renewable resources up to that amount only where there is no deterioration of their stock. Additionally, for wastes one should adopt

Sustainability of Materials 99

environment as a sink up to that amount only where there is no deterioration of their natural absorptive capacity.

6.3 HISTORY OF SUSTAINABILITY

In order to maintain our well-being, an importance of scientific understanding of the environment had been growing rapidly in 1960s. Because of the negative effects of internationally reported extreme events and local environmental problems, a parallel growth in public awareness has been made. On the night of 2–3 December 1984, a mixture of poisonous gases released from a Union Carbide pesticide manufacturing plant in Bhopal, India. There was a loss of thousands of lives and, in history, it turned-out as a worst industrial disaster. On the other hand, the BP oil well's offshore failure was seen in the Gulf of Mexico in 2010.

In agriculture, the unregulated usage of several pesticides like DDT has created serious health problems. In order to draw attention to the environment, a book titled "*Silent Spring*" [9] has played a role in raising public concern for the environment. During the 1960s, regarding toxic substances releasing to water, air, and land and the usage of pesticides, some stringent laws have been introduced by the governments to grow public awareness. Concerns for the environment were further raised and an attention to the issues of global development was further made by many articles in popular magazines, journals, and books. Another early book was *The Population Bomb* [10]. At that time, this book was considered to be a most influencing book since it raised serious issues. There is an arithmetical growth in the agricultural output, whereas a geometrical growth in the population is resulted. Therefore, the production of food will always fall behind than the demand. The problems faced by humanity were studied by a not-for-profit organization named the Club of Rome. The results of their study have been presented in the book *The Limits to Growth* [11]. In a world of abruptly increasing population, the implications of finite resources impact have been modeled in their study. Resource depletion, food production, pollution, industrialization, and population were the five examined variables. The collapse of the economic system would be led by the continuing growth of the global economy, as per the result of their study. Their study also resulted that through the combination of early alterations in technology, policy, and behavior, such collapses could be avoided.

At national and international levels, social and environmental issues have been raised time-to-time by the United Nations through international forums. The requirement of an alteration in the approach to development has been debated in 1972 by the UN Conference in Stockholm. In the Stockholm Declaration among the participating nations, a level of understanding and consensus has been achieved. In order to address the major issues facing the world, the Declaration has raised a set of principles. The natural resources and the human environment's accelerating deterioration have been addressed in 1983 during World Commission on Environment and Development. For social and economic development, the consequences of that deterioration were also been addressed herewith. After its chairperson Gro Harlem Brundtland, this commission was known as the Brundtland Commission. The term "without degrading the future generation's abilities, achieving our own needs as sustainable development" was first popularized in the report of Brundtland Commission.

At Rio de Janeiro, among all sectors of human endeavor for sustainable activity, 26 principles were provided in the United Nations Conference on Environment and Development in 1992. It is usually described as the Earth Summit. The precautionary principle is described by the statement of the principle 15 of the Rio Declaration. According to this principle, in order to prevent environmental degradation, for postponing cost-effective measures, the lack of full scientific certainty should not be adopted as a reason where there are dangers of irreversible or serious damage. According to precautionary principle, before the necessary protection of the environment is indicated by the scientific results, appropriative preventative measurements must be undertaken. According to the capabilities of the states, the precautionary approach must be broadly accepted in order to protect the environment.

In 2002 at Johannesburg, there was a United Nations World Summit on Sustainable Development. Both nongovernmental and business organizations' leaders were brought together in this summit. In comparison with the environmental issues, political, economic, and social issues were focused in greater details in the Johannesburg Declaration. Chronic, communicable, and endemic diseases, especially tuberculosis, malaria, and HIV/AIDS; xenophobia; incitement and intolerance to religious, ethnic, racial, and other hatreds; terrorism; trafficking in persons; illicit arms trafficking; natural disasters; corruption; organized crime; illicit drug problems; armed conflict; foreign occupation; malnutrition; and chronic hunger were included in their agreement to prevent the sustainable development of our people.

After the Earth Summit, in 1995, there was a formation of World Business Council for Sustainable Development (WBCSD). For sharing best practices, experiences, and knowledge and to explore sustainable development, a platform is provided by WBCSD to companies. In the varieties of forum working with inter-government, nongovernment, and government organizations, business positions can be advocated by the companies on these issues. It has approximately two hundred member companies. The principles of sustainability have been adopted by many organizations and companies. This is because to enhance their innovation scope; to describe potential cost savings; to modify access to investors; to retain and attract employees; to secure social license to operate; and to increase their reputation. For sustainable development, the business cases are often referred by them. The social and environmental harms can be caused by the companies; therefore for attaining sustainability, the business cases have grown. The measurements regarding contribution to sustainability have now been reported by many companies.

In response of increasing community expectations, principles of corporate social responsibility (CSR) have been adopted by some companies in recent years. In order to contribute and behave ethically to economic development, the continuing commitment by business is CSR. Here, the work-force and their family's life quality have been improving. The voluntary acceptance of ethics codes beyond and above legal requirements, sustainability standards, and social responsibilities by companies have been implied by CSR. If a company adopts CSR principles, then it means that international norms and environmental, social, and ethical standards will be ensured by the company. For the effect of products and activities of the company on

Sustainability of Materials 101

the community, employees, consumers, and environment, the company would accept its responsibilities. By eliminating harmful practices and by supporting and encouraging community development, company should promote public interest. Through operating with broad long-term perspectives, companies may get benefits in several intangible and tangible methods—as per CSR supporters. The maximization of the shareholder value, which is the primary role of business, is distracted by CSR—as per CSR criticisers. CSR is greenwashed or window-dressing which allows 'licence-to-operate'—as per others arguments.

6.4 METRICS FOR SUSTAINBILITY

New ways of thinking are requisite to meet with the requirements of the society without degrading the environment. For this instead of innovation's strong records, environmental problems should be addressed directly by the new solutions, human health, and the environment should be protected by manufacturing processes, and the needs of customer should be met by the products. With industrial chemistry, there is generally a critical relationship of greater public. More responsibly acts are performed by the industries to perceive the needs of the society. In order to ensure these acts, increasingly strict legislations are introduced by the governments all around the world. For the further development, an important issue is the quantification of the sustainability advantages of the industrial chemistry. In order to have new processes and products development, it is essential to use quantitative methods for validated and realistic estimation of innovative potentials of industrial chemistry. Within the chemicals industry, a crucial factor for the sustainable development is the link among metrics for sustainability and industrial chemistry. With an extensive and a powerful methodology, we should evaluate the development of industrial processes in combination with the latest developments and approaches of "Green Engineering" and "Green Chemistry" [12]. In impact assessment research as well as in new techniques and tools development, a key role has been played by the chemical industry. To research and development, a huge portion of its resources has been allocated consistently. The chemical industry is also imperative to promote understanding and communications among the academic communities. Through the evaluation of the product development industry's seeking ideas and the ideas generated by academia, we can bridge the gap among their ideas with an innovative way.

The existence of a service caused or of a given product can have some environmental impacts. Therefore, the valuations and investigations of such environmental impacts are particularly an important issue. Throughout the life cycle of a product, such impacts can be standardized and analyzed by adopting different approaches. The mass balance of system's inputs and outputs are evaluated by a method known as life-cycle assessment (LCA). Cradle-to-grave analysis and life-cycle analysis are the other names of LCA. Here, in accordance with the distinctive ecological areas, environmental impacts of inputs and outputs are converted and organized. Before starting an LCA, defining "functional unit" is one of the initial steps. The function, which a service or product shall be delivering, is related to the "functional unit". Among life-cycle equivalent stages, i.e., product A consumer stage and product B consumer stage, the comparisons are also allowed by LCA. For this, on the same assumptions and on the same databases, the Life-Cycle Inventory (LCI) relies as the basic data.

On a mass or volume basis, throughout the life cycle of a given product system, the quantifications and complications of inputs and outputs are pursued in LCI. Cubic meter of solid waste, kg of cadmium, kg of CO_2, etc. are its examples. The phase of LCA is the product-specific requirements (PSRs). Here, throughout the life cycle of a given product system, the quantifications and complications of inputs and outputs are involved. In addition to this, Environmental Product Declaration (EPD) will contain the specific technical data information. All inputs and outputs information are provided by the LCI analysis result. In the study, all the involved unit processes provided the result in the form of elementary flow. The flows which are going beyond the system boundary are included in the results of an LCI analysis. For life-cycle impact assessment (LCIA), the starting point has been provided then. The significance and magnitude of a product system's potential environmental impacts are aimed to evaluate and understand in the phase of LCA, called as LCIA. In order to attain recommendations and conclusions in consistent with the designated scopes and goals, the outcomes of either the impact assessment or the inventory analysis, or both, are combined in the phase of LCA, called as Life-Cycle Interpretation.

In the late 1990s, Roland Berger Consulting, Munich, developed a new approach by following above general methodology approaches. A life-cycle management tool was seen by this new methodology naming the Eco-Efficiency Analysis. Entire product LCAs, i.e., end-of-life ← marketing ← implementation ← designing ← concept development issues can be involved in it. Both environmental and economic aspects may be incorporated in this analysis method. Over entire life cycle, comprehensive evaluation of processes and products may also be conducted. A specific process or product's weaknesses and strengths are shown in the results of the method of eco-efficiency analysis. On environmental and costs impact, aggregated information has been presented in the results of this method.

The eco-efficiency analysis process steps are portrayed in Figure 6.2. There are several key stages through which every eco-efficiency analysis has to pass. The

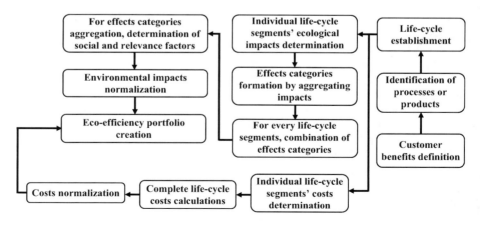

FIGURE 6.2 Eco-efficiency analysis process steps.

Sustainability of Materials 103

comparability of different studies and consistent quality has been ensured by this testing. LCA is adopted to determine the environmental impacts. National economic models or, in some instances, usual business models are used to calculate the economic data. In eco-efficiency analysis, there are three basic preconditions. First, both an economic and environmental assessment; second, the entire life cycle; and third, linking the identical described customer benefits in processes or products studies.

There are several defined and specific calculations which are used to work out the eco-efficiency analysis. These are inclusion of social aspects; analysis of business, sensitivities, scenarios, and weakness options; an eco-efficiency portfolio creation; economy versus ecology's relative importance determination; with societal factors, life-cycle analysis factors' weighting; for specific weighting, relevance calculations; over the whole life cycle, the determinations of effects on the safety and the health of the people; for all investigated processes or products, a particular life-cycle analysis calculations; and total cost calculations as per the viewpoints of the customer. German Technical Inspection Association has approved this methodology. In different studies of Association of Plastics Manufacturers in Europe (APME), this methodology has also been adopted in Freiburg, Germany. In order to provide a value for risk potential and toxicity potential, emissions and use of area, raw-material consumption and energy consumption, and analyses and assimilations of the data from the distinct production methods have been made so that each alternative's environmental position can be compared and calculated. Identically, the economic data may also be summarized and calculated from process evaluation or product application's life-cycle chain. With regard to capital investment, material utilization, and product design, better decisions can be led by this analysis. Landsiedel and Saling [13] and Saling et al. [14] have described the rationale behind this assessment tool. The decision-making processes can be supported by the metrics for sustainability, as shown by the practical/industrial examples.

6.5 SUSTAINABLE LOGISTICS: A FACTOR OF MODERN ECONOMICS

Due to the generation of environmental pollutions in consumers, sales outlets, warehouses, and supplying factories, logistics is apparently lied at the heart of the issues of the sustainable development providing the close relationship with transportation. In terms of fair trade, regional development, and employment, a higher contribution to social equity has also been made by logistics. Through efficient supply chain and quality improvement management, the enormous saving potentials contribute to the sustainable growth. In Europe, around 70% of the oil used is consumed in transport and therefore it is represented as greenhouse gas emissions primary generator. Imprint left by logistic infrastructure, noise, toxic exhaust emissions, energy consumption, and traffic congestion in urban areas are the other factors on the downside. Within the European Union, an expected increment of 60% in trade is the more alarming situation. The transport has a negative impact on the environment and it is a subject of much criticism. In order to meet the challenges of sustainable development, a great deal of expectation has risen by the logistic function. In optimizing and controlling distribution and production systems, a superior role has been played by this management discipline in the broadest sense. The economic vitality and business competitiveness of society in addition to reducing greenhouse gas emissions are also operating under this management. In the sustainable development, one of the capable contributing key players is this feature. The other two pillars of sustainability (social and economic) are connected with the well-known aspect of logistics. The purchasing power, employment, and economic growth are able to be influenced by this feature. Also, there is a reduction in the impacts of their activities on the environment. Through the adoption of "responsible practices" among logistic service providers, retailers, manufacturers, and suppliers and through the development of eco-technological solutions, the three aspects, such as environmental, social, and economic, can be reconciled by the sustainable logistics. In order to attain common objectives for the benefit of environment, society, and business, some collaborations and global projects are needed.

There are several endemic problems which the industry is facing today. Without making a progress in technology, it is difficult to solve these problems. In terms of sustainable development, counterproductive situations are generated by several individualistic practices. A loss of responsiveness, CO_2 emissions, and an explosion of logistic costs are entailed in short-term profitability considerations during the search for lowest production costs. In the direction of superior collaboration, responsibility, transparency, and balance among stakeholders, a breakthrough transformation has been called owing this alarming observation. In order to diminish supply-chain costs, infrastructure footprint, traffic congestion, and CO_2 emissions, vigorous steps changes are requisite. In general, human activities, such as environmental, social, and economic impacts, were highlighted during Grenelle Environment meeting in France in October 2007. From transport services, an assessment of emissions of greenhouse gas has already been advocated. The consequences for circulation of goods, warehousing, distribution networks, and organization of production are considerable. In the service of protection of the environment, social development, and economic growth, logistics have offered four essential drivers so that the challenges of sustainable development can be met and into concrete realities, their strategic objectives can be translated.

It is not difficult to attain this attractive vision. Based on the company's abilities to modernize their facilities between partners, resources, share information, coordinate activities, automate information exchange (ERP→ Enterprise Resource Planning and EDI→ Electronic Data Interchange), measure performance, cut down costs and lead-times, integrate the logistic process, and working methods, a long-haul and gradual approach is called in this vision. In order to make a link among strategic objectives and operational translations of companies, the concept of logistic drivers has been introduced herewith. Eco-logistic, logistic agility, logistic efficiency, and logistic reliability are the four essential drivers. In the service of sustainable development, several imperative advantages have been offered by logistics drivers. Through more mindful, agile, efficient, and reliable preservation of environmental and social balances, the expectations of environment, society, staff, customer, and share-holders can be reconciled by these logistics drivers.

In order to promote sustainable transport, European Sustainable Development Strategy, in 2006, has reaffirmed a series of measures such as technological

Sustainability of Materials

improvements, infrastructure developments, inter-modality development policy, and behavior. The development of more sustainable and efficient transport chains were encouraged in the highlights of Marco Polo programme in 2007 [15]. For the future supply chains, the requisite key changes were defined by Global Commerce Initiative (GCI) in 2016 to have fast-moving consumer goods industry [16]. Joint scorecard and business plan; efficient assets; identification and labeling; demand fluctuation management; reverse logistics; collaborative physical logistics; and in-store logistics are the seven sectors where several improvements are identified. Shopper interaction, shelf-ready products, and in-store visibility are required improvements in in-store logistics. Shared infrastructure, shared warehouse, and shared transport are required improvements in collaborative physical logistics. Returnable assets, packaging recycling, and product recycling are required improvements in reverse logistics. Monitoring, execution, and joint planning are required improvements in demand fluctuation management. Green buildings, switching modes, aerodynamic/ efficient vehicles, and alternative forms of energy are required improvements in efficient assets.

With an improvement in the on-shelf availability, there will be a lowering of around 25% of CO_2 emissions per pallet when this redesigned supply chain will be used, as per the estimation of GCI. GSI organizations in distinctive models (better traceability, cross-docking, stream consolidation, Forecasting and Replenishment, Collaborative Planning, Vendor-Managed Inventory, etc.), Voluntary Inter-industry Commerce Solutions (VICS), and Efficient Consumer Response (ECR) have successfully developed a strong collaboration among customer and supplier. In order to further optimize the use of resources, this development has been designed. Additionally, CO_2 emissions will be reduced significantly. New business opportunities and a future model leading to sustainable development will be resulted from these developments. Nowadays, tracing and tracking tools, Supply Chain Management software, and global communications standards (EDI, RFID, barcodes, product identifications) are available as certain supporting tools and technologies. In the physical supply chain, managements of social regulations, incentives and measures, organizational resources and design, capabilities, and investments are required to have new ways of working together which lead to a factor of modern economies. It shall help the small- and medium-sized industries to grow abruptly. With supporting regional development and purchasing power, the sharing of the gains and the needs of economic growth are the main objectives of this development.

6.6 BRIEF APPLICATIONS OF SUSTAINABILITY OF MATERIALS

In terms of carbon emissions, there must be a nonzero utilization and production of renewable materials so that we can have a desirable sustainable society. Both industrially and academically, much attention has been given to carbon materials with increasing applications in biomedicine, renewable energy conversion and storage, and CO_2 utilization. Because of the characteristics of easy synthesis, low cost, high abundant, and renewable, a potential advancement in the several applications has been exhibited by the sustainable carbon materials. Environmental pollution and energy consumption are the major issues with which the preparation processes of new carbon materials are suffering. Hence, it is requisite to have sustainable carbon material preparation methods which are of scalable, low-cost, economically and industrially attractive. A major feature of sustainability is the adoption of derivatives of biomass as a carbon materials precursor. Several researchers have summarized the synthetic strategy and emerging applications of sustainable carbon materials in greater details [17–21]. In order to produce sustainable carbon materials, several sustainable strategies and original intentions have been discussed by Lan et al. [18] in their review article. For sustainable carbon materials, further designing, significant, and insight guidelines have been provided in this review. In the biomedical and catalysis fields, the emerging applications have also been discussed. In manufacturing processes, carbon containing waste must be recycled. With the preparation of highly efferent carbon materials, there will be a considerable reduction in the unnecessary pollutions and wastes when sustainability approaches have been adopted. The dependence on natural resources, environmental pollution, and cost of carbon material preparation can be reduced considerably if the used carbon material is recycled.

It has been proven over time that during rehabilitation and construction of road pavements, the rate at which consumption and exploration of industrial and nonrenewable natural aggregates products have been made by the construction industry is resulting in non-sustainable and environmentally degrading materials. Inadequate disposal and generation of high-solid wastes are the major issues nowadays. Therefore, in order to integrate these solid wastes as alternative materials, several researchers have been performed a series of investigations in road maintenance and construction [22–26]. In both rigid and flexible pavements, the impacts of reusing these waste materials and the alarming rate of their generation have been understood in their studies. The selected waste materials benefits and drawbacks have been highlighted in the review study of Bamigboye et al. [26]. The economic sustainability implications, life cycle environmental and performances have also been highlighted in their studies. The industrial and municipal solid wastes can be recycled with a considerably improved rate if practical utilization of these materials are performed adequately. Consequently, certain performance criteria will be improved, maintenance and construction costs will be reduced, natural aggregates against depletion will be preserved, and requirements of land for landfills will be reduced effectively.

In residential, commercial, and industrial applications, the enhancement in profitability and efficiency of thermal systems has been shown when nanofluids are being adopting as an innovative alternative fluid solution. The lower costs, lower energy consumption, and reduction of environmental impact are resulted due to an enhancement in the thermal systems efficiency. Consequently, various imperative applications are resulted [27, 28]. Based on sustainability methodologies, environmentally and economically evaluation of nanofluids have been made in the recent years [29–32]. In order to have a sustainable technological development, overview of areas of opportunities and benefits of nanofluids have been given by Bretado-de los Rios et al. [29]. Sustainability assessments perspective of hybrid and simple nanofluids adopted in hybrid photovoltaic-thermal (PVT) solar systems, solar collectors, and heat exchangers have been focused in their study. Positive environmental and economic results can be implemented in a thermal system with stable nanofluids with the corrected selection of nanoparticles.

6.7 FUTURE PROSPECTS AND CHALLENGES

In industrialism and industrial society, the origin of modern environmental movement has established. An overarching commitment to growth for the material well-being, services produced, and the quantity of goods is defined as industrialism [33]. Several competing ideologies, especially fascism, communism, socialism, conservatism, and liberalism have been produced by industrial societies. Industrialism is committed by all of these ideologies irrespective of their differences. Highly suppressing or discounting environmental concerns are involved because of their identical attitude to the environment. Because of nonacceptance of industrialism, alternatives should be included to provide plausible responses to the sustainability challenge. Reformist or radical are the two varieties of the responses. Within the industrialism's present understanding, the responses which are seeking to work are the reformist responses. Between well-being, consumption, and growth, the responses which are seeking to break the nexus are the radical responses. One-dimensional matrix of possible responses is formed by them. In order to adopt Dryzek's terminology [33], responses can be imaginative or unimaginative (prosaic) when second dimension takes into account. Industrial society's politico-economic arrangements are accommodated by prosaic responses whereas politico-economic arrangements are seeking to redefine by the imaginative responses. Table 6.2 shows four responses to the sustainability challenges when two dimensions are putting together.

First response is green radicalism which is both imaginative and radical. The industrial society's basic structure is rejected by this response. For humans and societies, a series of alternative interpretations are being favored. Deep internal divisions and widely varying approaches are subjected for green radicalism due to its imaginative and radical nature. The Earth's stock has a certain limit for ecosystem services and resources. The growth of the population will eventually attain that limit, as per the assumption of survivalism response. Since, within the industrial-political economy, a redistribution of power is being sought by survivalism, this response is radical. Also, within the context of industrialism, it sees solution that makes it prosaic. Scientists, administrators, and other responsible elites have larger control on the existing systems. Hence, survivalism is prosaic and radical. The conflicts among economic and environmental values are seeking to be removed by sustainable development which is imaginative and reformist. In order to attain the limits, argument meaningless, development, and growth are sought to be redefined. Environmental protection and economic growth are looking as complementary in sustainable development. Through the adoption of market-type regulations or incentives (carbon

---

TABLE 6.2
Four Responses to the Sustainability Challenges

	Imaginative	Prosaic (Unimaginative)
Radical	Green Radicalism	Survivalism
Reformist	Sustainable Development	Environmental Problem-Solving

108　　　　　　　　　　　　　　　　　　Handbook of Sustainable Materials

tax, cap-and-trade schemes, etc.), public policy can be used to solve environmental problems, and the politico-economic status is accepted in environmental problem-solving response which is prosaic and reformist.

One of the subjective and complex concepts is sustainability. With an improvement in the awareness and understanding, the approaches toward sustainability are changing. This is because of not well development of the concept of sustainability. In order to keep debate at bay, it is often exploited by the decision makers in business, government, and other organizations at the industry sector, national, and international levels. Also, by figuring the lack of appropriate responses and agreement on the issues, they are delaying in taking necessary actions. In the sectors of climate change and global warming, this has been especially apparent. Due to coal combustion, the concepts of geosequestration and carbon-capture are being promoted by the coal companies as a solution to global warming. This is because of justifying industry's continuing growth. In order to solve the problem, a cheapest method of demonstrating commitment by many governments is the arranging of comparatively smaller amounts of funding for mitigation technologies and research into sequestration and capture. However, the business is allowed as usual. For short-term advantages, these strategies are used at the individual company levels. The organizational capacity is often weakened by this. Therefore, in order to shape business practices, new culture and knowledge are requisite to create and to result in practical sense of sustainability concepts.

The implementation of sustainable development can be substantially contributed by plasma-related research [34–36]. From forecasting space weather to health care and materials processing [37–40], powerful societal applications can be transformed from fundamental scientific research through plasma science and engineering. For a sustainable future, advancement and more exploration in plasma physics are required.

REFERENCES

1. Diamond J 2011 *Collapse: How Societies Choose to Fail or Succeed: Revised Edition* (Penguin)
2. Assessment M E 2005 *Ecosystems and Human Well-being: Wetlands and Water* (World Resources Institute)
3. Anastas P T and Warner J C 1998 Principles of green chemistry Green Chem. *Theory Pract.* **29**
4. Sanchez M C, Brown R E, Webber C and Homan G K 2008 Savings estimates for the United States Environmental Protection Agency's ENERGY STAR voluntary product labeling program *Energy Policy* **36** 2098–108
5. Hawkins P, Lovins A B and Lovins L H 1999 *Natural Capitalism: Creating the Next Industrial Revolution* (October 2000 ed.)
6. Elkington J and Rowlands I H 1999 Cannibals with forks: The triple bottom line of 21st century business *Altern. J.* **25** 42
7. Stiglitz J E 1974 Growth with exhaustible natural resources: The competitive economy *Rev. Econ. Stud.* **41** 139–52
8. Solow R M 1974 The economics of resources or the resources of economics *Classic Papers in Natural Resource Economics* (Springer) pp 257–76
9. Carson R 1962 *Silent Spring Houghton* (Mifflin: Boston, MA)
10. Ehrlich P R 1968 *The Population Bomb* (Ballantine Books: New York) pp 72–80

Sustainability of Materials

11. Meadows D H, Meadows D L, Randers J and Behrens W 1972 *The Limits to Growth* (Universe Books: New York, NY)
12. Shonnard D R, Kicherer A and Saling P 2003 Industrial applications using BASF eco-efficiency analysis: Perspectives on green engineering principles *Environ. Sci. Technol.* **37** 5340–8
13. Landsiedel R and Saling P 2002 Assessment of toxicological risks for life cycle assessment and eco-efficiency analysis *Int. J. Life Cycle Assess.* **7** 261–8
14. Saling P, Kicherer A, Dittrich-Krämer B, Wittlinger R, Zombik W, Schmidt I, Schrott W and Schmidt S 2002 Eco-efficiency analysis by BASF: The method *Int. J. Life Cycle Assess.* **7** 203–18
15. Sabathil G, Joos K and Kessler B 2008 *The European Commission: An Essential Guide to the Institution, the Procedures and the Policies* (Kogan Page Publishers)
16. Girouard B 2008 Supply Chain 2016 *MHD Supply Chain Solut.* **38**
17. Ghimire P P and Jaroniec M 2021 Renaissance of Stöber method for synthesis of colloidal particles: New developments and opportunities *J. Colloid Interface Sci.* **584** 838–65
18. Lan G, Yang J, Ye R, Boyjoo Y, Liang J, Liu X, Li Y, Liu J and Qian K 2021 Sustainable carbon materials toward emerging applications *Small Methods* **5** 2001250
19. Liang C, Chen Y, Wu M, Wang K, Zhang W, Gan Y, Huang H, Chen J, Xia Y and Zhang J 2021 Green synthesis of graphite from CO_2 without graphitization process of amorphous carbon *Nat. Commun.* **12** 1–9
20. Tan Z, Yang J, Liang Y, Zheng M, Hu H, Dong H, Liu Y and Xiao Y 2021 The changing structure by component: Biomass-based porous carbon for high-performance supercapacitors *J. Colloid Interface Sci.* **585** 778–86
21. Xu C, Chen J, Li S, Gu Q, Wang D, Jiang C and Liu Y 2021 N-doped honeycomb-like porous carbon derived from biomass as an efficient carbocatalyst for H_2S selective oxidation *J. Hazard. Mater.* **403** 123806
22. Gedik A 2020 A review on the evaluation of the potential utilization of construction and demolition waste in hot mix asphalt pavements *Resour. Conserv. Recycl.* **161** 104956
23. Nie X, Li Z, Yao H, Hou T, Zhou X and Li C 2020 Waste bio-oil as a compatibilizer for high content SBS modified asphalt *Pet. Sci. Technol.* **38** 316–22
24. Santos C R, Pais J C, Ribeiro J and Pereira P 2020 Evaluating the properties of bioasphalt produced with bio-oil derived from biodiesel production *Proceedings of the 9th International Conference on Maintenance and Rehabilitation of Pavements—Mairepav9* (Springer) pp 397–407
25. Situmorang Y A, Zhao Z, Yoshida A, Abudula A and Guan G 2020 Small-scale biomass gasification systems for power generation (< 200 kW class): A review *Renew. Sustain. Energy Rev.* 117 109486
26. Bamigboye G O, Bassey D E, Olukanni D O, Ngene B U, Adegoke D, Odetoyan A O, Kareem M A, Enabulele D O and Nworgu A T 2021 Waste materials in highway applications: An overview on generation and utilization implications on sustainability *J. Clean. Prod.* **283** 124581
27. Malik H K and Singh A K 2010 *Engineering Physics* (McGraw-Hill Education)
28. Malik H K 2021 *Laser-Matter Interaction for Radiation and Energy* (CRC Press)
29. Bretado-de los Rios M S, Rivera-Solorio C I and Nigam K D P 2021 An overview of sustainability of heat exchangers and solar thermal applications with nanofluids: A review *Renew. Sustain. Energy Rev.* **142** 110855
30. Iqbal A, Mahmoud M S, Sayed E T, Elsaid K, Abdelkareem M A, Alawadhi H and Olabi A G 2021 Evaluation of the nanofluid-assisted desalination through solar stills in the last decade *J. Environ. Manage.* **277** 111415
31. Rasheed A H, Alias H B and Salman S D 2021 Experimental and numerical investigations of heat transfer enhancement in shell and helically microtube heat exchanger using nanofluids *Int. J. Therm. Sci.* **159** 106547

32. Yakasai F, Jaafar M Z, Bandyopadhyay S and Agi A 2021 Current developments and future outlook in nanofluid flooding: A comprehensive review of various parameters influencing oil recovery mechanisms *J. Ind. Eng. Chem.* **93** 138–62
33. Dryzek J S 2021 *The Politics of the Earth* (Oxford University Press)
34. Dhawan R and Malik H K 2020 Sheath formation criterion in collisional electronegative warm plasma *Vacuum* **177** 109354
35. Dhawan R and Malik H K 2020 Sheath characteristics in plasma carrying finite mass negative Ions and Ionization at low frequency *Chinese J. Phys.* **66** 560–72
36. Dhawan R and Malik H K 2020 Behaviour of sheath in electronegative warm plasma *J. Theor. Appl. Phys.* **14** 121–8
37. Malik H K and Dhawan R 2020 Sheath structure in electronegative plasma having cold ions: An impact of negative ions' mass *IEEE Trans. Plasma Sci.* **48** 2408–17
38. Dhawan R, Kumar M and Malik H K 2020 Influence of ionization on sheath structure in electropositive warm plasma carrying two-temperature electrons with non-extensive distribution *Phys. Plasmas* **27** 63515
39. Kumar S, Dhaka V, Singh D K and Malik H K 2021 Analytical approach for the use of different gauges in bubble wakefield acceleration *J. Theor. Appl. Phys.* **15** 1–12
40. Dhawan R and Malik H K 2021 Modelling of electronegative collisional warm plasma for plasma-surface interaction process *Plasma Sci. Technol.* **23** 45402

7 Experimental Investigation on Durability Properties of Concrete Incorporated with Fly Ash

V. G. Meshram, A. M. Pande, and B. P. Nandurkar

CONTENTS

7.1 Introduction 111
    7.1.1 Permeable Voids of Fly Ash Concrete.. 112
        7.1.1.1 Mechanism of Permeable Voids 112
        7.1.1.2 Effect of Fly Ash on Permeable Voids
                in Concrete 112
    7.1.2 Acid Resistance of Fly Ash Concrete 113
        7.1.2.1 Mechanism of the Acid Attack 113
        7.1.2.2 Effect of Fly Ash on the Acid Attack 113
7.2 Experimental Methodology 114
    7.2.1 Materials 114
    7.2.2 Mix Design and Test Specimen Preparation 114
    7.2.3 Test Methods.. 114
7.3 Results and Discussion 121
    7.3.1 Compressive Strength Studies 121
    7.3.2 Permeable Voids Studies 123
    7.3.3 Acid Attack Studies 126
        7.3.3.1 Comparison of Percentage Weight Reduction 126
        7.3.3.2 Comparison of Percentage Compressive Strength
                Reduction 129
7.4 Conclusion 131
References.. 132

7.1 INTRODUCTION

The sustainable development that meets present-day requirements without compromising the future is need of the hour. In the construction industry, the use of pozzolanic materials in cement concrete has resulted in the reduction of the use of raw materials, CO_2 emission, disposal problems of industrial wastes, and concrete

DOI: 10.1201/9781003297772-7

production cost, thus meeting the requirements of sustainable development. Strength and durability are two important factors to consider when fly ash is used in concrete. Presently, the durability of concrete becomes a major concern due to two reasons. Firstly, the hostile conditions in which the concrete structures are built, and secondly, the use of added cementitious materials in concrete. Therefore, this chapter deals with the durability aspects of fly ash concrete. The durability aspect can be broadly divided into permeable voids and resistance against chemical attack. In the first part, permeable void studies are carried out and in the second part, chemical resistance studies are carried out.

7.1.1 PERMEABLE VOIDS OF FLY ASH CONCRETE

The structure of porosity is one of the most important variables impacting the strength and durability of concrete. The porosity is the percentage of the material's bulk volume occupied by voids. Porosity in concrete comes in a variety of forms. Gel pores with dimensions of 0.5–10 nm, mesopores with dimensions of 5–5000 nm, macropores caused due to insufficient compaction, paste-aggregate interfacial zones, and aggregate porosity are all sources of porosity in concrete [1–3].

7.1.1.1 Mechanism of Permeable Voids

In the preparation of concrete mix, water is an important ingredient. This water remains in the concrete mass as entrapped water. When the concrete is hardened, some part of this water is used for hydration cement, while the remaining part evaporates. This process creates a porous channel in concrete mass. The hydrated products of the cement paste fill some part of this porous channel. The remaining part of the porous channel comprises capillary voids. These capillary voids are responsible for the ingress of water. In the case of fly ash concrete, the surplus lime liberated by cement reacts with fly ash producing additional cementitious material. This glue-like substance seals off these capillary spaces while also lowering the likelihood of excess free lime leaching and so lowering concrete permeability [4–6]. The primary role of natural pozzolan hydration products is to reduce the pore connectivity. Therefore, the movement of the pores into the microstructure of the concrete structure becomes more difficult, and the concrete porosity is reduced [7]. According to the existing literature [8–10], the concrete blended with fly ash outperformed controlled concrete in terms of permeable voids.

7.1.1.2 Effect of Fly Ash on Permeable Voids in Concrete

Mixture proportions, fineness, shape, particle size distribution, and degree of hydration are all factors that govern the effect of fly ash on permeable voids. Termkhajornkit et al. [3], Cox and De Belie [9], and Sideris et al. [11] have demonstrated that the porosity and water absorption increase with fly ash amount in the mixture. Gencel et al. [12] have shown that when cement is replaced by Class F fly ash, the total cement content and the water-to-cement ratio are to be adjusted to obtain increased durability qualities. Chindaprasirt [10] and Nath and Sarker [13] have demonstrated that the porosity of blended concrete decreases with an increase in fly ash fineness. Termkhajornkit et al. [3], Cox and De Belie [9], and

Experimental Investigation on Durability Properties

Sinsiri et al. [14] have demonstrated that with the increase in the age of concrete, the porosity reduces. This can be attributed to the increase in the hydration of cementitious materials and the formation of secondary Calcium Silicate Hydrate (CSH).

7.1.2 ACID RESISTANCE OF FLY ASH CONCRETE

Present-day concrete is most susceptible to acid attack due to the acidic environment which is caused by industrial wastes, the degrading of agricultural products, and the formation of sulfuric acid in sewage system [15–17]. The acid attack causes weight loss, concrete fracture, and further concrete deterioration. The acid attack is broadly classified as the Sulfate attack and Chloride attack. Acid attack on concrete is influenced by various parameters, including acid concentration, aggregate type, and cement type [18].

7.1.2.1 Mechanism of the Acid Attack

It has been shown that Portland cement is more vulnerable to acid attacks due to the high proportion of calcium hydroxide [18]. Ordinary Portland cement having pH values, normally above 12.5, is highly alkaline. During the acid attack, the pH of the solution becomes lower than 12.6 and the equilibrium of the cement matrix is disturbed. Therefore, hydrated cement compounds are altered [16, 19]. In this process, the ettringite and gypsum are formed. Ettringite development promotes expansion and cracking in the hardened paste, whereas gypsum creation causes mass loss. Hydrochloric acid (HCl) attack causes the change in color of the concrete cubes, while cracks are seen under sulfuric acid (H_2SO_4) attack [20, 21].

7.1.2.2 Effect of Fly Ash on the Acid Attack

Under acid attack, the effect of fly ash is found to be beneficial in two ways. Firstly, the fly ash chemically reacts with free lime to create additional C-S-H gel. This gel closes the capillaries and reduces the ingress of sulfate ions [19, 22]. Secondly, with the reduction in cement content, the amount of Calcium Aluminate is also reduced and ettringite formation is prevented [23].

There are few publications available that investigate the effect of fly ash on concrete under acid attack. Murthi and Sivakumar [23] and Nath and Sarker [24] carried out a durability study for up to 6–8 months on concrete specimens immersing them in 5% H_2SO_4 and 5% HCl solution. They have shown that all plain concrete specimens were severely deteriorated. When the fly ash was used, it was found that the specimens were affected moderately and slightly by acid attack. Bremseth [25] has shown that the increase in the cementitious compounds and decrease in water demand and capillaries were observed using fly ash. Therefore, more dense concrete is produced, which reduces the void spaces contributing to the resistance against weak acids, salts, and sulfates. Turkel et al. [15] have shown that the incorporation of fly ash does not have every time positive effect on the durability characteristics of cement against acid attack.

Thus, on this background, more studies are required on the various parameters which affect the durability features of fly ash concrete. This chapter discusses the performance of fly ash concrete against permeable voids and acid attacks emphasizing on fly ash fineness.

7.2 EXPERIMENTAL METHODOLOGY

7.2.1 MATERIALS

The materials utilized in this study were all readily available in the vicinity of Nagpur, India. These were conventional Portland cement, Grade 43, conforming to IS 8112-1989 [28], fine and coarse aggregates conforming to IS 383-1970 [29], and Class F fly ash from hoppers 1 to 5 with varying fineness conforming to IS 3812 (Part 1):2003 [27]. Table 7.1 lists some of the physical properties of fly ash tested according to IS 1727 (1967) [26].

7.2.2 MIX DESIGN AND TEST SPECIMEN PREPARATION

The concrete mix design for M25 was done as per IS 10262-2009 [30]. The fly ash of 0, 12.5, 25, and 37.5% from hoppers 1 to 5 was used to replace cement.

For compressive strength and acid resistance test, the 150 mm. cubic specimen was used.

For Permeable void: The specimens were prepared as per ATSM C 642-97 [32]. According to this code, the volume and weight of each specimen should be at least 350 cm^3 and 800gm, respectively. Therefore, a 10 cm × 10 cm × 20 cm size specimen of M25 grade concrete is cast. This specimen having an approximate weight of 5–6 kg is cut into three pieces in such a way that each piece should not have observed cracks or crushed edges.

7.2.3 TEST METHODS

A. **Compressive Strength Test:** IS 516-1959 [31] was used for testing the compressive strength. The test was conducted after 7, 28, and 90 days of curing period. The test results are given in Table 7.2.

B. **Permeable Void Test:** ATSM C 642-97 [32] was used for carrying out the permeable void test. This test involves the following four steps:

1. **Oven Dry Weight (A):** To determine the oven dry weight, the specimen was kept in an oven and allowed to dry for 24 hours. The temperature of 100–110°C was maintained during this period. After 24 hours, the specimen was removed from the oven and kept in a desiccator

TABLE 7.1
Physical Properties of Fly Ash

S No.	Test Conducted	Test Result				
		Hopper 1	Hopper 2	Hopper 3	Hopper 4	Hopper 5
1	Blain's Fineness (Sq. m./kg)	229	320	474	536	607
2	Specific gravity	2.04	2.06	2.17	2.19	2.28

Experimental Investigation on Durability Properties

TABLE 7.2
Compressive Strength Test Results

		Compressive Strength (N/mm²)		
Hopper No.	Fly Ash (FA) %	7 Days	28 Days	90 Days
Control Mix	0	28.44	35.11	38.67
Hopper 1	12.5	19.55	31.11	45
	25	18.4	25.56	38
	37.5	22.44	24.22	28.22
Hopper 2	12.5	24.44	32	42.44
	25	20.88	29.11	33.55
	37.5	16.66	26	31.15
Hopper 3	12.5	23.55	32.88	39.78
	25	20.28	29.1	32.44
	37.5	22.22	24.88	30.22
Hopper 4	12.5	22	35.77	40.66
	25	25.33	38	42.44
	37.5	17.33	28	33.55
Hopper 5	12.5	27.84	38.44	42.21
	25	24.67	36.75	44.89
	37.5	19.34	29.77	36.67

for cooling. The temperature of 20–25°C was maintained during this cooling period. After cooling, the weight of the specimen was measured. This process was repeated until the constant weight value was attained. This value was called oven dry weight (A) (Figure 7.1).

2. **Saturated Mass after Immersion (B):** To determine the saturated mass, the specimen was immersed in water for 48 hours at an approximate temperature of 21°C. After 48 hours, the specimen was removed and the weight of the surface-dried sample was measured. This process was repeated until the constant weight value was attained. This value was called saturated mass after immersion (B) (Figure 7.2).

FIGURE 7.1 Oven dry weight (A).

FIGURE 7.2 Saturated mass after immersion (B).

3. **Saturated Mass after Boiling [C]:** To determine saturated mass after boiling, the specimen was allowed to boil. After 5 hours, the specimen was removed. The specimen was then cooled for 14 hours at 20–25°C temperature. The surface of the specimen was cleaned, and the weight was measured. This value was called saturated mass after boiling (C) (Figure 7.3).
4. **Immersed Apparent Mass [D]:** To determine immersed apparent mass, the specimen was suspended in water by means of a wire and the immersed apparent mass was measured. This value was called immersed apparent mass (D) (Figure 7.4).

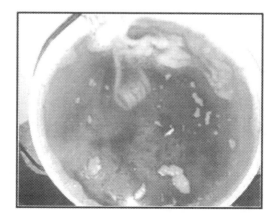

FIGURE 7.3 Saturated mass after boiling (C).

Experimental Investigation on Durability Properties

FIGURE 7.4 Immersed apparent mass (D).

The volume of permeable void is calculated by the following formula:

$$PV = [(\text{Apparent density} - \text{Bulk density}) / \text{Apparent density}] * 100$$

In general, the volume of permeable void = [(C − A)/(C − D)] *100

The test results of permeable voids are shown in Table 7.3.

C. **Acid Attack Test:** ATSM C 642-97 was used for carrying out the acid attack test. The 28 days water-cured concrete specimens were surface dried and the weight of the specimens was measured. After that, the specimens were separately immersed in 1% diluted H_2SO_4 solutions and 3% diluted HCl solutions for 30 days. To maintain the pH values, the solution was inspected frequently during the test period. After 30 days, the specimen was removed from the water and the surface was cleaned. The weight of the specimen was measured and the compressive strength was found. The percentage weight reduction and percentage compressive strength reduction were calculated. The test results of weight reduction are shown in Table 7.4, and the test results of compressive strength reduction are shown in Table 7.5 (Figures 7.5 and 7.6).

TABLE 7.3
Permeable Voids Test Results

Hopper No.	Age of Concrete (days)	FA %	A	B	C	D	PV
0	7	0	1194.67	1256.67	1260	750	12.81
0	28	0	1290.33	1351	1353	799.667	11.33
0	90	0	1133.33	1182.67	1185	742	11.66
1	7	12.5	1374	1462.67	1462	855.333	14.51
1	7	25	1089.67	1158.67	1161.33	661.333	14.33
1	7	37.5	1080	1158	1160.33	633.667	15.25
1	28	12.5	1243.33	1320	1321.33	768	14.1
1	28	25	1288	1358.33	1360.67	800.667	12.98

(*Continued*)

TABLE 7.3 (*Continued*)
Permeable Voids Test Results

Hopper No.	Age of Concrete (days)	FA %	A	B	C	D	PV
1	28	37.5	1163.33	1233.33	1234.33	714.333	13.65
1	90	12.5	1226.67	1283.33	1285.67	729	10.6
1	90	25	1085	1133.33	1136.67	640	10.4
1	90	37.5	1161.67	1218	1221	701	11.41
2	7	12.5	1143	1221.67	1223	663	14.29
2	7	25	1163.33	1247.33	1248.67	728.667	16.41
2	7	37.5	1114.33	1203.33	1205	691.667	17.66
2	28	12.5	1250	1316.67	1320	773.333	12.8
2	28	25	1284	1360	1361.67	791.667	13.63
2	28	37.5	1138.33	1207.67	1209.33	709.333	14.2
2	90	12.5	1194.33	1247.67	1250.67	730.667	10.83
2	90	25	1205	1262.67	1266.67	726.667	11.42
2	90	37.5	1221.67	1280	1283.67	730.333	11.2
3	7	12.5	1230	1300	1303.33	763.333	13.58
3	7	25	1092.67	1158.67	1161.33	668	13.92
3	7	37.5	1161	1224.33	1231.33	724.667	13.88
3	28	12.5	1278	1333.33	1339	805.667	11.44
3	28	25	1121	1176.67	1179.67	686.333	11.89
3	28	37.5	1229	1296.67	1298.33	758.333	12.84
3	90	12.5	1230.67	1283.33	1287.67	745.333	10.51
3	90	25	1245.33	1308.67	1309.67	766.667	11.85
3	90	37.5	1213	1271	1275.67	734.333	11.58
4	7	12.5	1241	1311.5	1313	803	14.12
4	7	25	1112	1177	1179	685.667	13.58
4	7	37.5	1179.5	1249.5	1252	727	13.81
4	28	12.5	1253	1314.33	1315.33	755.333	11.13
4	28	25	1216	1273	1275.5	773	11.84
4	28	37.5	1214	1280	1281.67	741.667	12.53
4	90	12.5	1260.67	1320	1322	762	10.95
4	90	25	1252.67	1306.67	1307.67	747.667	9.82
4	90	37.5	1263.5	1320	1325	755	10.79
5	7	12.5	1446.67	1522	1525.33	926	13.13
5	7	25	1198	1273	1274	764	14.9
5	7	37.5	1190	1261.67	1265	751.667	14.61
5	28	12.5	1306.67	1363.33	1367	810.333	10.84
5	28	25	1227	1281.5	1285	755	10.94
5	28	37.5	1257	1323.5	1325	775	12.36
5	90	12.5	1256	1307.33	1308.67	772	9.81
5	90	25	1189.33	1234.67	1235.67	705.667	8.74
5	90	37.5	1166	1220	1220.33	693.667	10.32

TABLE 7.4
Results of Weight Reduction in HCl and H$_2$SO$_4$ Solution

		HCl Solution			H$_2$SO$_4$ Solution		
Hopper No.	FA %	Weight before Immersion (gm)	Weight after Immersion (gm)	% Weight Reduction	Weight before Immersion (gm)	Weight after Immersion (gm)	% Weight Reduction
0	0	8870	8830	0.45	8740	8650	1.03
1	12.5	8600	8540	0.7	8620	8530	1.04
1	25	8870	8820	0.56	8710	8610	1.15
1	37.5	8450	8410	0.47	8150	8020	1.60
2	12.5	8650	8620	0.35	8600	8530	0.81
2	25	8700	8650	0.57	8550	8480	0.82
2	37.5	8380	8320	0.72	8020	7930	1.12
3	12.5	8380	8360	0.24	8900	8820	0.90
3	25	8540	8510	0.35	8430	8360	0.83
3	37.5	8100	8080	0.25	7760	7710	0.64
4	12.5	9210	9180	0.33	8670	8610	0.69
4	25	8950	8920	0.34	8700	8650	0.57
4	37.5	8320	8290	0.36	8320	8240	0.96
5	12.5	8560	8540	0.23	8790	8710	0.91
5	25	8910	8890	0.22	9300	9220	0.86
5	37.5	8480	8460	0.24	7900	7830	0.89

FIGURE 7.5 Effect of hydrochloric acid on concrete cubes.

TABLE 7.5
Results of Compressive Strength Reduction in HCl and H$_2$SO$_4$ Solution

		HCl Solution			H$_2$SO$_4$ Solution		
Hopper No.	FA %	Comp. Strength before Immersion (N/mm^2)	Comp. Strength after Immersion (N/mm^2)	% Comp. Strength Reduction	Comp. Strength before Immersion (N/mm^2)	Comp. Strength after Immersion (N/mm^2)	% Comp. Strength Reduction
0	0	35.11	31.55	10.14	35.11	32.44	7.6
1	12.5	31.105	28.88	7.15	31.105	28.94	6.96
1	25	25.55	23.55	7.83	25.55	24	6.07
1	37.5	24.22	21.77	10.12	24.22	21.84	9.83
2	12.5	31.998	28.88	9.74	31.998	30.22	5.56
2	25	29.108	25.77	11.47	29.108	27.11	6.86
2	37.5	25.995	22.66	12.83	25.995	23.11	11.1
3	12.5	32.88	30.22	8.09	32.88	30.67	6.72
3	25	34.89	31.56	9.54	34.89	32.14	7.88
3	37.5	24.88	22.67	8.88	24.88	23.11	7.11
4	12.5	35.77	33.33	6.82	35.77	33.77	5.59
4	25	38	35.11	7.61	38	36.44	4.11
4	37.5	28	25.77	7.96	28	25.78	7.93
5	12.5	38.44	36	6.35	38.44	36.44	5.2
5	25	36.75	34.67	5.66	36.75	35.11	4.46
5	37.5	29.77	27.55	7.46	29.77	28.44	4.47

FIGURE 7.6 Effect of sulfuric acid on concrete cubes.

Experimental Investigation on Durability Properties

7.3 RESULTS AND DISCUSSION

7.3.1 COMPRESSIVE STRENGTH STUDIES

Table 7.2 shows the results of compressive strength test.

The compressive strength results given in Table 7.2 are shown graphically through the Figures 7.7–7.12.

From Figures 7.7–7.9, it can be seen that the compressive strength of concrete increases with age. This trend is persistent for all replacement levels and all grades of fly ash.

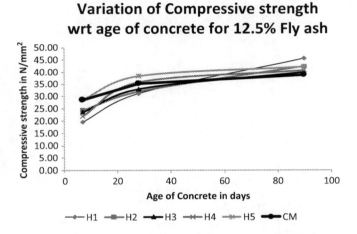

FIGURE 7.7 Relation between age of concrete and compressive strength for 12.5% FA.

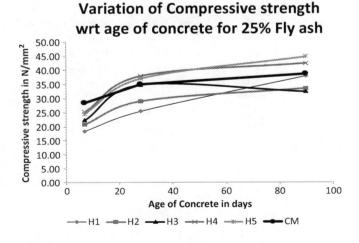

FIGURE 7.8 Relation between age of concrete and compressive strength for 25% FA.

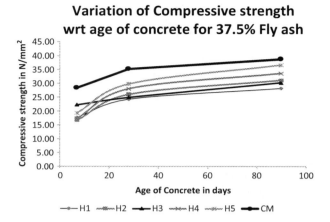

FIGURE 7.9 Relation between age of concrete and compressive strength for 37.5% FA.

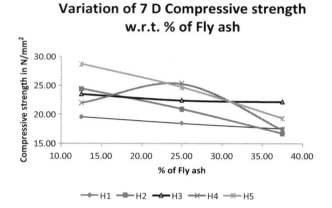

FIGURE 7.10 Relation between FA percentage and compressive strength for 7 days.

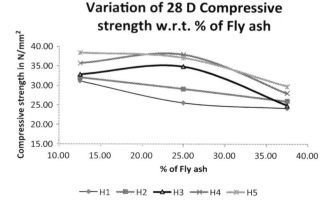

FIGURE 7.11 Relation between FA percentage and compressive strength for 28 days.

Experimental Investigation on Durability Properties 123

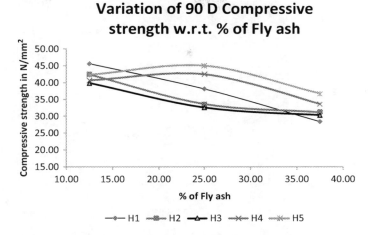

FIGURE 7.12 Relation between FA percentage and compressive strength for 90 days.

From Figures 7.10–7.12, it can be seen that the compressive strength of concrete decreases slightly as the proportion of fly ash increases from 12.5 to 37.5%. The compressive strength of finer fly ash, on the other hand, increases at a replacement level of 25%.

The finer fly ash performed better in terms of compressive strength development. It is also discovered that the compressive strength of finer fly ash increases more than that of coarser fly ash. The fineness of fly ash in concrete has the greatest impact on strength increase from 28 to 90 days of age. The 90-day compressive strength of fly ash is proportional to its fineness. This could be due to subsequent CSH production, which is more prominent in finer fly ash.

7.3.2 Permeable Voids Studies

Table 7.3 shows the results of permeable voids test. The permeable voids results given in Table 7.3 are shown graphically through the Figures 7.13–7.18.

From the graphs as shown in Figures 7.13–7.15, it is seen that

i. The permeable voids of fly ash concrete decrease with the increase in the age of the concrete.
ii. The 7 and 28 days permeable voids of plain concrete are as compared to fly ash concrete irrespective of the fly ash replacement levels and hoppers except hopper 5.
iii. The permeable voids for hoppers 3, 4, and 5 are on the lower side as compared to hoppers 1 and 2 at 7 and 28 days of age of concrete.
iv. At 90 days of the age of concrete, the permeable voids are found to decrease for all hoppers and all percentages of fly ash.

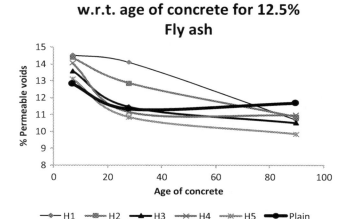

FIGURE 7.13 Relation between age of concrete and permeable voids for 12.5% FA.

From the graphs as shown in Figures 7.16–7.18, it is seen that

i. At the 7 and 28 days curing period, as the percentage of fly ash increases from 12.5 to 37.5, the permeable voids of fly ash concrete increase. However, at 25% fly ash, the 90 days permeable voids are less for finer fly ash.
ii. For hopper 5, it is also found that permeable voids are on the lower side at 28 days of the age of concrete. This shows that for fly ash from hopper 5 the pozzolanic reaction is faster as compared to fly ash from other hoppers.

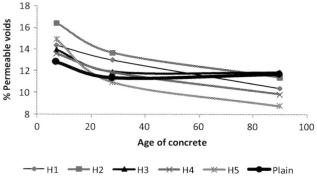

FIGURE 7.14 Relation between age of concrete and permeable voids for 25% FA.

Experimental Investigation on Durability Properties

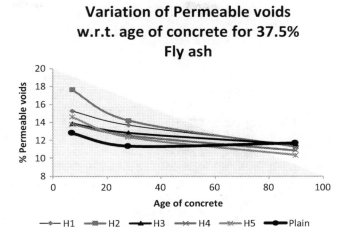

FIGURE 7.15 Relation between age of concrete and permeable voids for 37.5% FA.

Durability of concrete is normally linked to the formation of permeable voids. It is observed that the formation of permeable voids is restricted by the higher fineness of fly ash in concrete. Up to 7 days of the age of concrete, secondary C-S-H formation is not significant. The effect goes on increasing with the age of concrete. At 28 days of age, secondary C-S-H formation, as a part of pore refinement, is significant and the importance of fineness in reducing permeable voids is visible. At 90 days of the age of concrete, fly ash concrete has shown a reduction in permeable voids as compared to controlled concrete irrespective to the fineness of fly ash. This indicates the participation of all grades of fly ash in secondary C-S-H formation and partial reduction of permeable voids.

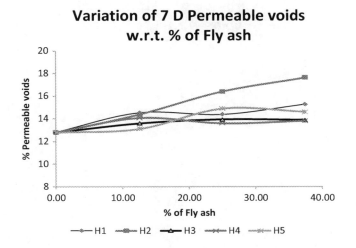

FIGURE 7.16 Relation between FA percentage and permeable voids for 7 days.

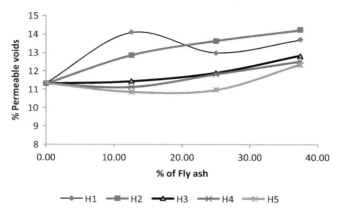

FIGURE 7.17 Relation between FA percentage and permeable voids for 28 days.

7.3.3 Acid Attack Studies

7.3.3.1 Comparison of Percentage Weight Reduction

Table 7.4 shows the results of weight reduction when immersed in HCl and H_2SO_4 solution.

The results of weight reduction given in Table 7.4 are shown graphically through the Figures 7.19–7.23.

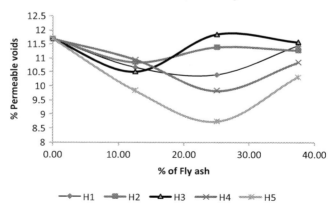

FIGURE 7.18 Relation between FA percentage and permeable voids for 90 days.

Experimental Investigation on Durability Properties

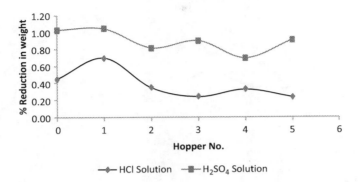

FIGURE 7.19 Relation between Hopper No. and percentage weight reduction for 12.5% FA.

From the graphs as shown in Figures 7.19–7.23, it is seen that

i. The percentage weight reduction for fly ash concrete is less as compared to controlled concrete.
ii. The percentage weight reduction is less up to 25% fly ash replacements and the upward trend is seen afterward.
iii. The percentage weight reduction is more in H_2SO_4 solution as compared to HCl solution though the percentage of HCl solution is more. This implies that the fly ash concrete offers more resistance against HCl than H_2SO_4.

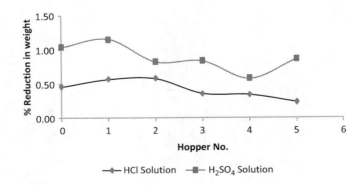

FIGURE 7.20 Relation between Hopper No. and percentage weight reduction for 25% FA.

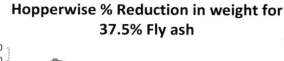
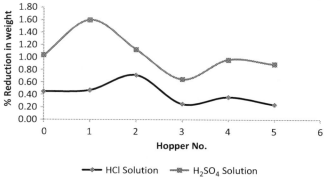

FIGURE 7.21 Relation between Hopper No. and percentage weight reduction for 37.5% FA.

iv. The percentage weight reduction of fly ash concrete cubes immersed in HCl as well as in H_2SO_4 solution is more for coarser ash for all percentages of fly ash.

v. In both immersion media, the loss of weight of fly ash concrete for hoppers 3, 4, and 5 is less as compared to hoppers 1 and 2, which are coarse. This implies that, at the curing period of 28 days, the effect of coarse fly ash is not seen in resisting the weight loss. However, the finer fly ash resists the

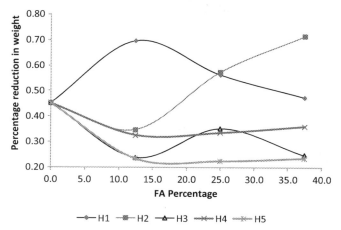

FIGURE 7.22 Relation between FA percentage and percentage weight reduction in HCl.

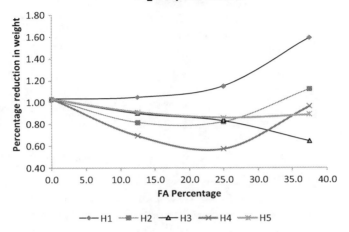

FIGURE 7.23 Relation between FA percentage and percentage weight reduction in H_2SO_4.

acid attack. Therefore, it can be inferred that the resistance to weight reduction depends on the fineness of fly ash and it increases with the increase in the fineness of fly ash.

7.3.3.2 Comparison of Percentage Compressive Strength Reduction

Table 7.5 shows the results of compressive strength reduction when immersed in HCl and H_2SO_4 solution.

The results of compressive strength reduction given in Table 7.5 are shown graphically through the Figures 7.24–7.28.

From the graphs as shown in Figures 7.24–7.28, it is seen that

i. The percentage reduction in compressive strength of fly ash concrete is less as compared to controlled concrete up to 25% percentage of fly ash replacements for all hoppers.
ii. For hoppers 1 and 2, the reduction in compressive strength was found to be more at 37.5% fly ash. This implies that the percentage reduction in compressive strength of fly ash concrete cubes immersed in HCl as well as in H_2SO_4 solution is more for coarser ash for all percentage of fly ash. Therefore, it can be inferred that the resistance of fly ash concrete to compressive strength reduction depends on the fineness of fly ash, and the resistance increases with the increase in fineness of fly ash.
iii. In HCl solution, the percentage reduction in compressive strength is more as compared to H_2SO_4 solution. Therefore, it can be inferred that the fly ash concrete offers more resistance against compressive strength reduction in H_2SO_4 than HCl.

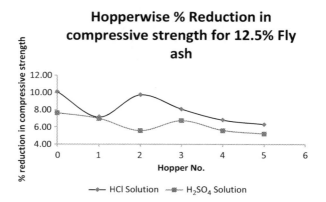

FIGURE 7.24 Relation between Hopper No. and percentage compressive strength reduction for 12.5% FA.

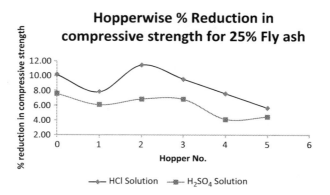

FIGURE 7.25 Relation between Hopper No. and percentage compressive strength reduction for 25% FA.

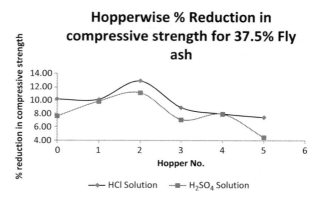

FIGURE 7.26 Relation between Hopper No. and percentage compressive strength reduction for 37.5% FA.

Experimental Investigation on Durability Properties

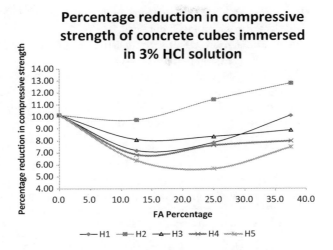

FIGURE 7.27 Relation between FA % and percentage compressive strength reduction in HCl.

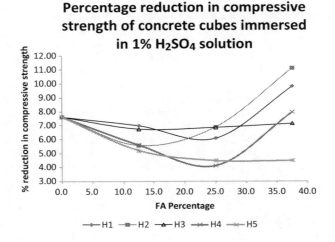

FIGURE 7.28 Relation between FA % and percentage compressive strength reduction in H_2SO_4.

7.4 CONCLUSION

The results of the investigations given above can lead to the following conclusions:

1. The compressive strength of fly ash concrete increases as compared to controlled concrete up to a replacement level of 25%.
2. The permeable voids of fly ash concrete decrease with the age of concrete due to secondary C-S-H formation.

3. The fly ash concrete offers more resistance against weight loss and strength reduction at a 25% fly ash replacement level.
4. The fly ash concrete is more effective in resisting weight loss and strength reduction against HCL as compared to H_2SO_4.
5. The resistance of fly ash concrete to weight reduction depends on the fineness of fly ash and it increases with the increase in the fineness of fly ash.
6. The fly ash up to a replacement level of 25% can safely be used in the concrete for the fulfillment of the strength and durability criteria of concrete.

REFERENCES

1. D.M. Roy et al., (1993), "Concrete microstructure – porosity and permeability", Strategic Highway Research Program National Research Council, Washington, DC
2. Rakesh Kumar and B. Bhattacharjee, (2003), Porosity, pore size distribution and in situ strength of concrete, Cement and Concrete Research Vol. 33, pp. 155–164
3. Pipat Termkhajornkit et al., (2009), Self-healing ability of fly ash–cement systems, Cement & Concrete Composites, Vol. 31, pp. 195–203
4. Fly Ash for Cement Concrete, (2007), Ash Utilization Division NTPC Limited, 2007
5. Seung Heun Lee et al., (2003), Effect of particle size distribution of fly ash–cement system on the fluidity of cement pastes, Cement and Concrete Research, Vol. 33, pp. 763–768
6. http://www.flyash.com/data/upimages/press/HWR_brochure_flyash.pdf "Fly Ash for Concrete"
7. Mauricio López and José Tomás Castro, (2010), Effect of natural pozzolans on porosity and pore connectivity of concrete with time, Revista Ingeniería de Construcción, Vol. 25, No. 3, pp. 419–431
8. N. Bouzoubaa, M.H. Zhang, and V.M. Malhotra, (2001), "Mechanical properties and durability of concrete made with high-volume fly ash blended cements using a coarse fly ash", International Centre for Sustainable Development of Cement and Concrete (ICON), CANMET/Natural Resources, Canada
9. K. Cox and N. De Belie, (2007), "Durability behavior of high-volume fly ash concrete", Magnel Laboratory for Concrete Research, Ghent University, Ghent, Belgium
10. Prinya Chindaprasirt, Chai Jaturapitakkul, and Theerawat Sinsiri, (2005), Effect of fly ash fineness on compressive strength and pore size of blended cement paste, Cement & Concrete Composites, Vol. 27, pp. 425–428
11. K.K. Sideris et al., (2001), Resistance of Fly Ash and Natural Pozzalan Blended Cement Mortars and Concrete ta Catenation, Sulfate Attack and Chloride fan Penetration, SP 199-Vol-I, Seventh CANMET/ACI International Conference on Fly Ash, Silica Fume, Slag and Natural Pozzolans in Concrete, pp. 275–293
12. Osman Gencel et al., (2012), Combined effects of fly ash and waste ferrochromium on properties of concrete, Construction and Building Materials, Vol. 29, pp. 633–640
13. Pradip Nath and Prabir Kumar Sarker, (2013), Effect of mixture proportions on the drying shrinkage and permeation properties of high strength concrete containing class F fly ash, KSCE Journal of Civil Engineering, Vol. 17, No. 6, pp. 1437–1445
14. Theerawat Sinsiri et al., (2010), Influence of fly ash fineness and shape on the porosity and permeability of blended cement pastes, International Journal of Minerals, Metallurgy and Materials, Vol. 17, No. 6, pp. 683–690
15. S. Turkel, B. Felekoglu, and S. Dulluç, (2007), Influence of various acids on the physico–mechanical properties of pozzolanic cement mortars, Sadhana, Vol. 32, Part 6. 2007, pp. 683–691

Experimental Investigation on Durability Properties 133

16. Ali Allahverdi et al., (2000), Acidic corrosion of hydrated cement based materials part 1. – mechanism of the phenomenon, Institute of chemical technology, Department of glass and ceramics, Technická 5, 166 28 Prague 6, Czech Republic

17. H. Siad et al., (2010), Influence of natural pozzolan on the behavior of self-compacting concrete under sulphuric and hydrochloric acid attacks, comparative study, The Arabian Journal for Science and Engineering, Vol. 35, No. 1B

18. Sumrerng Rukzon and Prinya Chindaprasirt (2008), Development of classified fly ash as a pozzolonic material, Journal of Applied Sciences, Vol. 8, No. 6, pp. 1097–1102

19. Shintaro Miyamoto et al., (2013), A Study on the Reaction Mechanism of Hardened Cement Chemically Damaged by Mixed Acids, Third International Conference on Sustainable Construction Materials and Technologies

20. J.K. Tishmack, J. Olek, and S. Diamond, (1999), Characterization of high-calcium fly ashes and their potential influence on ettringite formation in cementitious systems, Cement, Concrete, and Aggregates, CCAGDP, Vol. 21, No.1, pp. 82–92

21. Alireza Motamednia et al., (2013), Laboratory investigation of the durability of light-weight and normal concrete against acids (hydrochloric, sulfuric and lactic acid), Research Journal of Chemical and Environmental Sciences, Vol. 1, No. 3, pp. 20–25

22. K. Kruse, A. Jasso, K. Folliard, R. Ferron M. Juenger, and T. Drimalas, (2012), "Characterizing Fly Ash", Report No. FHWA/TX-13/0-6648-1; Published August 2013

23. P. Murthi and V. Sivakumar, (2008), Studies on acid resistance of ternary blended concrete, Asian Journal of Civil Engineering (Building and Housing) Vol. 9, No. 5, pp. 473–486

24. P. Nath and P. Sarker, (2011), Effect of Fly Ash on the Durability Properties of High Strength Concrete, The Twelfth East Asia-Pacific Conference on Structural Engineering and Construction

25. Sigrun Kjær Bremseth, (2009), "Fly ash in concrete: A literature study of the advantages and disadvantages", SINTEF Building and Infrastructure, COIN Project report 18 – 2009

26. IS 1727: 1967 (Reaffirmed 1999) "Methods of test for pozzolanic materials", Bureau of Indian Standards, New Delhi

27. IS 3812 (Part 1): 2003, "Pulverised fuel ash – specifications", Bureau of Indian Standards, New Delhi

28. IS 8112: 2013, "Indian standard ordinary Portland cement, 43 grade – specification (second revision)", Bureau of Indian Standards, New Delhi

29. IS 383: 1970, "Indian standard specification for coarse and fine aggregates from natural sources for concrete", Bureau of Indian Standards, New Delhi

30. IS 10262-1982, "Recommended guidelines for concrete mix design", Bureau of Indian standards, New Delhi

31. IS:516-1959 (Reaffirmed 1999), "Methods of tests for strength of concrete", Bureau of Indian Standards, New Delhi

32. ASTM C 642-1997: Standard test method for density, absorption, and voids in hardened concrete, "Annual book of ASTM standards", American Society for Testing and Materials

8 Investigation of Electrode Materials Used in Electrocoagulation Process for Wastewater Treatment

Mukul Bajpai, Surjit Singh Katoch,
Akhilesh Nautiyal, Rahul Shakya,
and Samriddhi Sharma

CONTENTS

Nomenclature 135
    Abbreviations 135
    Notations 136
8.1   Introduction 136
    8.1.1 General 136
    8.1.2 Conventional Wastewater Treatment Technologies 137
    8.1.3 Decentralized Wastewater Treatment System 137
8.2   Electrocoagulation (EC) Process 138
    8.2.1 Electrochemical and Chemical Reactions 139
    8.2.2 Critical Factors Affecting EC Efficiency 139
8.3   Application of EC in Various Kinds of Wastewater 141
8.4   Application of Different Electrode Materials Used in the
    Electrocoagulation Process 142
8.5   Conclusion 145
Reference 146

NOMENCLATURE

ABBREVIATIONS

BOD: Biological oxygen demand
CEZ: Cefazolin
COD: Chemical oxygen demand
DC: Direct current
DWTS: Decentralized wastewater treatment systems

DOI: 10.1201/9781003297772-8

EC: Electrocoagulation
ECs: Emerging contaminants
HSST: Hygiene, Sanitation, Sewage Treatment
IC: Initial concentration
IED: Inter-electrode distance
MLD: Million liters per day
MP-P: Monopolar-parallel
RBCs: Rotating biological contractors
RE: Removal efficiency
RO: Reverse osmosis
SS: Suspended solids
TSS: Total suspended solids
UASBR: Up-flow anaerobic sludge blanket reactor
WHO: World Health Organization
WWT: Water and wastewater treatment

NOTATIONS

$C_{electrode}$**:** Electrode dissolution
C_{energy}**:** Energy consumption
CD: Current density
F: Faraday's constant
I: Current
j: Actual density
M: Relative molar mass
n: Number of oxidations reacting electron
t_{EC}**:** Electrolysis time
U: Voltage
V: Volume of wastewater sample
W: Weight of dissolved electrode material

8.1 INTRODUCTION

8.1.1 GENERAL

Water demand and wastewater generation have reached record levels as a result of industrial expansion and increasing urbanization. A considerable amount of untreated residential and industrial wastewater is disposed of directly into water bodies due to a gap in the installed treatment capacity of wastewater, causing surface water quality to deteriorate (Breida et al. 2020). The issue is even more urgent in underdeveloped nations, where sustainable wastewater management and treatment are given less attention owing to financial restrictions. As per the report generated by ENVIS center on hygiene, sanitation, sewage treatment (HSST) systems, and technology in 2019, out of 61754 million liters per day (MLD) of sewage, only 22963 MLD is treated, leaving approximately 62.8% of sewage is untreated due to the lack of sewage treatment capacity of treatment plants (Ghosh & Sarkar 2020).

8.1.2 Conventional Wastewater Treatment Technologies

The goal of the wastewater treatment process is to eliminate inorganic and organic matter, poisonous substances, and other contaminants. As a result, the treated water quality is enhanced to meet WHO standards or the needs of the respective country's pollution control authority (Bajpai et al. 2019). So, the steps for treating wastewater depend on the type of waste and the quality of water that is needed after treatment. The wastewater treatment process is commonly divided into three stages: primary, secondary, and tertiary (Figure 8.1). The majority of large particles are removed using the first and second treatment processes, respectively. When the primary and secondary treatments fail to remove certain undesirable substances from treated water, the tertiary treatment is used as a polishing unit. The combination of physical, chemical, and biological processes is mainly used in most WWT. However, in the present scenario, traditional WWT technologies are inefficient in removing emerging contaminants (ECs) and micropollutants from wastewater. In this way, researchers are now focusing on new and innovative ways to clean up wastewater that has proven to be much more effective and reliable at getting rid of micropollutants and ECs.

8.1.3 Decentralized Wastewater Treatment System

To overcome the issue of freshwater demand, wastewater treatment for metropolitan areas has progressed significantly, but it is still insufficient to meet today's needs. The high maintenance and capital costs, as well as the huge area requirement, are the fundamental investments needed in conventional wastewater treatment units or plants. In most developing countries, the unsatisfactory functioning of these treatment plants is due to a lack of local competence, and insufficient funding. To address this issue, decentralized wastewater treatment systems (DWTS) are gaining

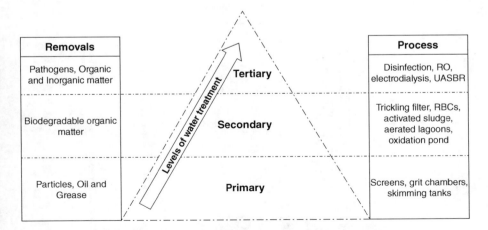

FIGURE 8.1 Flowchart shows the levels of conventional wastewater treatment.

popularity as a government project to improve sanitation and provide a non-potable water supply (Sathe & Munavalli 2019). Individual residences, industrial or institutional buildings, clusters of houses or companies, and entire communities can all benefit from DWTS, which includes a variety of strategies for wastewater collection, treatment, and disposal. In this way, the electrocoagulation (EC) process seems to be a much better way to clean up wastewater than the systems that have been used in the past.

8.2 ELECTROCOAGULATION (EC) PROCESS

EC is the process of dissolving, coagulating, and flocculating the pollutants in the presence of electricity (Mohamad et al. 2021; Tegladza et al. 2021). The term "electrolysis" is derived from the underlying theory of EC. Electrolysis simply implies "the splitting apart of material by electricity". Michael Faraday was the first to propose the electrolysis theory in 1820. In 1889, the EC approach was initially developed for municipal wastewater by the addition of salty water in it (Vik et al. 1984). Various steps occur concurrently in an EC process which is depicted in Figure 8.2: the *first stage* includes, the dissolution of the anode (generally Fe and Al) with H_2O; In the *second stage*, H_2O is hydrolyzed in H_2 gas on the cathode surface; the *third stage*, free electrons cause destabilization of solids and oils; the *fourth stage*, large and dense flocs comprising of solids, oil and grease, and formation of contaminants occurs; and *finally*, flocs are removed from water via filtration. The whole process takes place in an ionic solution, which allows the flow of ions between the electrodes. The cations flow toward the cathode and anions flow toward the anode when the current is applied. At the electrodes, the anions are oxidized, which makes the cations smaller (Bajpai et al. 2022; Shahedi et al. 2020; Sivaranjani et al. 2021; Wagle et al. 2020).

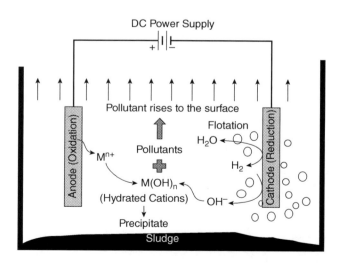

FIGURE 8.2 Pollutant removal mechanism in an EC process.

Investigation of Electrode Materials Used in Electrocoagulation Process — 139

8.2.1 ELECTROCHEMICAL AND CHEMICAL REACTIONS

EC process imposes a potential that is nobler than the reversible potential of the anode, resulting in rapid corrosion (Mohammed-Ridha 2021; Sahu & Dhanasekaran 2021; Syam Babu et al. 2020). As a result, it was hypothesized that the suspension's pH has a huge role in chemical reactions. When Fe or Al anodes are used in water with a pH below 4.0, the reactions (8.1), (8.2), (8.3), and (8.4) occur simultaneously:

Anode:

$$Fe \rightarrow Fe^{2+} + 2e^{-1} \tag{8.1}$$

$$Al \rightarrow Al^{3+} + 3e^{-1} \tag{8.2}$$

Cathode:

$$8H^+ + 8e^{-1} \rightarrow 4H_2 \tag{8.3}$$

$$2H_2O + 2e^{-1} \rightarrow H_2 + 2OH^{-1} \tag{8.4}$$

The rate of the global reaction is affected by the way metals dissolve, the temperature, the pH, and how the solution is stirred. When the pH is less than 4, the ferrous oxide is soluble, and there is no protective covering on the metal's surface, so the dissolving process is accelerated. Similarly, the electro-produced cations of ferrous or aluminum may react with anions like phosphates, carbonates, oxalates, etc., or oxygen molecules created on the anode, creating insoluble products on the electrode's surface, which causes the dissolution rate to become passive.

8.2.2 CRITICAL FACTORS AFFECTING EC EFFICIENCY

The efficacy of EC in WWT is influenced by several factors. They are divided into three groups: To begin, there are the operating circumstances, namely applied current density (CD), type of power supply (AC or DC), mixing speed, and electrolysis time (t_{EC}). Second, the process parameters of water and wastewater, for example, pH, conductivity, initial pollutant concentration, and temperature effect; Finally, the EC setup's shape or geometry, as well as the electrode placement (Dadban Shahamat et al. 2021; Sravanth et al. 2020). The key elements that affect the EC process are summarized in Figure 8.3; their correlations highlight the vectors by which EC performance can be influenced.

- *pH of the solution*
  The pH of water has a significant role in the EC process. The adjustment of pH is done by using NaOH or H_2SO_4. The generation of monomeric species is controlled by the pH of the wastewater at an initial level. At pH levels of less than 6 and higher than 10, the efficiencies of the removal of pollutants decrease. This decrease is caused by the amphoteric performance of iron (II) and iron (III) hydroxides, which results in a soluble Fe^{2+} and Fe^{3+} cation (Bajpai & Katoch 2021; Eslami et al. 2021; Khadir et al. 2020).

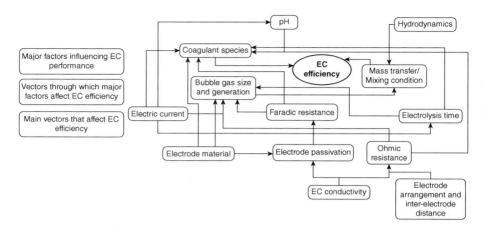

FIGURE 8.3 Factors affecting pollutant removal efficiency (RE) and dependent vectors (Bajpai et al. 2022).

- ***Current density***
  Current density (CD) can influence the performance of the EC process. High electrode dissolution occurs at higher current values. At higher current values, the rate of creation of Fe_2O_3 and polymeric metal composites is high, leading to higher pollutant removal due to the various monomeric species being formed (Adeogun et al. 2021). Theoretically, the quantity of dissolved metal is calculated by Faraday's law using equation (8.5) (Zini et al. 2020):

$$W = \frac{jtM}{nF} \tag{8.5}$$

  Here, W = weight of dissolved electrode material (grams of metal per cm²); j = actual density (Amperes per cm²); t = time (minutes); M = relative molar mass; n = number of oxidations reacting electron, and F = a constant of Faraday, 96.485 coulombs/mole.

  The quantity of metal that gets dissolved, is determined by atomic absorption, and then is correlated with the theoretical value, which is computed from Faraday's law, so a current efficiency can be calculated. Hence, this parameter is used to comment on the electrolytic process performance for different operating conditions.

- ***Electrolysis time***
  EC time is also a substantial operating aspect in an EC process. EC time controls monomeric species formation and concentration which controls pollutant removal and the consumption of electrode material. An anode creates a metal ion which is destabilizing agent during the electrochemical reaction (Criado et al. 2020). The dosage of metal ions is inefficient in destabilizing all the suspended and colloidal particles if the current is weak (Shahedi et al. 2020). Therefore, COD removal will not be high at a lower EC time.

Investigation of Electrode Materials Used in Electrocoagulation Process **141**

- ***Electrical conductivity***
  Increases in salt content enhance the ion concentration of the solution, lowering the resistance between the electrodes. NaCl has usually applied to electrolytes for increasing the electrical conductivity of wastewater to be treated. An increase in the salt concentration decreases the battery voltage at fixed CD, and therefore power utilization is reduced (Ghahrchi et al. 2021; Zaied et al. 2020).

- ***Electrode material***
  The majority of studies employ iron, stainless steel, and aluminum electrodes as sacrificial electrodes, with just a few studies using copper (Cu) and nickel (Ni) as anode electrodes. During EC, these multivalent valance metal electrodes form a metal hydroxide coagulant (Abdel-shafy et al. 2020; Zaied et al. 2020).

- ***Energy and electrode consumption***
  The economic study of the EC method was heavily influenced by energy and electrode use. Using equations (8.6) and (8.7), energy and electrode usage were evaluated (Bajpai et al. 2022; Shaker et al. 2021):

$$\text{Operating cost} = a.\ C_{\text{energy}} + b.\ C_{\text{electrode}} \tag{8.6}$$

$$C_{\text{energy}} = \frac{UIt}{V} \tag{8.7}$$

where, U = voltage; V = volume; I = current; t = time; $C_{\text{electrode}}$ = electrode dissolution; and C_{energy} = energy consumption.

8.3 APPLICATION OF EC IN VARIOUS KINDS OF WASTEWATER

- ***EC on textile wastewater:*** Using Al and Fe as electrode materials, the EC method was effectively applied to real textile wastewater. The Al electrode removed 18.6% COD, 42.5% TOC, 90.3–94.9% color, 83.5% turbidity, 64.7% TSS, under ideal conditions of 25 mA/cm² CD, and pH 5 having t_{EC} between 0 and 120 minutes (Bener et al. 2019).

- ***EC on tannery wastewater:*** The leather sector is dealing with the difficulties of removing various types of impurities generated by the manufacturing process of leather all over the world (Thirugnanasambandham & Sivakumar 2016). The recent EC study investigated removing chromium from real tannery wastewater. Under the ideal working conditions of CD: 13 mA/cm² and pH 7, maximum chromium removal was 100% (Genawi et al. 2020).

- ***EC on paper mill wastewater:*** An EC experiment was examined on Kraft paper mill wastewater using Fe as an electrode. 1.5 mA/cm² and electrolysis time: 25 minutes were the optimal values for maximum pollutant removal. Under ideal conditions, tannin/lignin removal rates were greater than 70% and color removal rates were greater than 95%, respectively (Wagle et al. 2020).

- ***EC on petroleum wastewater:*** Oil refineries utilize a big amount of water, which results in a large volume of effluent. The effectiveness of a combined

EC-photocatalytic method using ZnO nanoparticles for the removal of COD from petroleum effluent was examined. The COD removal rate was 47% after 120 minutes of t_{EC} under ideal conditions (initial COD conc: 1000 mg/L, ZnO conc: 80 g/m^2, irradiation power: 32W, CD: 20 mA/cm^2, pH: 8.5, and NaCl conc: 0.5g/L) (Keramati & Ayati 2019).

- **EC on dairy wastewater:** Dairy effluent is fatty, contains lactose (along with detergents), has a high biological oxygen demand (BOD), and frequently contains sanitary agents. The efficiency of an EC reactor in removing COD, phosphate, SS, and turbidity from dairy effluent was studied. The findings revealed that the RE of SS, COD, phosphate, and turbidity was 100%, 80%, 98%, and 100%, respectively, under optimal operating circumstances (CD: 0.65 A/m^2, charge loading: 0.59 F/m^3, and pH: 6) (Bassala et al. 2017).
- **EC on distillery wastewater:** The impact of direct current (DC) and alternating current (AC) on the EC process for color and COD removal from distillery industrial effluent was investigated. Maximum removal of COD: 95% and color: 100% were achieved with AC at the ideal condition of pH: 7, CD: 0.4A/dm^2, initial COD conc: 3000 mg/L, IED: 1 cm, pulse cycle: 0.45, and t_{EC}: 3.5 hours using Fe as electrode material (Asaithambi et al. 2021).
- **EC on the removal of pharmaceutical compounds:** EC experiment was conducted on Fe and Al electrodes to investigate oxytetracycline hydrochloride removal from synthetic water. The best CD using Fe and Al electrodes was 20 mA/cm^2, with 93.2% and 87.75% removal, respectively (Nariyan et al. 2017). Another EC research examined how to treat pharmaceutical wastewater that contained the antibiotic cefazolin (CEZ). At an optimal operating conditions (pH: 8, CD: 16 mA/cm^2, and CEZ conc: 25 mg/L), the maximum RE was 85.65% after a 40-minute equilibrium t_{EC} (Bajpai et al. 2021).
- **EC on real graywater:** A study was done to investigate the EC system for real graywater treatment and its reuse potential. It was observed that at an optimum operating condition (Voltage: 14 V, time: 47 minutes, and pH: 7.35), maximum removal for COD, TDS, turbidity, and chloride was 75.6%, 78.7%, 93.4%, and 63.2%, respectively (Bajpai et al. 2020).
- **EC on hospital wastewater:** The experiment was carried out to investigate the RE of COD and chloride from clinical wastewater using an EC procedure and a Fe electrode. At an optimum condition of pH: 7.41, current: 2.64 A, and t_{EC}: 41.31 minutes, maximum removal efficiencies of 92.81% and 71.23% for COD and chloride were attained (Bajpai & Katoch 2020).

8.4 APPLICATION OF DIFFERENT ELECTRODE MATERIALS USED IN THE ELECTROCOAGULATION PROCESS

The electrochemical reactions that will occur in the EC system are determined by the type of electrode material used. The EC unit has successfully used Al and Fe in the majority of previous studies (Kalla 2021; Sahu & Dhanasekaran 2021). Other than aluminum and iron, various other electrodes are successfully applied to EC processes such as titanium, copper, stainless steel, and magnesium. Table 8.1 summarizes a variety of electrodes installed in an EC process.

Investigation of Electrode Materials Used in Electrocoagulation Process 143

TABLE 8.1

Application of Fe, Al, Mg, SS, Ti, and Cu electrodes in an EC process

Electrode Materials	Country	Wastewater Type	Optimum Operating Condition	Pollutant RE	References
Iron (Fe)	Turkey	Slaughterhouse wastewater	MP-P arrangement; For TOC: CD = 22.7 mA/m^2, IED = 12.03 mm, t_{EC} = 78.95 mins and OC = 2.45 US \$/m^3. For color: CD = 23.08 mA/m^2, IED = 15.84 mm, t_{EC} = 80.77 mins and OC = 2.57 US \$/m^3	TOC = 94.77% Color = 99.32%	(Tanyol & Tevkur 2021)
	India	Metal complex dye (MCD) industrial effluent	CD = 89.45 A/m^2, pH = 5.83, IED = 0.7 cm, t_{EC} = 50 mins, ENC = 2.499 kWh/m^3, OC = 0.207 US\$/m^3	Chromium = 99.64%, TDS = 70.02%, COD = 83.03%, BOD = 81.38%, TS = 65.17%	(Patel & Parikh 2021)
	Qatar	Textile wastewater	CD = 75 A/m^2, pH = 7, t_{EC} = 45 mins	TSS = 55%, Turbidity = 82%, COD = 52%	(Mohammad et al. 2021)
	Egypt	Synthetic wastewater	Voltage = 12 V, t_{EC} = 90 mins, IED = 3 cm, ENC = 112.8 kWh/m^3	Urea = 51%	(Mamdouh et al. 2021)
	Morocco	Tannery Wastewater	pH = 7, applied current = 68 mA, t_{EC} = 15 mins, OC = 2.01 US\$/m^3	COD = 64%, TSS = 96%, Chromium = 99%	(Aboulhassan et al. 2018)
	Iran	Hospital wastewater	CD = 15 mA/m^2, pH = 7.5, IED = 1.58 cm, t_{EC} = 20 mins, ENC = 0.522 kWh/m^3	Ciprofloxacin = 99%	(Yoosefian et al. 2017)
Aluminum (Al)	Morocco	Groundwater	pH = 7, Voltage = 30 V, t_{EC} = 150 mins	Nitrate = 94.41%	(Amarine et al. 2020)
	UK	River water	CD = 6 A/m^2, pH = 6, IED = 0.5 cm, t_{EC} = 60 mins, IC = 100 mg/L, ENC = 4.34 kWh/m^3, OC = 0.503 US\$/m^3	Phosphate = 99%	(Hashim et al. 2019)
	Malaysia	Palm oil mil effluent (POME)	CD = 40.21 A/m^2, pH = 4.4, IED = 3 cm, t_{EC} = 45.67 mins	TSS = 100% Color = 96.8% COD = 71.3%	(Bashir et al. 2019)
	India	Municipal wastewater	CD = 40 A/m^2, pH = 7.4–8.5, IED = 1 cm, t_{EC} = 20 mins, ENC = 2.27 kWh/m^3	TDS = 49.78% Turbidity = 93.97% COD = 92.01%	(Nawarkar & Salkar 2019)

(Continued)

TABLE 8.1 (*Continued*)

Application of Fe, Al, Mg, SS, Ti, and Cu electrodes in an EC process

Electrode Materials	Country	Wastewater Type	Optimum Operating Condition	Pollutant RE	References
Magnesium (Mg)	Egypt	Cooling tower blowdown water	CD = 14.29 mA/cm^2, IED = 1 cm, t_{EC} = 60 mins, pH = 8, ENC = 1.81 kWh/m^3 OC = 0.88 US$/m^3	Hardness ion (Ca^{2+}, Mg^{2+}) = 51.80%, silica = 93.70%	(Abdel-shafy et al. 2020)
	Mexico	Industrial wastewater	CD = 201.5 mA/cm^2, pH = 7.12, IED = 1 cm, t_{EC} = 60 mins, OC = 0.08 US$/m^3	COD = 67.7%, Color = 97.5%, Turbidity = 96.3%	(Carmona-Carmona et al. 2020)
	India	Synthetic wastewater	CD = 0.2 mA/cm^2, pH = 7, IED = 0.5 cm, t_{EC} = 25 mins, ENC = 1.035 kWh/m^3	Arsenate = 97.9%	(Vasudevan et al. 2012)
Stainless steel (SS)	Egypt	Industrial electroplating rinsing wastewater	Voltage = 12 V, pH = 7.12, IED = 1 cm, t_{EC} = 60 mins, OC = 0.08 US$/m^3	Nickel = 98.9%, copper = 96.6%, zinc = 97.4%	
	India	Landfill leachate	Voltage = 12 V, pH = 9.8, IED = 1 cm, t_{EC} = 120 mins	COD = 73.5%, Color = 65%	(Bharath & Krishna 2019)
	Morocco	Textile wastewater	CD = 20 mA/cm^2, pH = 7, IED = 1.5 cm, t_{EC} = 6 mins, ENC = 5.84 kWh/m^3	COD = 45%, TOC = 51%	(Titchou et al. 2020)
	Australia	Synthetic water	CD = 0.45 mA/cm^2, pH = 7, IED = 1 cm, t_{EC} = 5 mins, OC = 0.240 US$/m^3	Arsenate = 92%	(Nguyen et al. 2021)
	Iraq	Synthetic water	CD = 20 mA/cm^2, pH = 4, IED = 1.5 cm, t_{EC} = 60 mins, OC = 0.613 US$/m^3, ENC = 3.21 kWh/m^3	Ciprofloxacin = 93.47% Levofloxacin = 88%	(Mohammed et al. 2021)
Titanium (Ti)	Turkey	Carwash wastewater	CD = 3 mA/cm^2, pH = 4, IED = 2 cm, t_{EC} = 40 mins, OC = 9.67 US$/m^3, ENC = 10.1 kWh/m^3	COD = 84%, Oil & grease = 82%, anionic surfactant = 99.3%	(Gönder et al. 2019)
	Egypt	Real printing wastewater	CD = 15 mA/cm^2, IED = 4 cm, t_{EC} = 90 mins, ENC = 1.7 kWh/m^3	COD = 46%, TDS = 9%	(Safwat 2020)

(Continued)

Investigation of Electrode Materials Used in Electrocoagulation Process 145

TABLE 8.1 (*Continued*)
Application of Fe, Al, Mg, SS, Ti, and Cu electrodes in an EC process

Electrode Materials	Country	Wastewater Type	Optimum Operating Condition	Pollutant RE	References
Copper (Cu)	India	Dyestuff industries effluent	CD = 68.87 A/m^2, IED = 0.7 cm, t$_{EC}$ = 50 mins, ENC = 3.32 kWh/m^3, OC = 0.277 US\$/m^3, pH = 5.83	Chromium = 99.7%	(Patel et al. 2021)
	Egypt	Real Printing wastewater	CD = 28 mA/m^2, pH = 12, IED = 4 cm, t$_{EC}$ = 90 mins, ENC = 0.86 kWh/m^3	COD = 67%, TDS = 24%	(Safwat et al. 2019)
	Egypt	Real wastewater from the sedimentation tank	CD = 5 mA/cm^2, pH = 9.2, IED = 4 cm, t$_{EC}$ = 90 mins, ENC = 24.7 kWh/m^3, OC = 0.8-3 US\$/m^3	Chromium = 98%	(Shaker et al. 2021)
	India	Sugar industry wastewater	CD = 178 A/m^2, pH = 6.5, IED = 2 cm, t$_{EC}$ = 120 mins	COD = 83%, Color = 90%	(Sahu & Dhanasekaran 2021)
	India	Rice grain-based distillery effluent	CD = 89.3 A/m^2, IED = 2 cm, t$_{EC}$ = 120 mins, ENC = 114.28 kWh/m^3, pH = 3.5	COD = 80%, Color = 65%	(Prajapati et al. 2016)

Note: BOD: Biochemical oxygen demand; CD: Current density; COD: Chemical oxygen demand; ENC: Energy consumption; IC: Initial concentration; IED: Inter-electrode distance; MP-P: Monopolar-parallel; OC: Operating cost; TDS: Total dissolved solids; t$_{EC}$: Electrolysis time; TOC: Total organic carbon; TS: Total solids; TSS: Total suspended solids.

8.5 CONCLUSION

The emergence of novel toxins in the ecosystem has hastened the study and application of numerous WWT methods. EC surpasses conventional WWT systems due to several benefits, including a simple design, quick initiation time, minimal sludge formation, ease of operation, lower maintenance requirements, and lower cost. Commercial, municipal, rural, and pharmaceutical waste can all benefit from EC treatment. EC also effectively removes colors, heavy metals, and micropollutants from a variety of wastewater.

In an EC process, a reaction mechanism is governed by the quality of the electrode material used, which directly affects the RE of the EC reactor. The study of the various literature suggested that the EC method was successfully established with an iron (Fe) electrode as an anode material for the removal of color up to 99.32%,

TOC 94.77%, a heavy metal such as chromium up to 99.64%, and antibiotics such as ciprofloxacin as high as 99%. Apart from the iron material, most of the studies reveal that the aluminum (Al) electrode as the anode is best for the removal of mineral content from the water, such as phosphate with 99%, nitrate with 94.41% RE, and total suspended solids, are removed up to 100%. The maximum removal of the turbidity of 96.3% and the arsenate content of 97.9% was obtained by using the magnesium (Mg) electrode as the anode material. It was found that using stainless steel (SS) electrodes, the removal of metals, such as Ni, Cu, and Zn, was highly reduced, up to 96% in an EC process. Various studies reveal that the EC plays a vital role in the removal of oil and grease with 82% and surfactant with 99.3% RE when a titanium (Ti) electrode is used in the EC process as an anode.

The performance of an EC process could be enhanced by developing novel electrode materials, using different configurations, and choosing economical design reactors. More study is needed to install the EC process on a larger scale and in continuous systems. Therefore, the EC process is a feasible wastewater treatment technique. More mathematical, technological, and practical research are needed for future use.

REFERENCES

Abdel-shafy HI, Shoeib MA, El-khateeb MA, Youssef AO, Hafez OM. 2020. Electrochemical treatment of industrial cooling tower blowdown water using magnesium-rod electrode. Water Resour Ind. 23:100121.

Aboulhassan MA, El Ouarghi H, Ait Benichou S, Ait Boughrous A, Khalil F. 2018. Influence of experimental parameters in the treatment of tannery wastewater by electrocoagulation. Sep Sci Technol. 53(17):2717–2726.

Adeogun AI, Bhagawati PB, Shivayogimath CB. 2021. Pollutants removals and energy consumption in electrochemical cell for pulping processes wastewater treatment: Artificial neural network, response surface methodology and kinetic studies. J Environ Manage [Internet]. 281(December 2020):111897. https://doi.org/10.1016/j.jenvman.2020.111897

Amarine M, Lekhlif B, Sinan M, El Rharras A, Echaabi J. 2020. Treatment of nitrate-rich groundwater using electrocoagulation with aluminum anodes. Groundw Sustain Dev. 11(March):100371.

Asaithambi P, Govindarajan R, Yesuf MB, Selvakumar P, Alemayehu E. 2021. Investigation of direct and alternating current–electrocoagulation process for the treatment of distillery industrial effluent: Studies on operating parameters. J Environ Chem Eng [Internet]. 9(2):104811. https://doi.org/10.1016/j.jece.2020.104811

Bajpai M, Katoch SS. 2020. Techno-economical optimization using Box–Behnken (BB) design for chemical oxygen demand and chloride reduction from hospital wastewater by electro-coagulation. Water Environ Res [Internet]: wer.1387. https://onlinelibrary.wiley.com/doi/abs/10.1002/wer.1387

Bajpai M, Katoch SS. 2021. Reduction of COD from real graywater by electro-coagulation using Fe electrode: Optimization through Box-Behnken design. Mater Today Proc [Internet]. 43, pp. 303–307. https://doi.org/10.1016/j.matpr.2020.11.667

Bajpai M, Katoch SS, Chaturvedi NK. 2019. Comparative study on decentralized treatment technologies for sewage and graywater reuse – A review. Water Sci Technol [Internet]. 80(11):2091–2106. https://iwaponline.com/wst/article/80/11/2091/72028/Comparative-study-on-decentralized-treatment

Bajpai M, Katoch SS, Kadier A, Ma P-C. 2021. Treatment of pharmaceutical wastewater containing cefazolin by electrocoagulation (EC): Optimization of various parameters using response surface methodology (RSM), kinetics and isotherms study. Chem Eng Res Des [Internet]. 176(December):254–266. https://doi.org/10.1016/j.cherd.2021.10.012

Bajpai M, Katoch SS, Kadier A, Singh A. 2022. A review on electrocoagulation process for the removal of emerging contaminants: Theory, fundamentals, and applications. Environ Sci Pollut Res [Internet]. (0123456789). https://doi.org/10.1007/s11356-021-18348-8

Bajpai M, Singh Katoch S, Singh M. 2020. Optimization and economical study of electrocoagulation unit using CCD to treat real graywater and its reuse potential. Environ Sci Pollut Res [Internet]. http://link.springer.com/10.1007/s11356-020-10171-x

Bashir MJ, Lim JH, Abu Amr SS, Wong LP, Sim YL. 2019. Post treatment of palm oil mill effluent using electro-coagulation-peroxidation (ECP) technique. J Clean Prod. 208:716–727.

Bassala HD, Kenne Dedzo G, Njine Bememba CB, Tchekwagep Seumo PM, Donkeng Dazie J, Nanseu-Njiki CP, Ngameni E. 2017. Investigation of the efficiency of a designed electrocoagulation reactor: Application for dairy effluent treatment. Process Saf Environ Prot [Internet]. 111:122–127. http://dx.doi.org/10.1016/j.psep.2017.07.002

Bener S, Bulca Ö, Palas B, Tekin G, Atalay S, Ersöz G. 2019. Electrocoagulation process for the treatment of real textile wastewater: Effect of operative conditions on the organic carbon removal and kinetic study. Process Saf Environ Prot [Internet]. 129:47–54. https://doi.org/10.1016/j.psep.2019.06.010

Bharath M, Krishna BM. 2019. Electrocoagulation treatment for landfill leachate using stainless steel electrode. Int J Eng Adv Technol. 9(1):2851–2855.

Breida M, Alami Younssi S, Ouammou M, Bouhria M, Hafsi M. 2020. Pollution of Water Sources from Agricultural and Industrial Effluents: Special Attention to NO 3 ⁻, Cr(VI), and Cu(II). In: Water Chem [Internet]. [place unknown]: IntechOpen. https://www.intechopen.com/books/water-chemistry/pollution-of-water-sources-from-agricultural-and-industrial-effluents-special-attention-to-no-sub-3-

Carmona-Carmona PF, Linares-Hernández I, Teutli-Sequeira EA, López-Rebollar BM, Álvarez-Bastida C, Mier-Quiroga M de los A, Vázquez-Mejía G, Martínez-Miranda V. 2020. Industrial wastewater treatment using magnesium electrocoagulation in batch and continuous mode. J Environ Sci Heal - Part A Toxic/Hazardous Subst Environ Eng. 56(3):269–288.

Criado SP, Gonçalves MJ, Ballod Tavares LB, Bertoli SL. 2020. Optimization of electrocoagulation process for disperse and reactive dyes using the response surface method with reuse application. J Clean Prod. 275.

Dadban Shahamat Y, Hamidi F, Mohammadi H, Ghahrchi M. 2021. Optimisation of COD removal from the olive oil mill wastewater by combined electrocoagulation and peroxone process: Modelling and determination of kinetic coefficients. Int J Environ Anal Chem [Internet]. 00(00):1–14. https://doi.org/10.1080/03067319.2021.1937615

Eslami A, Khavari Kashani MR, Khodadadi A, Varank G, Kadier A, Ma PC, Madihi-Bidgoli S, Ghanbari F. 2021. Sono-peroxi-coagulation (SPC) as an effective treatment for pulp and paper wastewater: Focus on pH effect, biodegradability, and toxicity. J Water Process Eng [Internet]. 44(December):102330. https://doi.org/10.1016/j.jwpe.2021.102330

Genawi NM, Ibrahim MH, El-Naas MH, Alshaik AE. 2020. Chromium removal from tannery wastewater by electrocoagulation: Optimization and sludge characterization. Water (Switzerland). 12(5).

Ghahrchi M, Rezaee A, Adibzadeh A. 2021. Study of kinetic models of olive oil mill wastewater treatment using electrocoagulation process. Desalin WATER Treat [Internet]. 211:123–130. http://www.deswater.com/DWT_abstracts/vol_211/211_2021_123.pdf

Ghosh K, Sarkar A. 2020. Sustainability study of in-situ hydroponic vetiver system for urban wastewater management in developing countries: A theoretical review. Asian J Water, Environ Pollut [Internet]. 17(4):97–103. https://www.medra.org/servlet/aliasResolver?alias=iospress&doi=10.3233/AJW200056

Gönder ZB, Balcıoğlu G, Kaya Y, Vergili I. 2019. Treatment of carwash wastewater by electrocoagulation using Ti electrode: Optimization of the operating parameters. Int J Environ Sci Technol. 16(12):8041–8052.

Hashim KS, Al Khaddar R, Jasim N, Shaw A, Phipps D, Kot P, Pedrola MO, Alattabi AW, Abdulredha M, Alawsh R. 2019. Electrocoagulation as a green technology for phosphate removal from river water. Sep Purif Technol [Internet]. 210(June 2018):135–144. https://linkinghub.elsevier.com/retrieve/pii/S1383586618313261

Kalla S. 2021. Use of membrane distillation for oily wastewater treatment - A review. J Environ Chem Eng [Internet]. 9(1):104641. https://doi.org/10.1016/j.jece.2020.104641

Keramati M, Ayati B. 2019. Petroleum wastewater treatment using a combination of electrocoagulation and photocatalytic process with immobilized ZnO nanoparticles on concrete surface. Process Saf Environ Prot [Internet]. 126:356–365. https://doi.org/10.1016/j.psep.2019.04.019

Khadir A, Negarestani M, Motamedi M. 2020. Optimization of an electrocoagulation unit for purification of ibuprofen from drinking water: Effect of conditions and linear/nonlinear isotherm study. Sep Sci Technol [Internet]. 56(8):1431–1449. https://doi.org/10.1080/01496395.2020.1770795

Mamdouh M, Safwat SM, Abd-Elhalim H, Rozaik E. 2021. Urea removal using electrocoagulation process with copper and iron electrodes. Desalin Water Treat. 213:259–268.

Mohamad HAED, Hemdan M, Bastawissi AAED, Bastawissi AEDM, Panchal H, Sadasivuni KK. 2021. Industrial wastewater treatment by electrocoagulation powered by a solar photovoltaic system. Energy Sources, Part A Recover Util Environ Eff. 00(00):1–12.

Mohammed SJ, M-Ridha MJ, Abed KM, Elgharbawy AAM. 2021. Removal of levofloxacin and ciprofloxacin from aqueous solutions and an economic evaluation using the electrocoagulation process. Int J Environ Anal Chem [Internet]. 00(00):1–19. https://doi.org/10.1080/03067319.2021.1913733

Mohammed-Ridha M. 2021. Optimization of levofloxacin removal from aqueous solution using electrocoagulation process by response. IRAQI J Agric Sci [Internet]. 52(1):204–217. https://jcoagri.uobaghdad.edu.iq/index.php/intro/article/view/1252

Nariyan E, Aghababaei A, Sillanpää M. 2017. Removal of pharmaceutical from water with an electrocoagulation process; effect of various parameters and studies of isotherm and kinetic. Sep Purif Technol [Internet]. 188:266–281. http://dx.doi.org/10.1016/j.seppur.2017.07.031

Nawarkar CJ, Salkar VD. 2019. Solar powered electrocoagulation system for municipal wastewater treatment. Fuel [Internet]. 237(September 2018):222–226. https://doi.org/10.1016/j.fuel.2018.09.140

Nguyen TTQ, Loganathan P, Dinh BK, Nguyen TV, Vigneswaran S, Ngo HH. 2021. Removing arsenate from water using batch and continuous-flow electrocoagulation with diverse power sources. J Water Process Eng. 41(March):102028.

Patel SR, Parikh SP. 2021. Chromium removal from industrial effluent by electrocoagulation: Operating cost and kinetic analysis. J Environ Treat Tech. 9(3):621–628.

Patel SR, Parikh SP, Prajapati AK. 2021. Copper electrode for the removal of chromium from dyestuff industries effluent by electrocoagulation: Kinetic study and operating cost. J Dispers Sci Technol. 0(0):1–11.

Prajapati AK, Chaudhari PK, Pal D, Chandrakar A, Choudhary R. 2016. Electrocoagulation treatment of rice grain based distillery effluent using copper electrode. J Water Process Eng [Internet]. 11:1–7. http://dx.doi.org/10.1016/j.jwpe.2016.03.008

Safwat SM. 2020. Treatment of real printing wastewater using electrocoagulation process with titanium and zinc electrodes. J Water Process Eng. 34(January): 101137.

Safwat SM, Hamed A, Rozaik E. 2019. Electrocoagulation/electroflotation of real printing wastewater using copper electrodes: A comparative study with aluminum electrodes. Sep Sci Technol. 54(1):183–194.

Sahu O, Dhanasekaran P. 2021. Electrochemical removal of contaminates from sugarcane processing industry wastewater using copper electrode. J Iran Chem Soc. 18(8):2101–2113.

Sathe SM, Munavalli GR. 2019. Domestic wastewater treatment by modified bio-rack wetland system. J Water Process Eng [Internet]. 28(January):240–249. https://doi.org/10.1016/j.jwpe.2019.02.010

Shahedi A, Darban AK, Taghipour F, Jamshidi-Zanjani A. 2020. A review on industrial wastewater treatment via electrocoagulation processes. Curr Opin Electrochem [Internet]. 22(June):154–169. https://doi.org/10.1016/j.coelec.2020.05.009

Shaker OA, Matta ME, Safwat SM. 2021. Nickel and chromium removal by electrocoagulation using copper electrodes. Desalin Water Treat. 213:371–380.

Sivaranjani, Gafoor A, Ali N, Kumar S, Ramalakshmi, Begum S, Rahman Z. 2021. Applicability and new trends of different electrode materials and its combinations in electro coagulation process: A brief review. Mater Today Proc [Internet]. 37(Part 2): 377–382. https://doi.org/10.1016/j.matpr.2020.05.379

Sravanth T, Ramesh ST, Gandhimathi R, Nidheesh PV. 2020. Continuous treatability of oily wastewater from locomotive wash facilities by electrocoagulation. Sep Sci Technol [Internet]. 55(3):583–589. https://doi.org/10.1080/01496395.2019.1567548

Syam Babu D, Anantha Singh TS, Nidheesh PV., Suresh Kumar M. 2020. Industrial wastewater treatment by electrocoagulation process. Sep Sci Technol [Internet]. 55(17): 3195–3227. https://doi.org/10.1080/01496395.2019.1671866

Tanyol M, Tevkur S. 2021. Removal of total organic carbon and color from slaughterhouse wastewaters using electrocoagulation process: Central composite design optimization. Desalin Water Treat. 223:227–234.

Tegladza ID, Xu Q, Xu K, Lv G, Lu J. 2021. Electrocoagulation processes: A general review about role of electro-generated flocs in pollutant removal. Process Saf Environ Prot [Internet]. 146:169–189. https://doi.org/10.1016/j.psep.2020.08.048

Thirugnanasambandham K, Sivakumar V. 2016. Removal of ecotoxicological matters from tannery wastewater using electrocoagulation reactor: Modelling and optimization. Desalin Water Treat. 57(9):3871–3880.

Titchou FE, Afanga H, Zazou H, Ait Akbour R, Hamdani M. 2020. Batch elimination of cationic dye from aqueous solution by electrocoagulation process. Mediterr J Chem. 10(1):1–12.

Vasudevan S, Lakshmi J, Sozhan G. 2012. Studies on the removal of arsenate from water through electrocoagulation using direct and alternating current. Desalin Water Treat. 48(1–3):163–173.

Vik EA, Carlson DA, Eikum AS, Gjessing ET. 1984. Electrocoagulation of potable water. Water Res [Internet]. 18(11):1355–1360. https://linkinghub.elsevier.com/retrieve/pii/0043135484900034

Wagle D, Lin CJ, Nawaz T, Shipley HJ. 2020. Evaluation and optimization of electrocoagulation for treating Kraft paper mill wastewater. J Environ Chem Eng [Internet]. 8(1):103595. https://doi.org/10.1016/j.jece.2019.103595

Yoosefian M, Ahmadzadeh S, Aghasi M, Dolatabadi M. 2017. Optimization of electrocoagulation process for efficient removal of ciprofloxacin antibiotic using iron electrode; kinetic and isotherm studies of adsorption. J Mol Liq. 225:544–553.

Zaied BK, Rashid M, Nasrullah M, Zularisam AW, Pant D, Singh L. 2020. A comprehensive review on contaminants removal from pharmaceutical wastewater by electrocoagulation process. Sci Total Environ [Internet]. 726:138095. https://doi.org/10.1016/j.scitotenv.2020.138095

Zini LP, Longhi M, Jonko E, Giovanela M. 2020. Treatment of automotive industry wastewater by electrocoagulation using commercial aluminum electrodes. Process Saf Environ Prot [Internet]. 142:272–284. https://doi.org/10.1016/j.psep.2020.06.029

9 Enhancing Mechanical Properties of Adobe by Use of Vernacular Fibers of Agave Americana and Eulaliopsis Binata for Sustainable Traditional Mud Houses of Himachal Pradesh

Vandna Sharma and Aniket Sharma

CONTENTS

9.1 Introduction 152
9.2 Materials and Methodology 153
    9.2.1 Materials 153
    9.2.2 Material Attributes 153
        9.2.2.1 Soil – Preliminary Tests 153
        9.2.2.2 Particle Size Distribution 154
        9.2.2.3 Proctor Compaction Test 155
        9.2.2.4 Consistency Limits 155
        9.2.2.5 Fibers 156
    9.2.3 Preparation of Test Samples 157
9.3 Experimental Investigation 158
    9.3.1 Proctor Compaction Tests for Three Soils and Different Proportions of Cement 159
    9.3.2 Strength Tests 159
9.4 Results and Discussion 165
    9.4.1 Unconfined Compressive Strength (UCS) Tests 165

DOI: 10.1201/9781003297772-9

9.4.2 Stress-Carrying Capacity	170
9.4.2.1 Analyzing Cubical and Cylindrical Specimens for Stress-Carrying Capacity	171
9.5 Conclusion	173
References	174

9.1 INTRODUCTION

The vernacular architectural style is essentially climate-responsive and has energy-conserving characteristics that provide enhanced thermal comfort at a negligible cost [1-4]. Vernacular wisdom or indigenous knowledge of the built environment has always been looked upon by academicians, scientists, and researchers as the environmentally most viable solution for issues of non-sustainable construction practices around the globe [5]. Studies have shown that indigenous architecture or local or vernacular architecture, as it is known, has magnificent climate-responsive features and strategies [6]. These strategies include design and planning considerations, use of local natural building materials, sociocultural aspects, adoption of self-reliance techniques for construction in rural areas, and thermal comfort considerations. Most importantly, vernacular architecture exhibits respect for natural & environmental resources and reserves [7]. In this very context, rural mud houses have always enchanted researchers for their excellent thermal performance, and sustainable low-cost solutions for mass housing, especially for those who cannot afford the luxuries of expensive housing [8]. Rural mud or earth houses involve the use of earth in many forms across the globe due to variations in the topographic, geographic, geological, and other microclimatic conditions. Earth is used in a variety of forms varying from the use of adobe bricks to the use of compressed earth blocks wattle & daub, cob and rammed earth, etc. [9]. Similarly, the use of adobe for making houses is also not new to the human race. People in rural Indian villages have ingeniously used adobe for the construction of mud houses. However, as per census 2011 reports [10], this vernacular mud architecture is on the verge of decline owing to many reasons like unavailability of masons well versed with the mud construction, status quotient, aesthetical preferences, structural constraints, difficulty to maintain, durability aspects, etc. Out of these varied problems, studies have shown that poor strength and durability have been the main propellers for the shift of people from mud architecture to more flexible and low-maintenance modern architectural practices. Modern architectural practices involve the use of cement concrete, bricks, etc., which are otherwise unsustainable, energy-consuming, and expensive as well [11].

With the growing concern to explore more sustainable solutions for mass housing, improving the properties of the earth has become a great research area by amalgamating natural and vernacular fibers. Many research works have shown the effectiveness of fibers in ameliorating the mechanical attributes of adobe when subjected to experimental investigations as per standard code provisions. Research work [12, 13] provides information that people residing especially in rural areas have their local wisdom of ingeniously incorporating vernacular materials like straw, cow dung, husk, or other local fibers, etc., in the soil to enhance its properties before using it as building material either as adobe or rammed earth or cob. In this context,

Enhancing Mechanical Properties of Adobe 153

stabilization as discussed by many researchers either by natural stabilizer or by synthetic stabilizer has proven to be very effective [14, 15]. Literature review in context to improvement in mechanical, thermal, and physical properties of adobe using natural fibers as reinforcement shows many successful studies, whereby the authors have benefited society with the propagation of modified adobe as sustainable building material [16]. Another study [17] has successfully shown that the inclusion ameliorates the physical and mechanical properties of adobe advantageously. Another similar study [18] proved that stabilization by plant and vegetable fibers improves the strength attributes of the soil. Research conducted by Kafodya and others [19] to ameliorate the mechanical attributes of adobe using sisal fibers indicated that fiber inclusion increased the tensile, shear, and compressive strength of the adobe considerably. On the parallel lines, the present research dealt with an exploration into the properties of the earth after reinforcing it with natural indigenous fibers of Agave Americana and Eulaliopsis Binata. The importance of this study was to address the shortcomings in rural adobe houses in the villages of Himachal Pradesh, the Northern state of India. This was done to mitigate the mechanical properties related problems in adobe namely compressive strength.

9.2 MATERIALS AND METHODOLOGY

9.2.1 MATERIALS

The experimental investigation included the use of the soil, cement, and indigenous fibers of Agave Americana and Eulaliopsis Binata. The investigation included the use of two types of soil samples brought from the village named Jandot located in District Bilaspur and the village named Kashmir located in District Hamirpur. The study areas were selected with the dominance of vernacular houses in the state of Himachal Pradesh. The soil was taken from the barren pasture/fields to avoid harm to agricultural land. Similarly, two types of native fibers indigenous to both the soil types; namely Agave Americana for the soil brought from Jandot village (Bilaspur) and Eulaliopsis Binata (Buggad) for the soil brought from Kashmir village (Hamirpur) were used for experimental investigation respectively. The cement of standard grade 43 with specific gravity of 3.10 was applied in the experimental exploration as a stabilizer.

9.2.2 MATERIAL ATTRIBUTES

9.2.2.1 Soil – Preliminary Tests

The soil from both locations was preliminarily tested for evaluating the size of grain particles, the content of water, liquid and plastic limit, specific gravity, optimum water content, and related maximum dry density (MDD) as per Indian code provisions [20-27] for information about attributes of soil. The outputs of the same have been reported in Table 9.1. The compaction curves obtained from Standard Proctor compaction tests were analyzed to find optimum moisture content (OMC). Varying proportions of 0.5%, 1.0%, 1.5%, and 2.0% of vernacular fibers of Agave Americana and Eulaliopsis Binata were cut manually in lengths of 30 mm as suggested by the

154   Handbook of Sustainable Materials

TABLE 9.1

Physical Attributes of the Soils Evaluated properties of Water content [21], liquid and plastic limit [22], specific gravity [23], optimum water content [21], and maximum dry density [24, 25] tested for soils of Jandot (Bilaspur) and Kasmir (Hamirpur) as per IS codes

Attribute Name	Jandot (Bilaspur)	Kashmir (Hamirpur)
Water content (%)	18.55	14.146
Liquid limit (%)	39.13	10.56
Plastic limit (%)	35.05	42.2
Plasticity index (PI)	−4.08	31.6
Optimum Moisture Content (OMC %)	14	14
Maximum Dry Density (MDD) (kN/m³)	15.65	16.51
Specific gravity (G)	2.92	2.18
Indian Standard classification	SC	SC

literature review and were used in the investigation. The natural fibers were mixed randomly in the samples. Properties of indigenous fibers of Agave Americana (Dwaraya) and Eulaliopsis Binata (Baggad) were also tested and reported in Table 9.2.

9.2.2.2 Particle Size Distribution

For particle size distribution of the soil of Jandot (Bilaspur) and Kashmir (Hamirpur), an initial soil composition check and grain size analysis (also called sieve analysis) were done as per Indian standards. It was found that the soil from both the locations used under consideration has the configuration of coarse grain particles near to 0.0156 for sand and near to 0.0005 for clay consistent with Indian Standard

TABLE 9.2

Physical Attributes of Fiber – Agave Americana Evaluated properties of fiber elongation, tensile strength, fiber shape, diameter, length, and color for fibers of Agave Americana and Eulaliopsis Binata

Property	Agave Americana	Eulaliopsis Binata
Elongation at break (%)	4.39	3.552
Tensile strength (N)	46.35	29.50
Shape	Triangular profile	Circular
Diameter (mm)	0.45	0.45
Length (mm)	100	100
Speed of Jogg (mm/min)	10	10
Color	Dark brown (dry fiber)	Light brown/golden

Enhancing Mechanical Properties of Adobe

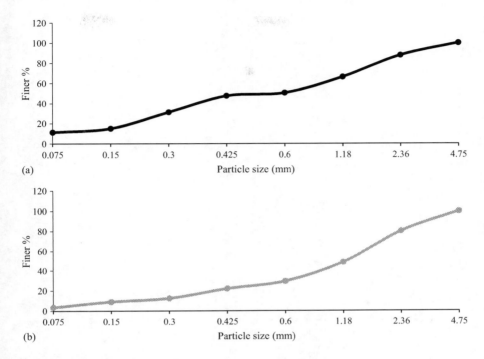

FIGURE 9.1 Particle size distribution curve of sand-clay soil for soils of (a) Jandot (Bilaspur) and (b) Kashmir (Hamirpur).

Grain size analysis conducted as per IS:2720-4,1985 yielded a particle size distribution curve of sand-clay soil for soils of Jandot (Bilaspur) and Kashmir (Hamirpur) showing soils as sand-clay soil according to Indian Standard classification of the soil.

classification of the soil. Investigation displayed sand-clay soil, assigned with symbol SC, consistent with the Indian Standard classification as provided in code provision. Further, this sand-clay soil used in the investigation is of low compressive strength as shown in Figure 9.1.

9.2.2.3 Proctor Compaction Test

The compaction characteristics of sand-clay soil of Jandot (Bilaspur) and Kashmir (Hamirpur) were found with standard proctor test as per procedure laid in Indian standards. In the compaction test, for the soil of Jandot (Bilaspur), OMC was calculated as 14% and related MDD was 15.65 kN/m^3. Similarly, for the soil of Kashmir (Hamirpur), OMC was 14% and related MDD was 16.51 kN/m^3 as reported in Table 9.1. The Proctor compaction test yielded the following curve for sand-clay soil as shown in Figure 9.2.

9.2.2.4 Consistency Limits

The consistency limits of sand-clay soil were determined as per laboratory procedure consistent with Indian standards. The specific gravity, plastic limit,

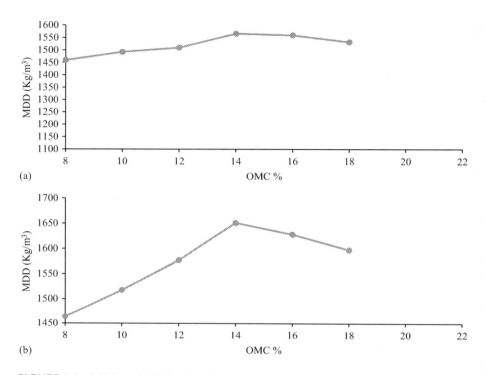

FIGURE 9.2 MDD vs OMC for the soil.

Figure 9.2 presents the calculation of MDD corresponding to OMC on the basis of Proctor compaction test conducted as per IS: 4332-3, 1967, and IS: 4332-10, 1969.

plasticity index, and liquid limit were determined as per the prescribed code. Tests disclosed that the soil (Jandot, Bilaspur) has a. natural water content of 2.11%, the specific gravity of 2.92%, plastic limit of 39.13%, liquid limit of 35.05%, plasticity index (PI) of 4.08, and the soil (Kashmir, Hamirpur) has natural water content of 10.61%, specific gravity of 2.18%, plastic limit of 10.56%, liquid limit of 42.20%, plasticity index (PI) of 31.6 as given in Table 9.1. Particle size distribution of both the soils was studied and found to be lying in the category of Sandy Clay (SC) in context to the Unified Soil Classification System having low compressive strength.

9.2.2.5 Fibers

Fibers of Agave Americana and Eulaliopsis Binata were tested for physical and engineering properties as shown in Figure 9.3. Literature review shows fiber inclusion serves a very crucial function in improving the strength attributes of adobe when used as reinforcement [28]. Further, research studies have proved the significance of fiber length in soil-fiber compositions. According to these studies when fibers of specific lengths/dimensions are included in the soil

Enhancing Mechanical Properties of Adobe

FIGURE 9.3 Fibers used in experimental investigation.

Addition of fibers of Agave Americana and Eulaliopsis Binata cut in lengths of 30 mm by weight as per IS code [26, 27].

compositions, it leads to considerable improvement in the strength attributes of the soil [29]. The study [29] showed the improvement in compressive and flexural strength of adobe by the inclusion of Angustifolia haw agave fiber of length 10–25 mm.

Therefore, physical attributes/properties of both the fibers of Agave Americana and Eulaliopsis Binata were checked and reported as shown in Table 9.2.

9.2.3 Preparation of Test Samples

Collected soil was first air-dried for 48 hours and then sieved. The soil and cement (used as a stabilizer as per IS codes) were first infused in the dry state and then water with fibers was supplemented (addition by weight and content decided as per IS codes). The samples were prepared with compositions such as soil, soil with cement, and vernacular fibers [30, 31]. Nearly 240 samples of both stabilized and unstabilized samples (120 cubical and 120 cylindrical) were prepared for each soil and fiber type at a maximum dry unit weight and related OMC values [25, 26] and [32-34]; thus achieved from standard compaction tests. Cylindrical specimens were developed with proportions of 38 mm diameter and 76 mm height [35], while cubical soil samples had preparatory dimensions of 190 mm × 90 mm × 90 mm as per Indian code provisions [25, 26, 32-34]. Specimens were designated as per the category and content. In total, 240 cubical and 240 cylindrical stabilized and unstabilized soil samples (in total for both soil types) were tested for maximum stress-carrying capacity and UCS testing respectively, as shown in Table 9.3. After preparation, the samples were first dried in the open air for a period of 24 hours, and then curing was completed for periods of 07, 14, 28, 56, and 90 days as per Indian code provisions [34]. The equipment used involved a Load Frame testing machine which has a maximum capacity of 50 kN with a persistent strain rate of 1.25 mm/min for UCS tests and a compression and flexure testing machine for stress-carrying capacity determination.

158   Handbook of Sustainable Materials

TABLE 9.3

Designations of Different Mix Specimens of Adobe for Compressive Strength Tests Stabilized and unstabilized samples with or without the use of fibers for Jandot soil designated by A and those for Kashmir soil designated by E prepared as per IS: 4332-1, 1967 for UCS Testing and stress-strain capacity evaluation

Sample Designations	Mix Specification	Cubical Samples (No.)	Cylindrical Samples (No.)
A1	Soil (Jandot, Bilaspur) (No Cement, No Fiber)	20 each	20 each
A2	Soil + 4% Cement + No Fiber		
A3	Soil + 4% Cement + 0.5% Agave Americana		
A4	Soil + 4% Cement + 1.0% Agave Americana		
A5	Soil + 4% Cement + 1.5% Agave Americana		
A6	Soil + 4% Cement + 2.0% Agave Americana		
A1	Soil (Kashmir, Hamirpur) (No Cement, No Fiber)		
E2	Soil + 3% Cement + No Fiber		
E3	Soil + 3% Cement + 0.5% Eulaliopsis Binata		
E4	Soil + 3% Cement + 1.0% Eulaliopsis Binata		
E5	Soil + 3% Cement + 1.5% Eulaliopsis Binata		
E6	Soil + 3% Cement + 2% Eulaliopsis Binata		
Total number of samples	240	240	

9.3 EXPERIMENTAL INVESTIGATION

The research work included studying the influence of added indigenous natural fibers over the unconfined compressive strength (UCS) of low compressive soil, compressed at an MDD and OMC. The surge in strength by adding up natural fibers in low compressive soil was quantified by conducting a set of UCS tests on the soil specimens prepared both with and without the addition of cement stabilizer and fiber reinforcement at OMC and related maximum dry unit weight [20-23, 26, 32]. Each sample composition was investigated for different curing durations. Nearly four test samples were tested for every individual composition according to Indian Codal provisions [34]. Verification tests were conducted conducive to establishing the validity, replication, and authenticity of experiments.

Enhancing Mechanical Properties of Adobe 159

9.3.1 Proctor Compaction Tests for Three Soils and Different Proportions of Cement

Proctor compaction tests were performed with the intent to find the OMC and MDD values; the results of which were further used in making cylindrical test specimens. These results were reported in Table 9.4 and in Figure 9.4.

Based on the analysis of MDD and related OMC for different soils and varied proportions of cement as given in Figure 9.4 and Table 9.4, it was therefore interpreted that with the addition of 4% cement in the soil of Jandot (Bilaspur); the value of MDD at given OMC value under consideration was over and above the comparable values of MDD for 3% and 5% added cement proportions for same soil. Similarly, in the case of the soil of Kashmir (Hamirpur), for addition of 3% cement in the soil, the value of MDD for the given OMC value was better than comparable values of MDD for 4% and 5% cement proportions added. Therefore, in case of the soil of Jandot (Bilaspur) for addition of 4% cement in the soil, and in case of the soil of Kashmir (Hamirpur) augmentation of 3% cement in the soil was used for stabilization purposes. Further using this proportion of stabilizers, samples for further experimental investigation were prepared.

9.3.2 Strength Tests

The compressive strength of all developed specimens was checked and measured as UCS. Tests were administered at a constant strain rate of 1.25 mm/min., consistent with Indian codal provisions [30, 31, 34]. UCS tests were administered after curing durations of 07, 14, 28, 56, and 90 days. Strength parameters for all mix compositions were evaluated in the form of unconfined compression strength. Nearly, 240 cylindrical mix compositions were subjected to test examination after 07, 14, 28, 56, and 90 days for gauging the level of improvement in low compressive clay soil by insertion of cement stabilizer, Agave Americana, and Eulaliopsis Binata fibers as reinforcement supplement. The stress-strain graphs were framed as shown in Figures 9.5 to 9.16, and analyzed for all mix compositions and compared.

As shown in Figure 9.5, the compressive strength for sample A1 (soil + 0% cement + 0% fiber) increased from 667 kN/m^2 to 692.8 kN/m^2, 745 kN/m^2, 921 kN/m^2, and 1120 kN/m^2 for test durations 07, 14, 28, 56, and 90 days, respectively.

TABLE 9.4

Maximum Dry Density for Different Soil-Cement Content MDD calculated corresponding to OMC from Proctor compaction tests as per [25, 33] for Jandot and Kashmir soils

	MDD(kN/m^3)	
Percentage of Cement Additive	**Jandot (Bilaspur)**	**Kashmir (Hamirpur)**
3% Cement	14.74	17.13
4% Cement	16.02	16.65
5% Cement	15.95	15.80

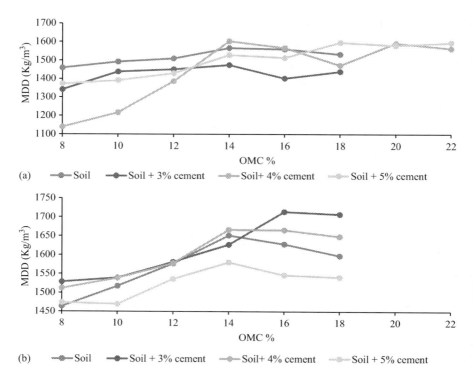

FIGURE 9.4 Moisture content vs dry density for different composites of the soil and cement.

Moisture content corresponding to dry density evaluated gave an appropriate percentage of cement to be added in the soil of Jandot (Bilaspur) as 4% and for Kashmir (Hamirpur) as 3%.

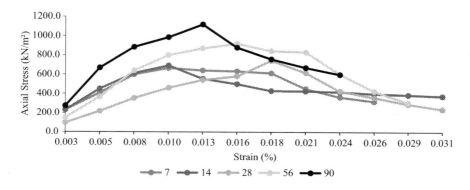

FIGURE 9.5 Development of strength in soil specimens: A1.

UCS Testing of specimen A1 (Jandot soil + no cement + no fiber) for curing durations of 07, 14, 28, 56, and 90 days was conducted as per IS: 2720-7, 1980.

Enhancing Mechanical Properties of Adobe

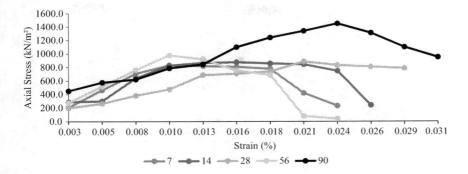

FIGURE 9.6 Development of strength in soil specimens: A2.

UCS Testing of specimen A2 (Jandot soil.+.4% cement + no fiber) for curing durations of 07, 14, 28, 56, and 90 days was conducted as per IS: 2720-7, 1980.

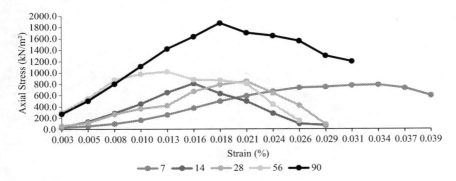

FIGURE 9.7 Development of strength in soil specimens: A3.

UCS Testing of specimen A3 (Jandot soil + 4% cement + 0.5% fiber) for curing durations of 07, 14, 28, 56, and 90 days was conducted as per IS: 2720-7, 1980.

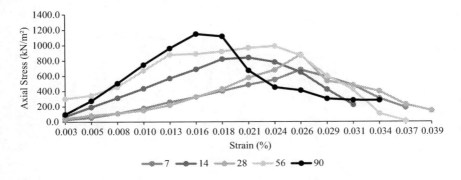

FIGURE 9.8 Development of strength in soil specimens: A4.

UCS Testing of specimen A4 (Jandot soil + 4% cement + 1.0% fiber) for curing durations of 07, 14, 28, 56, and 90 days was conducted as per IS: 2720-7, 1980.

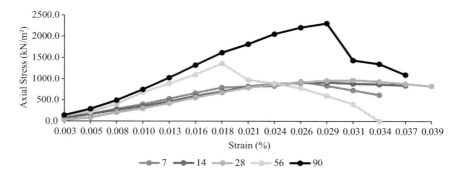

FIGURE 9.9 Development of strength in soil specimens: A5.

UCS Testing of specimen A5 (Jandot soil + 4% cement + 1.5% fiber) for curing durations of 07, 14, 28, 56, and 90 days was conducted as per IS: 2720-7, 1980.

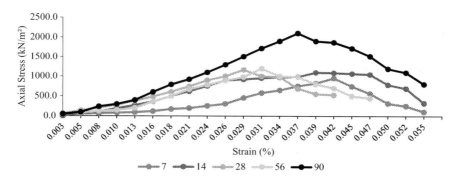

FIGURE 9.10 Development of strength in soil specimens: A6.

UCS Testing of specimen A6 (Jandot soil + 4% cement + 2.0% fiber) for curing durations of 07, 14, 28, 56, and 90 days was conducted as per IS: 2720-7, 1980.

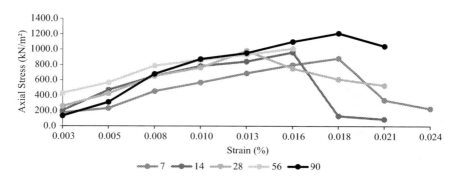

FIGURE 9.11 Development of strength in soil specimens: E1.

UCS Testing of specimen E1 (Kashmir soil + no cement + no fiber) for curing durations of 07, 14, 28, 56, and 90 days was conducted as per IS: 2720-7, 1980.

Enhancing Mechanical Properties of Adobe 163

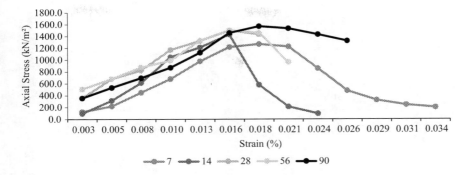

FIGURE 9.12 Development of strength in soil specimens: E2.

UCS Testing of specimen E2 (Kashmir soil + 3% cement + no fiber) for curing durations of 07, 14, 28, 56, and 90 days was conducted as per IS: 2720-7, 1980.

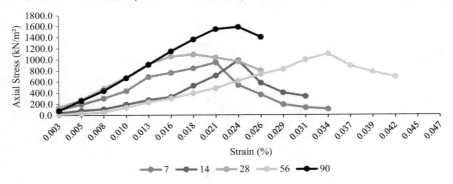

FIGURE 9.13 Development of strength in soil specimens: E3.

UCS Testing of specimen E3 (Kashmir soil + 3% cement + 0.5% fiber) for curing durations of 07, 14, 28, 56, and 90 days was conducted as per IS: 2720-7, 1980.

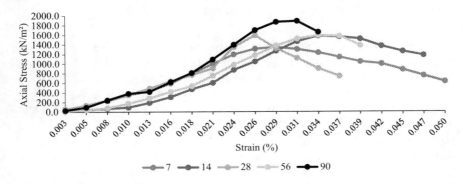

FIGURE 9.14 Development of strength in soil specimens: E4.

UCS Testing of specimen E4 (Kashmir soil + 3% cement + 1.0% fiber) for curing durations of 07, 14, 28, 56, and 90 days was conducted as per IS: 2720-7, 1980.

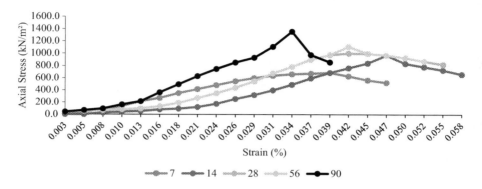

FIGURE 9.15 Development of strength in soil specimens: E5.

UCS Testing of specimen E5 (Kashmir soil + 3% cement + 1.5% fiber) for curing durations of 07, 14, 28, 56, and 90 days was conducted as per IS: 2720-7, 1980.

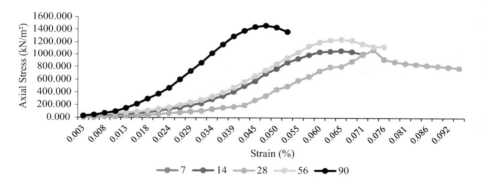

FIGURE 9.16 Development of strength in soil specimens: E6.

UCS Testing of specimen E6 (Kashmir soil + 3% cement + 2.0% fiber) for curing durations of 07, 14, 28, 56, and 90 days was conducted as per IS: 2720-7, 1980.

As shown in Figure 9.6, the compressive strength for sample A2 (soil + 4% cement + 0% fiber) increased from 834 kN/m^2 to 877 kN/m^2, 888 kN/m^2, 978 kN/m^2, and 1450 kN/m^2 for test durations 07, 14, 28, 56, and 90 days, respectively.

As exhibited in Figure 9.7, the compressive strength for sample A3 (Soil + 4% Cement + 0.5% Agave Americana) increased from 782.4 kN/m^2 to 811 kN/m^2, 848 kN/m^2, 1020 kN/m^2, and 1873 kN/m^2 for test durations 07, 14, 28, 56, and 90 days respectively.

As exhibited in Figure 9.8, the compressive strength for sample A4 (Soil + 4% Cement + 1.0% Agave Americana) increased from 682 kN/m^2 to 841 kN/m^2, 878 kN/m^2, 990 kN/m^2, and 1148.7 kN/m^2 for test durations 07, 14, 28, 56, and 90 days respectively.

As shown in Figure 9.9, the compressive strength for sample A5 (Soil + 4% Cement + 1.5% Agave Americana) increased from 935.9 kN/m^2 to 920.03 kN/m^2,

Enhancing Mechanical Properties of Adobe

970.7 kN/m², 1360 kN/m², and 2306.7 kN/m² for test durations 01, 14, 28, 56, and 90 days respectively.

As shown in Figure 9.10, compressive strength for sample A6 (Soil + 4% Cement + 2.0% Agave Americana) increased from 667 kN/m² to 1100 kN/m², 1170.0 kN/m², 1200 kN/m² and 2100 kN/m² for test durations 01, 14, 28, 56, and 90 days respectively.

As shown in Figure 9.11, the compressive strength for sample E1 (Soil + 0% Cement + 0% fiber) decreased from 883.5 kN/m² to 962.0 kN/m², 980.9 kN/m², 1011.0 kN/m², and 1723.1 kN/m² for test durations 07, 14, 28, 56, and 90 days respectively.

As shown in Figure 9.12, the compressive strength for sample E2 (Soil + 3% Cement + 0% fiber) increased from 1266.9 kN/m² to 1689.2 kN/m², 1699.3 kN/m², 1711.2 kN/m², and 1762.8 kN/m² for test durations 07, 14, 28, 56, and 90 days respectively.

As shown in Figure 9.13, the compressive strength for sample E3 (Soil + 3% Cement + 0.5% Eulaliopsis Binata) increased from 687.4 kN/m² to 710.5 kN/m², 1091.7 kN/m², 1100.2 kN/m², and 1581.9 kN/m² for test durations 07, 14, 28, 56, and 90 days respectively.

As shown in Figure 9.14, the compressive strength for sample E4 (Soil + 3% Cement + 1% Eulaliopsis Binata) increased from 1346.7 kN/m² to 1578.2 kN/m², 1588.1 kN/m², 1591.4 kN/m², and 1888.3 kN/m² for test durations 07, 14, 28, 56, and 90 days respectively.

As shown in Figure 9.15, the compressive strength for sample E5 (Soil + 3% Cement + 1.5% Eulaliopsis Binata) increased from 923.3 kN/m² to 850.3 kN/m², 999.7 kN/m², 1101.2 kN/m², and 1365.7 kN/m² for test durations 07, 14, 28, 56, and 90 days respectively.

As shown in Figure 9.16, the compressive strength for sample E6 (Soil + 3% Cement + 2% Eulaliopsis Binata) increased from 757.0 kN/m² to 1064.9 kN/m², 1078.1 kN/m², 1244.5 kN/m², and 1465.2 kN/m² for test durations 07, 14, 28, 56, and 90 days respectively.

9.4 RESULTS AND DISCUSSION

9.4.1 UNCONFINED COMPRESSIVE STRENGTH (UCS) TESTS

Experimental investigation including UCS tests for a series of curing duration was conducted in the laboratory under controlled circumstances to gauge the strength improvement of low-compressive clay soil after reinforcement with vernacular fibers of Agave Americana and Eulaliopsis Binata. The stress-strain curves were drawn, analyzed, and compared.

Analysis of results of UCS test for untreated and treated soil-fiber specimens have been illustrated in Figure 9.17 for Jandot soil using fibers of Agave Americana and in Figure 9.18 for Kashmir Soil using fibers of Eulaliopsis Binata respectively. The analysis is shown comparatively with each other in terms of early gain in strength and final gain in strength which further have been discussed in subsequent sections.

The analysis of gain in strength with fibers of Agave Americana for Jandot soil has been illustrated in Figure 9.17. Figures 9.17a and 9.17b show values for the UCS test for test durations of 07 and 14 days (designated by 7 and 14 days respectively).

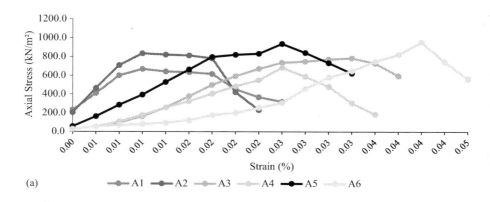

(a)

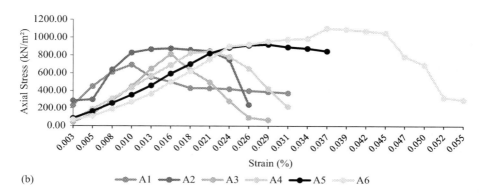

(b)

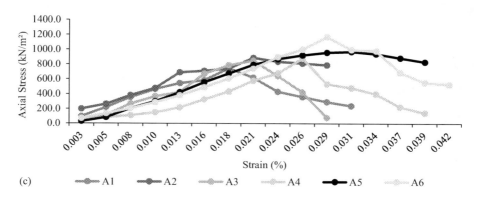

(c)

FIGURE 9.17 Compressive strength trends of Jandot soil, Bilaspur.

Trends in UCS for Jandot soil and Agave Americana fibers for curing durations of 07, 14, 28, 56, and 90 days show Specimens A5 and A6 having maximum UCS.

Enhancing Mechanical Properties of Adobe

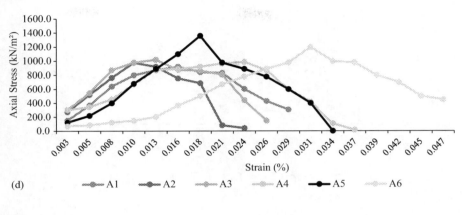

(d)

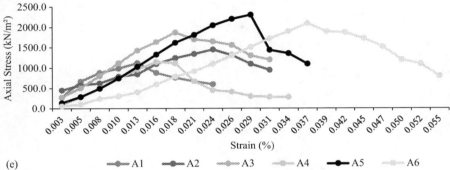

(e)

FIGURE 9.17 *(Continued)*

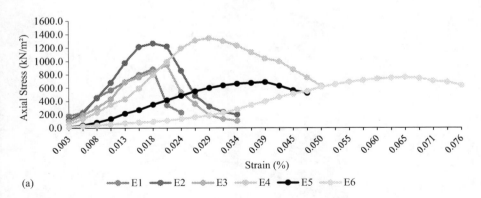

(a)

FIGURE 9.18 Compressive strength trends of Kashmir soil, Hamirpur.

Trends in UCS for Kashmir soil and Eulaliopsis Binata fibers for curing durations of 07, 14, 28, 56, and 90 days show Specimen E4, E2, and E6 having maximum UCS.

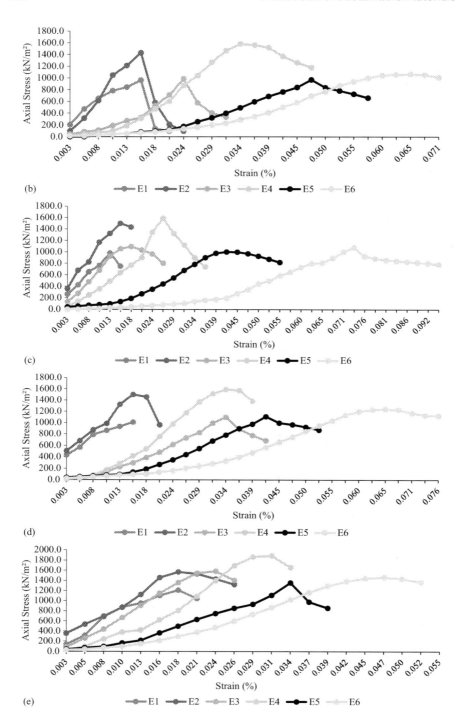

FIGURE 9.18 (*Continued*)

Enhancing Mechanical Properties of Adobe

Figure 9.17a gives the idea that test sample A5 has attained a maximum gain of 40% in strength, test sample A6 shows a gain of 30%, and test sample A3 shows a gain of 17.30% against the base strength of 667 kN/m^2 of soil. In the case of Figure 9.17b, test sample A6 shows a maximum gain of 58.77% in strength. Further, test sample A5 shows a gain of 32.79%, test sample A2 shows a gain of 26.58%, test sample A4 shows a gain of 21.39%, and test sample A3 shows a gain of 17.06% against a base strength of 692.8 kN/m^2 of soil. It is important to mention here that samples A6, A5, A4, and A3 refer to mixes with fiber reinforcement of 2%, 1.5%, 1%, and 0.5% of Agave Americana along with stabilization of 4% cement, and A2 correspond to mix with 3% cement stabilizer and no fiber reinforcement. Similarly, Figures 9.17c to 9.17e show values for the UCS test for test durations of 28, 56, and 90 days (represented by 28, 56, and 90 days respectively). Figure 9.17c gives the idea that for a test duration of 28 days; test sample A6 shows a maximum gain of 57% in strength. Further, test sample A5 shows a gain of 30.30%, test sample A2 shows a gain of 19.19%, test sample A4 shows a gain of 15.14%, and test sample A3 shows a gain of 13.8% against a base strength of 745 kN/m^2 of soil. Figure 9.17d gives the idea that for a test duration of 56 days; test sample A5 shows a maximum gain of 47.66% in strength, test sample A6 shows a gain of 30.29%, and test sample A3 shows a gain of 10.74% against base strength of 921 kN/m^2 of soil. Figure 9.17e gives the idea that for a test duration of 90 days; test sample A5 shows a maximum gain of 105.96% in strength, test sample A6 shows a gain of 87.5%, and test sample A3 shows a gain of 67.23% against base strength of 1120 kN/m^2 of soil.

Figures 9.18a and 9.18b show values for the UCS test for test durations of 07 and 14 days (designated by 07 days and 14 days respectively). Figure 9.18a gives the idea that test sample E4 has attained a maximum gain of 52.43% in strength, test sample E2 shows a gain of 43.36%, and E6 shows a gain of 11.29% against the strength of the soil as 883.5 kN/m^2. In Figure 9.18b, the test sample E4 shows a maximum gain of 64.05% in strength. Further, test sample E2 shows a gain of 48.44% and test sample E6 shows a gain of 10.69% against the strength of 962 kN/m^2 of the soil. It is important to mention here that samples, namely E6, E5, E4, and E3, refer to mixes with fiber reinforcement of 2%, 1.5%, 1%, and 0.5% of Eulaliopsis Binata along with stabilization of 3% cement and E2 correspond to mix with 3% cement stabilizer and no fiber reinforcement.

Similarly, Figures 9.18c to 9.18e show values for the UCS test for test durations of 28, 56, and 90 days (represented by 28, 56, and 90 days respectively). Figure 9.18c gives the idea that for a test duration of 28 days; test sample E4 shows a maximum gain of 61.90% in strength. Further, test sample E2 shows a gain of 52.71%, test sample E3 shows a gain of 11.38%, test sample E6 shows a gain of 9.9% against a base strength of 980.9 kN/m^2 of soil. Figure 9.18d gives the idea that for a test duration of 56 days; test sample E4 shows a maximum gain of 57.40% in strength. Further, test sample E2 shows a gain of 48.48% and test sample E6 shows a gain of 23.09% against a base strength of 1011 kN/m^2 of soil. Figure 9.18e gives the idea that for a test duration of 90 days; test sample E4 shows a maximum gain of 55.85% in strength. Further, test sample E3 shows a gain of 30.56%, and test sample E2 shows a gain of 29.52% against the base strength of 1211.6 kN/m^2 of soil.

170   Handbook of Sustainable Materials

It can be interpreted from Figures 9.17 and 9.18 that the addition of 1.5% Agave Americana fiber to Jandot soil leads to 105.96% amelioration in compressive strength of soil and with the addition of 1.0% Eulaliopsis Binata fibers to Kashmir soil leads to 55.85% amelioration in compressive strength of soil for rural villages of districts Bilaspur and Hamirpur respectively. This is made possible by better adhesion of soil-fiber particles, which creates a strong bond between them leading to improved strength. Similar results have also been reported in other research studies [36-38]. During the lab investigation, it was seen that on increasing the fiber content (both for Agave Americana and Eulaliopsis Binata fibers) there is an initial surge in compressive strength for curing durations of 14 and 28 days. However, after curing durations of 56 and 90 days, it was seen that for soil samples reinforced with Agave Americana fibers; sample A5 (fiber content: 1.5%) has more final gain in strength than that sample A6 (fiber content: 2.0%). Similarly, for soil samples reinforced with Eulaliopsis Binata fibers; sample E4 (fiber content: 1.0%) has more final gain in strength than that of samples E5 (fiber content: 1.5%), and E6 (fiber content: 2.0%). Therefore, the curves show that the presence of fiber in the soil increases its compressive strength as compared with virgin soil samples; however, this increase is feasible only up to a specific limit. Beyond this specific limit, if the fiber proportion is further increased, the compressive strength value decreases. The cause is improper bonding between soil-fiber particles due to agglomeration of fibers at higher content, which has also been reported by Abd El-Baky and others [39]. Further, the breaking pattern was also found to be brittle for unreinforced and unstabilized soil samples. However, reinforced samples did not break all of a sudden since, due to the reorganization of inner forces within the soil-fiber specimens from soil to fibers, fibers bind together particles of soil [40]. Due to this, the split of fibers did not happen all at once even though the fiber-soil bond was ruptured. This study resulted in the optimum fiber content of Agave Americana and Eulaliopsis Binata fibers to be added beneficially to the soil for an increase in compressive strength which can be incorporated by local masons at the village level for making strong adobe bricks for rural mud houses.

9.4.2 STRESS-CARRYING CAPACITY

Stress-carrying capacity along with the respective compressive strength value was compared for cylindrical specimens and cubical specimens to gauge the better performance in the development of strength in context to geometrical shapes. In total, 240 cubical specimens of composition, as given in Table 9.3, were administered for tests for periods of 07, 14, 28, 56, and 90 days to assess the achievement in the strength of the specimens as exhibited in Figures 9.19 and 9.20 using compression and flexure testing machine. Figures 9.19 and 9.20 reflect that the stress-carrying capacity of specimens enhances with time span which further reflects enhancement in strength.

It is revealed from the graphs that: (i) specimens A1, A2, E1, and E2 have brittle and fragile conduct. They shatter into pieces all at once just when the cracks become visible after attainment of maximum load-carrying capacity, (ii) samples A3, A4, A5, A6, E3, E4, E5, and E6 show ductile and flexible conduct. Specimens did not shatter into pieces all at once just when the cracks become visible after the attainment of

Enhancing Mechanical Properties of Adobe

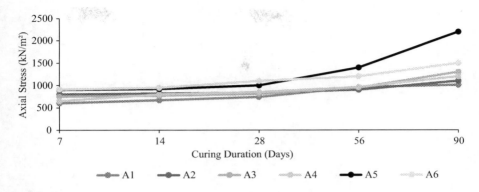

FIGURE 9.19 Gain in maximum stress-carrying capacity with time in cubical soil samples for Jandot soil, Bilaspur.

Stress-carrying capacity corresponding to UCS values calculated for 120 cubical specimens for Jandot soil, Bilaspur, with resepct to curing durations of 07, 14, 28, 56, and 90 days respectively.

maximum load-carrying capacity. Instead, this disruption was slowed with time, and specimens that had more fiber proportion reflected the highest slowdown in disruption and breaking, (iii) specimens A5 and A6 (of Agave Americana) revealed more ductility than specimens E4, E5, and E6 (of Eulaliopsis Binata).

9.4.2.1 Analyzing Cubical and Cylindrical Specimens for Stress-Carrying Capacity

The soil mix compositions (specimens) of both cubical and cylindrical shapes were equated and compared for stress-carrying capacity and related compressive strength

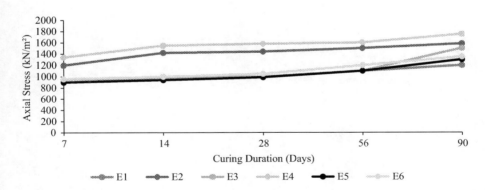

FIGURE 9.20 Gain in maximum stress-carrying capacity with time in cubical soil samples for Kashmir soil, Hamirpur.

Stress-carrying capacity corresponding to UCS values calculated for 120 cubical specimens for Kashmir soil, Hamirpur, with resepct to curing durations of 07, 14, 28, 56, and 90 days respectively.

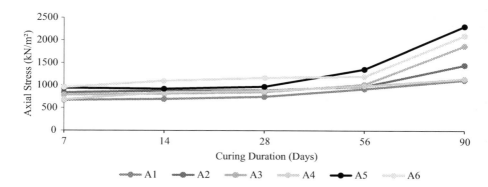

FIGURE 9.21 Gain in maximum stress-carrying capacity with time in cylindrical soil samples for Jandot soil, Bilaspur.

Stress-carrying capacity corresponding to UCS values calculated for 120 cylindrical specimens for Jandot soil, Bilaspur, with resepct to curing durations of 07, 14, 28, 56, and 90 days respectively.

characteristics. The results exhibited that although the values are proportionate for both types, still the value of strength for cylindrical specimens was more than that for cubical specimens as shown by Figures 9.21 and 9.22 respectively. The graphs reveal that the maximum stress-carrying capacity for cubical specimens is 2200 kN/m² and for cylindrical specimens is 2306.7 kN/m² for Jandot soil samples reinforced with fibers of Agave Americana. Similarly, the maximum stress-carrying capacity for cubical specimens is 1750kN/m², and for cylindrical specimens is 1888.3 kN/m² for Kashmir soil samples reinforced with fibers of Eulaliopsis Binata. It can be interpreted from Figures 9.21 and 9.22 that the maximum stress-carrying capacity of the

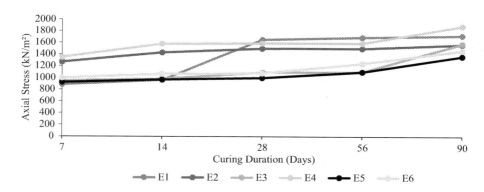

FIGURE 9.22 Gain in maximum stress-carrying capacity with time in cylindrical soil samples for Kashmir soil, Hamirpur.

Stress-carrying capacity corresponding to UCS values calculated for 120 cylindrical specimens for Kashmir soil, Hamirpur, with resepct to curing durations of 07, 14, 28, 56, and 90 days respectively.

Enhancing Mechanical Properties of Adobe 173

soil, which is also a synonym for its compressive strength, gets enhanced by adding up fibers of Agave Americana and Eulaliopsis Binata. Additionally, it also tells that soil specimens of fibers Agave Americana are stronger than soil specimens of fiber Eulaliopsis Binata.

9.5 CONCLUSION

The investigation involving a series of UCS and stress-carrying capacity experiments to check the amelioration in compressive strength of low-strength soils of rural villages of the state Himachal Pradesh, India, was carried out as per provisions of IS codes. The investigation involves the use of vernacular fibers of Agave Americana and Eulaliopsis Binata. The investigation showed a notable increment in the compressive strength of soil after augmenting it with native fibers. Results, in specific, have been summarized on the experimental investigation are as follows:

1. Soil samples without the addition of any stabilizer or fibers showed brittle behavior which is not safe from a seismic point of view and requires modification.
2. Soil samples even after the addition of cement showed brittle behavior which is again not safe as per seismic considerations and requires further material modification of adobe under study.
3. A significant increment in UCS values was seen in all samples reinforced with fibers as compared with virgin soil samples.
4. Soil samples when reinforced with fibers of Agave Americana and Eulaliopsis Binata along with the addition of cement as a stabilizer (as per IS code provisions) showed improvement in compressive strength by 105.95% for fibers of Agave Americana (against the base soil sample strength of 1120 kN/m^2) and 55.85% for fibers of Eulaliopsis Binata (against the base soil sample strength of 1211.6 kN/m^2) in addition to improved ductile behavior.
5. Soil samples reinforced with different proportions of fibers of Agave Americana and Eulaliopsis Binata did not break into pieces all at once even after the cracks became visible on reaching a maximum load-carrying capacity of samples.
6. Fiber-reinforced soil samples of soil Jandot (fiber – Agave Americana) showed more ductile behavior than fiber-reinforced soil samples of soil Kashmir (fiber – Eulaliopsis Binata). This is also evident from the tensile properties of both fibers. The tensile strength of the fibers of Agave Americana is more than that of Eulaliopsis Binata.
7. For fiber Agave Americana, significant gain in UCS was shown by samples A5 (soil + 4% cement + 1.5% Agave Americana), A6 (soil + 4% cement + 2.0% Agave Americana), A2 (soil + 4% cement + no fiber), A3 (soil + 4% cement + 0.5% Agave Americana). While for fiber Eulaliopsis Binata, significant gain in UCS was shown by samples E4 (soil + 3% cement + 1.0% Eulaliopsis Binata), E3 (soil + 3% cement + 0.5% Eulaliopsis Binata), E2 (soil + 4% cement + no fiber), E6 (soil + 4% cement + 2.0% Agave Americana).

8. UCS behavior of samples showed that gain in strength increases with fiber proportion but only up to a specific limit. Beyond this limit, if the fiber proportion is further increased, it will lead to a decrease in compressive strength. This is attributed due to the excessive presence of fibers which leads to the slipping of fibers in soil and prevents proper bonding of soil particles and fiber. This recommends the appropriate fiber proportion which can be augmented to the soil for enhancing its compressive strength for fieldwork.

An experimental investigation involving the upgrading of mechanical attributes of adobe by using natural vernacular fibers of Agave Americana and Eulaliopsis Binata showed that fiber addition would successfully lead to an amelioration in compressive strength of adobe in practice for rural mud houses for propagating sustainable housing solutions for masses of Himachal Pradesh.

REFERENCES

1. C. Alexander, S. Ishikawa & M. Silverstein. 1977. A Pattern Language: Towns, Buildings, Construction: Oxford University Press. Vol. 2.
2. P. Oliver. 1983. Earth as a Building Material Today, Oxford Art Journal, 5:2, 31–38.
3. P. Oliver. 1997. Encyclopedia of Vernacular Architecture of the World: Cambridge University Press.
4. T. Alastair, M. Marsh & Yask Kulshreshtha. 2021. The State of Earthen Housing Worldwide: How Development Affects Attitudes and Adoption, Building Research & Information, 50:5, 485–501. DOI: 10.1080/09613218.2021.1953369
5. Adriana Silva, Inês Oliveira, Vítor Silva, José Mirão & Paulina Faria. 2022. Vernacular Caramel´s Adobe Masonry Dwellings – Material Characterization, International Journal of Architectural Heritage, 16:1, 67–84. DOI: 10.1080/15583058.2020.1751343
6. Maria Philokyprou & Aimilios Michael. 2021. Environmental Sustainability in the Conservation of Vernacular Architecture – The Case of Rural and Urban Traditional Settlements in Cyprus, International Journal of Architectural Heritage, 15:11, 1741–1763. DOI: 10.1080/15583058.2020.1719235
7. F. De Filippi, R. Pennacchio, S. Torres & L. Restuccia. 2021. Anti-Seismic Retrofitting Techniques for Vernacular Adobe Buildings in Colombia: A Proposed Framework for Developing and Assessing Sustainable and Appropriate Interventions, International Journal of Architectural Heritage, 16:6, 923–939. DOI: 10.1080/15583058.2021.1992538
8. Cristiana Costa, Ângela Cerqueira, Fernando Rocha & Ana Velosa. 2019. The Sustainability of Adobe Construction: Past to Future, International Journal of Architectural Heritage, 13:5, 639–647. DOI: 10.1080/15583058.2018.1459954
9. João Luís Parracha, José Lima, Maria Teresa Freire, Micael Ferreira & Paulina Faria. 2021. Vernacular Earthen Buildings from Leiria, Portugal – Material Characterization, International Journal of Architectural Heritage, 15:9, 1285–1300. DOI: 10.1080/15583058.2019.1668986
10. Census of India, Ministry of Home Affairs. 2011. Housing, Household Amenities and Assets Database (ver. 2.0), Census Information.
11. Faris Karahan & Sanaz Davardoust. 2020. Evaluation of Vernacular Architecture of Uzundere District (Architectural Typology and Physical form of Building) in Relation to Ecological Sustainable Development, Journal of Asian Architecture and Building Engineering, 19:5, 490–501. DOI: 10.1080/13467581.2020.1758108

Enhancing Mechanical Properties of Adobe

12. Mohammad Sharif Zami. 2021. Barriers Hindering Acceptance of Earth Construction in the Urban Context of the United Kingdom, Architectural Engineering and Design Management. DOI: 10.1080/17452007.2021.1995314
13. Guru Kumar Manjappara Subramanian, M. Balasubramanian & A. Arul Jeya Kumar. 2021. A Review on the Mechanical Properties of Natural Fiber Reinforced Compressed Earth Blocks, Journal of Natural Fibers. DOI: 10.1080/15440478.2021.1958405
14. Sunil Kumar Meena, Raghvendra Sahu & Ramanathan Ayothiraman. 2021. Utilization of Waste Wheat Straw Fibers for Improving the Strength Characteristics of Clay, Journal of Natural Fibers, 18:10, 1404–1418. DOI: 10.1080/15440478.2019.1691116
15. Elhoussine Atiki, Bachir Taallah, Sadok Feia, Kamal Saleh Almeasar & Abdelhamid Guettala. 2021. Effects of Incorporating Date Palm Waste as a Thermal Insulating Material on the Physical Properties and Mechanical Behavior of Compressed Earth Block, Journal of Natural Fibers, DOI: 10.1080/15440478.2021.1967831
16. C. Babé, D. K. Kidmo, A. Tom, R. R. N. Mvondo, R. B. E. Boum & N. Djongyang. 2020. Thermomechanical Characterization and Durability of Adobes Reinforced with Millet Waste Fibers (Sorghum Bicolor), Case Studies in Construction Materials, 13: e00422.
17. Necmi Yarbasi. 2022. Effect of Freeze-Thaw on Compressive Strength of Clayey Soils Reinforced with Wool, Journal of Natural Fibers, 19:1, 382–393. DOI: 10.1080/15440478.2021.1875357
18. Kirupairaja Thanushan, Youganathan Yogananth, Pooraneswaran Sangeeth, Juthathatheu Gracian Coonghe & Navaratnarajah Sathiparan. 2021. Strength and Durability Characteristics of Coconut Fibre Reinforced Earth Cement Blocks, Journal of Natural Fibers, 18:6, 773–788. DOI: 10.1080/15440478.2019.1652220
19. I. Kafodya, F. Okontaa & P. Kloukinas. 2019. Role of Fiber Inclusion in Adobe Masonry Construction. Journal of Building Engineering, 26: 100904.
20. IS: 2720-4. 1985. Grain Size Analysis: Indian Standards. Bureau of Indian Standard, New Delhi.
21. IS: 2720-2. 1973. Methods of Test for Soils, Determination of Water Content. Bureau of Indian Standard, New Delhi.
22. IS: 2720-5. 1985. Determination of Liquid and Plastic Limit: Indian Standards. Bureau of Indian Standard. New Delhi.
23. IS: 2720-3-1. 1980. Method of Test for Soils – Determination of Specific Gravity. Bureau of Indian Standard. New Delhi.
24. IS: 4332-3. 1967. Method of Test for Stabilized Soils – Part 3: Test for Determination of Moisture Content – Dry Density Relation for Stabilize Sol Mixtures. Bureau of Indian Standard. New Delhi.
25. IS: 4332-10. 1969. Method of Test for Soils – Part 10: Methods of Test for Stabilized Soils. Bureau of Indian Standard. New Delhi.
26. IS 1725. 1982. Specification for Soil Based Blocks Used in General Building Construction. Bureau of Indian Standard. New Delhi.
27. IS 1498. 1970. Classification and Identification of Soils for General Engineering Purposes. Bureau of Indian Standard. New Delhi.
28. Lamia Guettatfi, Abdelmadjid Hamouine, Khedidja Himouri & Boudjemaa Labbaci. 2021. Mechanical and Water Durability Properties of Adobes Stabilized with White Cement, Quicklime and Date Palm Fibers, International Journal of Architectural Heritage, 1–15. DOI: 10.1080/15583058.2021.1959675
29. Magdaleno Caballero-Caballero, Fernando Chinas-Castillo, José Luis Montes Bernabé, Rafael Alavéz-Ramirez, María Eugenia Silva Rivera. 2018. Effect on Compressive and Flexural Strength of Agave Fiber Reinforced Adobes, Journal of Natural Fibers, 15(4): 575–585.

30. V. Sharma, H. K. Vinayak & B. M. Marwaha. 2015a. Enhancing Sustainability of Rural Adobe Houses of Hills by Addition of Vernacular Fiber Reinforcement, International Journal of Sustainable Built Environment, 4(2): 348–358.
31. V. Sharma, H. K. Vinayak & B. M. Marwaha. 2015b. Enhancing Compressive Strength of Soil Using Natural Fibers, Construction and Building Material, 93: 943–949.
32. IS: 2720-7. 1980. Method of Test for Soils – Part 7: Determination of Water Content-Dry Density Relation Using Light Compaction. Bureau of Indian Standard. New Delhi.
33. IS: 4332-1. 1967. Method of Test for Soils – Part 1: Method of Sampling and Preparation of Stabilized Soils for Testing. Bureau of Indian Standard. New Delhi, 1967.
34. IS: 2720-10. 1991. Method of Test for Soils – Part 10: Determination of Unconfined Compressive Strength. Bureau of Indian Standard. New Delhi.
35. IS 2110. 1980 (reaffirmed 2002). Code of Practice for In-situ Construction of Walls in Buildings with Soil-cement. Bureau of Indian Standard. New Delhi.
36. Teingteing Tan, Bujang B. K. Huat, Vivi Anggraini, Sanjay Kumar Shukla & Haslinda Nahazanan. 2021. Strength Behavior of Fly Ash-Stabilized Soil Reinforced with Coir Fibers in Alkaline Environment, Journal of Natural Fibers, 18:11, 1556–1569. DOI: 10.1080/15440478.2019.1691701
37. Bibekananda Sahoo, Bishnu Prasad Nanda & Alok Satapathy. 2022. Palm Leaf Fibers: A Potential Reinforcement for Enhancing Thermal Insulation Capability of Epoxy Based Composite, Journal of Natural Fibers, 1–13. DOI: 10.1080/15440478.2022.2028213
38. Ramanathan Ayothiraman, Raghvendra Sahu & Priyabrata Bhuyan. 2021. Strength and Deformation Behavior of Fine-Grained Soils Reinforced with Hair Fibers and Its Application in Pavement Design, Journal of Natural Fibers, 1–18. DOI: 10.1080/15440478.2021.1954129
39. M. A. Abd El-Baky, Mohamed A. Attia, Mostafa M. Abdelhaleem & Mohamad A. Hassan. 2022. Flax/basalt/E-glass Fibers Reinforced Epoxy Composites with Enhanced Mechanical Properties, Journal of Natural Fibers, 19(3): 954–968.
40. H. P. Singh & M. Bagra. 2013. Improvement in CBR Value of Soil Reinforced with Jute Fiber, International Journal of Innovative Research in Science, Engineering and Technology, 2(8): 3447–3452.

10 Green Synthesis of Cerium Oxide Nanoparticles and Its Recent Electrochemical Sensing Applications
A Study

Satyajit Das and Partha Pratim Sahu

CONTENTS

10.1 Introduction 177
10.2 Green Synthesis 178
      10.2.1 Plant-Mediated 179
      10.2.2 Fungus-Mediated 181
      10.2.3 Polymer-Mediated 182
      10.2.4 Nutrient-Mediated 182
10.3 Electrochemical Sensing Applications 183
      10.3.1 Electrochemical Sensor 183
10.4 Conclusion and Prospects 188
References 188

10.1 INTRODUCTION

Cerium is a rare earth metal that exists in two oxidation states (+3 and +4) and belong to the lanthanide series [1]. The oxide of cerium (CeO_2) represents the cubic fluorite structure and has oxygen deficiencies, to provide it with redox reaction sites [2].

The nanoparticles (NPs) with their excellent properties, like miniature size and shape, high surface-to-volume ratio, etc., have got much consideration in industrial, environmental, medical, and material science showing their significant growth [3, 4]. Likewise, CeO_2 NPs also have various applications such as catalyst [5, 6] sensors, sunscreen cosmetics, fuel cells with solid oxide, chemicals alteration in the body, antibacterial activity, drug delivery carriers, antiparasitic ointments, therapeutics agents, etc. [7-9]. Compared to metal oxide NPs, like TiO_2 and ZnO, CeO_2 NPs possess less toxicity while applying to cell activity [10]. CeO_2 NPs with AO

DOI: 10.1201/9781003297772-10

TABLE 10.1
Methods for the Synthesis of Cerium Nanoparticles by Means of Chemical and Biological/Green Processes

	Methods	Ref
Chemical synthesis	Sol–gel	[32]
	Mechanochemical	[33]
	Co-precipitation	[34]
	Pyrolysis method	[35]
	Sonochemical	[36]
	Solvothermal	[37]
Green/biosynthesis	Plant-mediated	[9, 27, 28, 38]
	Fungus-mediated	[29]
	Polymer-mediated	[39]
	Nutrient-mediated	[40]

properties have potential applications in daily human life as evaluated by various researchers [11, 12].

Normally, CeO_2 NPs are synthesized by hydrothermal methods, sonochemical technique, microwave-assisted, flame spray pyrolysis, sol-gel approach, co-precipitation, etc. [13-17], as described in Table 10.1. Although these methods are helpful in controlling the size and shape of NPs, these methods seem costly, use materials that are toxic, require high temperature and pressure, have lesser biocompatibility, and are not environment friendly [2, 8, 18, 19]. Recently green synthesis methods got much interest to researchers for the synthesis of nanomaterial. This synthesis process involves eco-friendly materials like plants, microorganisms, and other biological products which have the potential for biological reduction of metals and/or act as stabilizing agents [20].

Electrochemical sensors based on CNP have been studied by many researchers. CNP deposited on glassy carbon surface has been used for rapid detection of ethanol [21]. CNP when immobilized with oxidase enzymes tyrosinase, lactate oxidase, and glutamate oxidase improves the performance of biosensors for detecting dopamine [22], lactate [23], and glutamate [24] respectively. The cerium oxide nanoparticles (CNPs) doped with cobalt doped were used for the detection of hydrazine electrochemically. The hydroxyl radicals were identified with CNPs with graphene oxide (CeNP/GO) composite which were coated over screen-printed carbon electrode. Similarly, cerium oxide and reduced graphene oxide (CeO_2/RGO) composite were used for the detection of tryptophan electrochemically in food.

In this chapter, the green synthesis methods of CeO_2 NPs along with electrochemical sensing applications have been reviewed.

10.2 GREEN SYNTHESIS

The green synthesis of nanostructures has emerged as a simple, nontoxic, nonexpensive, and efficient approach for nanofabrication. The nontoxic nature of synthesis is environment-friendly also. In this direction, biosynthesis of inorganic

Green Synthesis of Cerium Oxide Nanoparticles

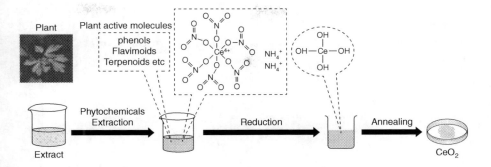

FIGURE 10.1 Process showing the formation of CeO$_2$ NPs using green synthesis.

nanostructures, such as oxide NPs, sulfide NPs, metallic NPs, and other nanostructures, have been reported for applications like drug delivery, antibacterial activity, gene therapy, DNA analysis, bio-sensing, and MRF imaging extract [20].

The plant extracts, micro-organisms, and biological products are used for biosynthesis of nanostructures, especially metallic NPs where phenolic groups, polyphenols, and terpenoids reduce particle sizes making them nanostructure forms [25]. Moreover, biosynthesis provides stabilities and bioavailability of NPs [26]. Recently CeO$_2$ NPs have been focused on antibacterial behavior with bacteria. Green synthesis of CeO$_2$ NPs is carried out by using plant extracts such as Acalypha indica [27], Petroselinum crispum [9, 28], and Gloriosa superba [9]. Figure 10.1 shows the green synthesis process for the preparation of CNPs. Fungal extracellular compound humicola sp. is also used for the synthesis of CeO$_2$ NPs [29]. Here, the plant extract stabilizes the NPs and acts as a capping agent [30], but produces NPs of large size. Apart from the plant extract, honey, egg white, and pectin are used for isotropic growth of nutrient-mediated biosynthesis of CeO$_2$ NPs [31]. Different methodologies for green/bio-synthesized cerium NPs are briefed in Table 10.2. The precursor for synthesis, morphologies of synthesized nanostructures, their particle sizes, and other applications as reported by different researchers are shown in Table 10.2

10.2.1 Plant-Mediated

The Gloriosa superba L. leaf was used for extraction of its phytochemicals as stated by Arumugam et al. and the extract was applied to synthesize CeO$_2$ NPs retaining their cubic structure. The particle size presented by the TEM image was 5 nm with the characteristics of a high surface area [9]. In the green synthesis process used by Sebastiammal et al., sweet basil plant leaves were washed with water and cut off into small pieces. Then the leaf extract was obtained by stirring well with water. The 50 mL of extract was added to the homogeneous aqueous solution of 0.5 M cerium (III) nitrate hexahydrate, and stirred at 80°C till the supernatant was evaporated. The product obtained was cachinnated for 2 hours at 600°C. The face-centered cubic crystal was confirmed by the X-ray diffraction pattern of CeO$_2$ NPs. The crystal size of 13 nm was reported with the agglomeration of the green synthesized CeO$_2$ NPs as observed in FESEM [38].

TABLE 10.2

Recently Applied Green/Biosynthesis Methods for Cerium Oxide Nanoparticles

Precursors	Source for Green/ Biosynthesis	Crystallization Temperature	Morphology, Particle Size	Other Applications/ Studies	Ref.
$CeCl_3$	Gloriosa superba	400°C	Spherical shaped, 5 nm	Antibacterial studies	[9]
Cerium (III) nitrate hexahydrate ($CeN_3O_9\cdot6H_2O$)	(Fungus) *Humicola* sp.	-	-	-	[29]
$CeCl_3$-$7H_2O$	Acalypha indica leaf extract	400°C	Spherical, 20–30 nm	Antibacterial	[30]
$Ce(NO_3)_3\,6H_2O$	Honey	200, 400, 600, and 800°C	Spherically shaped, 23 nm	Neurotoxicity effects	[31]
Cerium (III) nitrate hexahydrate	Sweet basil plant	600°C	-	Antibacterial studies	[38]
Cerium nitrate	Pectin	400°C	Polydispersed spherical, 40 nm	Antioxidant, antibacterial, cytotoxicity	[39]
Cerium (III) nitrate hexahydrate	Egg white	200, 400, 600, and 800°C	Spherically shaped, 25 nm	In-vitro cytotoxicity studies	[40]
$CeCl_3$	Prosopis juliflora	800°C	Spherically shaped, 15 nm	Antibacterial activity	[41]
Cerium(III) nitrate hexahydrate	Aloe vera leaf	600°C	Spherical, 63 nm	-	[42]
$Ce(NO_3)_3\,6H_2O$	Hibiscus sabdariffa flower extract	500°C	3.9 nm	-	[44]
$Ce(NO_3)_2$	Datura metel	500°C	Spherical, 5–15 nm	Antioxidant	[45]
$CeCl_3$	Calotropis procera	400°C	Spherical like, 7 nm, 21 nm	Antibacterial, photocatalytic degradation of methyl orange	[46]
$CeCl_3$	Aspergillus niger	350 and 400°C	Spherical shaped, 5–20 nm	Antibacterial activity	[47]
Cerium(III) nitrate hexahydrate	Starch	120–600°C	6 nm	Dose dependent toxicity, up to 175 mg/mL nontoxic effect of concentration	[48]

Green Synthesis of Cerium Oxide Nanoparticles

Arunachalam et al. reported the preparation of CNPs using ultrasound-assisted leaf extract of *Prosopis juliflora*. The aqueous solution of 0.1 M cerium chloride ($CeCl_3$) was produced and dropped into an equal volume of *P. juliflora* leaf extract. The presence of white precipitate in the solution indicated that hydroxide-mediated precipitation had occurred. The solution was heated in an oven at 80°C until it was half its original volume. A yellowish-brown precipitate happened to appear after the precursor was operated at a power of 800 W with the frequency 2450 MHz for 10 minutes in a microwave oven. It was annealed at 800°C for 2 hours and the formation of NPs was characterized. HRTEM scans revealed spherical-shaped particles averaging 15 nm in size. The concentrations of Ce^{3+} and Ce^{4+} in the produced CNPs were determined using XPS signals [41].

The CNPs synthesis by green approach was reported by means of Acalypha indica leaf extract. The analysis of the NPs was performed using XRD, TEM, Eto theDX, and revealed the size of NPs as 20–30 nm. Ce-O-Ce stretching was confirmed by FTIR spectra. An antibacterial study was also performed with a positive result [30]. The Cerium (III) nitrate hexahydrate was reduced by the plant extract of aloe vera leaf to prepare cerium NPs. The reduced sample was annealed at 600°C for 2 hours to obtain spherical-shaped particles of size 63 nm [42]. Cerium NPs were biologically reduced through aloe vera leaf extract where the as-synthesized and synthesized particles after annealing show the size of 7 nm and 12 nm respectively. FESEM shows the spherical shape of the nanostructures [43].

Thovhogi reported the use of Hibiscus sabdariffa extract in the synthesis of CNPs where leaf extract acts as a chelating agent. Calcination (500°C) of the reduced particles shows high crystallinity as confirmed by the analysis such as HRTEM, XRD, and XPS. The particle size histogram as measured through Image J software shows the size of the particles as 3.9 nm [44]. Yulizar et al. conveyed the green synthesis of NPs using the effect of weak base and capping agent of Datura metel L. leaves extract. The spherical-sized NPs of cerium oxides as confirmed by the XRD analysis have the face-centered cubic structure. The particle size as reported by the author was 5–15. The DPPH scavenging activity performed by the author achieved around 16.61% of radicle scavenging [45]. The research work here used Calotropis procera flower extract for the reduction of $CeCl_3$ in the process of synthesis of cerium nanostructures. The particle sizes of 7 nm and 21 nm were reported during XRD and HRTEM analysis respectively. Methyl orange dye for its photocatalytic degradation by nanoceria was studied which shows 98% of its degradation. While doing an antibacterial activity, the study shows important activity against bacteria (Gram-positive bacteria over Gram-negative) [46].

10.2.2 Fungus-Mediated

The biological routes of nanoparticle synthesis are always nontoxic, cheap, and eco-friendly as explained by many researchers. Here the author tried for the CNPs synthesis through microorganisms (fungus) *Humicola* sp. The extracted protein-covered NPs have transition states as established by the X-ray Photoemission Spectroscopy (XPS) analysis [29]. *Aspergillus niger* (*A. niger*) culture filtrate was used to make CNPs (CeO_2 NPs) as reported by Gopinath et al. The fungi, *A. niger*, was cultured and 100 mL of the fungal filtrate was added to 3.72 g of $CeCl_3.H_2O$. For 4–6 hours at 80°C, this solution was regularly stirred. After constant stirring in the flask, a white precipitate appeared which later turned yellowish brown. The precipitate was calcined for 2 hours at 350°C and 400°C. Finally, CeO_2 NPs nanopowders were obtained. The cubic structure and spherical form of the mycosynthesized CeO_2 NPs were identified by characterization, with particle sizes ranging from 5 to 20 nm. The NPs were tested for antibacterial efficacy against Gram-positive bacteria and two Gram-negative (G-) bacteria [47]. Darroudi et al. reported long chain starch facilitated green synthesis of cerium NPs using sol-gel method. The precursor salt cerium nitrate hexahydrate was mixed with starch and precipitated with ammonium hydroxide. The reduced sample was made crystalline by calcination at a temperature range of 120–600°C showing a crystal size of 6 nm. The cytotoxicity study performed by the author shows dose-dependent nontoxic effects in the range of 175 mg/mL [48].

10.2.3 POLYMER-MEDIATED

A simple green approach for monodispersed CNPs (CeO_2-NPs) fabrication was described by Kargar et al. With the use of agarose matrices which is nontoxic in nature, the CeO_2-NPs were grown through calcination at different temperature. The NPs were successfully characterized using FESEM, PXRD, FTIR, TGA/DTA, and UV–vis spectroscopy techniques. The authors explain *in vitro* cytotoxicity studies on L929 cells, a nontoxic effect in all concentrations (up to 800 µg/mL). The authors also believe that the NPs have viable applications in different fields of medicine [49].

Pectin, a nontoxic biopolymer extracted from the seeds of Citrus maxima was used for the CeO_2 NPs synthesis. The phytochemicals that exist in pectin help in the process of synthesis of the nanostructure cerium. The analyses (XRD, FESEM, and DLS) of the NPs revealed it to be spherically shaped with polydispersed having a particle size of less than 40 nm. The DPPH scavenging activity study of CeO_2 NPs showed a change in color indicating free radical scavenging activity. The antibacterial and cytotoxicity studies done by the authors also show positive results [39].

10.2.4 NUTRIENT-MEDIATED

The fresh egg white was used for green synthesis of CeO_2 with size being controlled. The low-cost materials with the bio-directed green synthesis method produce NPs of 25 nm as exhibited by the FESEM images. The calcination temperatures for the crystallization of particles were 200, 400, 600, and 800°C [40]. In preparing CeO_2 NPs of approximately 5.0 g, 12.6 g of $Ce(NO_3)_3$ $6H_2O$ was dissolved in 30 mL of distilled water and stirred for 30 minutes. For a transparent honey solution, 25 g of honey was dissolved in 50 mL of distilled water and swirled for 15 minutes at room temperature. Following that, the cerium nitrate solution was added to the honey solution, and the container was placed in an oil bath at 60°C. The light yellow color resin was obtained after 6 hours of stirring, annealed at temperatures of 200, 400, 600, and 800°C, and then the products were kept for 2 hours at the specified temperatures in the air to obtain CeO_2-NPs. The synthesized CeO_2-NPs were spherical in shape, and FESEM imaging indicated the formation of NPs in size of about 23 nm. The fluorite cubic structure of CeO$_2$-NPs was revealed by preferential orientation at (111) reflection plane as analyzed by XRD [31].

10.3 ELECTROCHEMICAL SENSING APPLICATIONS

The oxide of cerium (CeO$_2$) represents the cubic fluorite structure with oxygen deficiencies [2] that provide redox reaction sites of cerium NPs. This property results in it being highly conductive and hence making it to be suitably used in an electrochemical sensor. The other characteristics that allow CNP to use in electrochemical sensors include catalytic activity, surface coating, surface reactivity, etc. [50]. The functions, such as transduction, enhancement of electrochemical signals, catalyst, etc., are carried out by NPs while used in a sensor [51]. Besides, a transition of cerium in its two oxidation states Ce^{3+} and Ce^{4+} (Figure 10.2) improves the electrochemical sensing property of ceria NPs.

10.3.1 ELECTROCHEMICAL SENSOR

In electrochemical sensors, the sensing element is used for the recognition of the analyte, and an electrode is used as a transducer that converts the analyte recognition system to some measurable electrical output. The NPs used in the electrode surface increase its surface area and hence increase the performance of the sensor system.

Based on the property such as catalytic activity of CNPs, H$_2$O$_2$ in the living cell has been detected by means of an electrochemical sensing technique [52]. By the wet-chemical deposition, CNPs were synthesized. This inexpensive, highly sensitive, simple electrochemical sensor contains Au microelectrodes integrated with ceria NPs over the glass substrate. The sensor shows the limit of detection (LOD) of 20 nm with a 226.4 µA·cm^{-2}·µM^{-1} of sensitivity showing it to be promising in sensing along with the diagnosis. The sensitivity of CeO$_2$ based sensor for H$_2$O$_2$ detection was enhanced by about two times when integrated with single-walled carbon nanohorns (SWCNHs). The transition in oxidation states in cerium oxide catalyzed the electro-reduction of H$_2$O$_2$. Excellent selectivity and flexibility with reproducibility were shown in the sensor [53]. Jiang et al. reported electrochemical-based sensing through CeO$_2$ and the gold NPs (AuNPs) fabricated on glassy carbon electrodes (GCE) in determining the concentration of glucose. Here the cerium and gold NPs facilitate the system to simulate electron transfer of enzyme and glucose oxidase.

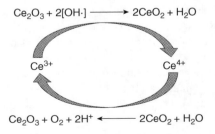

FIGURE 10.2. Transition of oxidation states of cerium.

The sensor exhibited reproducibility, selectivity, and stability with LOD of 2.86 × 10^{-3} mM [54]. Glucose was also quantified using an electrochemical sensor where C/CeO_2 was coated over the surface GCE. LOD of 0.8 mM and 1.8–2.0 mM range for glucose detection was observed with this C/CeO_2-based device [55]. Li et al. reported nonenzymatic glucose detection. The author used the electrode constructed based on nanorods of CeO_2 implanted in the matrix of nickel hydroxide ($Ni(OH)_2$). The response of the composite materials was improved by the effect called synergetic effect between the materials CeO_2 and $Ni(OH)_2$. For the sensor, the detection limit of 1.13 μM was reported and detection was in the range of 2 μM to 6.62 mM and a 594 μA mM^{-1} cm^{-2} of sensitivity. Less than a 5-second response time was exhibited by the sensor Senor [56].

Lactate acts a key role in the human body, especially the sports persons where lactate is related to hypoxia, heart problems, and muscle fatigue of the body. Hence, determining lactate concentration in the human body is important in the diagnosis of the same. Nesakumar et al. fabricated an electrochemical sensor with nicotin-amide adenine dinucleotide (NADH) and lactate dehydrogenase (LDH) with nano-structures of $-CeO_2$ over GCE bioelectrode for determining lactate concentration in blood samples of human. Cerium NPs were synthesized using a hydroxide-mediated approach. NADH and LDH were immobilized on GCE where CeO_2 NPs act as an interface. The amperometric study was carried out using three electrode systems (working electrode as $NADH/LDH/NanoCeO_2/GCE$, the electrode as Ag/AgCl (reference) with 0.1 M KCl and Pt wire (counter electrode)). From 0.2 to 2 mM linearity having a 4-second reaction time was reported for the sensor [23].

CNPs with graphene oxide (CeNP/GO) composite were coated over screen-printed carbon electrode. The electrode was used for hydroxyl radicals detection. Following the Fenton reaction, the interaction of composite with the analyte cyclic voltammetry (CV) was used. Different sizes of Cerium NPs were examined with different concentrations of cerium NPs on the composite. LOD of 0.085 mM for 8 nm cerium NPs was reported [57]. CNPs altered graphite paste electrode ($CeO_2@GP$) was used for formalin (FAL) detection in mushrooms. The electrochemical parameters of the fabricated sensor were examined by CV. It was further tested with differential pulse voltammetry (DPV). The quantification range for the sensor was 25 μM–1 mM with LOD of 0.1 μM. The electrochemical device exhibited good repeatability, selectivity, reproducibility, and stability [58]. Lavanya et al. worked on copper-doped cerium NPs coated on GCE ($Cu-CeO_2/GCE$) for the detection of purine (GU and AD) and pyrimidine which are essential DNA constituents. Voltammetric techniques were used to study the electrolytic activities of DNA using the buffer solution (PBS, pH 7.0). The mediator was not being employed. In the concentration range of 0.1–500 μM for AD the purine and pyrimidine bases exhibited linear characteristics. Similarly, for GU, TY, and CY, it showed linear characteristics in the range of 1–650 μM, 1–300 μM, and 1–250 μM respectively. LOD values for GU, AD, CY, and TY of 0.021, 0.031, 0.024, and 0.038 μM respectively [59]. The nanoceria modified Carbon Paste Electrode (CPE) exhibited a strong peak when cyclic voltammograms were done in comparison to bare CPE when used for detection of dopamine. Increasing dopamine showed that anodic peak current also increased simultaneously [22].

Green Synthesis of Cerium Oxide Nanoparticles

Tryptophan in food was detected by an electrochemical method where the electrode was loaded with cerium oxide (CeO_2) and reduced graphene oxide (RGO) composite. The CV and DPV were used for electrochemical analysis in which the electrode showed a high degree of selectivity and reproducibility. The sensor attained a linear range of 0.2–25 µM and a LOD of 80 nm [60].

The 5-aminosalicylic acid (Mesalamine, MES) which is anti-inflammatory was being electrochemically sensed by cerium oxide/tin oxide (CeO_2/SnO_2) composite-based electrode. The GCE modified by CeO_2/SnO_2 NPs revealed an exceptional electrocatalytic activity in terms of higher anodic peak current and lower peak potential. The LOD of 0.006 µM and range (linear) of 0.02–1572 µM were achieved by the electrochemical sensor [61]. Anh et al. developed a microelectrode coated with CeO_2 NPs and polypyrrole (PPy) with cholesterol oxidase (ChOx) immobilized on it for detection of cholesterol. This ChOx/CeO_2 NPs/PPy/electrode showed linearity in the range from 50 to 500 mg/dL. The sensitivity of $5:7 \times 10^{-6}$ mA/mg.dL^{-1} was archived for the microelectrode [62]. Tributyrin was detected using lipase-immobilized CeO_2 deposited in Indium tin oxide-coated glass electrode. The CV results exhibited good linearity (50–500 mg/dL) with good shelf life. LOD for the sensor was 32.8 mg/dL and the bioelectrode specified a high affinity of lipase toward tributyrin [63]. GCEs modified with cerium NPs – carbon nanotubes were used for finding out dopamine (DA), ascorbic acid, uric acid, and acetaminophen. The presence of CNT rises the area (surface) of the electrode and that in turn grows the electron transfer. Electrochemical impedance spectroscopy (EIS) and DPV were used for electrochemical analysis which showed a linear behavior for all the analytes. LOD for DA is 3.1 nm, for uric acid 2.4 nm, for ascorbic acid 2.6 nm, and 4.4 nm LOD were observed [64]. In the process of detecting hydroquinone (HQ) and catechol (CC) carbon (mesoporous) and cerium NPs composites (MC–CeNPs) adapted GCEs were fabricated. The developed sensor showed high sensitivity and selectivity [65].

By coprecipitation method, cerium oxide with cobalt doped were prepared for detection of hydrazine electrochemically. The oxidation states transition, ionic mobility, and vacancy in oxygen make the CeO_2-Co a prominent electrocatalytic oxidant. CV was equivalent to impedance spectra and the sensor shows a lower LOD value of 7.2 µM. The detection by the sensor was in the range from 7.18 to 1000 µM and exhibited good sensitivity, selectivity, and reproducibility in fast-detecting hydrazine [66]. The food-borne mycotoxin was detected by CNPs with chitosan (CH)-based biocomposite. The nanocomposite was coated onto ITO glass substrate. The electroactive surface area of the electrode is increased with the use of nanoceria which consequences in the high loading of r-IgGs. BSA/r-IgGs/CH-NanoCeO$_2$/ITO. The linearity was enhanced to 0.25–6.0 ng/dL with the LOD of 0.25 ng/dL. The device showed good sensitivity of 18 µA/ng dL^{-1} cm^{-2} and fast response [67]. The Carbon Paste Electrode-CMCPE, which was modified by cerium oxide, was used as an electrochemical sensor for caffeine detection. CV and DPV analysis reveal sensitive quantification of caffeine. The range of detection is 5–80 µM. The device also showed good reproducibility and high catalytic activity. 0.036 µM of LOD was found for the device [68]. A composite (nano) of copper–ceria ($CuO–CeO_2$) was formed using the calcination process of copper/cerium Cu(II)/Ce(III) for the detection of insecticide malathion [69].

186 Handbook of Sustainable Materials

Toxic methyl orange was detected by GCE electrode based on CeO_2/graphene oxide (GO) – polylactic (PLA) composite. CV analysis and impedance measurement for the composite were studied which imply it to be a suitable electrode material that helps in the process of removing MO dye from wastewater [70]. The redox reaction Fe $(CN)_6^{3-/4-}$ was used in the process of screen-printed carbon electrode (SPCE) incorporated with polyacrylic acid that is coated with nanoceria (PAA–CNPs). The PAA–CNPs on the electrode surface create an oxidative effect which is high for Fe $(CN)_6^{4-}$ was observed [71]. Saraf et al. fabricated an electrochemical transistor for the quantification of limonin, a biomolecule responsible for delayed bitterness in citrus fruits. The ultralow concentration of limonin was detected by the transistor which is functionalized with cerium NPs. The detection limit of 10 nm was perceived as having a high sensitivity of 10 μA/μM. An abnormally high concentration of Limonin causes greening disease in citrus fruits and hence detection and reduction of this biomolecule in an early stage will improve the harvesting of the same [72]. The comparative study for the recent electrochemical sensors that use cerium NPs has been listed in Table 10.3.

TABLE 10.3
Recent Applications of Cerium Oxide Nanoparticles in Electrochemical Sensing

Electrode/ Electrode with Alteration	Target Element	Synthesis Method	Characteristics	Ref.
Au-CeO$_2$	H$_2$O$_2$	Cerium oxide NSs were prepared by a wet-chemical deposition method	LOD: 20 nm and sensitivity: 226.4 μA·cm^{-2} μM^{-1}	[52]
SWCNHs with CeO$_2$	H$_2$O$_2$	-	Excellent selectivity, flexibility, reproducibility, stable	[53]
CeO$_2$ + (Au NPs) on GCE electrode	Glucose	-	Detection limit: 2.86×10^{-3} mM linear range: 0.02–0.6 mM	
C/CeO$_2$-GCE	Glucose	Solvothermal	LOD: 0.8 mM with range between 1.8 and 2.0 mM	[55]
NADH/LDH/ Nano-CeO$_2$/GCE	Lactate	Hydroxide-mediated approach		[23]
CeO$_2$ and Ni(OH)$_2$	Detection of glucose nonenzymatic	-	Detection limit: 1.13 μM, range: 2.0 μM to 6.620 mM, sensitivity: 594 microA. mM^{-1} cm^{-2}	[56]

(Continued)

Green Synthesis of Cerium Oxide Nanoparticles

TABLE 10.3 (*Continued*)
Recent Applications of Cerium Oxide Nanoparticles in Electrochemical Sensing

Electrode/ Electrode with Alteration	Target Element	Synthesis Method	Characteristics	Ref.
Screen-printed carbon electrode-CeNP/GO	OH group	Precipitation method	LOD: 0.085 mM	[57]
CeO$_2$@GP	Formalin	Sol-gel	Detection range: 25 µM–1 mM, LOD of 0.1 µM. good repeatability, selectivity, reproducibility and stability	[58]
Cu-CeO$_2$/GCE)	Purine and pyrimidine	-	LOD values for GU, AD, CY, and TY of 0.021, 0.031, 0.024, and 0.038 µM respectively	[59]
The nanoceria modified Carbon Paste Electrode (CPE)	Dopamine	-	Oxidation peak +0.2 V, +0.13 V reduction peak for nanoceria modified CPE	[22]
CeO$_2$/RGO	Tryptophan	Hydrothermal	Range: 0.2–25 µM, LOD: 80 nm	[60]
(CeO$_2$/SnO$_2$)-GCE	5-aminosalicylic acid		Detection limit of 0.006 µM, linear range 0.02–1572 µM	[61]
ChOx/CeO$_2$ NPs/ PPy	Cholesterol	-	Sensitivity of 5.7 × 10^{-6} mA/ mg.dL−1	[62]
GCE-CeO$_2$-CNT	DA, uric acid, ascorbic acid, and acetaminophen	-	LOD (DA): 3.1 nm, for DA, LOD (ascorbic acid): 2.6 nm LOD (uric acid): 2.4 nm, and LOD (acetaminophen): 4.4 nm	[64]
Co-CeO$_2$	Hydrazine	Coprecipitation process	LOD: 7.2 lM. Detection range : 7.18–1000 lM good sensitivity, selectivity, reproducibility	[66]
CH-NanoCeO$_2$/ITO	Mycotoxin	-	Linearity 0.25–6.0 ng/dL, LOD: 0.25 ng/dL. sensitivity: 18 µA/ng dL^{-1} cm^{-2} with fast response	[67]
Cerium Oxide Modified Carbon Paste Electrode-CMCPE	Caffeine	-	Range of detection: 5–80 µM, LOD: 0.036 µM	[68]

LOD: Limit of detection.

10.4 CONCLUSION AND PROSPECTS

Compare to the conventional process of synthesis of CNPs, green synthesis approaches are safer, eco-friendly, and low-cost. The plant active materials, such as phenols, flavonoids, terpenoids, etc., help in the bioreduction process during the synthesis of NPs. The biosynthesized cerium NPs were characterized for the study of their properties like morphology, size, shape, crystallinity, etc. using different analysis methods such as UV-VIS, FTIR, Raman spectroscopy, photoluminescence, SEM, FESEM, TEM, HERTEM, XRD, XPS, TGA, and DTA/DTG. The CeO_2 NPs show much attraction while using electrochemical sensing as compared to other metal oxide NPs due to having features such as oxygen vacancies, nontoxicity, mechanical strength, stability, etc. This chapter described recently applied biosynthesis methods for the green synthesis of CNPs. It also incorporates other activities/functions like antioxidant and antibacterial properties of nanoceria. The chapter listed the recent development of electrochemical sensors based on CNPs. Though other nanomaterials are applied for electrochemical sensing, the biocompatibility of CNPs immobilizes the biomolecules through electrostatic interaction requiring low isoelectric points. CNPs thus applied as electrochemical sensors enhance the performance of the sensor by increasing its parameters like sensitivity, reproducibility, etc. Focus can be given to studying the nanotoxicological information of the particles. Many biomedical applications can be performed with these NPs based on the study done by the researchers. The electrochemical sensing applications studied by many researchers can have the potential to extend their application in the field of agriculture also.

Funding:

This review did not receive any specific grant from any funding agencies.

Conflicts of interest:

Authors declare no conflict of interest in doing this review.

Acknowledgments:

The authors would like to extend their gratitude to the Department of Electronics and Communication Engineering for providing facilities to prepare this review.

REFERENCES

1. Korsvik, C., et al., *Superoxide dismutase mimetic properties exhibited by vacancy engineered ceria nanoparticles*. Chemical Communications, 2007. (10): pp. 1056–1058.
2. Singh, K.R., et al., *Cerium oxide nanoparticles: properties, biosynthesis and biomedical application*. RSC Advances, 2020. **10**(45): pp. 27194–27214.
3. Rico, C.M. and S. Majumdar, *J. l. Gardea-Toressdey*. Journal of Agriculture And Food Chemistry, 2011. **59**(8): pp. 3485–3498.
4. Fard, J.K., S. Jafari, and M.A. Eghbal, *A review of molecular mechanisms involved in toxicity of nanoparticles*. Advanced Pharmaceutical Bulletin, 2015. **5**(4): p. 447.
5. Trovarelli, A., *Catalytic properties of ceria and CeO_2-containing materials*. Catalysis Reviews, 1996. **38**(4): pp. 439–520.
6. Wang, C.-H. and S.-S. Lin, *Preparing an active cerium oxide catalyst for the catalytic incineration of aromatic hydrocarbons*. Applied Catalysis A: General, 2004. **268**(1-2): pp. 227–233.

Green Synthesis of Cerium Oxide Nanoparticles

7. Das, S., et al., *Cerium oxide nanoparticles: applications and prospects in nanomedicine*. Nanomedicine, 2013. **8**(9): pp. 1483–1508.

8. Liying, H., et al., *Recent advances of cerium oxide nanoparticles in synthesis, luminescence and biomedical studies: a review*. Journal of Rare Earths, 2015. **33**(8): pp. 791–799.

9. Arumugam, A., et al., *Synthesis of cerium oxide nanoparticles using Gloriosa superba L. leaf extract and their structural, optical and antibacterial properties*. Materials Science and Engineering: C, 2015. **49**: pp. 408–415.

10. George, S., et al., *Use of a rapid cytotoxicity screening approach to engineer a safer zinc oxide nanoparticle through iron doping*. ACS Nano, 2010. **4**(1): pp. 15–29.

11. Soren, S., et al., *Antioxidant potential and toxicity study of the cerium oxide nanoparticles synthesized by microwave-mediated synthesis*. Applied Biochemistry and Biotechnology, 2015. **177**(1): pp. 148–161.

12. Lee, S.S., et al., *Antioxidant properties of cerium oxide nanocrystals as a function of nanocrystal diameter and surface coating*. ACS Nano, 2013. **7**(11): pp. 9693–9703.

13. Terribile, D., et al., *The synthesis and characterization of mesoporous high-surface area ceria prepared using a hybrid organic/inorganic route*. Journal of Catalysis, 1998. **178**(1): pp. 299–308.

14. Lee, J.-S. and S.-C. Choi, *Crystallization behavior of nano-ceria powders by hydrothermal synthesis using a mixture of H_2O_2 and NH_4OH*. Materials Letters, 2004. **58**(3-4): pp. 390–393.

15. Yin, L., et al., *Sonochemical synthesis of cerium oxide nanoparticles—effect of additives and quantum size effect*. Journal of Colloid and Interface Science, 2002. **246**(1): pp. 78–84.

16. Hayes, B.L., *Recent advances in microwave-assisted synthesis*. Aldrichimica Acta, 2004. **37**(2): pp. 66–77.

17. Oh, H. and S. Kim, *Synthesis of ceria nanoparticles by flame electrospray pyrolysis*. Journal of Aerosol Science, 2007. **38**(12): pp. 1185–1196.

18. Nadeem, M., et al., *The current trends in the green syntheses of titanium oxide nanoparticles and their applications*. Green Chemistry Letters and Reviews, 2018. **11**(4): pp. 492–502.

19. Srikar, S., et al., *Green synthesis of silver nanoparticles: a review*. Green and Sustainable Chemistry, 2016. **6** (2016): pp. 34–56.

20. Kulkarni, N. and U. Muddapur, *Biosynthesis of metal nanoparticles: a review*. Journal of Nanotechnology, 2014. **2014**.

21. Khan, S.B., et al., *Exploration of CeO_2 nanoparticles as a chemi-sensor and photocatalyst for environmental applications*. Science of the Total Environment, 2011. **409**(15): pp. 2987–2992.

22. Njagi, J., et al., *Amperometric detection of dopamine in vivo with an enzyme based carbon fiber microbiosensor*. Analytical Chemistry, 2010. **82**(3): pp. 989–996.

23. Nesakumar, N., et al., *Fabrication of lactate biosensor based on lactate dehydrogenase immobilized on cerium oxide nanoparticles*. Journal of Colloid and Interface Science, 2013. **410**: pp. 158–164.

24. Andreescu, D., et al., *Applications and implications of nanoceria reactivity: measurement tools and environmental impact*. Environmental Science: Nano, 2014. **1**(5): pp. 445–458.

25. Xiao, Z., et al., *Plant-mediated synthesis of highly active iron nanoparticles for Cr (VI) removal: investigation of the leading biomolecules*. Chemosphere, 2016. **150**: pp. 357–364.

26. Milani, N., et al., *Dissolution kinetics of macronutrient fertilizers coated with manufactured zinc oxide nanoparticles*. Journal of Agricultural and Food Chemistry, 2012. **60**(16): pp. 3991–3998.

27. Giraldo, J.P., et al., *Plant nanobionics approach to augment photosynthesis and bio-chemical sensing.* Nature Materials, 2014. **13**(4): pp. 400–408.
28. Korotkova, A.M., et al., *"Green" synthesis of cerium oxide particles in water extracts petroselinum crispum.* Current Nanomaterials, 2019. **4**(3): pp. 176–190.
29. Khan, S.A. and A. Ahmad, *Fungus mediated synthesis of biomedically important cerium oxide nanoparticles.* Materials Research Bulletin, 2013. **48**(10): pp. 4134–4138.
30. Kannan, S. and M. Sundrarajan, *A green approach for the synthesis of a cerium oxide nanoparticle: characterization and antibacterial activity.* International Journal of Nanoscience, 2014. **13**(03): pp. 1450018.
31. Darroudi, M., et al., *Food-directed synthesis of cerium oxide nanoparticles and their neurotoxicity effects.* Ceramics International, 2014. **40**(5): pp. 7425–7430.
32. He, H.-W., et al., *Synthesis of crystalline cerium dioxide hydrosol by a sol-gel method.* Ceramics International, 2012. **38**: pp. S501–S504.
33. Song, H., et al., *Aerosol-assisted MOCVD growth of Gd_2O_3-doped CeO_2 thin SOFC electrolyte film on anode substrate.* Solid State Ionics, 2003. **156**(3-4): pp. 249–254.
34. Godinho, M., et al., *Room temperature co-precipitation of nanocrystalline CeO_2 and Ce0. 8Gd0. 2O1. 9−δ powder.* Materials Letters, 2007. **61**(8-9): pp. 1904–1907.
35. Hu, J.-d., et al., *Preparation and characterization of ceria nanoparticles using crystalline hydrate cerium propionate as precursor.* Materials Letters, 2007. **28**(61): pp. 4989–4992.
36. Jimmy, C.Y., L. Zhang, and J. Lin, *Direct sonochemical preparation of high-surface-area nanoporous ceria and ceria–zirconia solid solutions.* Journal of Colloid and Interface Science, 2003. **260**(1): pp. 240–243.
37. Zhang, H., et al., *Nano-CeO_2 exhibits adverse effects at environmental relevant concentrations.* Environmental Science & Technology, 2011. **45**(8): pp. 3725–3730.
38. Sebastiammal, S., et al., *Chemical and sweet basil leaf mediated synthesis of cerium oxide (CeO_2) nanoparticles: antibacterial action toward human pathogens.* Phosphorus, Sulfur, and Silicon and the Related Elements, 2022. **197**(3): pp. 237–243.
39. Patil, S.N., et al., *Bio-therapeutic potential and cytotoxicity assessment of pectin-mediated synthesized nanostructured cerium oxide.* Applied Biochemistry and Biotechnology, 2016. **180**(4): pp. 638–654.
40. Kargar, H., H. Ghazavi, and M. Darroudi, *Size-controlled and bio-directed synthesis of ceria nanopowders and their in vitro cytotoxicity effects.* Ceramics International, 2015. **41**(3): pp. 4123–4128.
41. Arunachalam, T., M. Karpagasundaram, and N. Rajarathinam, *Ultrasound assisted green synthesis of cerium oxide nanoparticles using Prosopis juliflora leaf extract and their structural, optical and antibacterial properties.* Materials Science-Poland, 2017. **35**(4): pp. 791–798.
42. Priya, G.S., et al., *Biosynthesis of cerium oxide nanoparticles using Aloe barbadensis miller gel.* International Journal of Scientific and Research Publication, 2014. **4**(6): pp. 199–224.
43. Sebastiammal, S., et al., *Green synthesis of cerium oxide nanoparticles using aloevera leaf extract and its optical properties.* Songklanakarin Journal of Science & Technology, 2021. **43**(2).
44. Thovhogi, N., et al., *Nanoparticles green synthesis by Hibiscus sabdariffa flower extract: main physical properties.* Journal of Alloys and Compounds, 2015. **647**: pp. 392–396.
45. Yulizar, Y., et al., *Datura metel L. Leaves extract mediated CeO_2 nanoparticles: synthesis, characterizations, and degradation activity of DPPH radical.* Surfaces and Interfaces, 2020. **19**: pp. 100437.
46. Muthuvel, A., et al., *Green synthesis of cerium oxide nanoparticles using Calotropis procera flower extract and their photocatalytic degradation and antibacterial activity.* Inorganic Chemistry Communications, 2020. **119**: pp. 108086.

Green Synthesis of Cerium Oxide Nanoparticles

47. Gopinath, K., et al., *Mycogenesis of cerium oxide nanoparticles using Aspergillus niger culture filtrate and their applications for antibacterial and larvicidal activities.* Journal of Nanostructure in Chemistry, 2015. **5**(3): pp. 295–303.
48. Darroudi, M., et al., *Green synthesis and evaluation of metabolic activity of starch mediated nanoceria.* Ceramics International, 2014. **40**(1): pp. 2041–2045.
49. Kargar, H., F. Ghasemi, and M. Darroudi, *Bioorganic polymer-based synthesis of cerium oxide nanoparticles and their cell viability assays.* Ceramics International, 2015. **41**(1): pp. 1589–1594.
50. Hartati, Y.W., et al., *Synthesis and characterization of nanoceria for electrochemical sensing applications.* RSC Advances, 2021. **11**(27): pp. 16216–16235.
51. Huang, H. and J.-J. Zhu, *The electrochemical applications of rare earth-based nanomaterials.* Analyst, 2019. **144**(23): pp. 6789–6811.
52. Alizadeh, N., et al., *Intrinsic enzyme-like activities of cerium oxide nanocomposite and its application for extracellular H_2O_2 detection using an electrochemical microfluidic device.* ACS Omega, 2020. **5**(21): pp. 11883–11894.
53. Bracamonte, M.V., et al., *H_2O_2 sensing enhancement by mutual integration of single walled carbon nanohorns with metal oxide catalysts: the CeO_2 case.* Sensors and Actuators B: Chemical, 2017. **239**: pp. 923–932.
54. Jiang, L., et al., *A non-enzymatic nanoceria electrode for non-invasive glucose monitoring.* Analytical Methods, 2018. **10**(18): pp. 2151–2159.
55. Meng, F., et al., *The synthesis of carbon/cerium oxide composites clusters with the assistance of the glucaminium-based surfactant and their electrochemical performance in the glucose monitoring.* Journal of Alloys and Compounds, 2017. **713**: pp. 125–131.
56. Li, Y., et al., *CeO_2 nanorods embedded in $Ni(OH)_2$ matrix for the non-enzymatic detection of glucose.* Nanomaterials, 2017. **7**(8): pp. 205.
57. Duanghathaipornsuk, S., et al., *The effects of size and content of cerium oxide nanoparticles on a composite sensor for hydroxyl radicals detection.* Sensors and Actuators B: Chemical, 2020. **321**: pp. 128467.
58. Nag, S., et al., *A simple nano cerium oxide modified graphite electrode for electrochemical detection of formaldehyde in mushroom.* IEEE Sensors Journal, 2021. **21**(10): pp. 12019–12026.
59. Lavanya, N., J.N. Claude, and C. Sekar, *Electrochemical determination of purine and pyrimidine bases using copper doped cerium oxide nanoparticles.* Journal of Colloid and Interface Science, 2018. **530**: pp. 202–211.
60. Zhang, J.-W. and X. Zhang, *Electrode material fabricated by loading cerium oxide nanoparticles on reduced graphene oxide and its application in electrochemical sensor for tryptophan.* Journal of Alloys and Compounds, 2020. **842**: pp. 155934.
61. Sukanya, R., et al., *Ultrasound treated cerium oxide/tin oxide (CeO_2/SnO_2) nanocatalyst: a feasible approach and enhanced electrode material for sensing of anti-inflammatory drug 5-aminosalicylic acid in biological samples.* Analytica Chimica Acta, 2020. **1096**: pp. 76–88.
62. Anh, T.T.N., et al., *Cerium oxide/polypyrrole nanocomposite as the matrix for cholesterol biosensor.* Advances in Polymer Technology, 2021. **2021**.
63. Solanki, P.R., et al., *Nanostructured cerium oxide film for triglyceride sensor.* Sensors and Actuators B: Chemical, 2009. **141**(2): pp. 551–556.
64. Iranmanesh, T., et al., *Green and facile microwave solvent-free synthesis of CeO_2 nanoparticle-decorated CNTs as a quadruplet electrochemical platform for ultrasensitive and simultaneous detection of ascorbic acid, dopamine, uric acid and acetaminophen.* Talanta, 2020. **207**: pp. 120318.

65. Liu, D., et al., *Mesoporous carbon and ceria nanoparticles composite modified electrode for the simultaneous determination of hydroquinone and catechol.* Nanomaterials, 2019. **9**(1): pp. 54.

66. Ansari, A.A. and M. Alam, *Electrochemical sensitive detection of hydrazine through cobalt-doped cerium oxide nanostructured platform.* Journal of Materials Science: Materials in Electronics, 2021. **32**(10): pp. 13897–13905.

67. Kaushik, A., et al., *Cerium oxide-chitosan based nanobiocomposite for food borne mycotoxin detection.* Applied Physics Letters, 2009. **95**(17): pp. 173703.

68. Santhosh, B., et al., *Electrochemical investigation of caffeine by cerium oxide nanoparticle modified carbon paste electrode.* Journal of the Electrochemical Society, 2020. **167**(4): pp. 047503.

69. Xie, Y., et al., *A CuO-CeO$_2$ composite prepared by calcination of a bimetallic metal-organic framework for use in an enzyme-free electrochemical inhibition assay for malathion.* Microchimica Acta, 2019. **186**(8): pp. 1–9.

70. Mohammad, F., T. Arfin, and H.A. Al-lohedan, *Enhanced biosorption and electrochemical performance of sugarcane bagasse derived a polylactic acid-graphene oxide-CeO$_2$ composite.* Materials Chemistry and Physics, 2019. **229**: pp. 117–123.

71. Iglesias-Mayor, A., et al., *Nanoceria quantification based on its oxidative effect towards the ferrocyanide/ferricyanide system.* Journal of Electroanalytical Chemistry, 2019. **840**: pp. 338–342.

72. Saraf, N., et al., *Microsensor for limonin detection: an indicator of citrus greening disease.* Sensors and Actuators B: Chemical, 2019. **283**: pp. 724–730.

11 Self-lubricating Hybrid Metal Matrix Composite toward Sustainability

Sweta Rani Biswal and Seshadev Sahoo

CONTENTS

11.1 Introduction 193
11.2 Self-Lubricating Hybrid Metal Matrix Composite.................................. 195
     11.2.1 Tribological Aspects of Graphite and Graphene Reinforced Self-lubricating Hybrid Metal Matrix Composite 198
     11.2.2 Tribological Aspects of Dichalcogenides Reinforced Self-lubricating Hybrid Metal Matrix Composite........................ 199
     11.2.3 Tribological Aspects of Hexagonal Boron Nitride Reinforced Self-lubricating Hybrid Metal Matrix Composite.....200
11.3 Fabrication Route for Self-Lubricating Hybrid Metal Matrix Composite...204
11.4 Sustainability of Self-Lubricating Hybrid Metal Matrix Composite, its Future Prospective, and Challenges 205
11.5 Conclusion 206
References 207

11.1 INTRODUCTION

Global warming a serious environmental concern occurs from the depletion of natural resources in the earth's crust. That results in rapid and dramatic shifting toward a new world of energy and environmental sustainability for the paradigm of human scientific discovery and the advancement of human civilization. These necessities are not just a suitable and prompt response, but also one that can protect humanity against natural disasters. Environmental constraints and regulations are needs of the hour which currently pose a threat to the field of environmental engineering science. As a result, humankind is currently undergoing a period of intense scientific reflection and extensive scientific innovation. Thus, energy and environmental sustainability are critical to humanity's success today. The torchbearers of this treatise's scientific paradigm are green tribology's scientific success and deep scientific promise, as well as sustainability. The term "green tribology" was coined to describe all of the accomplishments in the study of zero-wear phenomenon and mechanism of super-antifriction, along with the creation of lubricants for practical applications [1, 2].

DOI: 10.1201/9781003297772-11

The new field of tribology which is been introduced as "Green tribology" is predicted to play a significant role in global scenarios to find the key solutions to resolve numerous problems such as environmental pollution, energy conservation, and minimize the use of environmental resources. Three important core topics of green tribology are (1) biomimetics (imitating living nature to address complicated human problems) and self-lubricating materials/surfaces; (2) biodegradable and environmental friendly lubrication and materials; and (3) renewable and/or sustainable energy resources [3]. Many biological materials offer incredible features like superhydrophobicity, self-cleaning, self-healing, high adhesion, reversible adhesion, high mechanical strength, antireflection, and so on, which is impossible to match by traditional engineering approaches. Solid lubricants are one among these self-lubricating materials, which offer self-lubricating behavior due to their structure and hierarchical multiscale organization. These structures have the ability to expand and adapt for providing natural self-healing or repair mechanisms due to minor frictional damages. Understanding their lubricating mechanisms has been a critical aspect in ensuring their sustainability in the field of automotive application. As a result, the biomimetic approach to green tribology holds a lot of promise.

In the era of formation of a new-age material, it is the time for the industry to abandon old technology and encourage innovation [4]. Development of materials, which creates their lubrication through the reinforcement of solid lubricant, popularly known as self-lubricating hybrid metal matrix composites (SLHMMCs). Solid lubricants [5–7], like 2D-molybdenum disulfide (MoS_2), 2D-hexagonal boron nitride (h-BN), 1D-graphene, and 2D-graphite, are used in severe circumstances like vacuum, radiation, pressure, and temperature to resolve the problem of direct application of lubricants. In these conditions, maintaining a homogeneous layer of solid lubricants between the moving parts is difficult. Self-lubricating composite (SLC) materials have been developed to address these problems. SLHMMCs materials are hybrid composite materials that also integrate a hard phase (normally ceramic) to improve the matrix's mechanical properties, such as strength and stiffness, in addition to soft reinforcement (solid lubricant) to induce self-lubrication. Metal and its alloy of aluminum, copper, nickel, magnesium, and silver are commonly utilized for the production of SLHMMCs. Matrix as well as reinforcements have impacts on the mechanical and tribological characteristics of SLHMMCs [8].

Considering the different aspects of SLHMMCs, the focus has been given to study the self-lubricating behavior of SLHMMCs as a function of matrix, reinforcements, its type, and concentration in the composite. Self-lubricating metal matrix composites (SLHMMCs) can be made using various fabrication techniques which have a greater influence on mechanical and tribological properties, i.e., discussed briefly in this chapter. This chapter highlights key points on tribological properties in association with SLHMMCs for their sustainability, eco-friendliness, and cost-effectiveness. Eventually, results and findings are offered to aid the researchers in their future endeavors. Moreover, this chapter focuses on the different aspects of SLHMMCs toward sustainability in the field of automotive sector as economical and energy-efficient solution. This research confirms the existence of a second generation of materials in the field of tribology.

11.2 SELF-LUBRICATING HYBRID METAL MATRIX COMPOSITE

In 21st century, SLHMMCs are popular as an advanced engineering material. These are not only dominating in the field like defense, marine, and sports but also captivate the major market of automobile and aerospace industries. The main characteristics of these materials are their excellent combination of properties such as ductility with strength, high wear resistance with self-lubricating property, toughness with high strength, and high superior thermal properties [9]. Lightweight is considered the most significant property in addition to enhanced mechanical properties with cost-effectiveness for automotive and aerospace applications. The three basic elements of SLHMMCs are matrix, primary reinforcement such as hard particle addition, and secondary reinforcement, i.e., solid lubricants. These elements are not only helping to enhance their mechanical and electrical properties but also help to increase the life of mechanical systems by enhancing their tribological properties. The combination of these elements induces the combination of properties that raise the demand for SLHMMCs. These composites are manmade composites with the addition of more than one reinforcement in matrix, and they combinedly contribute their best attributes. Here, a strong structural matrix is carefully mixed with primary and secondary reinforcement to provide a lubricating phase.

The basic mechanism behind these SLHMMCs is represented through schematic representation, as shown in Figure 11.1, which describes the role of solid lubricant. The mechanism explains that whenever there is interaction between contacting surfaces, some materials get removed from interacting surface which as a result exposes the solid lubricant inside the metal matrix composite. Once the solid lubricant gets exposed to the rubbing surface that will create a lubrication film and provides lubrication, for example, gray cast iron which constitutes uniformly distributed graphite flakes acts as a classic example of a self-lubricating composite. For efficient lubrication, solid lubricant must have to adhere strongly to the contacting surfaces, which will enhance the product life. Also, these solid lubricants are soft in nature which will help to reduce friction but, in contrary, degrade the strength of the materials. To provide strength without compromising its tribological phenomenon, researchers have added hard particles as the primary reinforcement. The lubricating phase can be incorporated in a variety of ways. The most efficient and simple solution to maintain continuous lubrication is by introducing uniformly distributed solid lubricant particles or fibers scattered throughout the matrix. The quality of these sorts of composites is determined by the qualities of the particular matrix, primary

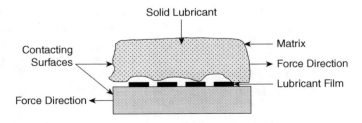

FIGURE 11.1 Schematic representation of self-lubricating mechanism.

reinforcement and solid lubricant, solid lubricant concentration, its distribution in matrix, and the interacting mechanism of matrix and its lubricating phase. Looking into the contribution of each element in SLHMMC, it is important to understand the individual contribution of each key element. As per the research, matrix is the base of the composite. So, it is important to understand the aspect of all available matrices. The popular matrices used for SLHMMCs are aluminum (Al) [6, 10, 11], copper (Cu) [12–14], magnesium (Mg) [15, 16], nickel (Ni) [17], etc.

A study on tribological properties of different matrices in addition to solid lubricants reveals the impact of Al, Mg, Ni, and Cu matrices in self-lubricating composites. The influence of tribological properties is analyzed through the calculation of "coefficient of friction (COF)" in variation with different wear parameters such as temperature, load, reinforcement concentration, speed, etc. The wear behaviors of those composites reveal that these matrices have good performance in wear resistance when added with reinforcement and also have a good impact in reference to tribological parameters variation. Kumar et al. [18] have found the fall of COF value in aluminum matrix composite at different reinforcement conditions, which is better than the wear characteristics of aluminum alloy. Investigation on nickel matrix by Zhu et al. [19] revealed that the addition of graphite particles improves friction and wear characteristics dramatically compared to pure nickel, at different temperature conditions. Similarly, a study on Cu matrix by Jamwal et al. [20] outperformed pure copper at different loading conditions, whereas, AZ91 magnesium alloy matrix, examined by Girish [21], has demonstrated the better wear characteristics and decreasing trend as compared to magnesium alloy. The studies on different matrices have suggested that Al, Mg, Cu, and Ni [13] are more compatible matrices for the solid lubricant for tribological applications.

Considering the huge impact of matrices on the tribological characteristics, aluminum and magnesium alloys as matrices are generally preferred as sustainable materials which can provide high-impact properties like low cost at low density with high strength-to-weight ratio [6, 22–25]. These are extensively used in advanced lightweight application which not only reduces the overall weight, but it also helps to decrease the fuel consumption efficiency and helps to increase greenhouse emissions for automotive and aerospace applications. As per the study, reduction of weight by 10% in lightweight vehicles like car or mini truck can reduce 6.9% of fuel consumption [26]. The study on self-lubricating composite [27] also reveals that CO_2 production will get reduced by 2000 kg in the mean lifetime of a vehicle. This characteristic of SLHMMCs provides sustainability in market as a substitute to other lubrication and also creates a huge demand for lightweight such as Mg- and Al-based composites with outstanding mechanical and tribological qualities in addition to additional benefits of higher fuel efficiency with lower pollution emissions. Aluminum and magnesium composites and hybrid composite materials are reinforced with diverse reinforcement materials, such as SiC [28, 29], Al_2O_3 [30, 31], and B_4C [9, 32], as primary reinforcement to meet the requirement for improvement of mechanical and tribological qualities. As a primary reinforcement, silicon carbide (SiC) at 3.18 g/cm^3 density and alumina (Al_2O_3) at 3.9 g/cm^3 density are popular for automotive and aerospace applications [13]. Particle addition of SiC and Al_2O_3 has a higher density than pure aluminum which makes the composite denser with the increase in reinforcement composition. Even modest amounts of ceramic reinforcements, like SiC [33] or Al_2O_3 [31], can increase mechanical as well as tribological characteristics significantly. Despite the benefits, preparing aluminum-ceramic composites is complex, while hard particle size, weight fraction, and concentration are having a significant changes in composite, the interface of matrix and reinforcing phase became more crucial. Due to weaker bonding, reinforcement is easily removed from matrix and treated as third-body abrasives which result in increasing wear rate. Furthermore, adding ceramic particles to the matrices improves the composite's hardness and makes machining more difficult and also does not solve the problem of frictional losses in frictional mechanical systems. In such cases, multiple reinforcements can be used to alleviate such difficulties. Since ceramic reinforcements are stronger than any other type of reinforcement, they are employed as the principal reinforcement in the production of hybrid composites [34]. The secondary reinforcements, i.e., solid lubricants, on the other hand, act as soft particle addition which makes the component machinable, as well as helps to reduce the frictional loss by inducing self-lubricating property in the hybrid composite, which helps to save fuel and provides sustainability to the mechanical system [35]. In addition to this, these characteristics of materials help to save money, life, and weight of mechanical systems. To provide self-lubrication, generally different types of solid lubricants are used and among them, some key types are "structural lubricants", "mechanical lubricants", "soaps", and "chemically active lubricants". The main focus of this chapter lies on the structural solid lubricant reinforced hybrid metal matrix composite. The most promising structural lubricants are one- and two-dimensional carbon atoms such as graphite (Gr) and graphenes (GNPs), two-dimensional MoS_2, tungsten disulfide, and h-BN, which are used for excellent wear resistance phenomenon. These layered structure materials consist of atoms with strong covalent bonds in the same plane and weaker atomic bonds known as Van der Waals bonds in the succeeding planes. Such designs allow consecutive planes to delaminate at relatively low-stress levels during rubbing action between surfaces that provides a continuous layer of lubricating film on the contacting surfaces, which helps to reduce wear during sliding as discussed in the earlier solid lubrication mechanism [35–37].

Looking into the significant impact of self-lubricating materials in the field of tribology, this chapter provides a deep review on different solid lubricants' self-lubricating behavior for their sustainability. The main emphasis is given to the tribological properties of carbon atoms of both one-dimensional (1D) graphene and two-dimensional (2D) graphite that are studied along with 2D-dichalcogenides and h-BN, and agro-waste materials reinforced self-lubricating hybrid composite are also investigated. Their sustainability in automotive and aerospace industries is adjudged through the wear performance under different environmental conditions. In this chapter, various parameters such as reinforcement content, shape, size, and fabrication method [24, 38], and properties evaluation parameters, i.e., applied load, sliding velocity [39], and service temperature conditions [9, 19, 40] are discussed to analyze the tribological characteristics.

11.2.1 Tribological Aspects of Graphite and Graphene Reinforced Self-lubricating Hybrid Metal Matrix Composite

As discussed earlier, graphite is one of the most common structural solid lubricants because of the 2D hexagonal system. The structure consists of carbon atoms that are linked tightly together in sheets but have low bond strength between sheets. Thus, graphite possesses low shearing strength under frictional force, which makes it the most promising solid lubricant for SLHMMC. Review on wear characteristics of Al and Mg matrix composites [9, 16, 26] have suggested that wear resistance gets enhanced within the specific limits with the addition of SiC, Al_2O_3, and B_4C. The increase in most significant wear parameters, like normal load, provides more contacting surfaces and leads to higher frictional loss due to high COF in hybrid composites. Abrasive wear dominates the wear mechanism for low normal loads. This study also advocates the contribution of different fabrication methods adopted in a graphite reinforced Al and self-lubricating aluminum hybrid composite and a self-lubricating magnesium hybrid composite where the liquid fabrication technique, stir casting (45%), dominates the fabrication technique followed by powder metallurgy (PM) (27%) and others. The most promising primary reinforcement is SiC (63%) with graphite followed by Al_2O_3 (16%) for Al- and Mg-based SLHMMCs.

The graphite reinforced self-lubricating hybrid composite (Al-SiC-Gr) [28] through the PM fabrication technique has been evaluated through parameters like applied loads at different sliding velocities to cover different sliding distance to analyze the wear performance. The hybrid composite shows self-lubricating behavior by forming Gr-rich tribolayers and helps to enhance its tribological properties, which can be a good substitute to Al MMCs in automotive parts. The wear mechanism of this hybrid composite is dominated by delamination and oxidative wear and forms smaller wear debris with the increase in SiC weight concentration. Likewise, for electrical applications like brushes (electrical sliding parts) in electrical generators, graphite-reinforced with Cu matrix composites are the most important self-lubricating materials having better electrical and thermal properties [41]. Self-lubricating Cu matrix hybrid composites with different reinforcements have been studied and their tribological performances with wear parameters are outstanding as compared to single reinforced Cu matrix composite [13, 42, 43]. The effects of CNTs and GNPs of Cu-based hybrid matrix composites have been thoroughly investigated. The inference made through this analysis is the tribological performance of hybrid Cu matrix is similar to Al- and Mg-based self-lubricating hybrid composites. In addition to the evaluation of tribological performance of Cu-based hybrid composite, Zhan et al. [44] have also studied the wear behavior in variation with temperature range (373–723 K). The dry sliding wear characteristics of single-reinforced and hybrid-reinforced Cu composites were examined and analyzed. The results conveyed that the hybrid composite has outperformed the single reinforced Cu composite at high temperature (723 K), with a more stable and lower friction coefficient. Cu-based hybrid composite might be used in high-temperature sliding wear applications. Further, investigation on self-lubricating behavior of hybrid reinforcements has been done on aluminum composite with variation in temperature [40]. The study reveals that the temperature and load are primary sources of change in wear resistance.

Self-lubricating Hybrid Metal Matrix Composite toward Sustainability 199

Vast research on graphite-reinforced hybrid composite has revealed that micronsize graphite particles [6] are generally used as solid lubricant in metal matrix composite, but the evolution of graphite in finer particle form, e.g., 1D multi-layer graphene (MLG) and carbon nanotubes (CNTs) open up new options to enhance the tribological performance of hybrid composite [45, 46]. Graphene (GPs) is having sheet-like morphology with tensile strength 130 (GPa), thermal conductivity (5.3 × 10^3 $Wm^{-1}K^{-1}$) and elastic modulus (0.5–1 TPa) [47]. Furthermore, graphite particles can be substituted by CNTs and GNPs, like hybrid reinforcements [48–51]. Graphene [49, 52, 53], as solid lubricant, has significant tribological benefits at very low addition (0.1–0.8 wt.%), but higher addition of Graphene (1.2 wt.%) may also cause severe wear loss due to agglomeration. Graphene has also a stronger interfacial bonding than CNTs and graphite [54] and has least tribological attributes [55].

A comparative study on Al-SiC-Gr and Al-SiC-GNPs [56] has been done by Zhang et al. The reinforcement of nano GNPs as solid lubricant indicates the improvement in wear resistance and friction coefficient. GNPs and SiC nanoparticles form core-shell structure results in less wear scars. Uniform distribution of GNPs in matrix is the major reason for the impressive tribological properties, which helps to create GNPs-rich self-lubricating films in the hybrid composite. A significant number of studies on tribological performance [18, 19, 36, 57] have also been undertaken on graphite and graphene-based Mg hybrid composites, which have a similar result as compared to other matrices which can be a good alternative to aluminum matrix self-lubricating composite.

11.2.2 Tribological Aspects of Dichalcogenides Reinforced Self-lubricating Hybrid Metal Matrix Composite

Two-dimensional (2D) transition metal dichalcogenides (TMDs) are the alternative to graphite in the field of tribology for dry and vacuum environmental conditions. They have outstanding lubricating characteristics due to their crystal structure, which are termed "frictionless", and because of low friction coefficient (0.001). Pure TMDs have a high-friction coefficient but are very sensitive in humid environment. TMDs can be made into different forms such as powder, burnished films, thin films, inorganic fullerene-like particles, or nanotubes for tribological applications [58]. TMDs are basically "MX_2 type", where "M" and "X" stand for transition metal atoms such as Mo, W, etc., and chalcogen atoms such as S, Se, or Te, respectively. Here, M atom of one layer is sandwiched between two layers of X atoms. These materials are popularly known as 2D materials [59]. Molybdenum, tungsten, and niobium disulfides, as well as diselenides, make up the TMD family [60]. MoS_2 is the most common TMD due to its low cost and outstanding tribological characteristics in both air and vacuum environment. However, when compared to MoS_2 or graphite, WS_2 has a greater oxidation temperature and thus has a wider effective lubricating range.

Many authors have reported the potentiality of dichalcogenides as a solid lubricant where graphite cannot be used for self-lubrication. As per the reports, MoS_2-reinforced Al-based hybrid composite exhibits enhanced wear and corrosion resistance properties in addition to CeO_2 [61], B_4C [62, 63], and SiC [64, 65]. As reported by Furlan et al., the value of COF gets reduced maximum by 63% [66].

From the wear characteristics study, Kumar et al. have found that [63] the better tribological performance of MoS_2 reinforced Al-based hybrid composite at varying temperature conditions. The study indicates that the gradual increase in wear rate is due to sliding velocity variation at a particular temperature, whereas it gradually decreases as the temperature rises from 50°C to 100°C at an interval of 25°C. It is related to the creation of an oxide layer on the surface at higher temperatures due to MoS_2 particles (solid-lubrication) addition as a solid lubricant which can resist wear and improve thermal stability. The wear mechanism reveals the presence of widest and deepest wear tracks at low temperatures whereas wear tracks dimensions are much smaller at elevated temperatures. Particulate microfracture is the reason for abrasive wear which is dominant wear mechanism. The aluminum hybrid composites are having smaller and smoother worn surface which results in better wear resistance as suggested by Raja et al. [62]. Likewise, the wear characteristics of MoS_2 hybrid reinforcement on Al7075 matrix also have better tribological characteristics due to smaller depth of deformation as compared to aluminum alloy with smoother wear surface which have a tremendous potential in automotive sector [64]. Due to outstanding lubricating characteristics at high loads and at temperatures up to 1100°C, for aluminum-reinforced hybrid composites, MoS_2 performs outstandingly as a dry solid lubricant. MoS_2 creates a continuous solid lubricating transfer film over the contact surfaces, which improves wear resistance [65]. There are a very few works reported in [14, 64] other matrices which have described the wear phenomenon in addition to MoS_2. Nautiyal et al. [14] have studied the Cu-based hybrid matrix composites and their wear characteristics in addition to MoS_2 solid lubricant fabricated through PM method. The hybrid combination shows better tribological properties. They have also mentioned to optimize the wear properties through the reinforcement concentration in matrix by controlling processing parameters. PM processed $AA7068/MoS_2/WC$ hybrid composite shows smaller wear debris with self-lubricating property due to uniform distribution of WC [67].

As like MoS_2, there are very less efforts in the invention of WS_2 reinforced hybrid composite. Only a few literatures have shown the self-lubricating behaviors and its automotive and aerospace applications. Wear mechanism of WS_2-reinforced hybrid composite is dominated by ploughing and delamination at higher percentage of WS_2 [68]. The study on WS_2 reinforcement in steel-based hybrid composite [69] also reveals that there is better tribological performance in a wide range of temperatures from room temperature (RT) to 800°C. WS_2 provides excellent lubricating characteristics at low temperatures from RT to 800°C when reinforced with ZNO in steel matrix, at minimum value COF, i.e., 0.19.

11.2.3 Tribological Aspects of Hexagonal Boron Nitride Reinforced Self-lubricating Hybrid Metal Matrix Composite

Hexagonal boron nitride reinforced hybrid composite is the substitute to generally used graphite and MoS_2 [70]. Like 2D dichalcogenides, BN also has a hexagonal sandwiched structure which can be easily sheared off during rubbing action. The most important characteristic of BN is the better self-lubricating behavior in humid conditions too where the performance of dichalcogenides and graphite is very poor [71].

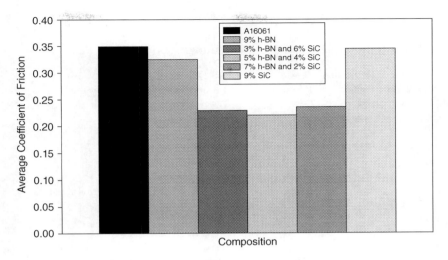

FIGURE 11.2 Wear performance of Al-SiC-hBN hybrid composite.

Its exceptional characteristic is the balance between lubricating properties and thermal stability. So, introduction of hexagonal BN provides a special platform for generating a new type of self-lubricating hybrid composite, which helps to sustain in the automotive industries. As reported by Sahoo et al. Al-hBN-SiC exhibits more wear resistance in comparison to single reinforced composite [72]. The wear characteristics of Al-hBN-SiC hybrid composite are represented in Figure 11.2. This figure also describes its tribological performance of different composites. Images of worn surfaces depict the presence of the narrow grooves in the parallel direction of sliding as described by the author [72]. Addition of SiC as primary reinforcement restricts the friction. Moreover, h-BN as solid lubricant creates a thin layer of solid lubrication and helps to reduce frictional agents in the composite [73].

Moreover, addition of BN in aluminum alloy with Al_2O_3 forms a hybrid composite which has reduced cutting forces by inducing self-lubricating nature through BN particles. Such types of composites are the future generation composites that can widely be used for machining operations by eliminating the vast use of external lubrication [74]. Some researchers have also reported the influence of h-BN on Cu-based hybrid composite. Tribological study on h-BN reinforced copper hybrid composite investigated by Chand et al. [75] in addition to another solid lubricant, i.e., graphite whose wear performance have represented with respect to wear load. The findings reveal the superiority of graphite as compared to h-BN as solid lubricant. A few studies on tribological aspects of SLHMMCs are discussed in Table 11.1.

Although there is a huge number of research, as listed in Table 11.1, based on synthetic self-lubricating composite, some have found the alternative to it. From the discussion on synthetic materials reinforced self-lubricating hybrid composite, impact on different aspects, like mechanical, machining, and tribological, gives a clear image of the significance of synthetic solid lubricant. The study found that synthetic solid lubricant-based hybrid composites require high cost-efficient fabrication route and technology, which is the main reason behind high-production cost, density,

TABLE 11.1
Tribological Performance of Self-Lubricating Hybrid Composite

Composition	Fabrication Method	*Wear Parameters	Tribological Performance	Applications	Ref.
Al + SiC + GNSs	Powder method (PM)	20 N, 43.96 mm/s, 43.96 m	GNSs-rich micro tribofilms reduced the wear rate as low as 0.0015 mm³/Nm for hybrid combination	Automotive industry	[56]
Fe-Si-C matrix + h-BN + graphite	PM	7 N at 10 min intervals	Decreasing h-BN proportion in enhanced tribological properties	Automotive parts	[76]
TiC-C/Cu	PM	15 N, 100 rpm, 25 m	Abrasive wear reason for COF reduction by 16% (CNTs) and 6% (graphene)	Electrical sliding parts	[41]
Cu-SiC-Gr	Solid state mixing	(10, 20, 30, 40, 50, 60 N), (500, 1000, 1500, 2000, 2500, 3000 m), (0.5, 1.0, 1.5, 2.0, 2.5 m/s)	Oxidative wear and delamination wear give better wear properties for hybrid composites	Bearings and brushes for engines, generators/motors, automobile electrical sliding parts	[77]
Al-Si alloy + B₄C + GNPs	Semi-PM	10–20–40 N loads, 0.08 m/s, 10 mm and 500 m	Adhesive and slight abrasive mechanisms were observed. 40% increase in wear resistance	Exhaust valves, bearing materials, and turbine blades	[47]
(Al-SiC(n) CNTs) and (Al-SiC(n) glassy carbon)	PM	10 N, 0.1 m/s, 100 m at 25°C up to 250 m at 450°C	Improved friction behavior at all temp	Automotive and aerospace parts	[78]
Mg + SiC + WS₂	PM, Spark plasma sintered	1, 4 N, at 22.5 mm/s, 4.5 mm, for 60 min, at room temperature and 100°C	Wear track is smoother, narrower in both temperature conditions Maximum COF reduced by 54.5%	Automobile, aerospace, defense, and telecommunication industries	[79]
Cu hybrid composite	Friction stirs processing	30 N, 1 m/s, 1500 m	Combination of adhesive and abrasive wear with low coefficient of friction	Electrical sliding parts	[42]

(Continued)

Self-lubricating Hybrid Metal Matrix Composite toward Sustainability

TABLE 11.1 (*Continued*)
Tribological Performance of Self-Lubricating Hybrid Composite

Composition	Fabrication Method	*Wear Parameters	Tribological Performance	Applications	Ref.
Mg + SiC + reduced graphene oxide	PM	5, 10, 15 N, 0.5, 1.0,1.5 m/s 500, 1000, 1500 m	Reinforcement helps to increase the coefficient of friction. Wear mechanism dominates with plastic deformations and delamination wear	Automobile, Mineral industry and aerospace applications	[80]

* Wear parameters: load, sliding speed, sliding distance, time.

and natural resources depletion. To eliminate the problem due to two synthetic reinforcements, some researchers have found replacement in agricultural or industrial waste derivatives along with synthetic reinforcement. A variety of agricultural (agro) and industrial wastes were turned into ashes, and their viability for use as reinforcing phase material was investigated [81, 82]. Fly ash (FA) as an industrial waste derivative is the most promising reinforcement in this category. It shows high hardness, tensile strength, and compressive strength with an ability to enhance wear resistance and machinability of hybrid composites at low cost and low density [83–86]. As a result, FA has been used in the creation of hybrid MMCs as a complement to synthetic ceramic particles. There are also several agro-waste reinforcements available, but the properties obtained by agro-waste reinforced composites are inferior as compared to synthetic reinforced composites [87]. The comparative tribological performance analysis has been done on agro-reinforced hybrid composite in addition to bamboo leaf ash. This study has reported that the 2 and 3 wt.% BLA hybrid reinforcement show improved wear resistance and is because of less debris on the frictional surface in 2 and 3 wt.% BLA hybrid composites [88].

For a sustainable solution to liquid lubricant, SLHMMCs are the most promising replacement of conventional MMCs. These materials are used in various sliding parts such as pistons, cylinders, clutches, and brake shoes in different automotive systems. Due to their better electrical properties along with tribological properties, these are also used in thrust washers, trusses, propeller shafts, printed circuit boards, transformers, ship hulls, and turbine blades, which fall under various multidisciplinary sectors. Mg and Al matrices are suitable choices for tougher, stronger, and lighter materials to be used in aerospace and automotive industries in addition to CNTs, GNPs, graphite, dichalcogenides, etc. [25, 40] These materials also have potential to balance strength to ductility. Hybrid combination of CNTs and GNPs along with Cu matrix provides high mechanical properties along with high thermal and electric conductivity best suited for electrical applications. For high-temperature applications, MoS_2 and WS_2 are good alternatives to graphite particles, CNTs, and GNPs, as hybrid reinforcements in different matrices. They have excellent self-lubricating properties in vacuum conditions. During their service life, however, tribological and wear degradation is unavoidable, resulting in significant financial losses. As a result, improvement in wear resistance is critical for increasing the sustainability of self-lubricating hybrid composite. The addition of different combinations of hybrid reinforcements with different concentrations and particle sizes provides the solution for better tribological performances.

11.3 FABRICATION ROUTE FOR SELF-LUBRICATING HYBRID METAL MATRIX COMPOSITE

Different fabrication routes are adopted for different combinations of hybrid MMCs to exhibit good overall properties. Among them, both ex-situ and in-situ have been advised for fabrication. The manufacturing processes for Al- and Mg-based HMMCs are classified as (i) solid-state – PM [30, 32, 36], microwave sintering [89], (ii) liquid state – infiltration method or squeeze casting [90], stir casting [69, 77], and (iii) semisolid state – compo casting [91]. Stir casting was determined to be the most straightforward and cost-effective method for fabricating Al- and Mg-based HMMCs in addition to graphite family-based solid lubricants [92]. Furthermore, stir-casting process factors, such as preheating, stirring speed, and stirring time, were critical in the production of high-quality HMMCs. The apparent non-wettability is the most difficult aspect of employing a liquid metallurgical technique for preparation [24, 92]. Agglomeration of particles, wettability, and the development of undesirable phases, which are some of the issues associated with the liquid metallurgical process, can readily be avoided with the solid-state technique [24, 93–95].

Solid state fabrication method PM has been highly recommended for hybrid combination of composite with higher volume proportion of reinforcements. This approach has also been used to construct hybrid reinforced with nanosized particles successfully [93]. PM product doesn't require post-fabrication machining. Solid lubricants, such as graphite, h-BN, MoS_2, WS_2, etc., which are reactive to high-temperature molten metal can be added easily and homogeneously into ceramic-reinforced matrices through PM. The schematic diagram of PM process is illustrated in Figure 11.3. This process provides a capacity to customize material's microstructure, and hence its qualities and performance are one of its benefits. There is a huge impact on microstructural changes due to the different fabrication conditions, and the choice of solid lubricant is referred to in different matrices. As a result, the final characteristics of reinforcement and matrix will be heavily influenced by the phases created during the fabrication process.

The critical analysis of current fabrication techniques suggests that there is a need for development of novel processing methods and hybrid composite formulations in order to sustain in this industry. To resolve these issues, novel fabrication techniques, i.e., named as additive manufacturing processes, are invented to produce self-lubricating hybrid MMCs with an optimum quantity of reinforcements [96, 97]. This technological development in fabrication industry will overcome the hybrid MMCs' strength-ductility trade-off and enhance its sustainability.

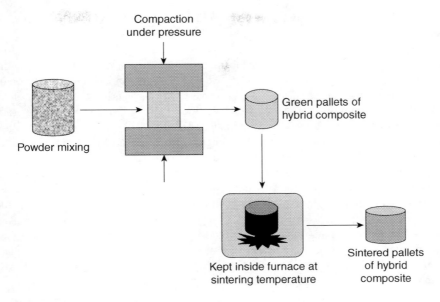

FIGURE 11.3 Schematic representation of powder metallurgy process.

11.4 SUSTAINABILITY OF SELF-LUBRICATING HYBRID METAL MATRIX COMPOSITE, ITS FUTURE PROSPECTIVE, AND CHALLENGES

Hybrid metal matrix composites with self-lubricating characteristics cover a unique gap in minimizing wear to obsolete liquid lubricants which are unfeasible or ineffective in a vacuum, space technology, or vehicle transportation. SLHMMC with agro waste and industrial waste in self-lubricating composite clears the gateway toward captivating the potential market as a new-age engineering material and provide the sustainability. The discovery of new-age technology improves the properties of these self-lubricating composites and opens the path toward sustainability. The use of self-lubricating material with graphite and MoS_2 is applied in adverse conditions by forming a continuous lubricating film on the surfaces due to self-lubricating phenomenon. In high-vacuum applications, nuclear reactors, h-BN, and dichalcogenides are good alternatives for avoiding contamination due to lubricating oils or greases. These types of self-lubricating hybrid composites help to eradicate the limitations like sustainability in extreme environmental condition of liquid lubricant. These materials have also solved the issues of solid lubricant coating such as oxidation, poor bonding, and low lifetime due to self-lubricating phenomenon. These are now famous for their characteristics of green or environmental friendly tribology, sustainability, and energy efficiency solution. Whereas oil and grease-based lubricants always produce harmful pollutants for the environment. Controlling the fabrication parameters of SHMMCs, i.e., stirring speed, pouring temperature, solidification time, etc., hybrid composite can fill the gaps and help to produce highly efficient, sustainable, eco-friendly, and energy efficiency solutions for aerospace and automobile industries with variable hybrid reinforcements.

The study reveals the fact that the main key self-lubricating hybrid composites are made up of graphite solid lubricant with aluminum matrix because the combination of properties, like higher wear resistance and mechanical properties, is obtained at a very low cost. However, various synthetic materials, such as lightweight magnesium, copper, nickel, steel, SiC, and Al_2O_3, are being studied by researchers in addition to dichalcogenides and boron nitrides which are quite important in vacuum and high-temperature conditions. Despite the lower friction coefficient, the high concentration of solid lubricants (up to 40% by volume) affects matrix continuity, impeding mechanical characteristics, and hence limiting the application of composites. Some high-performance sintered Al-based self-lubricating composites have been developed in the previous decade as a possible approach for balancing reduced friction coefficient with increased mechanical strength and wear resistance. With the humongous variation in reinforcements having synthetic to agro-waste/industrial-waste materials to induce better mechanical, structural, electrical, and tribological properties, SLHMMC can now be termed as "sustainable and eco-friendly material".

11.5 CONCLUSION

This chapter is written for professionals and university students who want a fundamental understanding of self-lubricating materials and their mechanism from various aspects of automotive applications. A comprehensive collection of outcomes from a variety of aspects, including mechanical, chemical, materials, and manufacturing, toward its sustainability, is included in this chapter. This chapter will be very valuable to professionals involved in the development and application of self-lubricating materials in comprehending the multidisciplinary knowledge. In the realm of SLHMMCs, a comprehensive list of references has been incorporated in this chapter which is a great source of information for the researchers working in the field of tribology. This is a valuable resource in the field of SLHMMCs. In particular, wear behavior under different lubrication conditions is addressed through this chapter. The thought of the studies is to expand the usage of these composites with optimum and sustainable utilization. Improved tribological performance in comparison to the unreinforced alloy has been established for the new generation of hybrid composites that entail the utilization of agricultural and industrial waste derivatives. Furthermore, the extent of enhancement in self-lubricating characteristic of hybrid AMCs comprising agricultural and industrial waste derivatives over self-lubricating materials with synthetic reinforcement needs to be investigated further toward sustainability. As these types of composites have the potential to reduce the cost by 50% by replacing the synthetic reinforcements. More agro-waste/industrial-waste reinforced hybrid composite should be investigated toward self-lubrication. More research interest has to be made for technological development in manufacturing process area to establish platform for fabrication to install different combinations of solid lubricants for hybrid composite at low cost and to make it environmentally friendly and pollution free.

REFERENCES

1. Zhang, S.W. 2013. Green tribology: Fundamentals and future development. Friction, 1(2):186–194.
2. Kuzharov, S. 2014. The concept of wearlessness in modern tribology. Izvestiya vuzov: North Caucasian region. Series: Engineering Sciences, 177:23–31.
3. Van Minh, N., Kuzharov, A., Huynh, N. and Kuzharov, A., 2020. Green Tribology. In: Tribology. Intech Open.
4. Thompson, R., 2021. Aluminium matrix composites: A sustainable solution. Reinforced Plastics.
5. Furlan, K.P., de Mello, J.D.B. and Klein, A.N. 2018. Self-lubricating composites containing MoS2: A review. Tribology International, 120:280–298.
6. Omrani, E., Afsaneh, D.M., Pradeep, L.M. and Pradeep, K.R. 2016. Influences of graphite reinforcement on the tribological properties of self-lubricating aluminum matrix composites for green tribology, sustainability, and energy efficiency— A review. The International Journal of Advanced Manufacturing Technology, 83(1–4):325–346.
7. Zhang, Y. and Chromik, R.R. 2018. Tribology of Self-Lubricating Metal Matrix Composites. In: Self-Lubricating Composites. Springer, Berlin, Heidelberg: 33–73.
8. Sahoo, S. 2021. Self-lubricating composites with 2D materials as reinforcement: A new perspective. Reinforced Plastics, 65(2):101–103.
9. Khatkar, S.K., Suri, N.M. and Kant, S. 2018. A review on mechanical and tribological properties of graphite reinforced self-lubricating hybrid metal matrix composites. Reviews on Advanced Materials Science, 56(1):1–20.
10. Idusuyi, N. and Olayinka, J.I. 2019. Dry sliding wear characteristics of aluminium metal matrix composites: A brief overview. Journal of Materials Research and Technology, 8(3):3338–3346.
11. Dorri Moghadam, A., Schultz, B.F., Ferguson, J.B., Omrani, E., Rohatgi, P.K. and Gupta, N. 2014. Functional metal matrix composites: Self-lubricating, self-healing, and nanocomposites—An outlook. Jom, 66(6):872–881.
12. Hidalgo-Manrique, P., Lei, X., Xu, R., Zhou, M.Y., Kinloch, I.A. and Young, R.J. 2019. Copper/graphene composites: A review. Journal of Material Science, 54(19):12236–12289.
13. Zhou, M.Y., Ren, L.B., Fan, L.L., Zhang, Y.W.X., Lu, T.H., Quan, G.F. and Gupta, M. 2020. Progress in research on hybrid metal matrix composites. Journal of Alloys and Compounds, 838:155274.
14. Nautiyal, H., Kumari, S., Tyagi, R., Rao, U.S. and Khatri, O.P. 2021. Evaluation of tribological performance of copper-based composites containing nano-structural 2D materials and their hybrid. Tribology International, 153:106645.
15. Al-maamari, A.E.A., Iqbal, A.A. and Nuruzzaman, D.M. 2020. Mechanical and tribological characterization of self-lubricating Mg-SiC-Gr hybrid metal matrix composite (MMC) fabricated via mechanical alloying. Journal of Science: Advanced Materials and Devices, 5(4):535–544.
16. Singh, N. and Belokar, R.M. 2021. Tribological behavior of aluminum and magnesium-based hybrid metal matrix composites: A state-of-art review. Materials Today: Proceedings, 44:460–466.
17. Sajjadnejad, M., Haghshenas, S.M.S., Badr, P., Setoudeh, N. and Hosseinpour, S. 2021. Wear and tribological characterization of nickel matrix electrodeposited composites: A review. Wear, 486:204098.
18. Kumar, N. and Irfan, G. 2021. Mechanical, microstructural properties and wear characteristics of hybrid aluminium matrix nano composites (HAMNCs)—Review. Materials Today: Proceedings, 45:619–625.

19. Zhu, S., Cheng, J., Qiao, Z. and Yang, J. 2019. High temperature solid-lubricating materials: A review. Tribology International, 133:206–223.
20. Jamwal, A., Seth, P.P., Kumar, D., Agrawal, R., Sadasivuni, K.K. and Gupta, P. 2020. Microstructural, tribological and compression behaviour of copper matrix reinforced with Graphite-SiC hybrid composites. Materials Chemistry and Physics, 251:123090.
21. Girish, B.M., Satish, B.M., Sarapure, S., Somashekar, D.R. and Basawaraj. 2015. Wear behavior of magnesium alloy AZ91 hybrid composite materials. Tribology Transactions, 58(3):481–489.
22. Ravindran, S., Mani, N., Balaji, S., Abhijith, M. and Surendaran, K. 2019. Mechanical behaviour of aluminium hybrid metal matrix composites—A review. Materials Today: Proceedings, 16:1020–1033.
23. Malaki, M., Xu, W.W., Kasar, A.K., Menezes, P.L., Dieringa, H., Varma, R.S. and Gupta, M. 2019. Advanced metal matrix nanocomposites. Metals, 9(3):330.
24. Rana, V., Kumar, H. and Kumar, A. 2022. Fabrication of hybrid metal matrix composites (HMMCs)—A review of comprehensive research studies. Materials Today: Proceedings.
25. Ravishankar, B., Nayak, S.K. and Kader, M.A. 2019. Hybrid composites for automotive applications—A review. Journal of Reinforced Plastics and Composites, 38(18): 835 –845.
26. Monteiro, W.A., Buso, S.J. and da Silva, L.V. 2012. In: Monteiro, W.A. (ed.) New Features on Magnesium Alloys. Intech, Rijeka.
27. Stojanović, B., Babić, M., Veličković, S. and Blagojević, J. 2016. Tribological behavior of aluminum hybrid composites studied by application of factorial techniques. Tribology Transactions, 59(3):522–529.
28. Mosleh-Shirazi, S., Akhlaghi, F. and Li, D.Y. 2016. Effect of graphite content on the wear behavior of Al/2SiC/Gr hybrid nano-composites respectively in the ambient environment and an acidic solution. Tribology International, 103:620–628.
29. Kumar, C.R., Malarvannan, R. and JaiGanesh, V. 2020. Role of SiC on mechanical, tribological and thermal expansion characteristics of B4C/Talc-reinforced Al-6061 hybrid composite. Silicon, 12(6):1491–1500.
30. Kanthavel, K., Sumesh, K.R. and Saravanakumar, P. 2016. Study of tribological properties on Al/Al2O3/MoS2 hybrid composite processed by powder metallurgy. Alexandria Engineering Journal, 55(1):13–17.
31. Liu, J., Huang, X., Zhao, K., Zhu, Z., Zhu, X. and An, L. 2019. Effect of reinforcement particle size on quasistatic and dynamic mechanical properties of Al-Al2O3 composites. Journal of Alloys and Compounds, 797:1367–1371.
32. Çelik, Y.H. and Seçilmiş, K. 2017. Investigation of wear behaviors of Al matrix composites reinforced with different B4C rate produced by powder metallurgy method. Advanced Powder Technology, 28(9):2218–2224.
33. Wozniak, J., Kostecki, M., Cygan, T., Buczek, M. and Olszyna, A. 2017. Self-lubricating aluminium matrix composites reinforced with 2D crystals. Composites Part B: Engineering, 111:1–9.
34. Zivic, F., Busarac, N., Milenković, S. and Grujovic, N., 2021. General Overview and Applications of Ceramic Matrix Composites (CMCs). Reference Module in Materials Science and Materials Engineering.
35. Kumar, R. and Antonov, M. 2021. Self-lubricating materials for extreme temperature tribo-applications. Materials Today: Proceedings, 44:4583–4589.
36. Biswal, S.R. and Sahoo, S. 2020. Fabrication of WS2 dispersed Al-based hybrid composites processed by powder metallurgy: Effect of compaction pressure and sintering temperature. Journal of Inorganic and Organometallic Polymers and Materials, 30(8):2971–2978.

37. Biswal, S.R. and Sahoo, S. 2022. Structural and mechanical properties of a novel Al-Al$_2$O$_3$-WS$_2$ hybrid composites. Materials Letters, 307:131017.
38. Prabhu, T.R., Varma, V., Vedantam, S. 2014. Effect of reinforcement type, size, and volume fraction on the tribological behavior of Fe matrix composites at high sliding speed conditions. Wear, 309:247–255.
39. Prabhu, T.R. 2015. Effects of solid lubricants, load, and sliding speed on the tribological behavior of silica reinforced composites using design of experiments. Mater Des, 77:149–160.
40. Mahaviradhan, N. and Sivaganesan, S. 2021. Tribological analysis of hybrid aluminum matrix composites for high temperature applications. Materials Today: Proceedings, 39:669–675.
41. Sadeghi, N., Aghajani, H. and Akbarpour, M.R. 2018. Microstructure and tribological properties of in-situ TiC-C/Cu nanocomposites synthesized using different carbon sources (graphite, carbon nanotube and graphene) in the Cu-Ti-C system. Ceramics International, 44(18):22059–22067.
42. Akbarpour, M.R., Alipour, S., Safarzadeh, A. and Kim, H.S. 2019. Wear and friction behavior of self-lubricating hybrid Cu (SiC+ x CNT) composites. Composites Part B: Engineering, 158:92–101.
43. Nautiyal, H., Kumari, S., Khatri, O.P. and Tyagi, R. 2019. Copper matrix composites reinforced by rGO-MoS$_2$ hybrid: Strengthening effect to enhancement of tribological properties. Composites Part B, 173:106931.
44. Zhan, Y.Z. and Zhang, G.D. 2006. The role of graphite particles in the high-temperature wear of copper hybrid composites against steel. Material and Design, 27(1):79–84.
45. Moghadam, A.D., Omrani, E., Menezes, P.L. and Rohatgi, P.K. 2015. Mechanical and tribological properties of self-lubricating metal matrix nanocomposites reinforced by carbon nanotubes (CNTs) and grapheme—A review. Composites Part B: Engineering, 77:402–420.
46. Giubileo, F., Di Bartolomeo, A., Iemmo, L., Luongo, G. and Urban, F. 2018. Field emission from carbon nanostructures. Applied Sciences, 8(4):526.
47. Polat, S., Sun, Y., Çevik, E., Colijn, H. and Turan, M.E., 2019. Investigation of wear and corrosion behavior of graphene nanoplatelet-coated B4C reinforced Al–Si matrix semi-ceramic hybrid composites. Journal of Composite Materials, 53(25):3549–3565.
48. Azarniya, A., Safavi, M.S., Sovizi, S., Azarniya, A., Chen, B. and Madaah Hosseini, H.R. 2017. Metallurgical challenges in carbon nanotube-reinforced metal matrix nanocomposites. Metals, 7(10):384.
49. Chen, X.F., Tao, J.M., Liu, Y.C., Bao, R., Li, F.X., Li, C.J. and Yi, J. H. 2019. Interface interaction and synergistic strengthening behaviour in pure copper matrix composites reinforced with functionalized carbon nanotube-graphene hybrids. Carbon, 146:736–755.
50. Zhou, M.Y., Ren, L.B., Fan, L.L., Tun, K.S., Gupta, M., Zhang, Y.W.X., Lu, T.H. and Quan, G.F. 2019. Achieving ultra-high strength and good ductility in AZ61 alloy composites containing hybrid micron SiC and carbon nanotubes reinforcements. Materials Science and Engineering: A, 768:138447.
51. Zhang, X., Li, S. F., Pan, D., Pan, B. and Kondoh, K. 2018. Microstructure and synergistic strengthening efficiency of CNTs-SiCp dual-nano reinforcements in aluminium matrix composites. Composites Part A, 105:87–96.
52. Kumar, H.P. and Xavior, M.A. 2016. Fatigue and wear behavior of Al6061—Graphene composites synthesized by powder metallurgy. Transactions of the Indian Institute of Metals, 69(2):415–419.
53. Akbarpour, M.R., Sadeghi, N. and Aghajani, H., 2022. Nano TiC-Graphene-Cu composites fabrication by a modified ball-milling method followed by reactive sintering: Effects of reinforcements content on microstructure, consolidation, and mechanical properties. Ceramics International, 48(1):130–136.

54. Christou, A., Stec, AA., Ahmed, W., Aschberger, K. and Amenta. V. 2016. A review of exposure and toxicological aspects of carbon nanotubes, and as additives to fire retardants in polymers. Critical Reviews in Toxicology, 46(1):74–95.
55. Singh, P.B., Sidhu, S.S. and Payal, H.S. 2016. Fabrication and machining of metal matrix composites: A review. Materials and Manufacturing Processes, 31(5): 553–573.
56. Zhang, J.S., Yang, S.F., Chen, Z.X., Wu, H., Zhao, J.W. and Jiang, Z.Y. 2019. Graphene encapsulated SiC nanoparticles as tribology-favoured nanofillers in aluminium composite. Composites Part B, 162:445–453.
57. Jayabharathy, S. and Mathiazhagan, P. 2020. Investigation of mechanical and wear behaviour of AZ91 magnesium matrix hybrid composite with TiO_2/graphene. Materials Today: Proceedings, 27:2394–2397.
58. Rosenkranz, A., Costa, H.L., Baykara, M.Z. and Martini, A. 2021. Synergetic effects of surface texturing and solid lubricants to tailor friction and wear—A review. Tribology International, 155:106792.
59. Eftekhari, A. 2017. Tungsten dichalcogenides (WS_2, WSe_2, and WTe_2): Materials chemistry and applications. Journal of Materials Chemistry A, 5(35):18299–18325.
60. Wei, Z., Li, B., Xia, C., Cui, Y., He, J., Xia, J.B. and Li, J. 2018. Various structures of 2D transition-metal dichalcogenides and their applications. Small Methods, 2(11):1800094.
61. Maji, P., Nath, R.K., Paul, P., Meitei, R.B. and Ghosh, S.K., 2021. Effect of processing speed on wear and corrosion behavior of novel MoS_2 and CeO_2 reinforced hybrid aluminum matrix composites fabricated by friction stir processing. Journal of Manufacturing Processes, 69:1–11.
62. Raja, T., Prabhakaran, R., Kumar, D.P. and Sathish, D. 2021. Mechanical and tribological characteristics of AL7075/MWCNT, B_4C & MoS_2 hybrid metal matrix composites. Materials Today: Proceedings.
63. Kumar, N.S., Suresh, R. and Shankar, G.S. 2020. High temperature wear behavior of Al2219/n-B_4C/MoS_2 hybrid metal matrix composites. Composites Communications, 19:61–73.
64. Umanath, K., Palanikumar, K., Sankaradass, V. and Uma, K. 2021. Optimization of wear properties on AA7075/SiC/MoS_2 hybrid metal matrix composite by response surface methodology. Materials Today: Proceedings, 46:4019–4024.
65. Rouhi, M., Moazami-Goudarzi, M. and Ardestani, M. 2019. Comparison of effect of SiC and MoS_2 on wear behavior of Al matrix composites. Transactions of Nonferrous Metals Society of China, 29(6):1169–1183.
66. Furlan, K.P., da Costa Gonçalves, P., Consoni, D.R., Dias, M.V.G., de Lima, G.A., de Mello, J.D.B. and Klein, A.N. 2017. Metallurgical aspects of self-lubricating composites containing graphite and MoS_2. Journal of Materials Engineering and Performance, 26(3):1135–1145.
67. Lakshmipathy, J., Rajesh Kannan, S., Manisekar, K. and Vinoth Kumar, S. 2017. Effect of Reinforcement and Tribological Behaviour of AA 7068 Hybrid Composites Manufactured through Powder Metallurgy Techniques. In: Applied Mechanics and Materials. Trans Tech Publications Ltd.: 867:19–28.
68. Singh, N., Mir, I.U.H., Raina, A., Anand, A., Kumar, V. and Sharma, S.M. 2018. Synthesis and tribological investigation of Al-SiC based nano hybrid composite. Alexandria Engineering Journal, 57(3):1323–1330.
69. Essa, F.A., Zhang, Q. and Huang, X., 2017. Investigation of the effects of mixtures of WS_2 and ZnO solid lubricants on the sliding friction and wear of M50 steel against silicon nitride at elevated temperatures. Wear, 374:128–141.
70. Prajapati, A.K., Omrani, E., Menezes, P.L. and Rohatgi, P.K. 2018. Fundamentals of Solid Lubricants. In: Menezes, P., Rohatgi, P., Omrani E. (eds) Self-Lubricating Composites. Springer, Berlin, Heidelberg.

Self-lubricating Hybrid Metal Matrix Composite toward Sustainability 211

71. Daniel, S.A.A., Sakthivel, M., Gopal, P.M. and Sudhagar, S. 2018. Study on tribological behaviour of Al/SiC/MoS$_2$ hybrid metal matrix composites in high temperature environmental condition. Silicon, 10:2129–2139.

72. Sahoo, S., Samal, S. and Bhoi, B., 2020. Fabrication and characterization of novel Al-SiC-hBN self-lubricating hybrid composites. Materials Today Communications, 25:101402.

73. Torres, H., Ripoll, M.R. and Prakash, B. 2018. Tribological behaviour of self-lubricating materials at high temperatures. International Materials Reviews, 63(5):309–340.

74. Kannan, C., Ramanujam, R. and Balan, A.S.S. 2018. Machinability studies on Al 7075/BN/Al$_2$O$_3$ squeeze cast hybrid nanocomposite under different machining environments. Materials and Manufacturing Processes, 33(5):587–595.

75. Chand, S. and Chandrasekhar, P. 2020. Influence of B$_4$C/BN on solid particle erosion of Al6061 metal matrix hybrid composites fabricated through powder metallurgy technique. Ceramics International, 46(11):17621–17630.

76. Hammes, G., Mucelin, K.J., da Costa Gonçalves, P., Binder, C., Binder, R., Janssen, R., Klein, A.N. and de Mello, J.D.B. 2017. Effect of hexagonal boron nitride and graphite on mechanical and scuffing resistance of self-lubricating iron-based composite. Wear, 376:1084–1090.

77. Sundararajan, S., Rajadurai, J.S. and Antony Vasanthakumar, C. 2017. Impact of solid self-lubricant and hard ceramic reinforcements on tribological characteristics of microwave-sintered Cu hybrid composites processed by solid-state mixing technique. Proceedings of the Institution of Mechanical Engineers, Part J: Journal of Engineering Tribology, 231(5):523–542.

78. Hekner, B., Myalski, J., Valle, N., Botor-Probierz, A., Sopicka-Lizer, M. and Wieczorek, J. 2017. Friction and wear behavior of Al-SiC (n) hybrid composites with carbon addition. Composites Part B: Engineering, 108:291–300.

79. Zhu, J., Qi, J., Guan, D., Ma, L. and Dwyer-Joyce, R. 2020. Tribological behaviour of self-lubricating Mg matrix composites reinforced with silicon carbide and tungsten disulfide. Tribology International, 146:106253.

80. Kavimani, V., Prakash, K.S. and Thankachan T. 2019. Experimental investigations on wear and friction behaviour of SiC@r-GO reinforced Mg matrix composites produced through solvent-based powder metallurgy. Composites Part B, 162:508–521.

81. Chen, M., Fan, G., Tan, Z., Xiong, D., Guo, Q., Su, Y., Zhang, J., Li, Z., Naito, M. and Zhang, D. 2018. Design of an efficient flake powder metallurgy route to fabricate CNT/6061Al composites. Materials & Design, 142:288–296.

82. Selvam, J.D., Dinaharan, I. and Rai, R.S. 2021. Matrix and reinforcement materials for metal matrix composites.

83. Singh, B., Grewal, J.S. and Sharma, S. 2022. Effect of addition of flyash and graphite on the mechanical properties of A6061-T6. Materials Today: Proceedings, 50:2411–2415.

84. Rajesh, K., Mahendra, K.V., Mohan, N., Sachit, T.S. and Akshay Prasad, M. 2020. Studies on mechanical and dry sliding wear behaviour of graphite/flyash reinforced aluminium (Al6Mg) MMCs. Materials Today: Proceedings, 27:2434–2440.

85. Ashwini, M.V., Patil, S. and Robionson, P. 2021. Evaluation of mechanical and tribological properties of AL7075 hybrid metal matrix composite reinforced with fly ash and graphite. Materials Today: Proceedings, 45:311–317.

86. Devadiga, U. and Fernandes, P. 2021. Taguchi analysis for sliding wear characteristics of carbon nanotube-flyash reinforced aluminium nanocomposites. Heliyon, 7(2):e06170.

87. Ravi Kumar, K., Pridhar, T. and Sree Balaji V.S. 2018. Mechanical properties and characterization of zirconium oxide (ZrO$_2$) and coconut shell ash (CSA) reinforced aluminium (Al 6082) matrix hybrid composite. Journal of Alloys and Compound, 765:171–179.

88. Alaneme, K.K., Olubambi, P.A., Afolabi, A.S. and Bodurin, M.O., 2014. Corrosion and tribological studies of bamboo leaf ash and alumina reinforced Al-Mg-Si alloy matrix hybrid composites in chloride medium. International Journal of Electrochemical Science, 9:5663–5674.

89. Singh, H., Jain, P.K., Bhoi, N. and Pratap, S. 2018. Experimental Study Pertaining to Microwave Sintering (MWS) of Al-Metal Matrix Composite—A Review. In: Materials Science Forum. Trans Tech Publications Ltd.: 150–155.

90. Dmitruk, A., Zak, A., Naplocha, K., Dudzinski, W. and Morgiel, J. 2018. Development of pore-free Ti-Al-C MAX/Al-Si MMC composite materials manufactured by squeeze casting infiltration. Materials Characterization, 146:182–188.

91. Jose, J., Christy, T.V., Peter, P.E., Feby, J.A., George, A.J., Joseph, J. and Benjie N.M. 2018. Manufacture and characterization of a novel agro-waste based low-cost metal matrix composite (MMC) by compocasting. Material Research Express, 5(6):066530.

92. Ogawa, F. and Masuda, C. 2021. Fabrication and the mechanical and physical properties of nanocarbon-reinforced light metal matrix composites: A review and future directions. Materials Science and Engineering: A, 820:141542.

93. Patil, O.M., Khedkar, N.N., Sachit, T.S. and Singh, T.P. 2018. A review on effect of powder metallurgy process on mechanical and tribological properties of Hybrid nano composites. Materials Today: Proceedings, 5(2):5802–5808.

94. Thankachan, T., Prakash, K.S. and Kavimani, V. 2019. Investigating the effects of hybrid reinforcement particles on the microstructural, mechanical and tribological properties of friction stir processed copper surface composites. Composites Part B, 174:107057.

95. Samal, P., Vundavilli, P.R., Meher, A. and Mahapatra, M.M. 2020. Recent progress in aluminum metal matrix composites: A review on processing, mechanical and wear properties. Journal of Manufacturing Processes, 59:131–152.

96. Yu, W.H., Sing, S.L., Chua, C.K., Kuo, C.N. and Tian, X.L. 2019. Particle-reinforced metal matrix nanocomposites fabricated by selective laser melting: A state-of-the-art review. Progress in Material Science, 104:330–379.

97. Dadbakhsh, S., Mertens, R., Hao, L., Humbeeck, J.V. and Kruth, J.P. 2019. Selective laser melting to manufacture "in situ" metal matrix composites: A review. Advanced Engineering Materials, 21(3):1801244.

12 E-Plastic Waste—A Sustainable Construction Material
An Indian Perspective

Sweta Sinha and Ateeb Hamdan

CONTENTS

12.1 Introduction 213
12.2 Challenges 215
    12.2.1 Population Growth 215
    12.2.2 Construction Industry and Waste Generation 215
    12.2.3 E-plastic Waste Generation 218
12.3 E-Plastic and its Incorporation in Concrete 220
    12.3.1 Feasible Possibilities 220
    12.3.2 Strength Characteristic of E-plastic Concrete 221
        12.3.2.1 Compressive Strength 224
        12.3.2.2 Splitting Tensile Strength 225
        12.3.2.3 Flexural Strength Test 226
12.4 Result and Discussion 227
12.5 Conclusion 229
References 229

12.1 INTRODUCTION

The twenty-first century saw a wholesome representation of climate activists, environmentalists, and economists emphasizing the importance of sustainable development. This constant effort to bring attention of the masses has been existing since 1970s with many environmentalists' discussions being focused on ecological and sustainable development [1]. The world experiencing an expedient development has certainly provided a better standard of living globally. However, human activities generate waste [2] and its generation has increased by many folds when compared to prehistoric period [3]. An increase in volume of waste also increases the bulk variety of waste [4], making its recycling and disposal far complex. The environment now is unable to absorb the waste produced without having a degrading impact on itself.

Figure 12.1 illustrates the general classification of waste that can be done based on their physical state, source generation, degradability, and its environmental impact.

DOI: 10.1201/9781003297772-12

213

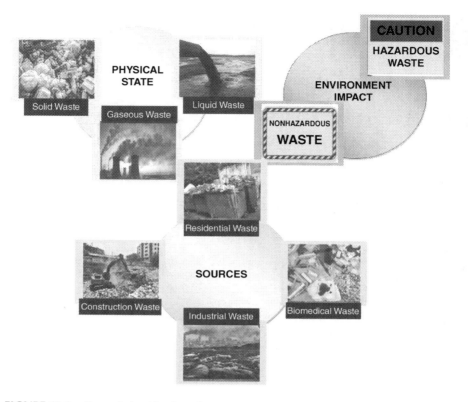

FIGURE 12.1 General classification of waste.

Although these wastes pose threat to the environment, nonbiodegradable (plastic and metals) and hazardous (biomedical and E-waste) wastes are of significant concern. These wastes persist in the environment long enough and are difficult to reuse and dispose of. Thus, an approach for effective management of these wastes for sustainable development is a present necessity.

India constitutes 17% of the world population and contributes to 7% of world emissions. The nation has also committed its cooperation to tackle the climate change by achieving the ambitious targets of COP 26 Summit in Glasgow till 2030. That includes, reducing total projected carbon emissions by one billion tons, consuming 50% of the country's energy needs through renewable sources, and reducing carbon intensity of its economy by more than 45%. For upholding all these commitments, India needs productive ways through which it can reduce carbon dependency. 39% of global carbon emission is generated by construction industry globally and this is evident in India too. Population-oriented solutions, interventions at policy level by the government, mobilizing, and improving individual and community capacities toward carbon reduction waste segregation and low nonbiodegradable consumption are paths forward for the nation toward achieving sustainable outcomes [5]. However, achieving such a farfetched goal constitutes a function of time, the interval of which is invariable.

E-Plastic Waste   215

Thus, the prime focus of this chapter is to investigate a possible divergent solution that not only deals with a huge amount of nonbiodegradable wastes, plastic being the most evident but also reduces the carbon footprints of the country. The efficient reuse and utilization in sectors that shall be befitting and even reap some monetary value shall be a step toward sustainability.

12.2 CHALLENGES

12.2.1 Population Growth

According to the United Nations, the world's population growth is expected to increase by 2 billion from 7.7 billion to 9.7 billion in 2050. While 61% live in Asia, Africa is predicted to be the fastest continent and has had the highest rate of population growth. India and China remain at the top in terms of population, currently being the two most populous countries in the world. Considering the world population prospects of 2019, it is suggested that India (1.33 billion) will overtake China (1.44 billion) by 2027 due to decrease of latter population by 2.2% in the country (between 2019 and 2050) and consequently of its one-child policy. It is evident that rapid economic development around the globe has resulted in a stable population growth and development. Fertility rate, life longevity increasing urbanization, and inter/intra migration are cited to be dominating factors for such a dramatic growth that shall portray an outreaching implication for future generations.

Figure 12.2 depicts the present and predicted population of the three most populous countries (e.g., China, India, and the USA) and their urban growth which absorbs the maximum resources of state. It is also to be noted that, apart from being densely populated states, they are manufacturing, producing, and consuming centers of the world. They are ranked foremost in plastic, E-waste, radioactive, and biomedical amongst many other waste generators globally. Addition to it, China and India are among the top countries consuming and burning fossil fuels like coal and oil; thereby, significantly contributing to the greenhouse gas (GHG) emissions resulting in global warming. Although population growth has a positive impact on economic development and is in support of the population-driven economic growth [6]. It often causes several obstacles in sustainable advancements in terms of environmental pollution, waste generation, and ecological instability.

12.2.2 Construction Industry and Waste Generation

The exponential increment in population has no sign of decline. Restricting its growth is unethical and immoral, at the same time it contradicts the notion of development for all. Although developed countries have certain ways to manage, this expeditious growth of developing countries, like India, having a large percentage of the world population is facing several hindrances to endure its progressive positive growth largely due to its financial constraints.

According to the UN world population report, although China and the US relish on a large urban population, India experiences fastest rate of urbanization in the world [7]. This influx is mainly due to inter- and intra-migration to urban centers for

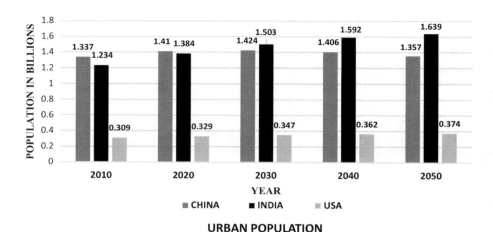

FIGURE 12.2 (a) Population of China, India, and the USA (in billion). (b) Urban population of China, India, and the USA (in billion).

better livelihood opportunities. This apparent change has had a coupling effect on the rapid and massive urban development to meet substantial accommodation requisites, majority of which are unplanned.

To accommodate this colossal influx, the construction industry of India is relied upon. India's construction industry is 3rd largest in the world and is one of the pioneers in boosting the economy of the country by contributing more than 10% of India's GDP [8] (Figure 12.3). However, since construction market is a great contributor in carbon emission, this positive in turn contradicts the global pretext of COP 26, which is to reduce carbon emission intensity negatively.

Coherent to rise of construction market, many natural resources, like sea sand, mountain rocks, and cement, which are important building materials, are being excessively consumed [9] resulting in the depletion of natural resources. Moreover, a large market implies a wide range of assembling/dismantling and use/disposal of construction materials which have resulted in exponential production of construction waste like huge debris and deformed concrete structures that are polluting the environment incessantly (Figure 12.4).

E-Plastic Waste

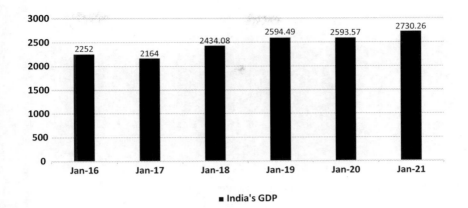

FIGURE 12.3 Construction-Market contributions in India's GDP (in billion India rupees).

In addition to this, cement, which is an integral constituent of construction material, is being consumed at a concerning level. In 2019, India consumed 337 million tons of cement. By 2025, this number is expected to rise to 550 million tons [10]. India's massive cement industry contributes to about 8% of global emissions. Causing CO_2 emission of 276 kg CO_2/ton of cement produced [11]. Not only does the construction process deplete natural resources, but it also emits GHG, which contributes to global climate change and ozone depletion. Although COVID-19 pandemic and the subsequent lockdown slowed down the cement market in India (Figure 12.5) and reduced the overstressed ecosystem, the statistics suggest that the market has bounced back and is much capable than before. As positive as it can be for the economy of the nation, it poses a potential threat of ecological degradation.

Schemes and projects focusing on infrastructure development, such as Smart City, PM Gram Sadak Yojana, Urban Transport and Metro Rail projects, National Housing Development Program, and Housing For All, are primary factors for this growth. Environmental protection is of utmost importance in reducing carbon dioxide produced by the construction and cement industries. Therefore, it is imperative to introduce and adopt alternatives to conventional construction materials.

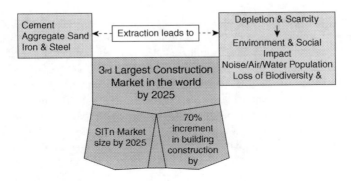

FIGURE 12.4 Construction scope-growth-impacts in India.

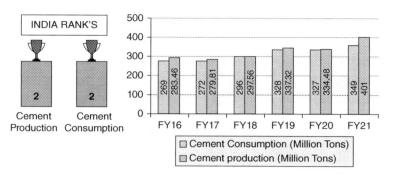

FIGURE 12.5 Cement production and consumption profile of India.

12.2.3 E-PLASTIC WASTE GENERATION

Rapid development has irrespectively brought in digital revolution in the first and third world countries. Benefits from the digitalization are on the rise through electronic appliances and advanced gadgets, like Android phones, smartwatches, and TVs, which act as a medium. This massive influx in smart technological appliances has positively impacted various sectors, like education, health care, administration, etc., and assisted in overall economic development of the country.

India, which is also experiencing rapid Internet penetration, has reported a sweeping adoption and consumption of these sophisticated instruments. However, these digital appliances have an average service life of 5–7 years, after which their disposal is an optimum choice. This consequentially resulted in soaring record generation of electronic equipment wastes, also known as Waste Electrical and Electronic Equipment (WEEE).

The European Union (EU) has defined E-waste or WEEE "as electrical or electronic equipment that is waste, including all components, subassemblies and consumables that are part of the product at the time of discarding" (E-WASTE Report, 2020). India in its E-waste (Management) Rules of 2016 – issued by the Central Pollution Control Board (CPCB) states that "electrical and electronic equipment (EEE), whole or in part, discarded as waste by consumers (individual or bulk) as well as rejects from manufacturing, refurbishment and repair process" (EWS), 2019, are to be considered as E-waste.

According to the Associated Chambers of Commerce and Industry of India (ASSOCHAM), India produces 2.7 million tons of E-waste annually, of which 68% is electronic gadgets (TVs and Computers), 14% is telecommunications devices, 8% is electrical equipment, 7% is medical equipment, and 5% is others [12] (Figure 12.6).

Many different sources of E-waste can be identified, including households, large consumers like government offices, commercial establishments, and manufacturers. These sources are generally more hazardous than municipal waste. Mercury, arsenic, cadmium, and lead are toxic materials present in E-waste (Figure 12.7), which must be handled carefully to avoid negative impacts on human health, whereas gold, platinum, silver, and copper are nontoxic materials that can be reused [12]. As a result, conventional disposal processes are not a viable and eco-friendly solution [13].

E-Plastic Waste

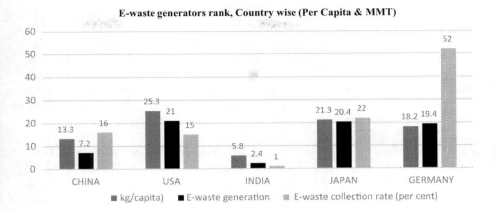

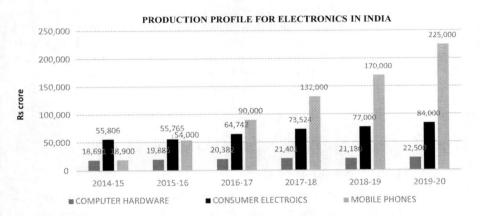

FIGURE 12.6 (a) E-waste generators rank, country wise (per capita and MMT); (b) Production profile for electronics in India. Source: CSE2020.

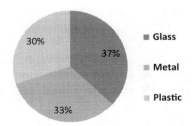

FIGURE 12.7 Components of E-waste. Source: ASSOCHAM, 2018.

Thus, several mitigation practices and scientific disposal procedures are strictly followed by government entities for disposal of E-waste.

Plastic represents a major component of WEEE mainly as a result of its advantages over other nonbiodegradable counterpart wastes like glass and metal. Plastic is used in electronics and in general at an astronomical rate. The main reasons for this are durability, low cost, shape and workability, as well as low density and higher strength-to-weight ratio. According to recent statistics, 3.4 million tons of plastic waste were generated in India in 2019–2020, an increase of more than 1 million tons from 2018–2019 [14].

At present several millions of tons of plastic waste generated, frequently found their way into landfills, oceans, rivers, coasts, beaches, and every year around the world [15]. The present statistics suggest that 25% of these nonbiodegradable wastes are recycled around the world indicating that plastic recovery and recycling continue to be insufficient and its issues remain neglected. The decrement in production of plastic is uncertain and its threat to the global environment is evident. However, it is to be noted that metals present in E-waste are generally recyclable and have significant commercial value and are reused, while plastic that is a major component of WEEE is nondegradable and having no significant resale value continues persisting in the environment [16]. Thus, a befitting solution can be an effective implementation of 3Rs, i.e., Recycling, Reducing, and Reusing as recycling of any or all sorts of waste is a step forward toward sustainability that shall result in conserving natural resources.

12.3 E-PLASTIC AND ITS INCORPORATION IN CONCRETE

12.3.1 Feasible Possibilities

A recent study shows that 80% of Indian households possess electronic appliances like desktops, televisions, printers, etc. Almost all Indians own mobile phones [17]. Consumption of electronic instruments is at its peak, similar to cement and aggregate in the concrete industry for urban development. Since plastic polymers of E-waste are largely left out or neglected and the complexity, it poses in terms of degradation is concerning for nature. Therefore, a synergization of waste plastic in a concrete matrix seems to be a productive approach. Several studies and experiments have been conducted previously for observing the effective and satisfactory results of using plastic in concrete manufacturing. Many studies have considered replacement of Fine Aggregate (FA) or Coarse Aggregate (CA) with plastic as its alternative replacer. Although replacement of waste material in concrete is beneficial from an environmental perspective, such end-products portray distinct engineering (e.g., mechanical and thermal) properties. Since properties of plastic distinctly differ from natural aggregates.

The use of plastic in electronic equipment has grown incessantly to reduce total weight and make products more cost-efficient. Plastics are divided into two types: thermosetting and thermoplastic. While the latter is easily recyclable without much alteration in its property and constitutes to 78% of the total plastic [18], thermosets pose a challenge for reuse due to their brittle tendency. Plastic wastes from

E-Plastic Waste

TABLE 12.1
Different Type of Plastic Used Commonly (Source: [19, 20])

Polyethylene terephthalate (PET)	Acrylonitrile butadiene styrene (ABS)	Polystyrene (PS)
Polyurethane (PE)	Polylactic acid (PLA)	Polypropylene (PP)
High-density polyethylene (HDPE)	Polyvinyl chloride (PVC)	Low-density polyethylene (LDPE)/Linear low-density polyethylene (LLDPE)

electronic gadgets (E-plastic) are generally composed of thermoplastic polymers, either in combination or independent. However, due to its existence in multiple forms added with varied polymers fractions when used in electrical equipment, it poses a major impediment in the recycling process since it makes a complex stream to recycle (Table 12.1).

The CPCB in its survey of 2015 depicts that 15342.6 tons of plastic waste are generated per day in India. An enormous percentage of which is low-density polyethylene (LDPE), since it has low-recycling value [18]. However, LDPE has good ductility and is impact resistant. Polyethylene terephthalate (PET) also prevailing in huge quantity at various disposal sites also has good strength, ductility, stiffness, and hardness.

Many scientists have attempted to successful amalgamation of plastic garbage into concrete. Inferences from these experimental studies suggested proportion of plastic waste kept limited to less than 2% of the overall weight of concrete [21]. Numerous researches have also been conducted to investigate the use of high-density polyethylene (HDPE), polyvinyl chloride (PVC), polypropylene (PP), and polyethylene terephthalate (PET) as viable cement mortar and concrete replacements [12]. These observations were similar to PET bottles used by Jo et al. [22] and Sadrmomtazi [23], where conventional FA and additional superplasticizers were replaced to boost endmost shear strength.

However, electronic appliances' waste components having plastic polymers are mainly composed of polyurethane (PE), acrylonitrile butadiene styrene (ABS), polypropylene (PP), and polystyrene (PS), which account for approximately 70% [13, 20] of such HDPE, PVC, and LDPE, which exist in the environment in large volume that are also used in electronic instruments like cable casing, cover, etc. Although a limited study on E-plastic reuse in concrete has been conducted, inferences from successful experimental studies observed that, at a certain increase in percentage of E-plastic, strength of design mix decreases. However, these decrements were less distinct to some extent compared to the reuse of conventional plastic in conventional concrete mix. Table 12.2 illustrates the relevant research and the conclusive results which were obtained.

12.3.2 STRENGTH CHARACTERISTIC OF E-PLASTIC CONCRETE

The amalgamation of waste plastic with concrete can only reap benefit if it is competent enough to bear heavy loads and capable to be used in stress zones of heavy structures.

TABLE 12.2

Properties of Concrete with E-Plastic in Previous Studies and Their Inference

References	Form of E-Plastic	Type of E-plastic	Replacement % of E-Plastic	Size of E-plastic particle use (mm)	Water-cement Ratio	Grade of Concrete	Optimum Result at	Remarks	Properties
[24]	Crushed, grinded to particle size	–	4, 8, 12, 16, 20, 24 By weight of coarse aggregate	1.18–2.36	0.55	M20	12% (E-plastic) & 10% (Flyash)	10% Flyash by weight of cement	
[25]	Crushed	Nonmetallic	5, 10, 15, 20 By weight of coarse aggregate	4.75, 10, 20	0.50	M20 & M25	5–6% ideal & 10% allowable		OPC 43 grade cement 60 specimens
[26]	Crushed	–	10, 20, 30 By weight of cement	5–20	0.50	M20	10%		OPC 53 grade cement
[27]	Shredded (gravel size)		5, 10, 15 By Weight of Coarse aggregate	20	0.50	M20	15% (E-plastic) & 10% (Flyash)	10% Flyash by weight of cement	
[12]	Chips & Flakes	HIPS	10, 20, 30, 40, 50 By volume of coarse aggregate	6–12	0.53, 0.49, 0.45	–	20%	–	486 concrete specimens
[28]	Fibers cable outer casing	Polyvinyl Chloride (PVC)	0.6, 0.8, 1 By weight of cement	35 & 4 (diameter)	0.4	M30	0.6%	Silica powder 10% by volume of the cement	

(Continued)

TABLE 12.2 (*Continued*)

Properties of Concrete with E-Plastic in Previous Studies and Their Inference

References	Form of E-Plastic	Type of E-plastic	Replacement % of E-Plastic	Size of E-plastic particle use (mm)	Water-cement Ratio	Grade of Concrete	Optimum Result at	Remarks	Properties
[29]	Shredded wire insulation	Polyvinyl Chloride (PVC)	0.4, 0.6, 0.8, 1	30, 40, 50 & 4 (diameter)	0.5	Mix ratio is 1:1.31:3.10			OPC 53 grade
[30]			10, 20 By volume of coarse aggregate	20	0.53	M20	10%		OPC 53 grade
[13]	Angular		10, 20 By volume of coarse aggregate & fine aggregate		0.5	M20	10% (fine & coarse both)	Replacement of Ground Granulated Blast Furnace Slag (GGBS) by 50% weight of cement	OPC 53 grade
[16]	Grinded	ABS	40, 50, 60	9.5–19	0.53	–			
[33]	Crushed	PHB (polyhydroxybutyrate)	5, 10, 15 By weight of coarse aggregate	100	0.46	M30	20%		OPC 53 grade
[20]	Pellets	ABS & PP	5, 10, 15, 20 By weight of coarse aggregate	3–6	0.45	M40	5%		OPC 53 grade

Thus, a test of its efficiency and its possible adoption in construction market as a sustainable alternative material for conventional concrete is a prerequisite. Strength test, like Compressive, Tensile, and Flexure, is thus analyzed.

12.3.2.1 Compressive Strength

Compressive strength (CS) is a property of concrete that determines its capacity to carry loads and is an important aspect of reinforced concrete design [31]. The concrete's durability is determined by its CS. The CS of the specimen was calculated by using the formula:

$$F_{ck} = P_c / A$$

P_c = Failure load in compression, in KN
A = Loaded area of the specimen, in mm^2

Lakshmi and Nagan [24] in their study observed that no significant increase in strength was noticed with replacement of E-plastic as CA in cement. Although at 4% and 8% replacement, durable strength was less but close to the control mix (M20). At 20% replacement, maximum strength reduction was observed. Successful replacement, however, was observed at 12% of E-plastic with 10% addition of flyash in cement, which made the concrete denser, giving result of 29.79 MPa, even higher than the CS of conventional mix 28.79MPa. Manatkar and Deshmukh [25] in their research work analyzed the durability of M20 and M25 concrete. The outcome observed from the research work was that 5% replacement of CA by E-plastic waste tends to have similar or near to strength when compared to its design mix while 15% replacement resulted in a maximal decrease of strength. Manjunath [26] tested the shear strength with 10, 20, and 30% replacement of CA and FA by E-plastic waste, wherein the author observed that 10% replacement of waste plastic in mix yielded satisfactory CS. Akram et al. [27] obtained a similar result when they replaced CA with shredded plastic aggregates partially, and obtained a 34% decline in CS at 15% of E-waste incorporation. However, after addition of 10% of flyash, the CS resembled to the strength of control specimen even at 15% replacement where a maximum decrease was observed. Kumar and Baskar [12] performed an experiment that comprised of 486 specimens manufactured by partial removal of CA and substitution of high impact PS (HIPS) plastic, which is often found in computers, by volume of 10, 20, 30, 40, and 50% for varied W/C (water cement) ratios (0.53, 0.49, and 0.45). They discovered that HIPS as aggregates preserves at least 50% strength under all test settings. The volumetric replacement of HIPS from CA demonstrated a linear decremented relationship, indicating that as HIPS concentration rose, CS decreased. As a result, the paper suggests utilization of such concrete in non-load bearing buildings such as partition walls and in earthquake-prone locations due to lightweight characteristic of mix consisting HIPS aggregate.

However, in the research study of Kurup and Kumar [32], they transformed outer casing of electrical wire into fibers. The addition of E-waste fibers resulted in an improvement in CS when present in 0.8% of the weight of cement. In their research, they indicated that E-plastic waste fibers of various lengths and shapes, particularly those of small size, will produce better outcomes than those of greater size. This

E-Plastic Waste

analysis is subsequently supported by the findings of Gull and Balasubramanian [29], where they discovered a 55.5% improvement in strength when E-plastic waste fiber of 3 cm length was employed [20]. Experimental inferences were also noticeable in their analysis of M40 specimen. They obtained a significant increase at 5 and 10% replacement of ABS and PP plastic mixed in equal proportions [30]. It indicated a marginal increase in CS up to 10% of E-plastic waste replacement and decrement trend in furthermore replacement which was similar in both 7 and 28 days tests. Similar inferences for CS in CA replacement by E-plastic were provided by Gavhane et al. [13], wherein he evaluated CA and FA CS at 7, 14, and 28 days for 10 and 20% replacement. A comparable result for FA was also obtained by Subramani and Pugal [33] who discovered that replacing 15% of the coarse material with recycled E-plastic resulted in an increase of 8% in CS when compared to the generic (control concrete) concrete mix.

Multiple research studies on E-waste incorporation in concrete, its influence on properties, and its behavioral reaction show that E-plastic waste can replace up to 30% of CA [28]. The partial substitution of conventional CA with recycled E-plastic waste has resulted in better physical and mechanical qualities of concrete mixes to some extent.

12.3.2.2 Splitting Tensile Strength

An important property of concrete is its tensile strength. Structural loads make concrete vulnerable to tensile cracking. Concretes, which are much prone to tensile failure, thus need reinforcement in the form of steel to counter the tension force. They are usually much lower than its CS, roughly 10–15% of CS. The split tensile strength of the specimen was calculated as:

$$F_{ct} = 2 * P / \pi * L * D$$

P = Maximum load in Newton applied to the specimen
L = Length of specimen, in mm
D = Cross-sectional dimension of specimen, in mm

Manjunath [26] reported that, after 28 days of split, tensile strength of concrete remained 4.8, 5.4, 3.8 N/mm^2 at replacement of 10, 20, 30% when compared to conventional mix having 4.9 N/mm^2. Akram et al. [27] tested split tensile strength for different proportions of E-plastic waste with or without 10% flyash addition at the end of 7, 14, and 28 days. The conclusion obtained indicated that up to 10% E-plastic waste addition by weight of CA, the split tensile strength of E-plastic aggregates replacement concrete was greater than the control concrete, but beyond that, a drop in split tensile strength was noted. The results of Kumar and Baskar [12] were analyzed and portrayed. HIPS replacement with split-tensile strength on different curing days for three distinct W/C ratios. The observations distinctly presented decrement in strength with increase in HIPS ratio. However, under splitting tension stress, control specimens were distorted in two halves post-suffering brittle rupture, but specimens containing HIPS aggregates did not experience brittle failure and were never divided into two halves. This suggested that HIPS concrete specimens failed in a ductile way due to the HIPS aggregate's capacity to endure substantial elastic deformation before breaking. Even when substantially broken, the failed specimens were able to bear quantifiable post-failure stresses. Noticeably the percentage curve for split-tensile strength reduction percentage of HIPS replacement, as well as the percentage curve for W/C ratios, resulted in a nearly linear relationship with the same slope. The percentage drop in tensile strength at 50% replacement after 28 days was 47.99, 47.89, and 40.48% for 0.53, 0.49, and 0.45 W/C, respectively. Despite the fact that the concrete specimens containing HIPS had unique properties, increasing the HIPS ratio resulted in a decrease in strength.

Gull and Balasubramanian [29] conducted their split tensile strength for 13 different mix proportions of Ordinary Portland Cement (OPC) + FA + CA and various plastic varying 0.4, 0.6, 0.8, and 1 for variable sizes of 3, 4, 5 cm concluded for their split tensile strength after 28 days varying from minimum 3.38 N/mm^2 for 13th mix (OPC + FA + CA + 1%, 3 cm)–3.6 N/mm^2 with mixes 9th (OPC + FA + CA + 1%, 4 cm) and 10th (OPC + FA + CA + 0.4%, 3 cm) having the maximum. Muthadhi et al. [30] also presented split tensile strength results of concrete from their experiment, which was that up to 20% replacement E-plastic waste aggregate at 28 days, splitting tensile strength shows 2.91 MPa in conventional mix and 3.43 MPa, 2.68 MPa in 10% and 20% E-plastic. Gavhane et al. [13] acquired 1.91 N/mm^2 as the tensile strength observation for 10% replacement at 28 days. This finding was consistent with E-plastic CS behavior. Integration of E-plastic waste marginally boosts the splitting tensile of concrete up to 10% and lowers above 10%. Subramani and Pugal [33] discovered that replacing 15% of the coarse material with recycled E-plastic resulted in increase of 3%, in split tensile strength when compared to the generic control concrete.

12.3.2.3 Flexural Strength Test

The bending strength or transverse rupture strength of concrete or the maximum stress in a material just before its bending evaluation is known as Flexural strength. The test indirectly analyzes concrete tensile strength by assessing the capacity of an unreinforced concrete beam or slab to sustain bending failure. The flexural strength of all concrete mixes is shown as a modulus of rupture for each specimen (MR) in MPa or psi. The specimen's flexural strength was calculated as follows:

$$F_b = P * 1 / b * d^2,$$

if a > 20.0 cm for 15 cm specimen
P = maximum Load (KN)
1 = Supported length (cm)
b = Width of specimen (cm) and d = Failure point depth (cm)

Manjunath [26] casted beams of dimensions $150 \times 150 \times 700$ mm for investigating flexural behavior of their altered specimens. The specimens are cured for 28 days in pure water before being evaluated on universal testing equipment (UTM). The conventional mix had a flexural strength of 4.35 N/mm^2 and flexural strengths of 4.4, 4.3, and 2.5 N/mm^2 for 10, 20, and 30% E-plastic incorporation. Gull and Balasubramanian [29] conducted an experiment using waste aggregate sizes of 5, 4, and 3 cm having replacement percentage of 0.4, 0.6, 0.8, and 1%. For these 13 variable specimens having varying permutation and combination of OPC, FA, CA, and PWA, they obtained flexural strength ranging from a minimum 4.5 N/mm^2 at 0% replacement for 5, 4, and 3 cm to a maximum 7 at 1% of 3 cm size (mix 13th). It is noticeable that mix 13th had a minimum split tensile strength while had maximum flexural strength, which is roughly double. Gavhane et al. [13] reported flexural strength of 2.74 N/mm^2 at a replacement percentage of 10% at 28 days, which was comparatively less when compared the conventional mix having 3.41 N/mm^2. Subramani and Pugal [33] discovered that replacing 15% of the coarse material with recycled E-plastic resulted in increase of 5% flexural strength original mix. However, on further replacement resulted in a loss of strength. He also noticed an oversupply of water in the mix, indicating that E-plastic trash had a low-water absorption rate when compared to CA. The experiment of Kumar and Baskar [12] was rather significant wherein beam specimens were loaded on the flexural testing machine having a capacity of 100 KN at a constant loading rate of 180 (kg/min). The conducted-experimental inferences of flexural strength depicted that with an increase in replacement percentage of HIPS content for 0.53, 0.49, 0.45 W/C ratios, the flexural strength decreases. The decrease percentage in flexural strength at 28 days with 50% replacement was observed to be as 41.58, 37.38, and 36.99% for 0.53, 0.49, and 0.45 W/C ratios, respectively. Although HIPS plastic resulted in a decrease in overall strength, it enhances the flexural behavior at ultimate load due to HIPS fibrous nature, i.e., even the control specimens of concrete have yielded and broken into two pieces at ultimate load. HIPS plastics were able to delay the failure of specimens post reaching the ultimate load benchmark and resisted further propagation of crack.

12.4 RESULT AND DISCUSSION

The potential for E-plastic embedded concrete is vast and withholds several new possibilities. The end products can be used successfully in non-load-bearing structures, like constructing potta cabins or partition walls. They are also being used to construct paver blocks, kerbs, finishing tiles, bricks, and even considered to be used as pavement material in roads, where freight weight is considerably low, i.e., school and college roads.

Post review of several articles and research papers, which have successfully integrated E-plastic in concrete, the following inferences can be derived:

I. E-Plastic material has a high-abrasion resistance. This material can be used to construct gutters, manhole covers, pipes of low-pressure flow, etc.

II. Impact strength is excellent, so it can be used as a pedestal for machines in workshops and colleges to absorb shock caused by impacts.

III. Incorporation of E-plastic waste as CA does not affect the workability of concrete, in fact when compared to the W/C ratio of conventional mix E-plastic concrete depicted higher workability.

IV. According to results, concrete containing E-plastic showed comparable resistance to aggressive chemical attack conditions like exhibiting better resistance to sulfate attack.

V. When replaced by E-plastic, volume of CA decreases. This results in reduced density in comparison to control mix, making it feasible for manufacturing lightweight structures, like temporary cabins at the construction sites. However, this change influences the strength of the concrete.

However, E-plastic utilized as a coarse and FA as an alternative in partial proportion for traditional aggregates resulted in a negative-linear relationship, i.e., with increase in E-plastic content decline in strength was obtained. Replacement of coarse and FA up to 10% resulted in a comparable strength with regards to design mix and was ideally observed in many studies. Further replacement of E-plastic in conventional concrete illustrated a decline in strength (compressive/tensile/flexural) with variation for 7, 14, and 28 days results. With respect to the strength of concrete, following opinions can be made:

I. When used as conventional aggregate, E-plastics in concrete mixes for a given weight-to-cement ratio, E-plastic waste reduces the density, CS, and tensile strength of the concrete. However, this was evident only after 10% replacement, i.e., below it according to the slump test results, all concrete mixes containing E-plastic waste were quite workable in their fresh state.
II. The decrease in strength of the specimen may be attributed to a lack of bond strength between the cementitious matrix of concrete and the E-plastic aggregate. Another factor is the hydrophobic nature of plastic which also results in improper bonding. To address this issue, post-material classification and before beginning of grinding process, replacers (E-plastic) must be processed in accordance with the texture requirement to increase the bond strength of materials with the cement.
III. The shape, texture, and size of aggregate also affected the strength properties of the concrete specimen. While Kumar and Baskar [12] observed that stress applied resulted in cracking at the interface between flaky and smooth textured HIPS aggregates and cement paste which caused specimens to fail at lower loads. Mainly because of poor bond strength between the E-plastic and cement pastes. However, Gull and Balasubramanian [29] depicted an increase in compressive, tensile, and flexural strength with varying sizes of E-plastic waste.
IV. Reduction in split tensile strength was influenced by the properties of the interfacial transition zone. As evident by Muthadhi et al. [30], the smooth surface of the plastic particles and the free water accumulated at the surface of plastic aggregate resulted in a weaker bond between the PVC particles and the cement paste.
V. Also a difference in specific gravity of the E-plastic aggregate and the conventional concrete materials resulted in a non-homogenous specimen. The E-plastic tends to move upward when compacted, resulting in a higher accumulation near the top layer surface. Causing subsequent reduction in concrete strength at higher replacement of E-plastic.
VI. A significant advantage of E-plastic concrete specimen is its increased ductility. E-plastic waste reduces the brittle failure of conventional mix.

E-Plastic Waste 229

The block on application of load initially compressed rather than failing abruptly during the test. This depicted the ability of concrete to significantly deform before failure. The gained ductility in the modified specimen is useful in seismic areas where it provides some time for clearance and in areas of harsh weather like contraction and expansion, or thaw and freeze.

VII. Addition of flyash, silica powder, and E-waste fiber when added to the concrete mix results in a slight increase in shear strength. It also improved the ductility behavior and reduced the brittle failure tendency of concrete.

12.5 CONCLUSION

India lacks a proper waste management sector, adequate mechanisms, and organizations to effectively counter the menace posed by the construction market and E-waste generation. Amalgamation of these hazardous and nonbiodegradable matters in the waste stream has caused environmental degradation and poses a serious threat to human health. Improper management and irresponsible attitudes toward the nuisance of such waste have only exposed the underlying issues. A treatment or waste-to-energy plant similar to those in developed countries might not be the best option for India, since affordability, population, and adaptability are all factors, which are different from those found in 1st world nations. Hence, the utilization of waste plastics, especially the plastics obtained from E-waste that has seen exponential waste generation in recent times, as aggregates in concrete, can offer a solution through which the majority of the waste plastics can be recycled, thereby increasing the possibilities of converting the plastics into housing and construction-related products. For this purpose, a detailed review of research literature was done in this chapter wherein integration of E-plastics in concrete has been done and its viability was tested.

Apart from the recycling of waste it will also reduce the rapid depletion of river sand and natural aggregates and ensure that the landfill remains free from nonbiodegradable and toxic substances of E-waste thereby reducing the already overstressed capacity of existing dumping grounds. However, in contrast to other streams of waste plastic (household plastics), a mere fraction of research has been conducted for E-plastics. Thus, utilization of such alternative replacements in large quantities must be analyzed and a feasible relation between waste E-plastic and its successful incorporation in concrete needs to be tested, wherein several behavioral characteristics of E-plastic concrete in different weather conditions and leaching shall be studied in the future.

REFERENCES

1. Y. Song and H. Zhang, "Research on Sustainability of Building Materials," *IOP Conf. Ser. Mater. Sci. Eng.*, vol. 452, p. 022169, Dec. 2018, doi: 10.1088/1757-899X/452/2/022169.
2. P. H. Brunner and H. Rechberger, "Waste to Energy – Key Element for Sustainable Waste Management," *Waste Manag.*, vol. 37, pp. 3–12, Mar. 2015, doi: 10.1016/j.wasman.2014.02.003.
3. E. Amasuomo and J. Baird, "The Concept of Waste and Waste Management," *J. Manag. Sustain.*, vol. 6, no. 4, p. 88, Nov. 2016, doi: 10.5539/jms.v6n4p88.

4. S. E. Vergara and G. Tchobanoglous, "Municipal Solid Waste and the Environment: A Global Perspective," *Annu. Rev. Environ. Resour.*, vol. 37, no. 1, pp. 277–309, Nov. 2012, doi: 10.1146/annurev-environ-050511-122532.
5. E. C. Simmons and M. Sanders, "Building Sustainable Communities for Sustainable Development: An EVIDENCE-BASED Behavior Change Intervention to Reduce Plastic Waste and Destructive Fishing in Southeast Asia," *Sustain. Dev.*, p. sd.2296, Feb. 2022, doi: 10.1002/sd.2296.
6. F. Furuoka and Q. Munir, "Is Population Growth Beneficial or Detrimental to Economic Development? New Evidence from Pakistan," *J. Popul. Soc. Stud.*, vol. 18, no. 2, Jan. 2010, Accessed: Feb. 20, 2022. [Online]. Available: https://www.researchgate.net/publication/235935626_Is_Population_Growth_Beneficial_or_Detrimental_to_Economic_Development_New_Evidence_from_Pakistan
7. C. Chauhan, "Urbanisation in India faster than rest of the world | Latest News India - Hindustan Times," Jun. 28, 2007. Accessed: Feb. 19, 2022. [Online]. Available: https://www.hindustantimes.com/india/urbanisation-in-india-faster-than-rest-of-the-world/story-IdmQ4BSqxEZe874AprzfnL.html
8. S. Nihas, K. Barlish, and D. Kashiwagi, "Construction Industry Structure in India," Sep. 2013.
9. Md. Masuduzzaman, S. K. S. Amit, and Md. Alauddin, "Utilization of E-Waste in Concrete and Its Environmental Impact – A Review," in *2018 International Conference on Smart City and Emerging Technology (ICSCET)*, Mumbai, Jan. 2018, pp. 1–4. doi: 10.1109/ICSCET.2018.8537301.
10. S. B. Hegde, "Sustainability in Indian Cement Industry – Challenges and Avenues," *NBM&CW Infra Construction & Equipment Magazine*, Dec. 17, 2021. https://www.nbmcw.com/product-technology/construction-chemicals-waterproofing/concrete-admixtures/sustainability-in-indian-cement-industry-challenges-and-avenues.html (accessed Feb. 01, 2022).
11. S. Bharadwaj, D. Tewari, and B. Natarajan, "Emission Reduction Approaches for the Cement Industry – Alliance for an Energy Efficient Economy," *Alliance for an Energy Efficient Economy (AEEE)*, Feb. 04, 2021. https://aeee.in/emission-reduction-approaches-for-the-cement-industry/ (accessed Nov. 13, 2021).
12. K. Senthil Kumar and K. Baskar, "Development of Ecofriendly Concrete Incorporating Recycled High-Impact Polystyrene from Hazardous Electronic Waste," *J. Hazard. Toxic Radioact. Waste*, vol. 19, no. 3, p. 04014042, Jul. 2015, doi: 10.1061/(ASCE)HZ.2153-5515.0000265.
13. A. Gavhane, D. Sutar, S. Soni, P. Patil, and R. M. D. S. S. of E. Pune, "Utilisation of E – Plastic Waste in Concrete," *Int. J. Eng. Res.*, vol. V5, no. 02, p. IJERTV5IS020538, Feb. 2016, doi: 10.17577/IJERTV5IS020538.
14. PTI, "Over 34 lakh tonnes of plastic waste generated in FY 2019-20: Government – The New Indian Express," *The New Indian Express*, Jul. 19, 2021. Accessed: Nov. 13, 2021. [Online]. Available: https://www.newindianexpress.com/nation/2021/jul/19/over-34-lakh-tonnes-of-plastic-waste-generated-in-fy-2019-20-government-2332396.html
15. D. K. A. Barnes, F. Galgani, R. C. Thompson, and M. Barlaz, "Accumulation and Fragmentation of Plastic Debris in Global Environments," *Philos. Trans. R. Soc. B Biol. Sci.*, vol. 364, no. 1526, pp. 1985–1998, Jul. 2009, doi: 10.1098/rstb.2008.0205.
16. M. Sabau and J. R. Vargas, "Use of E-Plastic Waste in Concrete as a Partial Replacement of Coarse Mineral Aggregate," *Comput. Concr.*, vol. 21, no. 4, pp. 377–384, Apr. 2018, doi: 10.12989/CAC.2018.21.4.377.
17. S. Sun, "India: device ownership 2020," *Statista*, Apr. 28, 2021. https://www.statista.com/statistics/1228293/india-device-ownership/ (accessed Nov. 14, 2021).

18. K. Monish, J. J. Jesuran, and S. Kolathayar, "A Sustainable Approach to Turn Plastic Waste into Useful Construction Blocks," in *Smart Technologies for Sustainable Development*, vol. 78, S. K. Shukla, S. Chandrasekaran, B. B. Das, and S. Kolathayar, Eds. Singapore: Springer Singapore, 2021, pp. 55–62. doi: 10.1007/978-981-15-5001-0_5.
19. A. J. Babafemi, B. Šavija, S. C. Paul, and V. Anggraini, "Engineering Properties of Concrete with Waste Recycled Plastic: A Review," *MDPI*, no. 3875, p. 26, Oct. 2018, doi: https://doi.org/10.3390/su10113875.
20. Anand and A. Hamdan, "Impact of Partial Replacement of Coarse Aggregate with Electronic Plastic Waste on Compressive Strength of Concrete," *Mater. Today Proc.*, vol. 56, pp. 143–149, 2022, doi: 10.1016/j.matpr.2021.12.573.
21. L. Gu and T. Ozbakkaloglu, "Use of Recycled Plastics in Concrete: A Critical Review," *Waste Manag.*, vol. 51, pp. 19–42, May 2016, doi: 10.1016/j.wasman.2016.03.005.
22. B. W. Jo, S. K. Park, and J. C. Park, "Mechanical Properties of Polymer Concrete Made with Recycled PET and Recycled Concrete Aggregates," *Constr. Build. Mater.*, vol. 22, no. 12, pp. 2281–2291, Dec. 2008, doi: 10.1016/j.conbuildmat.2007.10.009.
23. A. Sadrmomtazi, S. Dolati-Milehsara, O. Lotfi-Omran, and A. Sadeghi-Nik, "The Combined Effects of Waste Polyethylene Terephthalate (PET) Particles and Pozzolanic Materials on the Properties of Self-Compacting Concrete," *J. Clean. Prod.*, vol. P4, no. 112, pp. 2363–2373, 2016, doi: 10.1016/j.jclepro.2015.09.107.
24. R. Lakshmi and S. Nagan, "Investigations on Durability Characteristics of E-Plastic Waste Incorporated Concrete," *ASIAN J. Civ. Eng.*, vol. 12, no. 6, pp. 773–787, Apr. 2011.
25. P. Manatkar and G. P. Deshmukh, "Use of Non-Metallic E-Waste as a Coarse Aggregate in a Concrete," *Int. J. Res. Eng. Technol.*, vol. 04, pp. 242–246, Mar. 2015, doi: 10.15623/ijret.2015.0403040.
26. B. T. A. Manjunath, "Partial Replacement of E-plastic Waste as Coarse-Aggregate in Concrete," *Procedia Environ. Sci.*, vol. 35, pp. 731–739, Jan. 2016, doi: 10.1016/j.proenv.2016.07.079.
27. A. Akram, C. Sasidhar, and K. M. Pasha, "E-Waste Management by Utilization of E-Plastics in Concrete Mixture as Coarse Aggregate Replacement," *ISSNOnline 2319-8753ISSN Print 2347-6710 Int. J. Innov. Res. Sci. Eng. Technol.*, vol. 4, no. 7, p. 9, Jul. 2015.
28. A. R. Kurup and K. Senthil Kumar, "Effect of Recycled PVC Fibers from Electronic Waste and Silica Powder on Shear Strength of Concrete," *J. Hazard. Toxic Radioact. Waste*, vol. 21, no. 3, p. 06017001, Jul. 2017, doi: 10.1061/(ASCE)HZ.2153-5515.0000354.
29. I. Gull and M. Balasubramanian, "IJETT – A New Paradigm on Experimental Investigation of Concrete for E-Plastic Waste Management," *Int. J. Eng. Trends Technol. – IJETT*, vol. 10, no. 4, 2014, doi: 10.14445/22315381/IJETT-V10P234.
30. A. Muthadhi, A. M. Basid, R. Madivarma, J. S. Kumar, and R. Raghuvarman, "Experimental Investigations on Concrete with E-Plastic Waste," vol. 8, no. 6, p. 5, 2017.
31. A. Masi, A. Digrisolo, and G. Santarsiero, "Analysis of a Large Database of Concrete Core Tests with Emphasis on within-Structure Variability," *Materials*, vol. 12, no. 12, p. 1985, Jun. 2019, doi: 10.3390/ma12121985.
32. A. R. Kurup and K. S. Kumar, "Novel Fibrous Concrete Mixture Made from Recycled PVC Fibers from Electronic Waste," *J. Hazard. Toxic Radioact. Waste*, vol. 21, no. 2, p. 04016020, Apr. 2017, doi: 10.1061/(ASCE)HZ.2153-5515.0000338.
33. T. Subramani and V. K. Pugal, "Experimental Study on Plastic Waste as a Coarse Aggregate for Structural Concrete," vol. 4, no. 5, p. 9, 2015.

13 Green Synthesis of Sustainable Materials
A Stride toward a Viable Future

Mandeep Kaur, Gagandeep Kaur,
and Divya Sareen

CONTENTS

13.1 Introduction 233
13.2 Green Synthesis of Polymers 234
    13.2.1 Use of Renewable Raw Materials 235
    13.2.2 Use of Sustainable Solvents/Methods 236
    13.2.3 Use of Carbon Dioxide (CO_2) as a Monomer 238
    13.2.4 Use of Industrial Wastes 239
13.3 Green Synthesis of Metal and Metal Oxide Nanoparticles 239
    13.3.1 By Using Microbes 241
        13.3.1.1 Bacteria 241
        13.3.1.2 Algae 241
        13.3.1.3 Yeast 242
        13.3.1.4 Fungi 242
    13.3.2 By Using Plants 242
13.4 Green Synthesis of Ceramics 244
    13.4.1 Agricultural Wastes as Raw Material 244
    13.4.2 Industrial Wastes as Raw Material 246
    13.4.3 Plant Extracts as Raw Material 249
13.5 Conclusion 250
References 251

13.1 INTRODUCTION

The concept of sustainability has developed from our growing obligation to maintain the available environmental resources along with a good quality of life over prolonged time. This necessitates the employment of materials, which can be produced in quantitative amounts without deprivation of the nonrenewable natural resources, i.e. sustainable materials synthesized using the green synthetic approaches. Such materials include polymers, ceramics, composites, nanomaterials, etc., all of which are either synthesized using biological/natural resources or are recyclable. These sustainable materials prove to be better alternatives to the existing, nonrenewable

DOI: 10.1201/9781003297772-13

sources-based scientific, technological, and industrial materials. They can be applied to a wide range of applications in a variety of fields, such as construction, agriculture, industry, science and technology, etc. But a significant challenge lies in the identification of such materials which are environmentally sustainable and cost-effective, along with maintaining the requisite physical, material, and chemical properties for their intended applications. Another important consideration concerning sustainable materials is the availability of a green approach to their synthesis. "Green Synthesis" works on the basic principle of minimization of pollution-causing waste production and use of safer as well as recyclable feedstock. The synthetic route should most often involve sustainable starting materials and involve energy-conserving reactions and processes, with minimal waste production. This chapter summarizes the "Green Synthesis" approaches to a few of the ubiquitous sustainable materials, i.e. polymers, metal and metal oxide nanoparticles, and ceramics, covering the recent developments made in this field, mainly in the last five years.

13.2 GREEN SYNTHESIS OF POLYMERS

Polymers have gained popularity and have acquired an important place in the life of every single human [1]. Based on the origin of raw materials, polymers may be broadly classified into natural polymers (e.g. starch, natural rubber, and wood) and synthetic polymers (e.g. polyethylene and teflon). The increasing dependency of mankind on polymers has led to a paradigm shift in the source of the raw materials used for procuring the polymers, from natural reserves to the chemical-based synthesis due to tailor-made properties that may be appended through chemical modifications. To synthesize polymers with the required strength and characteristics, this area of research has become one of the most expectant themes of investigation among researchers of academic institutes and industries. However, the exponential growth rate of the synthetic polymers has faced an unexpected resentment due to the drastic environmental degradation that these polymers have led to. The magnificent quality of synthetic polymers that has made them own the world, i.e. resistance toward any degradation in natural environment, has in fact resulted in the accumulation of the waste polymers in our ecosystem up to a level of disaster and the situation is predicted to be worse in the coming decade when hundreds of million metric tons of plastic waste will be accumulated in our environment [2]. Only coordinated global actions can encourage the detoxification caused by excessive build-up of the plastic waste that may include reducing the use of plastic, developing methods for recycling [3], and innovating new sustainable substitutes. Worldwide campaigning for discouraging the excessive use of plastics and new inventions in the field of sustainable polymers are the result of these actions.

Sustainable polymers are materials obtained from renewable feedstock (non-petroleum-based), generally with closed-loop life cycles [4]. These materials possess a wide range of applications ranging from their use as anti-corrosive [5], food packaging industry [6], nano sustained-release fertilizer [7], in addition to its well-known use in plastic, medicine, and semiconductor industry.

The "Green Synthesis" of these sustainable polymers is trusted to be a sole alternative for addressing the deteriorating polymer pollution being created worldwide

Green Synthesis of Sustainable Materials 235

FIGURE 13.1 Adoptable measures for green synthesis of sustainable polymers.

through the use of synthetic polymers and their chemical-based synthesis. The prime requirement for the green production of sustainable polymers is the use of renewable raw materials, consumption of lesser water, emission of lesser greenhouse gases, reduced quantity of waste production, and easy degradation of discarded polymers. This approach includes synthesizing polymers by chemically modifying the natural polymers, using safer solvents/catalysts, assessing renewable biomass, and agricultural/industrial waste as starting materials and devising strategies for its recycling and biodegradation (Figure 13.1).

13.2.1 Use of Renewable Raw Materials

Renewable feedstock, such as cellulose [8], lignin [9], sugar [10], etc., has been explored much to pave a route toward synthesis of sustainable polymers. Similarly, terpene-based monomers, e.g. polyisoprene, pinene, limonene, pinocarvone, myrcene, ocimene, and farnesene, have been used for the preparation of polymers due to the versatile structure and olefinic character, which makes them easy to functionalize. Monica et al. have elaborated the use of terpenes as raw materials for the fabrication of sustainable polymers including polycarbonates, polyesters, polyurethanes, and polyamides [11].

Likewise, mono-functionalized fatty acids (e.g. alcohol, amide, and ester derivatives of fatty acids) are processed to obtain monomers, which are then polymerized. Zhou et al. developed three strategies to synthesize thermoplastic polymers from renewable fatty acids using 1,1,3,3-tetramethylguanidine (TMG) as the catalyst [12]. TMG catalyzes the esterification of halogenated alcohols with acrylic acid or poly(meth)acrylic acid with 100% conversion efficiency. Moreover, TMG is insensitive toward the presence of moisture, oxygen, or radical initiators. To study the process, three strategies were developed using 4-vinylbenzyl chloride and poly(4-vinylbenzyl chloride) as the monomers to obtain poly(4-vinylbenzyl oleate) as the final product after subsequent polymerization and esterification reactions in varying order or performing both reactions in a concerted step as a single-pot synthesis. It was observed that carrying out the esterification reaction as the first step or both the reactions in a single step, exceeded 99% conversion efficiency, whereas proceeding the polymerization step prior to esterification reduced the conversion efficiency to 78%. This trend is owed to the fact that the rate of polymerization reaction is faster than esterification.

In addition to the use of natural raw materials, the methodology of synthesizing bio-based monomers has gained extensive focus in the field of green synthesis of sustainable materials. A recent research work has described the use of bio-based furfural, glycerol, and lactic acid as the raw materials for the synthesis of methacrylate monomers proceeding through a transesterification reaction [13]. The polymerization was carried out through a free-radical mechanism using 1,1-azobis cyclohexane carbonitrile as the initiator leading to the formation of a thermally stable copolymer that was reinforced to form an amorphous polymer composite. This composite was found to degrade in aqueous acidic and basic media up to 10% in 24 hours. As a result, it was deployed in the form of thin films as water filtration membranes. Likewise, Thakur in his recent paper has elaborated the production of bio-based succinic acid (by the use of microbes) as a monomer to synthesize a biodegradable polymer namely poly(butylene succinate) [14].

13.2.2 Use of Sustainable Solvents/Methods

The solvents are ranked as sustainable and thereby green based on the extent of harm they cause to humankind or the environment [15]. Whereas, for a method to be ascertained as green, it is assessed in terms of two mass-based metrics, namely atom economy and the environmental factor (E factor) [16]. Atom economy is the percentage of the ratio of molecular weights of the product and the reactants. A reaction with a higher atom economy is considered greener. Whereas, the E factor is a ratio of the total mass of the waste generated in the reaction over the mass of the required product. Thus, a larger E factor indicates higher waste production of the reaction indicating it as potentially hazardous toward the environment.

Hence, important key guiding rules toward sustainable polymer synthesis include incorporation of fewer steps and use of nonhazardous solvents. However, the synthesis of conjugated polymers mainly routes through the Stille or Suzuki C-C couplings, which involve the use of cryogenic conditions and hazardous metallic reagents as well as lead to the production of toxic by-products. Rather than choosing the C-C coupling reactions, generating the bond *via* straight functionalization of C–H bonds through direct arylation polymerization (DArP) in the presence of an appropriate solvent may prove as a more sustainable method. Pankow reported a DArP reaction providing high yield of poly[2,5-bis(2hexyldecyloxy) phenylenealt-(4,7-dithiophen-2-yl)benzo[c][1, 2, 5]thiazole] using cyclopentyl methyl ether (CPME) as the solvent as compared to the traditional solvent (Tetrahydrofuran) [17]. CPME proved to be a better solvent for DArP, resulting in the synthesis of polymers with negligible defects and larger average molecular weight, adding to the advantage of its zero waste production and non-radical

Green Synthesis of Sustainable Materials

initiator behavior. In a subsequent paper, Pankow et al. reported the use of CPME and anisole as solvents in the DArP reactions of ester-functionalized reactants [18]. The analysis of the polymers procured through this method has been used for the deduction of a structure-property relationship between the nature of the substituent appended and the defects observed in the synthesized polymers. It was affirmed that electron-withdrawing substituents enhance the activation of the adjacent C–H bonds resulting in the formation of crosslinks, i.e. branch defects in polymers, whereas electron-donating substituents reduce the defect formation in polymers.

Further, to increase the atom economy of the synthesis, curtailing the number of steps required for polymer production is an appreciable measure. An attempt for the same has been done for the synthesis of sustainable polyesters. Polyesters, such as poly(butylene terephthalate), are a class of sustainable polymers, which due to high-thermal stability and good mechanical properties are used as engineering plastics. However, to address the issue of its inherent brittleness, various ductile and flexible bifunctional monomers are incorporated into the structure through the two-step condensation-polymerization method [19]. Xu et al. have demonstrated an easy methodology for incorporation of functional groups into the polymer backbone through a green cascade polycondensation-coupling ring-opening polymerization method rather than opting for the traditional condensation-polymerization method [20]. This modification proved to end up as a high-atom economy method. The reaction involved the synthesis of poly(butylene-co-decylene terephthalate) copolyesters using bio-based 1,10-decanediol and cyclic oligo(butylene terephthalate) as monomers. Insertion of decanediol in poly(butylene terephthalate) led to enhanced toughness of the system for potential applications as sustainable polymers in plastic packaging, adding to the advantage of being a high-atom economy (87–99%) process due to lowering of the number of steps in the reaction.

Likewise, a recent publication described the cascade polycondensation-coupling ring-opening polymerization methodology for the production of poly(ethylene 2,5-furandicarboxylate)-*block*-poly(tetramethylene oxide), in place of the traditional condensation-polymerization technique that gave lower yields due to side reactions [21]. The route of rapid cascade polymerization decreased the occurrence of side reactions leading to procurement of polymer with high number-average molecular weight. Further, a structure-property relationship was deduced that depicted an increase in the strength of the polymer upon increasing the poly(ethylene 2,5-furandicarboxylate) content. The poly(ether ester)s obtained exhibited excellent crystallinity and biodegradability (controlled by the hard-soft ratio), thus producing thermoplastic elastomers possessing tailor-made properties.

Similarly, Zschech et al. have reported the synthesis of efficient polymeric materials adopting the electron-induced reactive processing methodology, i.e. initiating the reaction through continuous irradiation of high-energy electrons [22]. All temperature-dependent synthetic processes, i.e. melt mixing, heat transport, and chemical reaction, required a close check of temperature throughout the process because of the temperature-sensitive behavior of the chemical initiators being used during the production.

13.2.3 Use of Carbon Dioxide (CO_2) as a Monomer

CO_2, being one of the main gases responsible for the greenhouse effect, is a major contributor toward global warming. Thus, its consumption in the production of value-added materials, such as urea, methanol and cyclic organic carbonates to name a few, is an environmental benign process [23]. The organic carbonates are synthesized through the addition of CO_2 to epoxides. When the epoxides used are extracted from the renewable natural resources (such as cellulose, lignin, or vegetable oils), the attack of CO_2 ultimately ends up in the production of sustainable polycarbonates which can further yield polyurethanes upon reaction with diamines [24]. Bobbink has also reported the synthesis of polycarbonates using ionic liquids as the catalyst through the solvent-free CO_2-epoxide coupling reaction (i.e. ring opening upon attack of anion of the ionic liquid to be used as catalyst) resulting in the formation of an alkoxide, followed by insertion of CO_2 and ring-closing to obtain polycarbonates as product [25]. It was determined that the reactivity of reaction depends upon the structure of epoxides and epichlorohydrin was determined to be the most reactive when compared with propylene oxide or ethylene oxide (low boiling, thus requiring reduced pressure conditions) and internal epoxides, such as cyclohexene oxide or limonene oxide (due to lower reactivity of tertiary carbons), in the presence of homogeneous catalyst.

Similarly, Bai et al. routed the synthesis of carbonates, namely bis(4-formyl-2-methoxyphenyl)carbonate and bis(4-(hydroxymethyl)-2-methoxyphenyl)carbonate using vanillin-based epoxides and CO_2 as the precursors. These monomers may be used for the synthesis of poly(carbonate ester) oligomers possessing amide chains as substituents, prepared through Passerini reaction [26].

Muthuraj et al. reported the use of CO_2 for the synthesis of biodegradable polymers, i.e. polycarbonates and poly(propylene) carbonates as the potential feedstock for copolymer and polymer blend generation [27]. The prime requirement for the use of low-energy inert CO_2 molecule is the presence of a catalyst that may provide the required amount of activation energy for the reaction to occur. This has been done through the oxidative-coupling of CO_2 with unsaturated compounds using low-valency metal complexes or by increasing the electrophilicity of the carbonyl carbon [28]. Thus, the aliphatic polycarbonates obtained using CO_2 as the feedstock included latter's reaction with the high-energy oxirane moiety in the presence of a suitable catalyst.

Shaarani et al. reported a terpolymerization reaction between natural epoxidized soybean oil, propylene oxide, and CO_2 under high pressure in the presence of Co-Zn double metal cyanide as a heterogeneous catalyst to obtain a bio-based polymer poly(propylene carbonate) with a yield of 72%, which validated the replacement of petroleum-based epoxide with its natural counterpart [29].

Tang et al. have reported an altered reaction of CO_2 involving its insertion in polyethylenes, proceeding through copolymerization of ethylene with a lactone derived from CO_2 and butadiene [30]. Two methodologies were employed with each of them leading to differently substituted polyethylenes as products. The first approach routing through palladium-catalyzed coordination, followed by insertion of CO_2 resulted in the formation of polyethylenes appended with unsaturated lactones, whereas the

Green Synthesis of Sustainable Materials

second route following radical copolymerization process gave polyethylenes substituted with saturated bicyclic lactones proceeding through Michael addition.

13.2.4 Use of Industrial Wastes

The material scientists strive to generate sustainable and biodegradable polymers from renewable raw materials to minimize the environmental deterioration issues. However, if this can be done through the use of effluent waste of the industry, it will serve a dual benefit of saving the ecosystem from degradation and producing sustainable polymers.

Sulfur is a major by-product of hydrodesulfurization process carried out in petroleum refineries, leading to production of millions of tons of elemental sulfur in a year. The main consumption of elemental sulfur is done for the industrial production of sulfuric acid and to a comparative smaller extent in fertilizer and rubber industries [31]. Thus, a very large excess of elemental sulfur is left useless in the petroleum refineries. The synthesis of polymers from the waste sulfur may serve as a measure for sustainable production of polymers. The polymeric material from elemental sulfur with high-sulfur-content possessing polyenes as crosslinkers is synthesized through a simple method with high atom efficiency, known as inverse vulcanization [32]. Inverse vulcanization process has been postulated to address the issues of long reaction time and production of insoluble materials due to crosslinking [33]. In an attempt to further increase the sulfur content in the polymer, Duarte et al. reported the use of a renewable monomer possessing disulfide linkage [34]. Starting from this monomer to build the polymer involved incorporation of an additional S-atom between the two sulfurs of disulfide through an organocatalyst (1,5,7- triazabicyclo[4.4.0]dec-5-ene)-triggered sulfur exchange reaction between the polymer and elemental sulfur (S_8).

Sanvezzo et al. have reported the production of sustainable polymer composites through recycling of natural jute fiber and polypropylene collected from the waste produced in a carpet manufacturing industry [35]. The jute fibers act as a strengthening component in composites of polypropylene as a matrix. After processing of the material through thermo-pressing technique, it has been deployed as an automotive interior. Further, a three-component system possessing the polypropylene fibers as matrix along with industrial waste and nano-calcium carbonate particles as reinforcement material leads to the development of a strong polymer that can substitute both plastic as well as wooden stuff, indoor as well as outdoor.

13.3 GREEN SYNTHESIS OF METAL AND METAL OXIDE NANOPARTICLES

Due to an increase in demand for metal and metal oxide nanoparticles, large-scale synthesis *via* high-energy processes with diverse hazardous solvents has become necessary. But this pollutes the environment, thereby necessitating the development of environmentally acceptable green synthetic processes. Bioresources, such as plants, fungi, bacteria, algae, yeast, etc., can be used as a potential substitute to produce metal and metal oxide nanoparticles. In comparison to other methods, green synthesis has minimal side effects and is harmless for living beings and environment, making it the ideal way for generating metal and metal oxide nanoparticles [36]. Because of their huge surface area, high surface energy, and quantum confinement, nanoparticles possess a variety of exceptional optical, magnetic, and electrical properties [37]. Metallic nanoparticles and metal oxide nanoparticles can be synthesized in a variety of ways, including chemical, physical, and biological methods [38]. These numerous physicochemical and biological methods are classified based on two approaches: top-down approach and bottom-up approach. Nanoparticles are synthesized in the top-down approach by reducing the size of starting material [39]. On the other hand, in bottom-up approach, nanoparticles are formed from smaller units such as atoms and molecules [40].

The physical methods employed for the formation of nanoparticles have many shortcomings like requiring long time to establish thermal stability, wasting a lot of energy to achieve the temperature around the source material, and, in the case of tube furnaces, taking up a lot of space [41]. On the other hand, the use of harsh reducing agents and harmful solvents is the main drawback of chemical methods [42]. Thus, to overcome the above shortcomings of both physical and chemical processes, nontoxic and eco-friendly biological (green) methods are considered for the synthesis of metal and metal oxide nanoparticles. These biological (green) methods involve the bottom-up approach for the synthesis of nanoparticles in which plants and other microbes are used for the bioreduction of salts of metal into the metallic form. The synthesis of nanoparticles *via* green methods is carried out by the means of microbes such as bacteria, fungi, viruses, yeast, and plants (Figure 13.2).

Green synthesis process is affected by different factors such as temperature, pH, solvent system, and pressure. The phytochemicals present in plants, like phenols, ascorbic acids, carboxylic acids, amides, terpenoids, flavones, aldehydes and ketones, etc., are the key components which help in bioreduction of metal salts to metal nanoparticles [43-45]. The process of green synthesis is more efficient, simpler, economical, and environment-friendly and can easily be scaled up to perform larger operation [46]. The nanoparticles prepared by using green synthesis have specific morphology, shape, and size and are thus successfully used in various fields. The biosynthesized nanoparticles *via* green methods are free of toxic chemicals and are eco-friendly.

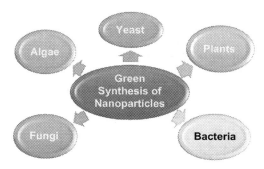

FIGURE 13.2 Green synthesis of metal and metal oxide nanoparticles *via* different routes.

Green Synthesis of Sustainable Materials   241

The extracellular and intracellular approaches have been used for the biosynthesis of nanoparticles in green methodology concept. The integration of metallic nanoparticles in intracellular [47] approach involves the use of metal-rich environment whereas the extracellular approach comprises of the use of bio-extracts procured from various sources. The characterization of synthesized nanoparticles is carried out using various analytical techniques such as UV-visible spectroscopy, which reveals the bioreduction of metal salt by observing the surface plasmon resonance peak. Scanning electron microscopy (SEM) and transmission electron microscopy (TEM) techniques give the information regarding morphology, size, and shape of the nanoparticles. The crystalline structure of the nanoparticles is determined with the help of X-ray powder diffraction (XRD).

13.3.1 By Using Microbes

13.3.1.1 Bacteria

Different types of bacteria are used to reduce the metal ions and for the synthesis of nanoparticles. The shape, size, and morphology of the nanoparticles can be altered by the use of bacteria. The synthesis of nanoparticles is carried out specifically using prokaryotic bacteria and actinomycetes [48]. The synthesis of metal and metal oxide nanoparticles using bacteria is carried out through two different approaches, namely extracellular and intracellular approaches. Extracellular biosynthesis of nanoparticles has the benefit of being a short process in comparison to intracellular process, as it does not require a downstream mechanism to collect nanoparticles from microorganisms. The reductase enzyme found inside bacteria's cells facilitates the reduction of inorganic metal ions to metal nanoparticles. *Deinococcus radiodurans* bacterial species have high antioxidant activity and are oxidative stress and radiation resistant [49]. The ionic form of this species made it a suitable candidate for the synthesis of gold nanoparticles and the nanoparticles thus obtained were more stable and had better antibacterial action. The silver nanoparticles of different morphologies and sizes are synthesized using various bacteria such as *Bacillus indicus*, *Pseudomonas antarctica*, *Klebsiella pneumonia*, and *Pseudomonas meridian*. *Pseudomonas fluorescens*, *Marinobacter pelagius*, *Plectonema boryanum* UTEX 485, and *Bacillus megaterium* D01 bacteria are used for reducing the gold metal ions to gold metal nanoparticles [50].

13.3.1.2 Algae

The formation of nanoparticles using algae as a reducing agent has been reported in some studies. The preparation of metal nanoparticles by means of algae involves three steps: (i) preparation of algal extract in water/organic solvent by heating for a specific time, (ii) preparation of solutions of metal salts of a particular strength, and (iii) incubation of both the above solutions under certain conditions followed by stirring. The synthesis of stable monodisperse gold nanoparticles with size 8–12 nm, confirmed by different characterization techniques, has been reported by the use of *Sargassum wightii* algae [51]. Similarly, *Sargassum wightii* algae were also used for the preparation of silver nanoparticles [52]. The synthesis of silver nanoparticles was also carried out by means of *Jania rubens* and *Ulva fasciata* algae [53]. Further, several other metallic nanoparticles have also been synthesized from various types of algae such as eukaryotic algae, marine algae, and marine cyanobacterium, *P. boryanum*, *Turbinaria conoides*, and *Jania rubens* [54]. In another work, the biological synthesis of silver nanoparticles has been demonstrated by using aqueous red algae Portieria *hornemannii* [55].

13.3.1.3 Yeast

Yeast is considered as the most preferred microbe for the green synthesis of nanoparticles owing to fast growth of their colonies even in common nutrient cultures and easy-to-control mass production. Moreover, nanoparticles are synthesized by microbes through the process of biomineralization, which is the most likely process in yeast to produce minerals for the rigorousness of their structure. In addition to this, yeast cells need no stabilizer for proceeding the encapsulation mechanism during the synthesis of nanoparticles. The intracellular formation of nanoparticles involves passage of metal salts into the cellular system through diffusion and removal of the salts from the extracellular space, followed by the reduction of the diffused metal salts using reducing agents (biomolecules), which are present in the cell or are transported into the cells [56].

On the contrary, the extracellular formation of the nanoparticles takes place at the surface of the microbe cells. Peiris has reported the synthesis of TiO_2 nanoparticles of less than 12 nm using *Saccharomyces cerevisiae* yeast cultured in sterile glucose solution (5% w/v) and using $TiCl_3$ as the metal salt. The nucleation of the nanoparticles was carried out by the macromolecules attached onto the porous surface of the yeast cells, which upon subsequent heat treatment led to the synthesis of porous lamellar structure containing TiO_2 nanoparticles and the yeast cells are removed [57].

13.3.1.4 Fungi

As fungi release a significant number of proteins, so these are used widely in the production of nanoparticles with certain sizes, structures, and characteristics. Fungi are used to make metal nanoparticles as a result of the presence of specific intracellular enzymes. Fungi are better at producing nanoparticles than other microbes because they grow faster and are easier to create and manage in the lab [58]. Further, many researchers reported that fungi is capable of producing the extracellular metabolites that help them to protect themselves from toxic compounds (such as metallic ions). The fungal mycelium is usually preserved in a salt solution during the production of metal nanoparticles. This prompts the fungus to create metabolites and enzymes necessary for its own survival. Using the catalytic effect of an extracellular enzyme and fungal metabolites, toxic metal ions are transformed into nontoxic metallic solid nanoparticles. The synthesis of gold nanoparticles has been reported using *Colletotrichum sp.* fungi whereas *Aspergillus fumigatus fresenius* fungi are known for the preparation of silver nanoparticles [55].

13.3.2 By Using Plants

Plant extracts can be used to effectively produce nanoparticles since they behave as chemical factories. The methods of synthesis of metal and metal oxide nanoparticles

Green Synthesis of Sustainable Materials

using plants are acquiring considerable attention because of their potential for bulk synthesis, elimination of the need for cell culture, and fabrication of sustainable nanoparticles with specified sizes and shapes. Moreover, they are cost-effective, simple, efficient, and feasible. The various parts of plants such as leaves, roots, stems, and fruits contain the phytochemicals, which are used in the synthetic process. The bioreduction of metal ions to metallic nanoparticles is carried out by means of a wide range of phytochemicals present in plants, like carboxylic acids, amides, aldehydes, sugars, ketones, flavones, and terpenoids [59].

Plants have mainly been used to synthesize gold [60] and silver [61] nanoparticles. Numerous plants are available for the synthesis of nanoparticles: mustard (*Brassica juncea*), lemon grass (*Cymbopogon flexuosus*), *Ginkgo biloba*, coriander (*Coriandrum sativum*), neem (*Azadirachta indica*), grape (*Vitis*), *Cydonia oblonga*, lemon (*Citrus limon*), tulsi (*Ocimum sanctum*), aloe vera (*Aloe barbadensis*), and oats (*Avena sativa*). The synthesis of various types of metallic and metal oxide nanoparticles using whole plants and parts of plants have been reported by many research groups and these reports are listed in Table 13.1.

TABLE 13.1
List of Metal and Metal Oxide Nanoparticles Using Whole Plants/Plant Parts

S. No.	Metal/Metal Oxide Nanoparticles	Whole Plants/Plant Parts	Size (in nm)	Shape	Ref. No.
1.	Silver	Reishi mushroom extract (*Ganoderma lucidum*)	15–22	Spherical	[61]
2.	Silver	Tomato (T), Onion (O), Acacia catechu (C) alone and mixed COT extracts	5–100	-	[62]
3.	Silver	Dried grass	15	–	[63]
4.	Silver	*Moringa oleifera* flowers	8	Spherical	[64]
5.	Silver	*Mentha aquatica* leaf extract	8	–	[65]
6.	Silver	*Parkia speciosa* leaf extract	31	–	[66]
7.	Gold	*Mimosa tenuiflora* bark extract	20	-	[67]
8.	Gold	*Clerodendrum inerme*	5.54	Spherical	[68]
9.	Gold	*Moringa oleifera* leaf extract	14–30	Crystalline	[69]
10.	Gold	Cinnamon bark extract	35	-	[70]
11.	Copper	*Jatropha curcas* leaf extract	12	-	[71]
12.	Copper	*Celastrus paniculatus* leaf extract	2–10	-	[72]
13.	Iron	*Eucalyptus robusta* leaf extract	8	-	[73]
14.	Iron oxide	*Hibiscus rosa-sinensis*	65	Spinel	[74]
15.	Platinum	*Peganum harmala* seed	22.3	Spherical	[75]
16.	Palladium	*Peganum harmala* seed	20.3	Spherical	[75]
17.	Zinc oxide	*Cayratia pedate*	52.25	-	[76]
18.	Titanium oxide	Jasmine flower extract	31–42	Spherical	[77]
19.	Titanium oxide	Lemon peel extract	80–140	Spherical	[78]

13.4 GREEN SYNTHESIS OF CERAMICS

Ceramics can broadly be defined as inorganic, nonmetallic, or metalloid substances having mixed (i.e. both covalent and ionic) bonding. They have gained significance because they offer distinctive characteristics of being thermally stable, chemically inert, brittle, and heat and electricity insulators. Thus, ceramics are finding new applications in a range of modern consumer products, varying from bricks, tiles, electronics, tableware, glass, sanitaryware to artificial teeth and bones, and nuclear power. Their demand is bound to increase manifold with increasing application possibilities such as in structural engineering. Traditionally, these ceramics have been largely manufactured from natural clay, thereby affecting ecology by depleting a natural resource. Alternatively, alumina, silica, silicon carbide, titania, zirconia, etc. are used as raw materials, but these are expensive and pollute the environment.

Another worrisome scenario is the ecological imbalance being created due to the enhanced industrial and agricultural production to meet the needs of the growing population. It is a double-edged situation, generating large amounts of hazardous wastes and pollution on one hand and exhausting the natural reserves on the other. Efficient recycling and utilization of the produced wastes is the current need of the time of all economies to be sustainable.

The waste recycling is of interest particularly in case of fabrication of ceramics. Recently, considerable progress has been made in employing wastes as alternative substitutes for natural raw materials for potential large-scale green synthesis of sustainable ceramics. These alternatives mainly include plant extracts, especially the fruit extracts and waste produced from different industrial processes such as oil production, water treatment, paper processing, etc. as well as agriculture waste [79]. Such an approach renders double benefits, by not only consuming the harmful waste and resolving its disposal problem, but also affording alternate, sustainable raw materials for manufacturing, thereby conserving natural resources. The conversion of wastes to marketable high-value-added ceramics at much lower costs is thus of high economic and ecological benefit. The wastes have been classified into three categories depending upon their role in ceramic production; fuel wastes which affect the sintering process, fluxing wastes which can form liquid phases to alter the sintering temperatures, and property-affecting wastes which modify the ceramic properties [80]. Some examples of wastes employed in preparation of ceramics include coal fly ash [81], blast furnace slag [82], rice husk ash [83], bottom ash [84], glass waste [85], paper processing residues [86], petroleum waste [87], water-treatment sludge [88], waste marble powder [89], and polished tile waste [90] (Figure 13.3). In the subsequent subsections, representative synthetic methodologies and approaches for green synthesis of ceramics using different kinds of wastes will be discussed. The emphasis will be on work done in the last five years.

13.4.1 AGRICULTURAL WASTES AS RAW MATERIAL

Bagasse is an extensively produced agricultural and forestry waste from sugarcane, with relatively high silicon content. A very recent paper reported the employment of bagasse as the silicon source to produce lithium silicate-based ceramics having

Green Synthesis of Sustainable Materials 245

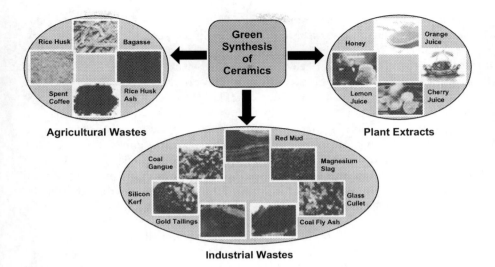

FIGURE 13.3 Raw materials used in green synthesis of ceramics.

applicability in CO_2 sorption with superior performance [91]. The bagasse was pretreated before use, in order to reduce the impurity elements present in it, which can cause side reactions during the ceramic preparation, thereby hindering the sorbent activity. Out of the different kinds of pretreatment methods (direct calcination, water/acid washing followed by calcination, and water/acid washing after calcination) employed, HCl acid washing after calcination gave the best results. It effectively removed the residual metals, such as Fe, Ca, etc., along with water-soluble impurities, e.g. K, P, S, etc., thus significantly increasing the silicon content up to nearly 93% (~2 times the original content). In addition, the pretreatment created a significant micro/mesoporous structure in the otherwise dense ash, being highly favorable for porosity and subsequent CO_2 sorption. This pretreated ash was mixed with lithium oxalate solution (lithium source) in the ratio 4.2:1 and heated to dryness, followed by calcination. The product obtained was ground and sieved to yield Li_4SiO_4-based ceramics.

Another abundantly available agricultural waste is rice husk (nearly 134 million tons produced per year in the world) which is mainly burnt or disposed to landfill or discharged into water bodies. It is a good amorphous silica waste that can be utilized as an alternative raw material for the manufacture of lightweight ceramics. It has been used as silica precursor in the production of variety of ceramic materials, such as glass ceramics [92], bricks [93], glazes [94], pigments [95], etc. Sharifikolouei et al. reported a new method wherein rice husk was initially utilized for adsorption of heavy metals and the resultant rice husk obtained was employed to produce porous glass ceramic [96]. Removal of toxic heavy metals such as lead and cadmium from industrial wastewaters is a grave environmental challenge. Rice husk was employed as a green adsorbent for these metals, thereby effecting wastewater treatment. Vitrified bottom ash, obtained from municipal solid waste and foaming agent (calcium carbonate), were mixed in the ratio 40:60 to prepare a porous glass matrix that was subsequently mixed with 70% by volume of Pb- and Cd-loaded rice husk. The mixture was heated to finally yield the porous glass ceramic. This work enabled effective utilization of two waste sources as well as provided a safe option for immobilization of heavy metals on rice husk followed by encapsulation inside glass ceramic structure. Another research work reported the use of rice husk ash in conjunction with another silica waste, i.e. glass cullet (broken/refuse glass) waste to form cellular glass granules, which were then coalesced to obtain foamed, lightweight ceramics [97]. Rice husk and glass cullet were mixed to form a paste using an alkali solution followed by heating to form solid specimens. The solid was then crushed, sieved, and heated to produce the cellular granular material. These cellular granules were subsequently mixed with clay powder and pressed to obtain raw ceramic bricks. Finally, sintering of the ceramic bricks caused coalescence of glass granules and consequent formation of pores inside the matrix to yield porous ceramics.

A novel approach involved using spent coffee grounds waste as the pore-forming agent in sacrificial phase technique to fabricate macro-porous alumina ceramics [98]. Waste coffee grounds serve as cheap, eco-friendly, and widely available pore-former material, and moreover produce harmless by-products during ceramic preparation. The alumina powder and coffee ground waste were mixed, to which an aqueous solution of polyvinyl alcohol was gradually added to form a blended paste. The pastes obtained were first air-dried and heated to give hardened masses, which were ground and sieved to give dry powder. The powder was converted into cylindrical form and sintered to give the porous alumina ceramic. These ceramics exhibited reasonably high porosities and mechanical strength and the porosity increased with increasing concentration of coffee grounds in the mixture. Such properties make these ceramics as potential candidates for applications in filters, catalyst supports, lightweight building materials, and thermal/electrical insulation.

13.4.2 Industrial Wastes as Raw Material

Red mud is a by-product of alumina production by the Baeyer process, produced around two times the amount of alumina, leading to generation of around 120 million tons per year. Though it is largely nontoxic but is highly alkaline and thus its permeation in soil and groundwater upon disposal results in environmental pollution. Similarly, tailings are remaining oxides generated as wastes during gold mining. It is estimated that production of 1 ton of gold generates about 1 megaton of solid waste. Center for Sustainable Resource Processing project has assessed tailings and red mud as primary and secondary wastes of serious concern owing to their excessive production and potential disposal risk. A number of works have been reported for their use in preparation of construction material [99], glass fiber [100], bricks [101], and ceramics [102]. Kim and Park devised a synthetic process to utilize these mining waste and smelting by-products in the production of wear-resistant glass ceramics [103]. Gold tailings and red mud were mixed with waste limestone and ferronickel slag in the ratio 25:30:15:30 by weight to make a homogeneous mixture which was air heated. The resulting molten oxide was solidified and the solid sample was then again heated, followed by cooling to generate cuboidal glass ceramic. The obtained

Green Synthesis of Sustainable Materials

ceramic is similar to fused cast basalt in chemical composition as well as crystalline phase, with a relatively higher compressive strength, and can thus be used as the latter's replacement.

Silicon kerf waste has been reported as a raw material for obtaining high-quality porous silicon carbide (SiC) ceramics [104], which are valuable for their high thermal stability and conductivity along with low thermal expansion and hence wide applicability in chemical, electronic, and aerospace industry. Silicon kerfs are Si/SiC containing tiny chips or shavings of the metal generated during preparation of silicon wafers. The growing demand of silicon wafer in applications such as solar photovoltaics necessitates its increased production, eventually generating larger silicon kerf waste to be addressed. This work described the synthesis of a series of SiC ceramics from silicon kerf waste with varying pore structures. The kerf was initially calcined to remove polyethylene glycol and water impurities and subsequently mixed with activated carbon and polyvinyl alcohol. The homogeneous mixture was then modeled, pressed, and sintered in a furnace. Different ceramic samples, e.g. SA, SN, and SC, were prepared by introducing different gases, namely argon, nitrogen, and carbon embedded, respectively into the furnace during sintering process. The variation of sintering atmosphere resulted in differing properties of SiC ceramics. The silicon kerfs have also been reported in the preparation of composite ceramics [105].

Coal being the main energy source, its mining is done worldwide on an extensive scale leading to massive production of coal gangue (around 10–15% of coal production). Its efficient disposal is thus imperative to minimize the harmful environmental effects. The gangue can act as an effective source of silica and alumina. Similarly, aluminum-containing refractory solid waste is also produced in large amounts during industrial processes which is an environmental hazard. Liu et al. exploited both these industrial wastes to synthesize mullite ceramics, employing the conventional solid-state reaction procedure [106]. Mullite, with chemical composition as $3Al_2O_3.2SiO_2$, is an important class of ceramic having favorable properties such as high strength, creep resistance, low conductivity, and high chemical and thermal stability. They are thus prospective candidates for applications in mechanics and advanced structural ceramics. The mixture of coal gangue and high alumina refractory waste, with the primary components as kaolinite, quartz, and alumina, were crushed, sieved, ball-mixed, and calcined. The obtained powders were pressed into bars and the resulting green bodies were sintered. It led to the formation of needle-like mullite powders. The phase-forming temperature of mullite obtained by this method was lower than by using pure oxide as raw materials, which is advantageous as porosity decreases with increasing temperature. In another work, low-density ceramic proppants were prepared by partially replacing the usually used low-grade bauxite raw material with coal gangue, in combination with another industrial solid waste, i.e. magnesium slag, as a sintering additive [107]. Magnesium slag is the residue obtained from magnesium smelting and causes soil hardening and water pollution when directly discharged. It is mainly composed of CaO, along with SiO_2 and unreduced MgO. This procedure enabled the decreasing of sintering temperature, along with reducing the production cost. Bauxite, gangue, and slag were homogeneously mixed in the ratio 77:20:3 and subsequently mixed with water to form spherical green bodies. These green bodies were then dried, sieved, and sintered. They were finally cooled and sieved to yield the proppant particles. The low-cost proppants thus obtained are lightweight and satisfy the requirements of perfect fracturing materials, thereby overcoming the conventional disadvantages of ceramic proppants being costlier and of high density.

Another widespread solid waste produced from coal is the coal fly ash, generated from its combustion in thermal power industry. It is a major environmental pollutant responsible for decreasing the air quality. Hence, its comprehensive use is desirable. The main components of coal fly ash include SiO_2, Al_2O_3, CaO, and Fe_2O_3. Coal ash has been reported as raw material for the fabrication of mullite ceramics [108], ceramic membranes [109], etc. Chen et al. have described a method for conversion of coal ash into a stable and porous aluminum titanate (Al_2TiO_5)-mullite composite ceramic using starch consolidation casting [110]. TiO_2 and AlOOH were used as additives to fabricate three phases, i.e. solid solution of $Al_{2(1-x)}Fe_{2x}TiO_5$, reinforced mullite and liquid anorthite. The formation of anorthite liquid phase was beneficial for reducing the sintering temperature and improving the flexural strength. The procedure involved wet-mixing of coal fly ash, AlOOH and TiO_2 in distilled water, followed by addition of corn starch (pore-forming and body-forming material), ethanol, and polyvinyl alcohol (binder). The pH was adjusted between 4–5 which aided in attaining homogeneity in the suspensions, which were then heated and solidified. Finally, the ceramic bodies were sintered to yield the composite ceramics. The conventional aluminum titanate ceramics, though having remarkable thermal- and corrosion-resistant properties, suffer from both poor mechanical strength and thermal stability in the range of 750–1300°C, thus limiting their applicability. This methodology provided an alternate approach to fabricate aluminum titanate ceramics, sans its disadvantages. The research group of Ma et al. has been actively involved in recycling coal fly ash to synthesize different kinds of ceramics [111]. In one of their works, they reported the use of coal ash to prepare porous $SiC-Al_2O_3$ ceramics by carbothermal reduction reaction under inert atmosphere [112]. Commercial SiC powder was used as an additive to improve the mechanical properties of the prepared ceramics. The homogenous mixture powder of coal ash and activated carbon in the ratio 100:58 in ethanolic medium was mixed with different amounts of SiC and phenolic resin was added as binder. The green samples were then pressed and subsequently sintered under argon atmosphere. The samples were then cooled and lastly air-heated to remove any left carbon. The ceramic sample containing 10% by weight SiC was found to have the densest microstructure and highest compressive strength. Much higher amounts of SiC conversely reduced the thermal shock resistance and mechanical strength. The same research group, in another work, utilized the coal fly ash to fabricate porous mullite ceramics in combination with bauxite using the reaction synthesis process [113]. Bauxite helped in neutralizing the excess silica present in coal fly ash. The effect of increasing SiC amounts in the ceramics was similar as in the previous case, with 10% by weight composition giving the best results. In addition, increasing the firing temperature enhanced the compressive strength and thermal shock resistance of the ceramics.

A new methodology has been designed by Monich et al. to manufacture dense glass ceramics from municipal solid waste [114]. The solid waste, upon plasma gasification, yields a synthetic gas from its organic fractions while its nonmetallic

Green Synthesis of Sustainable Materials

inorganic fraction undergoes vitrification upon fast cooling. This vitrified residue known as Plasmastone is upcycled to prepare glass ceramics by cold pressing and sinter crystallization. Kaolin clay was added as binder. Plasmastone was dried and milled to give powder, which was then mixed with soda-lime glass. This mixture was thereafter added to the suspension of kaolin clay in distilled water, followed by thorough mixing and overnight drying. The composition of mixture in weight percentage was 45% Plasmastone, 45% soda-lime glass, and 10% kaolin clay. The product obtained was milled, sieved, and pressed. The resulting green samples were sintered. The glass ceramics obtained by this approach exhibited desirable properties such as low water absorption and leachability and superior mechanical properties, which were comparable to commercial ceramic tiles. It also had beneficial environmental impact as all the heavy metals got incorporated into the glass crystals, leading to their very limited leaching.

13.4.3 PLANT EXTRACTS AS RAW MATERIAL

M. Salavati-Niasari and coworkers have been involved in green synthesis of ceramic nanostructures using juice extracts of plants as fuel, which are both nonhazardous and biodegradable. One of their works described the preparation of $Dy_2Ce_2O_7$ ceramic nanostructures from the juice of *Punica granatum* and metal nitrates [115]. It has been stated to be the first report of using *Punica granatum* juice as the fuel. This juice is rich in fructose and glucose, which act as reductants, as well as helps in determining the size and morphology of the nanostructures during preparation owing to their steric hindrance. The procedure involved slow addition of *Punica granatum* juice to the distilled water solution of dysprosium nitrate pentahydrate and ceric ammonium nitrate. The resultant mixture was evaporated to form a viscous gel, which was dried and subsequently heated to give the product. The synthesized ceramic nanostructures displayed outstanding photocatalytic efficiency for destroying Methylene Blue and Erythrosine pollutants under visible light. Another of their work involved the use of cherry or orange juice as natural surfactants for the preparation of $TbFeO_3$ ceramic nanostructures [116]. The ceramic nanostructures were fabricated using green sonochemistry method. The aqueous solutions of terbium nitrate hexahydrate and ferric nitrate nonahydrate were mixed and the surfactant (cherry/orange juice) was added to it. Thereafter, tetraethylenepentamine was added to the solution and it was sonicated and centrifuged. After washing with distilled water and ethanol and subsequent drying, the resulting powder was calcined to yield the final nanostructures. The use of cherry juice led to the formation of sponge-like $TbFeO_3$ nanostructures, whereas orange juice formed highly pure and crystallized, dense nanostructures with small and uniform size and shape. These nanostructures displayed appreciable photocatalytic activity for degradation of Methyl Orange, Acid Blue 92, Acid Black 1, and Acid Brown 214 contaminants under UV light and were thus applied in water treatment. The nanostructure prepared using orange juice as surfactant exhibited the best degradation efficiency which has been claimed to be the highest reported so far.

Khirade et al. have also been actively working in this field of green synthesis using natural fuel. One of their papers discussed the synthesis of spherical and uniform, multifunctional $Ba_{1-x}Sr_xTiO_3$ ceramic nano-powders using lemon juice as the fuel *via* sol-gel combustion method [117]. Lemon juice is a rich source of L-ascorbic acid, commonly known as vitamin C, and it helps in carrying out the synthetic process at low temperature. Also, its use helps in overcoming the disadvantage of release of hazardous gases during ignition in the usual sol-gel combustion method with chemical fuel. Two solutions were prepared; yellow-colored mixture solution of tetrabutyl titanate (in ethanol) and lemon juice (in deionized water) maintained at pH ~8, and mixture solution of barium nitrate and strontium nitrate in deionized water. Both these solutions were mixed homogenously and pH set at 7, after which it was heated till the formation of a viscous gel. Further heating led to evaporation of all water molecules from gel and eventually its ignition. The resulting coarse powders were ground to uniform powders and calcined to obtain highly pure, crystalline nano-ceramics. Similar green approach was used by this research group to fabricate $BaZr_{1-x}Ce_xO_{3-x/2}$ nano-ceramics using honey as the firing agent [118]. Honey is mainly composed of glucose and fructose and its use as fuel prevents the formation of toxic vapors in the combustion step. These cerium-incorporated barium zirconate ceramics produced *via* honey-assisted sol-gel combustion process are potential candidates for applications in electronics such as for solid-oxide fuel cells and microwave applications.

Juniperus phoenicea, commonly called Arar, has been a component of traditional medicine since a long time and is nowadays being investigated for its exceptional antibacterial and anticancer activities. Its fruit extract, being rich in a variety of functional groups, can act as both capping and reducing agent and hence can be potentially utilized for the synthesis of functional materials. A $YBa_2Cu_3O_7$ ceramic superconductor has been synthesized from aqueous fruit extract of Arar [119]. The aqueous solution of fruit extract of *Juniperus phoenicea* was gradually added to the aqueous solution of mixed metal acetate (prepared by mixing yttrium acetate, barium acetate, and copper acetate in water) to form dried brown-colored precipitate. These were made to stand overnight at 80°C, followed by calcination in order to eliminate any remaining volatile substances. The resulting powders were ground, pelletized, and sintered and finally cooled to give superconducting ceramics.

13.5 CONCLUSION

The plethora of usability has led to the prodigious production of materials (polymers, metal and metal oxide nanoparticles, and ceramics). However, this increased production has posed a serious threat to the environment due to the toxic reagents being used for the synthesis and the waste generated post-synthesis. To limit the discharge of the hazardous materials being thrown into the environment, green methodologies are developed for the production of the materials as well as recycling of the waste. The approaches involve the use of bio-based raw materials, green solvents, and the use of industrial/agricultural wastes for the production of sustainable materials. In addition, novel alternative methodologies, such as adopting high-atom economy approach, may be beneficial for the stride toward a sustainable future. Materials of desirable properties (such as size, shape, composition, strength, average molecular

Green Synthesis of Sustainable Materials

mass, crystallinity, etc.) may be obtained through these green methodologies, adding a further advantage.

Although enormous research is being done toward green synthesis of materials, yet their realization on a large scale is limited. A lot of work needs to be done in order to transfer the technology for commercial production. The bridging of the gap between academic research and its industrial utility will steer the development toward a viable future.

REFERENCES

1. Inal, S., Rivnay, J., Suiu, A.-O., Malliaras, G.G. and I. McCulloch. 2018. Conjugated polymers in bioelectronics. Acc. Chem. Res. 51:1368–76.
2. Lau, W.W.Y., Shiran, Y., Bailey, R.M. et al. 2020. Evaluating scenarios toward zero plastic pollution. Science. 369:1455–61.
3. Khui, P.L.N., Rahman, R., Jayamani, E. and K.B. Bakri. 2020. Advances in sustainable polymer composites. 267–82. Woodhead Publishing Series in Composites Science and Engineering.
4. Miller, S.A. 2014. Sustainable polymers: Replacing polymers derived from fossil fuels. Polym. Chem. 5:3117–8.
5. Verma, C. and M.A. Quraishi. 2021. Gum Arabic as an environmentally sustainable polymeric anticorrosive material: Recent progresses and future opportunities. Int. J. Biol. Macromol. 184:118–34.
6. Andrade, M.S., Ishikawa, O.H., Costa, R.S., Seixas, M.V.S., Rodrigues, R.C.L.B. and E.A.B. Moura. 2022. Development of sustainable food packaging material based on biodegradable polymer reinforced with cellulose nanocrystals. Food Packag. Shelf Life. 31:100807.
7. Tan, H., Zhang, Y., Sun, L. et al. 2021. Preparation of nano sustained-release fertilizer using natural degradable polymer polylactic acid by coaxial electrospinning. Int. J. Biol. Macromol. 193:903–14.
8. Shaghaleh, H., Xu, X. and S. Wang. 2018. Current progress in production of biopolymeric materials based on cellulose, cellulose nanofibers, and cellulose derivatives. RSC Adv. 8:825–42.
9. Kai, D., Tan, M.J., Chee, P.L., Chua, Y.K., Yap, Y.L. and X.J. Loh. 2016. Towards lignin-based functional materials in a sustainable world. Green Chem. 18:1175–200.
10. Warlin, N., Gonzalez, M.N.G., Mankar, S. et al. 2019. A rigid spirocyclic diol from fructose-based 5-hydroxymethylfurfural: Synthesis, life-cycle assessment, and polymerization for renewable polyesters and poly(urethane-urea)s. Green Chem. 21:6667–84.
11. Monica, F.D. and A.W. Kleij. 2020. From terpenes to sustainable and functional polymers. Polym. Chem. 11:5109–27.
12. Zhou, J., Wu, M., Peng, Q. et al. 2018. Highly efficient strategies toward sustainable monomers and polymers derived from fatty acids *via* tetramethylguanidine promoted esterification. Polym. Chem. 9:2880–6.
13. Techie-Menson, R., Rono, C.K., Etale, A., Mehlana, G., Darkwa, J. and B.C.E. Makhubela. 2021. New bio-based sustainable polymers and polymer composites based on methacrylate derivatives of furfural, solketal and lactic acid. Mater. Today Commun. 28:102721.
14. Thakur, S., Chaudhary, J., Singh, P., Alsanie, W.F., Grammatikos, S.A. and V.K. Thakur. 2022. Synthesis of bio-based monomers and polymers using microbes for a sustainable bioeconomy. Bioresour. Technol. 344:126156.
15. Prat, D., Hayler, J. and A. Wells. 2014. A survey of solvent selection guides. Green Chem. 16:4546–51.

16. Phan, S. and C.K. Luscombe. 2019. Recent advances in the green, sustainable synthesis of semiconducting polymers. Trends Chem. 1:670–81.
17. Pankow, R.M., Ye, L., Gobalasingham, N.S., Salami, N., Samal, S. and B.C. Thompson. 2018. Investigation of green and sustainable solvents for direct arylation polymerization (DArP). Polym. Chem. 9:3885–92.
18. Pankow, R.M., Yea, L. and B.C. Thompson. 2019. Influence of an ester directing-group on defect formation in the synthesis of conjugated polymers via direct arylation polymerization (DArP) using sustainable solvents. Polym. Chem. 10:4561–72.
19. Lavilla, C., Ilarduya, A.M., Alla, A., Martin, M.G., Galbis, J.A. and S. Guerra. 2012. Bio-based aromatic polyesters from a novel bicyclic diol derived from D-mannitol. Macromolecules. 45:8257–66.
20. Xu, S., Wu, F. and Li, Z. et al. 2019. A green cascade polymerization method for the facile synthesis of sustainable poly(butylene-co-decylene terephthalate) copolymers. Polymer. 178:121591.
21. Li, J., Tu, Y., Lu, H., Li, X., Yang, X. and Y. Tu. 2021. Rapid synthesis of sustainable poly(ethylene 2,5-furandicarboxylate)-*block*-poly(tetramethylene oxide) multiblock copolymers with tailor-made properties via a cascade polymerization route. Polymer. 237:124313.
22. Zschech, C., Mathias, P., Muller, M.T., Wiessner, S., Wagenknecht, U. and U. Gohs. 2020. Continuous electron-induced reactive processing—A sustainable reactive processing method for polymers. Radiat. Phys. Chem. 170:108652.
23. Perez-Fortes, M., Bocin-Dumitriu, A. and E. Tzimas. 2014. CO_2 utilization pathways: Techno-economic assessment and market opportunities. Energy Procedia. 63:7968–75.
24. Bobbink, F.D., Muyden, A.P.V. and P.J. Dyson. 2019. En route to CO_2-containing renewable materials: Catalytic synthesis of polycarbonates and non-isocyanate polyhydroxyurethanes derived from cyclic carbonates. Chem. Commun. 55:1360.
25. Bobbink, F.D., Vasilyev, D., Hulla, M. et al. 2018. Synthesis of carbonates and related compounds incorporating CO_2 using ionic liquid-type catalysts: State-of-the-art and beyond. ACS Catal. 8:2589–94.
26. Bai, D., Chen, Q., Chai, Y. et al. 2018. Vanillin derived a carbonate dialdehyde and a carbonate diol: Novel platform monomers for sustainable polymers synthesis. RSC Adv. 8:34297–303.
27. Muthuraj, R. and T. Mekonnen. 2018. Recent progress in carbon dioxide (CO_2) as feedstock for sustainable materials development: Co-polymers and polymer blends. Polymer. 145:348–73.
28. Sakakura, T., Choi, J.-C. and H. Yasuda. 2007. Transformation of carbon dioxide. Chem. Rev. 107:2365–87.
29. Shaarani, F.W. and J.J. Bou. 2017. Synthesis of vegetable-oil based polymer by terpolymerization of epoxidized soybean oil, propylene oxide and carbon dioxide. Sci. Total Environ. 598:931–6.
30. Tang, S., Zhao, Y. and K. Nozaki. 2021. Accessing divergent main-chain-functionalized polyethylenes via copolymerization of ethylene with a CO_2/butadiene-derived lactone. J. Am. Chem. Soc.143:17953–7.
31. Worthington, M.J.H., Kucera, R.L. and M. Justin. 2017. Chalker green chemistry and polymers made from sulphur. Green Chem.19:2748.
32. Chung, W.J., Griebel, J.J., Kim, E.T. et al. 2013. The use of elemental sulfur as an alternative feedstock for polymeric materials. Nat. Chem. 5:518–24.
33. Zhang, Y., Pavlopoulos, N.G., Kleine, T.S. et al. 2019. Nucleophilic activation of elemental sulfur for inverse vulcanization and dynamic covalent polymerizations. J. Polym. Sci., Part A: Polym. Chem. 57:7–12.
34. Duarte, M.E., Huber, B., Theato, P. and H. Mutlu. 2020. The unrevealed potential of elemental sulfur for the synthesis of high sulfur content bio-based aliphatic polyesters. Polym. Chem. 11:241–8.

Green Synthesis of Sustainable Materials

35. Sanvezzo, P.B. and M. C. Branciforti. 2021. Recycling of industrial waste based on jute fiber-polypropylene: Manufacture of sustainable fiber-reinforced polymer composites and their characterization before and after accelerated aging. Ind. Crops Prod. 168:113568.
36. Singh, J., Dutta, T., Kim, K.-H., Rawat, M., Samddar, P. and P. Kumar. 2018. Green synthesis of metals and their oxide nanoparticles: Applications for environmental remediation. J. Nanobiotechnology. 16:84.
37. Hussain, I., Singh, N.B., Singh, A., Singh, H. and S.C. Singh. 2016. Green synthesis of nanoparticles and its potential application. Biotechnol. Lett. 38:545–60.
38. Ijaz, I., Gilani, E., Nazir, A. and A. Bukhari. 2020. Detail review on chemical, physical and green synthesis, classification, characterizations and applications of nanoparticles. Green Chem. Lett. Rev. 13:223–45.
39. Meyers, M.A., Mishra, A. and D.J. Benson. 2006. Mechanical properties of nanocrystalline materials. Prog. Mater. Sci. 51:427–556.
40. Mukherjee, P., Ahmad, A., Mandal, D. et al. 2001. Fungus mediated synthesis of silver nanoparticles and their immobilization in the my celial matrix: A novel biological approach to nanoparticle synthesis. Nano Lett. 1:515–19.
41. Kawasaki, M., and N. Nishimura. 2006. 1064-nm laser fragmentation of thin Au and Ag flakes in acetone for highly productive pathway to stable metal nanoparticles. Appl. Surf. Sci. 25:2208–16.
42. Tarasenko, N.V., Butsen, A.V., Nevar, E.A. and N.A. Savastenko. 2006. Synthesis of nanosized particles during laser ablation of gold in water. Appl. Surf. Sci. 252:4439–44.
43. Jayappa, M.D., Ramaiah, C.K., Kumar, M.A.P. et al. 2020. Green synthesis of zinc oxide nanoparticles from the leaf, stem and in vitro grown callus of Mussaenda frondosa L.: Characterization and their applications. Appl. Nanosci. 10:3057–74.
44. Pillai, A.M., Sivasankarapillai, V.S., Rahdar, A. et al. 2020. Green synthesis and characterization of zinc oxide nanoparticles with antibacterial and antifungal activity. J. Mol. Struct. 1211:128107.
45. Shabaani, M., Rahaiee, S., Zare, M., and S.M. Jafari. 2020. Green synthesis of ZnO nanoparticles using loquat seed extract: Biological functions and photocatalytic degradation properties. LWT 134:110133.
46. Iravani, S., Korbekandi, H., Mirmohammadi, S.V. and B. Zolfaghari. 2014. Synthesis of silver nanoparticles: Chemical, physical and biological methods. Res. Pharm. Sci. 9:385–406.
47. Seralathan, J., Stevenson, P., Subramaniam, S. et al. 2014. Spectroscopy investigation on chemo-catalytic, free radical scavenging and bactericidal properties of biogenic silver nanoparticles synthesized using Salicornia brachiata aqueous extract. Spectrochim. Acta A, Mol. Biomol. Spectrosc. 118:349–55.
48. Singh, J., Dutta, T., Kim, K.H. Rawat, M., Samddar, P. and P. Kumar. 2018. "Green" synthesis of metals and their oxide nanoparticles: Applications for environmental remediation. J. Nanobiotechnol. 16:1–24.
49. Singh, A., Gautam, P.K., Verma, A. et al. 2020. Green synthesis of metallic nanoparticles as effective alternatives to treat antibiotics resistant bacterial infections: A review. Biotechnol. Rep. 25:e00427.
50. Zhang, D., Ma, X.-L., Gu, Y., Huang, H. and G.-W. Zhang. 2020. Green synthesis of metallic nanoparticles and their potential applications to treat cancer. Front. Chem. 8:799.
51. Singaravelu, G., Arockiamary, J.S., Kumar, V.G. and K. Govindaraju. 2007. A novel extracellular synthesis of monodisperse gold nanoparticles using marine alga, Sargassum wightii Greville. Colloids Surf. B: Biointerfaces. 57:97–101.

52. Govindaraju, K., Kiruthiga, V., Kumar, V.G. and G. Singaravelu. 2009. Extracellular synthesis of silver nanoparticles by a marine alga, Sargassum wightii Grevilli and their antibacterial effects. J. Nanosci. Nanotechnol. 9:5497–501.
53. Golinska, P., Wypij, M., Ingle, A.P., Gupta, I., Dahm, H. and M. Rai. 2014. Biogenic synthesis of metal nanoparticles from actinomycetes: Biomedical applications and cytotoxicity. Appl. Microbiol. Biotechnol. 98:8083–97.
54. Nair, G.M., Sajini, T. and B. Mathew. 2022. Advanced green approaches for metal and metal oxide nanoparticles synthesis and their environmental applications. Talanta Open. 5:100080–89.
55. Fatima, R., Priya, M., Indurthi, L., Radhakrishnan, V. and R. Sudhakaran. 2020. Biosynthesis of silver nanoparticles using red algae *Portieria hornemannii* and its antibacterial activity against fish pathogens. Microb. Pathog.138:103780.
56. Ma, G., Zhao, Z. and H. Liu. 2016. Yeast cells encapsulating polymer nanoparticles as Trojan particles via in situ polymerization inside cells. Macromolecules. 49:1545–51.
57. Peiris, M.M.K., Guansekera, T.D.C.P., Jayaweera, P.M. and S.S.N. Fernando. 2018. TiO_2 nanoparticles from baker's yeast: A potent antimicrobial. J. Microbiol. Biotechnol. 28:1664–70.
58. Soltys, L., Olkhovyy, O., Tatarchuk, T. and M. Naushad. 2021. Green synthesis of metal and metal oxide nanoparticles: Principles of green chemistry and raw materials. Magnetochemistry. 7:145–79.
59. Kumar, I., Mondal, M. and N. Sakthivel. 2019. Green synthesis of phytogenic nanoparticles. In micro and nano technologies. Shukla, A.K., Iravani, S., Eds.; Elsevier: Amsterdam, The Netherlands, 37–73.
60. Satpathy, S., Patra, A., Ahirwar, B. and M.D. Hussain. 2020. Process optimization for green synthesis of gold nanoparticles mediated by extract of Hygrophila spinosa T. Anders and their biological applications. Phys. E Low-Dimens. Syst. Nanostruct. 121:113830.
61. Aygun, A., Ozdemir, S., Gulcan, M., Cellat, K. and F. Sen. 2020. Synthesis and characterization of Reishi mushroom-mediated green synthesis of silver nanoparticles for the biochemical applications. J. Pharm. Biomed. Anal. 178:112970.
62. Chand, K., Cao, D., Fouad, D.E. et al. 2020. Green synthesis, characterization and photocatalytic application of silver nanoparticles synthesized by various plant extracts Arab. J. Chem. 13:8248–61.
63. Khatami, M., Sharifi, I., Nobre, M.A.L., Zafarnia, N. and M.R. Aflatoonian. 2018. Waste-grass-mediated green synthesis of silver nanoparticles and evaluation of their anticancer, antifungal and antibacterial activity. Green Chem. Lett. Rev. 11:125–34.
64. Bindhu, M.R., Umadevi, M., Esmail, G.A., Al-Dhabi, N.A. and M.V. Arasu. 2020. Green synthesis and characterization of silver nanoparticles from Moringa oleifera flower and assessment of antimicrobial and sensing properties. J. Photochem. Photobiol. B Biol. 205:111836.
65. Nouri, A., Yaraki, M.T., Lajevardi, A., Rezaei, Z., Ghorbanpour, M. and M. Tanzifi. 2020. Ultrasonic-assisted green synthesis of silver nanoparticles using Mentha aquatica leaf extract for enhanced antibacterial properties and catalytic activity. Colloids Interface Sci. Commun. 35:100252.
66. Ravichandran, V., Vasanthi, S., Shalini, S., Shah, S.A.A., Tripathy, M. and N. Paliwal. 2019. Green synthesis, characterization, antibacterial, antioxidant and photocatalytic activity of Parkia speciosa leaves extract mediated silver nanoparticles. Results Phys. 15:102565.
67. Rodriguez-Leon, E., Rodriguez-Vazquez, B.E., Martínez-Higuera, A. et al. 2019. Synthesis of gold nanoparticles using *Mimosa tenuiflora* extract, Assessments of Cytotoxicity, Cellular Uptake, and Catalysis. Nanoscale Res. Lett. 14:334.

Green Synthesis of Sustainable Materials

68. Khan, S.A., Shahid, S. and C.-S. Lee. 2020. Green synthesis of gold and silver nanoparticles using leaf extract of Clerodendrum inerme. Characterization, Antimicrobial, and Antioxidant Activities. Biomolecules. 10:835.
69. Boruah, J.S., Devi, C., Hazarika, U. et al. 2021. Green synthesis of gold nanoparticles using an antiepileptic plant extract: *In vitro* biological and photo-catalytic activities. RSC Adv. 11:28029–41.
70. ElMitwalli, O.S., Barakat, O.A., Daoud, R.M., Akhtar, S. and F.Z. Henari. 2020. Green synthesis of gold nanoparticles using cinnamon bark extract, characterization, and fluorescence activity in Au/eosin Y assemblies. J. Nanopart. Res. 22:309.
71. Ghosh, M.K., Sahu, S., Gupta, I. and T.K. Ghorai. 2020. Green synthesis of copper nanoparticles from an extract of *Jatropha curcas* leaves: Characterization, optical properties, CT-DNA binding and photocatalytic activity. RSC Adv. 10:22027–35.
72. Mali, S.C., Dhaka, A., Githala, K.G. and R. Trivedi. 2020. Green synthesis of copper nanoparticles using *Celastrus paniculatus* Willd. leaf extract and their photocatalytic and antifungal properties. Biotechnol. Rep. 27:e00518.
73. Vitta, Y., Figueroa, M., Calderon, M. and C. Ciangherotti. 2019. Synthesis of iron nanoparticles from aqueous extract of Eucalyptus Robusta Sm and evaluation of antioxidant and antimicrobial activity. Materials Science for Energy Technologies. 3:97–103.
74. Razack, S.A., Suresh, A., Sriram, S. et al. 2020. Green synthesis of iron oxide nanoparticles using Hibiscus rosa-sinensis for fortifying wheat biscuits. SN Appl. Sci. 2:898.
75. Fahmy, S.A., Fawzy, I.M., Saleh, B.M., Issa, M.Y., Bakowsky, U. and H.M.E.-S. Azzazy. 2021. Green synthesis of platinum and palladium nanoparticles using Peganum harmala L. Seed Alkaloids: Biological and computational studies. Nanomaterials. 11:965.
76. Jayachandran, A., Aswathy, T.R. and A.S. Nair. 2021. Green synthesis and characterization of zinc oxide nanoparticles using Cayratia pedata leaf extract. Biochem. Biophys. Rep. 26:100995.
77. Aravind, M., Amalanathan, M. and M.S.M. Mary. 2021. Synthesis of TiO_2 nanoparticles by chemical and green synthesis methods and their multifaceted properties. SN Appl. Sci. 3:409.
78. Nabi, G., Ain, Q.-U., Tahir, M.B., Riaz, K.M., Iqbal, T. and M. Rafique. 2020. Green synthesis of TiO_2 nanoparticles using lemon peel extract: Their optical and photocatalytic properties. Int. J. Environ. Anal. Chem. 102:434–42.
79. Hossain, Sk.S. and P.K. Roy. 2020. Sustainable ceramics derived from solid wastes: A review. J. Asian Ceramic Societies. 8:984–1009.
80. Vieira, C.M.F. and S.N. Monteiro. 2009. Incorporation of solid wastes in red ceramics—An updated review. Revista Materia. 14:881–905.
81. (a) Luo, Y., Ma, S., Zheng, S., Liu, C., Han, D. and X. Wang. 2018. Mullite-based ceramic tiles produced solely from high-alumina fly ash: Preparation and sintering mechanism. J. Alloy Comp. 732:828–37. (b) Han, G., Yang, S., Peng, W. et al. 2018. Enhanced recycling and utilization of mullite from coal fly ash with a flotation and metallurgy process. J. Clean. Prod. 178:804–13.
82. Ozturk, Z.B. and E. Gultekin. 2015. Preparation of ceramic wall tiling derived from blast furnace slag. Ceram. Int. 41:12020–26.
83. (a) Liu, X., Chen, X., Yang, L., Chen, H., Tian, Y. and Z. Wang. 2016. A review on recent advances in the comprehensive application of rice husk ash. Res. Chem. Intermed. 42:893–913. (b) Prasara-A, J. and S.H. Gheewala. 2017. Sustainable utilization of rice husk ash from power plants: A review. J. Clean. Prod. 167:1020–8.
84. Carrasco-Hurtado, B., Corpas-Iglesias, F.A., Cruz-Perez, N., Terrados-Cepeda, J. and L. Perez-Villarejo. 2014. Addition of bottom ash from biomass in calcium silicate masonry units for use as construction material with thermal insulating properties. Constr. Build. Mater. 52:155–65.

85. Silva, R.V., de Brito, J., Lye, C.Q. and R.K. Dhir. 2017. The role of glass waste in the production of ceramic-based products and other applications: A review. J. Clean. Prod. 167:346–64.
86. Sutcu, M. and S. Akkurt. 2009. The use of recycled paper processing residues in making porous brick with reduced thermal conductivity. Ceram. Int. 35:2625–31.
87. Pinheiro, B.C.A. and J.N.F. Holanda. 2013. Reuse of solid petroleum waste in the manufacture of porcelain stoneware tile. J. Environ. Manage. 118:205–10.
88. Geraldo, R.H., Fernandes, L.F.R. and G. Camarini. 2017. Water treatment sludge and rice husk ash to sustainable geopolymer production. J. Clean. Prod. 149:146–55.
89. Sutcu, M., Alptekin, H., Erdogmus, E., Er, Y. and O. Gencel. 2015. Characteristics of fired clay bricks with waste marble powder addition as building materials. Constr. Build. Mater. 82:1–8.
90. Ke, S., Wang, Y., Pan, Z., Ning, C. and S. Zheng. 2016. Recycling of polished tile waste as a main raw material in porcelain tiles. J. Clean. Prod. 115:238–44.
91. Li, Z., Liu, W., Yao, S., Yang, Y., Li, Q. and S. Zhou. 2021. Synthesis of waste bagasse-derived Li_4SiO_4-based ceramics for cyclic CO_2 capture: Investigation on the effects of different pretreatment approaches. Ceram. Int. 47:28744–53.
92. Andreola, F., Martin, M.I., Ferrari, A.M. et al. 2013. Technological properties of glass-ceramic tiles obtained using rice husk ash as silica precursor. Ceram. Int. 39:5427–35.
93. Andreola, F., Lancellotti, I., Manfredini, T., Bondioli, F. and L. Barbieri. 2018. Rice husk ash (RHA) recycling in brick manufacture: Effects on physical and microstructural properties. Waste & Biomass Val. 9:2529–39.
94. Bondioli, F., Barbieri, L., Ferrari A.M. and T. Manfredini. 2010. Characterization of rice husk ash and its recycling as quartz substitute for the production of ceramic glazes. J. Am. Ceram. Soc. 93:121–6.
95. Bondioli, F., Andreola, F., Barbieri, L., Manfredini, T. and A.M. Ferrari. 2007. Effect of rice husk ash (RHA) in the synthesis of $(Pr,Zr)SiO_4$ ceramic pigment. J. Eur. Ceram. Soc. 27:3483–8.
96. Sharifikolouei, E., Baino, F., Galletti, C., Fino, D. and M. Ferraris. 2019. Adsorption of Pb and Cd in rice husk and their immobilization in porous glass-ceramic structures. Int. J. Appl. Ceram. Technol. 0:1–8.
97. Ketov, A., Korotaev, V., Rudakova, L., Vaisman, I., Barbieri, L. and I. Lancellotti. 2021. Amorphous silica wastes for reusing in highly porous ceramics. Int. J. Appl. Ceram. Technol. 18:394–404.
98. Alzukaimi, J. and R. Jabrah. 2019. The preparation and characterization of porous alumina ceramics using an eco-friendly pore-forming agent. Int. J. Appl. Ceram. Technol. 16:820–31.
99. Li, Y.C., Min, X.B., Ke, Y. et al. 2018. Utilization of red mud and Pb/Zn smelter waste for the synthesis of a red mud-based cementitious material. J. Hazard. Mater. 344:343–9.
100. Kim, Y., Kim, M., Sohn, J. and H. Park. 2018. Applicability of gold tailings, waste limestone, red mud, and ferronickel slag for producing glass fibers. J. Clean. Prod. 203:957–65.
101. Kim, Y., Lee, Y., Kim, M. and H. Park. 2019. Preparation of high porosity bricks by utilizing red mud and mine tailing. J. Clean. Prod. 207:490–7.
102. Wang, W., Chen, W. and H. Liu. 2019. Recycling of waste red mud for fabrication of sic/mullite composite porous ceramics. Ceram. Int. 45:9852–7.
103. Kim, Y. and H. Park. 2020. A value-added synthetic process utilizing mining wastes and industrial byproducts for wear-resistant glass ceramics. ACS Sustainable Chem. Eng. 8:2196–204.

Green Synthesis of Sustainable Materials

104. Ren, X., Ma, B., Qian, F. et al. 2021. Green synthesis of porous SiC ceramics using silicon kerf waste in different sintering atmospheres and pore structure optimization. Ceram. Int. 47:26366–74.

105. (a) Zhou, L., Li, Z. and Y. Zhu. 2019. Porous silica/mullite ceramics prepared by foam-gelcasting using silicon kerf waste as raw material. Mater. Lett. 239:67–70. (b) Liu, W., Tian, J., Xiang, D., Mao, R. and B. Wang. 2020. Fabricating superior thermal conductivity SiC–AlN composites from photovoltaic silicon waste. J. Clean. Prod. 274:122799. (c) Sun, Q., Yuan, L., Jin, E., Wen, T., Tian, C. and J. Yu. 2021. Transformation of polysilicon cutting waste into SiC/α-Si_3N_4 composite powders via electromagnetic induction heating. Ceram. Int. 47:12269–75.

106. Liu, Y., Lian, W., Su, W., Luo, J. and L. Wang. 2019. Synthesis and mechanical properties of mullite ceramics with coal gangue and wastes refractory as raw materials. Int. J. Appl. Ceram. Technol. 00:1–6.

107. Hao, J., Ma, H., Feng, X., Gao, Y., Wang, K. and Y. Tian. 2018. Low temperature sintering of ceramic proppants by adding solid wastes. Int. J. Appl. Ceram. Technol. 15:563–8.

108. (a) Zhu, J. and H. Yan. 2017. Fabrication of an A356/fly-ash-mullite interpenetrating composite and its wear properties. Ceram. Int. 43:12996–3003. (b) Zhang, J., Li, H., Li, S. et al. 2018. Mechanism of mechanical–chemical synergistic activation for preparation of mullite ceramics from high-alumina coal fly ash. Ceram. Int. 44:3884–92.

109. Wei, Z., Hou, J. and Z. Zhu. 2016. High-aluminum fly ash recycling for fabrication of cost-effective ceramic membrane supports. J. Alloy Compd. 683:474–80.

110. Chen, Z., Xu, G., Du, H., Cui, H., Zhang, X. and X. Zhan. 2019. Realizable recycling of coal fly ash from solid waste for the fabrication of porous Al_2TiO_5-Mullite composite ceramic. Int. J. Appl. Ceram. Technol. 16:50–8.

111. (a) Ma, B.-Y., Li, Y., Cui, S.-G. and Y.-C. Zhai. 2010. Preparation and sintering properties of zirconia-mullite-corundum composites using fly ash and zircon. Trans. Nonferrous Met. Soc. China, 20:2331–5. (b) Ma, B.-Y., Sun, M.-G., Ding, Y.-S., Yan, C. and Y. Li. 2013. Fabrication of β-Sialon/ZrN/ZrON composites using fly ash and zircon. Trans. Nonferrous Met. Soc. China. 23:2638–43. (c) Ma, B., Ren, X., Yin, Y. et al. 2017. Effects of processing parameters and rare earths additions on preparation of Al_2O_3-SiC composite powders from coal ash. Ceram. Int. 43:11830–7.

112. Yin,Y., Ma, B., Hu, C. et al. 2019. Preparation and properties of porous SiC-Al_2O_3 ceramics using coal ash. Int. J. Appl. Ceram. Technol. 16:23–31.

113. Ma, B., Su, C., Ren, X. et al. 2019. Preparation and properties of porous mullite ceramics with high-closed porosity and high strength from fly ash via reaction synthesis process. J. Alloy Compd. 803:981–91.

114. Monich, P.R., Vollprecht, D. and E. Bernardo. 2020. Dense glass-ceramics by fast sinter-crystallization of mixtures of waste-derived glasses. Int. J. Appl. Ceram. Technol. 17:55–63.

115. Zinatloo-Ajabshir, S., Salehi, Z. and M. Salavati-Niasari. 2018. Green synthesis of $Dy_2Ce_2O_7$ ceramic nanostructures using juice of Punica granatum and their efficient application as photocatalytic degradation of organic contaminants under visible light. Ceram. Int. 44:3873–83.

116. Mehdizadeh, P., Orooji, Y., Amiri, O., Salavati-Niasari, M. and H. Moayedi. 2020. Green synthesis using cherry and orange juice and characterization of $TbFeO_3$ ceramic nanostructures and their application as photocatalysts under UV light for removal of organic dyes in water. J. Clean. Prod. 252:119765.

117. Khirade, P.P., Vinayak, V., Kharat, P.B. and A.R. Chavan. 2021. Green synthesis of $Ba_{1-x}Sr_xTiO_3$ ceramic nanopowders by sol-gel combustion method using lemon juice as a fuel: Tailoring of microstructure, ferroelectric, dielectric and electrical properties. Optical Materials 111:110664.
118. Khirade, P.P., Raut, A.V., Alange, R.C., Barde, W.S. and A.R. Chavan. 2021. Structural, electrical and dielectric investigations of cerium doped barium zirconate ($BaZrO_3$) nano-ceramics produced via green synthesis: Probable candidate for solid oxide fuel cells and microwave applications. Physica B. 613:412948.
119. Hamadneh, I., Al-Mobydeen, A., Al-Dalahmeh, Y. et al. 2022. Green synthesis of a $YBa_2Cu_3O_7$ ceramic superconductor using the fruit extract of *Juniperus phoenicea*. Ceram. Int. 48:12654–9.

14 Green Materials
Synthesis and Characterization

Rajeev Kumar and Jyoti Chawla

CONTENTS

14.1 Introduction 259
14.2 Synthesis of Materials 260
14.3 Reducing Chemical Waste Using Optimization Studies...........................263
    14.3.1 Response Surface Methodology 263
14.4 Characterization Techniques..265
    14.4.1 Fourier Transform Infrared (FTIR) 265
    14.4.2 X-ray Diffraction 265
    14.4.3 Scanning Electron Microscopy...266
    14.4.4 Energy Dispersive X-ray Analysis...266
    14.4.5 Brunauer, Emmett, and Teller (BET) Analysis............................267
    14.4.6 UV-visible Spectroscopy..267
    14.4.7 Nuclear Magnetic Resonance 267
    14.4.8 Mass Spectroscopy 268
    14.4.9 Atomic Adsorption Spectroscopy (AAS) 268
    14.4.10 Inductive Coupled Plasma Mass Spectroscopy (ICP-MS)...........269
14.5 Conclusion 269
References..269

14.1 INTRODUCTION

Natural resources are the fundamental resources for the development and economic growth of any country. These are the basic essential sources for income, employability, and well-being. Natural resources may be renewable or nonrenewable resources. Demands of raw materials have been increased by rapid industrialization in all over the world. Extraction of raw materials from natural resources has been increasing day by day. More than 100 billion metric tons raw materials may be extracted from natural resources by 2030. Population of the world increasing day by day requiring more food, more energy, more industries, and more materials placing extra load on the natural resources [1].

Materials are the prime requirement for any type of development in the world. Many types of natural and synthetic materials have been used in engineering and scientific field as building blocks. Single-use materials have been creating a lot of

DOI: 10.1201/9781003297772-14   **259** problems toward environment as well as human beings. It is now more important to explore the materials in terms of sustainable ways. Sustainable materials solve the long-term environmental problems and make the planet healthier. Sustainable materials are the materials that can be developed or produced in sufficient amount without depleting the natural nonrenewable resources and ensure the fulfillment of future requirements. There has been more discussion and research on the replacement of traditional materials by sustainable materials in all of the applied fields and there is an unstoppable momentum in the development of sustainable materials in solid, liquid, and gas phases. Researchers and designers have been collectively working for the development and production of sustainable materials with affordable cost as well as lesser carbon footprint [2-5]. Traditional or conventional non-recycled materials pose threat to human beings as well as environment, whereas sustainable or green or alternative materials can be recycled and environment friendly [6-9].

Conventional materials used traditionally for making structures are not feasible option for environment as well as human beings as lot of chemicals has been used for processing of these conventional materials creating lot of waste and other pollutants. Hazardous waste and other pollutants were not noticed earlier because of healthier environment. Current consumptions of materials are not completely in sustainable way and environment friendly. Most of the countries have been working in this direction to protect environment and improve resource productivity. The aim of this chapter is to explore the various green routes adopted by the researchers to synthesize the useful materials and identify the opportunities and challenges associated with the green synthetic routes.

14.2 SYNTHESIS OF MATERIALS

Materials play key roles in all applied fields. Materials may be classified mainly in four main groups (Figure 14.1).

Metallic materials are made from one or more metallic elements, such as copper, iron, silver, nickel, platinum, etc., and with some nonmetallic elements such as carbon, nitrogen, oxygen, etc. Polymers are natural or synthetic materials made from large number of molecules called as macromolecules. Ceramics are nonmetallic oxide, carbide or nitride materials having brittle nature, hard, and strong in compression, e.g. bricks and porcelain. Composites are advanced materials made by combination of two or more materials.

They may occur naturally or be synthesized. Synthesis of materials is an art in terms of atom economy and reaction yield. There are lots of conventional and non-conventional methods which are presently available for synthesis of materials.

Sustainable synthesis in terms of green chemistry was introduced by scientific community in 1991 to reduce the hazardous waste and increase the reaction yield for the protection of environment and human beings. Twelve principles are required for synthesis of materials in sustainable ways (Figure 14.2).

Twelve green chemistry principles listed in Figure 14.2 act as guiding force for synthesis of new green materials or sustainable materials. These principles have been applied simultaneously to get the more precise results. These principles are very simple and highly useful for synthesis of materials, but suitable knowledge of

Green Materials

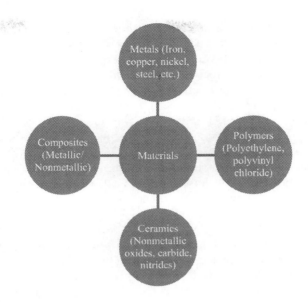

FIGURE 14.1 Classification of materials.

interaction of chemicals with environment is required for synthesis and modification of materials using greener approach [7,8]. A lot of efforts have been done by the researchers to avoid or reduce the waste and hazardous materials from the synthetic method and in this regard green synthesis plays a scientific role toward sustainable approaches.

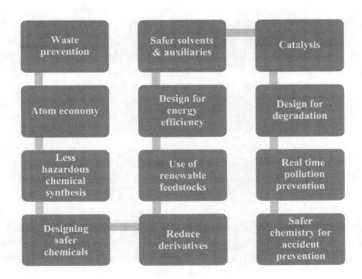

FIGURE 14.2 Twelve green chemistry principles.

Recently electrochemistry is widely used for the synthesis of materials. It includes electron exchange between electrodes and chemicals present in the solutions in the form of redox transformations. Waste generating reagents and hazardous compounds can be replaced by other electrochemical redox procedures which enhance sustainability and selectivity of the products [10,11].

Heterocycles compounds and their derivatives have been widely used in medicinal field. To avoid waste and hazardous materials during synthesis of these compounds, researchers have proposed safer and economical methods. Synthesis of quinoxalines in solvent-free/green solvent medium was described in sustainable ways [12]. Biomass resources were used in one pot sustainable synthesis of 2-furanylquinazolines and 2-furfurylidene derivatives from xylan, xylose, and fructose [13]. Cyrene (dihydrolevoglucosenone), a dipolar aprotic green solvent, was prepared through pyrolysis of biomass with cellulose [14]. It has good biodegradability and no NO_x or $SO_x$was produced. Silver-decorated graphene-based nanocomposite was synthesized in sustainable way by using *Kigelia Africana* stem extract [15]. Carbon quantum dots were prepared in sustainable ways using shrimp shell in one step hydrothermal method [16]. It is low-cost procedure used to prepare materials with having high fluorescent properties, bioimaging nature, and antibacterial properties. Silver (Ag) nanowires were prepared by green and sustainable way by using tannic acid as reducing, capping, and stabilizing agent [17]. In this method, also three light conditions, i.e., dark, partially illuminated, and fully illuminated, were applied to get the change in morphology of silver nanowires. Nickel-catalyzed benzazoles and purines were synthesized by sustainable methods [18]. It was acceptor less dehydrogenative coupling and borrowing hydrogen approach. Benzimidazole, benzothiazole, benzoxazole, and purine five-membered heterocycle compounds were synthesized using sustainable method. Natural adsorbents were synthesized and modified using sustainable electrochemical for adsorption of dyes and heavy metals from water [11]. One pot synthesis of honeycomb biomass adsorbent based on oxidized corn starch gelatin was proposed in sustainable and economical way using gelatin extracted from leather solid wastes. Silver nanoparticles were prepared by using microwave-assisted method for sustainable photocatalytic hydrogen evolution [19]. Green material 'corn starch', as stabilizing agent and glycerol, was used for synthesis of Ag nanoparticles by microwave irradiation and applied on to TiO_2 nanoparticles for photo-generation of H_2.

Copper-oxide nanoparticles were synthesized from bark extract of Thespesia populnea using a simple procedure [20]. Bark powder of Thespesia populnea was mixed with deionized water was boiled and then cooled at room temperature followed by filtration. After that 0.56 g of copper acetate was added and stirred magnetically for 5 minutes. The color of copper acetate solution under stirred conditions after 20 minutes changed blue to green and finally brown (CuO nanoparticles).

Silver-Zirconia composite was prepared by green synthetic method [10]. Sauropus androgynus extract was applied as green material for synthesis of Silver-Zirconia composite. $AgNO_3$ and $Zr(NO_3)_4$ were dissolved into Sauropus androgynus plant extract with stirring for 60minutes. Photo-induced reactions can be controlled by covering the reaction mixture with aluminum paper and change in color (the light green to deep yellowish brown) confirmed the formation of composite. M-MOF-74

Green Materials 263

for Cu, Co, and Zn were synthesized using sustainable methods [21]. Many such type of greener approaches have been proposed by different groups. However, more studies are required for implementation of these processes on large scale.

14.3 REDUCING CHEMICAL WASTE USING OPTIMIZATION STUDIES

Optimization is a mathematical approach to optimize the data in terms of one or more possible constraints. It is used to find the best result by optimizing the conditions to minimize the undesired products. The main objective of optimization techniques is to get the best results with lesser number of experiments, which help to reduce the chemical use and overall cost of the reaction. In case where outcome is affected by multiple variables, large numbers of experiment are required to optimize the data. It is too costly and requires too much chemicals to get the high yield of the products by repeating the experiments in different conditions. Using different experimental designs, desired outcome can be achieved with the help of results predicted by mathematical models [22-24]. Factorial analysis and response surface methodology (RSM) can be applied to optimize the data.

RSM is a statisticalmethod used to find the optimal values for synthesis of materials to get the high yield. Real and coded values for correlation can be expressed as follows:

$$X_i = \frac{x_i - x_o}{\Delta x} \tag{14.1}$$

X_i represents dimensionless value, x_i represents real value of the independent variable, and x_o represents real value at the center point of an independent variable, and Δx represents the step change value. Behavior of any system can be evaluated by using second-order polynomial model using RSM [25].

Box-Behnken design (BBD) and central composites design (CCD) (3^k factorial analysis) have been applied to optimize the conditions for second-order polynomial model. Regression equation is used to get the optimal values for best results. There are lot of other types of optimization techniques are also available, i.e., taguchi, heuristic search technique, tabu search, iterative mathematical search technique, and artificial neural network modeling for the optimization of the input variables.

14.3.1 RESPONSE SURFACE METHODOLOGY

RSM was first introduced by George E. P. Box & K. B. Wilson [25]. It is used to optimize the process in terms of yield and other factors for the synthesis of materials. Relationship between independent variables (input variables) and resulting output can be easily correlated by using this statistical method. It's a multivariable factor statistical tool. Design Expert® software was used by the researcher for evaluation of the data inputs. It works on designing the experiments with minimum numbers of trials to get the optimal results. CCD and BBD are widely used for designing of experiment for optimal values using Design Expert® software. The aim of this optimization is to optimize the input data to attain maximum output data. A number of designed runs in terms of experiments are conducted to find the changes in output values.

Water treatment and power generation networks were designed for achieving less or zero waste emission [26]. Decomposition-based strategy was applied for optimization for low-pollutant emissions and low-energy consumption. Nanocomposites having graphene oxide in the polymer network of chitosan and sodium alginate containing inclusion complexes of β-cyclodextrin with amlodipine besylate were prepared using microwave-assisted sustainable and eco-friendly method and optimized various parameters for synthesis including the amount of used solvent, time, ratio of backbones, amount of graphene oxide, and pH using RSM-CCD to get the maximum percentage swelling of the material [27].

Ethylhexyl palmitate and ethylhexyl stearate were synthesized using Novozym enzymes in solvent-free method [28]. Synthetic conditions were optimized using RSM. Optimized results showed that, at optimum conditions, more than 85% conversion was achieved in one hour. Also vacuum system and N_2 bubbling in the reaction system were used to improve the conversion.

Magnetic Moringa oleifera an efficient coagulant was synthesized using sonication-assisted method and optimized the conditions using RSM method [29]. Conditions were optimized in terms of sonication time, nanoparticles composition, and sonication temperature for removal of suspended materials from water. RSM results showed 1 wt% of MNPs mass fraction, 2.35 minutes of sonication time, and 50°C sonication temperature as the optimal conditions for preparation of an efficient coagulant. Removal of total suspended solid (83.3%), chemical oxygen demand (85.2%), and color (28.1%) was observed with optimum conditions.

Ag nanoparticles were synthesized using Aspergillus oryzae NRRL447 proteins in sustainable way [30]. Central composite design was used to set the various parameters, such as pH of solution, reaction time, cell-free filtrate ratio, and silver nitrate concentration, for desired silver nanoparticles formation. Central composite design in RSM results shows that 12.25 pH of solution, 108 minutes reaction time, and 2.25 mM silver nitrate concentration were best conditions to get the optimum value of silver nanoparticles.

Natural polysaccharide-based hydrogel of kappa-carrageenan and tamarind kernel powder crosslinker were synthesized in sustainable methods [31]. Central composite design with RSM was used for optimization of the synthetic parameters. Concentration of crosslinker, solvent amount, ratio of backbones, time, temperature, and pH were selected as active variables for optimization the reaction conditions using RSM. The ratio 1:1 of backbones, 30°C temperature, and 10 mL amount of solvent were found to be best conditions for maximum percentage swelling.

Synthetic parameters of inverse vulcanized copolymers from rubber seed oil were optimized by RSM [32]. Initial sulfur content of the reactant was found to be the most influential parameter. Different variables are optimized for maximum conversion of the sulfur in the respective product. Such type of optimization studies further helps to develop best sustainable synthetic routes for the preparation of materials.

Green Materials 265

14.4 CHARACTERIZATION TECHNIQUES

Characterization is the next step to be adopted post synthesis and all the conventional methods as discussed in this section may be used for characterization of green materials.

14.4.1 FOURIER TRANSFORM INFRARED (FTIR)

FTIR spectroscopy is perfect tool for identification of the types of bonds and presence of various groups in the synthesized materials. It involves measuring absorption or transmittance of IR radiation due to interaction of IR radiation with materials. Bonds and functional groups are identified due to absorption or transmitted IR radiation by the materials. Every bond and functional groups vibrate at particular wave number which can be measured using FTIR spectrometer.

Magnetic Moringa oleifera was characterized using FTIR [29]. FTIR spectrum of the product indicated sharp strong band at 3431 cm^{-1} due to proteins' characteristic vibration. Cerium-based nanoparticles were synthesized using Moringa oleifera leaf extract and FTIR spectrum of showed some characteristic peaks at correspond to N-O stretching, H-O-H bond, and OH stretching [33]. FTIR absorption spectrum of synthesized cerium oxide nanoparticles before calcination showed that the intensity of OH stretches and N-O stretch of nitrate is still strong around 1300 cm^{-1}. However, a new peak was observed at a wavelength of 453 cm^{-1} after calcinations. Also, a shift was observed in O-Ce-O bond peak at 556 cm^{-1} with high intensity.

Aloe vera extract was used to synthesize spinel lithium titanate (GSLTO) nanoparticles [34]. The FTIR pattern of the chemically synthesized spinel lithium titanate (LTO) showed strong peaks due to asymmetric stretching of Li_2CO_3 in LTO. The same characteristics bands were shifted to in GSLTO. Bending and stretching vibration of –OH group was observed for both LTO and GSLTO.

Thespesia populnea tree extract was used to synthesize copper-oxide nanoparticles [20]. FTIR analysis was used to identify various groups in the materials. Bark extract of Thespesia populnea contains phenols, alcohols (2980 cm^{-1}), alkanes (1638 cm^{-1}), amine (2865 cm^{-1}), alkanes (1453 cm^{-1}), and aliphatic amines (1053 cm^{-1}). Phenol and flavonoids were adsorbed at the surface of copper-oxide nanoparticles confirmed by the FTIR analysis.

14.4.2 X-RAY DIFFRACTION

X-ray diffraction is used to identify the crystalline or amorphousnature of the materials. It's a constructive interference of material and monochromatic X-rays. Intensity of X-rays (diffracted) is monitored and recorded as the function of rotation of at particular angles. Crytallinity of the materials depends on the intensity of peaks in the spectrum.

Graphene-silver nanocomposites were synthesized using green material Kigelia Africana stem extract [15]. Graphene peak was found at $2\theta = 26°$ along with silver peaks of 21.3 nm average particle size in XRD spectrum.

XRD spectrum of synthesized cerium oxide nanoparticles using green method showed some specific peaks in 2θ that confirmed the formation of cerium oxide nanoparticles [33].The XRD pattern of the chemically synthesized spinel lithium titanate (LTO) and green synthesized spinel lithium titanate (GSLTO) showed characteristics strong peaks at 2θ which was due to specific planes of the cubic spinel structure [33]. There were some emerged low-intensity minor peaks due to $Li_2Ti_3O_7$. Sharp and strong peaks in XRD spectrum confirmed the formation of crystalline LTO and GSLTO with the calculated average crystallite size 37 nm of LTO and 42 nm of GSLTO respectively. Nayak et al. (2022) have synthesized green silver-zirconia composite and cubic shape of silver and tetragonal shape of zirconia nanoparticles were confirmed by XRD analysis [10].

14.4.3 Scanning Electron Microscopy

Scanning electron microscopy is an analytical tool for surface characteristics of the nanomaterials. Chemical composition, size, and shape of the material can be easily estimated by using this technique. Various types of signals, i.e., backscattered electron, secondary electron imagine, and X-rays, are helpful to identify the surface characteristics of materials. From 1 cm to 5 microns area for 50–100 nm particles may be magnified in the range of 20–30000 X by conventional SEM technique.

The size of silver nanoparticles after 320 days of aging was characterized using SEM technique with diameters from 2.9 to ~44 nm [19]. SEM spectrum of synthesized cerium oxide nanoparticles using green method showed that cerium oxide nanoparticles have spherical shape with a uniform size but undergo agglomeration with time [33].

Size and morphology of synthesized copper-oxide nanoparticles were identified using SEM analysis that confirmed high-density poly-dispersed spherical and irregular form of 61–69 nm particle size [20]. Average size of copper nanoparticles was found to be in range of 41–55 nm and variation in the size was due to presence of various naturally derived constituents with different redox properties in the extract. Particle size distribution showed that copper-oxide nanoparticles have overall size ranges from 50–80 nm.

14.4.4 Energy Dispersive X-ray Analysis

Energy dispersive X-ray is used to identify the elements present in the material as well as composition. In this technique, emitted X-ray counts are observed from the solid materials due to interaction of incident X-rays with electron. A number of peaks may be observed in the EDX spectrum due to presence of various elements in the materials.

Cerium oxide nanoparticles were sustainably synthesized using Moringa oleifera leaf extract in 5 hours [33]. EDX analysis showed specific high peaks of Ce and O at 20 keV. Copper-oxide nanoparticles were synthesized using bark extract of Thespesia populnea tree in green method [20]. EDX study was used to identify the composition of copper-oxide nanoparticles. EDX spectrum confirmed the presence of copper (48.33%) in the materials with very small amount of chloride (0.95%).

Green Materials 267

Silver-zirconia composite and chitosan biopolymer (binder) were applied for fabrication of electrode [10]. Composition of various elements was confirmed by EDX study. Peaks at 0.2 keV, 0.5 keV, and 2.0 keV confirmed the elemental zirconium and oxygen. Additionally, small peaks at 1.2 keV and 1.3 keV represent zinc and magnesium respectively (plant extract materials). Peak at 3 keV for Ag along with zirconia peaks was observed in EXD analysis of Ag-zirconia composite.

14.4.5 BRUNAUER, EMMETT, AND TELLER (BET) ANALYSIS

BET analysis is used to evaluate the surface characteristics in terms of area and pores size of the particles. The name BET was given on the name of scientists who developed this method for analysis of particles of the materials [35]. Surface characteristics in terms of external surface area and pores size volumes of the particles can be calculated using adsorption isotherm (t-plot method) under nitrogen atmosphere. BJH method can be used for pore size distributions using N_2 isotherms. The BET-specific surface area of low-cost honeycomb biomass adsorbent based on oxidized corn starch gelatin was found to be 30.48 m^2/g [11].

M-MOF-74 for Cu, Co, and Zn were synthesized sustainable, and porosity and surface areas of these all M-MOF-74 were calculated using BET method [21]. Calculated BET surface area was found to be maximum for Cu and minimum for Zn-based metal organic framework. N_2 isotherm of Co-MOF-74 was also recorded at low range p/p0 (107–101) in order to check microporous nature of MOF having pore diameter (9.3 Å).

14.4.6 UV-VISIBLE SPECTROSCOPY

UV-visible spectroscopy is very important technology to identify conjugation and various functional groups in the materials. Different types of isomers of the materials can also be determined by using this tool. In this technique, ultraviolet and visible radiation (200–800 nm) interact with the sample to get peaks in the spectrum. Each peaks represents a particular type of transition correspond to a particular function group.

Graphene-silver nanocomposites were synthesized using Kigelia Africana stem extract in sustainable and graphene peak was found at λ_{max} of 268 along with silver peaks at 406 nm in UV-visible spectrometer [15]. Copper-oxide nanoparticles synthesized by using bark extract of Thespesia populnea tree showed absorbance at 390 nm in UV-visible spectrum that confirms the formation of copper-oxide nanoparticles [20].

14.4.7 NUCLEAR MAGNETIC RESONANCE

Nuclear magnetic resonance is used to identify the number of protons, structural isomers, and other isomers in the materials. In this technique, radiowave interacts with the sample in the presence of external magnetic field. Those materials which have spin quantum number (I>0) exhibit NMR spectrum.

Starch and non-isocyanate polyurethanes were used as green materials for synthesis of green hybrid materials [36]. The poly-hdroxyurethanes (PHUs) were prepared by ethylene carbonate and different diamines, such as 1,2-diaminoethane (EDA), 1,4-diaminobutane (BDA), 1,6-diaminohexane (HDA), at molar ratio of 2:1. The products of the respective diamines were labeled as PHU2, PHU4, and PHU6 respectively. Structure of these polyurethanes was verified based on position of hydrogen in NMR spectrum ((500 MHz, DMSO-d6). PHU2 was yellowish-white color powder having the positions of various proton at: δ (ppm) = 3.02 (CH_2NH, 4H), 3.52 (CH_2OH, 4H), 3.94 (CH_2O, 4H), 4.70 (OH, 2H), 6.72 (NH, 0.4H), and 7.09 (NH, 1.6H). However, PHU4 was white color powder having the positions of various proton at 1.37 (CH_2, 4H), 2.96 (CH_2NH, 4H), 3.52 (CH_2OH, 4H), 3.94 (CH_2O, 4H), 4.69 (OH, 2H), 6.73 (NH, 0.4H), and 7.09 (NH, 1.6H). Furthermore, PHU6 was white color powder having the positions of various proton at 1.22 (CH_2, 4H), 1.37 (CH_2, 4H), 2.95 (CH_2NH, 4H), 3.52 (CH_2OH, 4H), 3.93 (CH_2O, 4H), 4.68 (OH, 2H), 6.69 (NH, 0.4H), and 7.05 (NH, 1.6H).

14.4.8 MASS SPECTROSCOPY

Mass spectroscopy is an important tool for identification of material based on their mass. Different types of mass spectroscopy techniques are available for different types of materials. A number of ionization techniques are available for determination of mass of the materials. Atmospheric pressure chemical ionization, electrospray ionization, electron ionization, fast atom bombardment, field ionization, laser desorption, plasma desorption, matrix-assisted laser desorption ionization, thermospray ionization, etc. have been widely applied techniques for determination of the mass of the materials. Selectivity of the technique is based on the nature and structure of materials. The fast atomic bombardment has been widely applied for identification of the complex materials due to excessive fragmentation. Furthermore, matrix-assisted laser desorption ionization time of flight mass spectrometry, in a matrix consisting of 2,5-dihydroxy benzoic acid (DHB) dissolved in polar solvent, has also been widely used for identification of supramolecular complexes. In soft ionization method such as fast atomic bombardment electrospray ionization, there is little fragmentation and predominantly the molecular ions peaks which help to analyze the material very easily and quickly. Fourier transform ion cyclotron resonance, time of flight, ion trap, and quadrupole are widely used as mass analyzers in mass spectrometers. Fragments of various mass-to-charge ratios (m/z) are analyzed by the detector and converted into electric signals [37-39]. Materials mass can be calculated based on ions peaks with m/z ratio and their intensity of various fragments in the spectrum.

14.4.9 ATOMIC ADSORPTION SPECTROSCOPY (AAS)

AAS is an analytical tool for determination of concentration of various elements in the liquid. AAS is based on the principle that atoms or ions have tendency to absorb light of specific wavelength. In this technique, particular elements absorbed

Green Materials

the light at a specific wavelength between 190–900 nm and emit a specific wavelength. Concentration of atoms or ions is directly proportional to the amount of light absorbed/transmitted at specific wavelength. More than 60 elements can be detected by using AAS technique. A number of atomization techniques, such as flame, graphite furnace, glow-discharge atomizing systems, hydride-generating atomizers, and cold vapor atomization, are available for conversion of metal into atoms. Flame AAS(FAAS) has been widely applied for determination of concentration of elements in parts per million (ppm) or parts per billion (ppb) ranges. In FAAS, metals are injected in the form of spray at high temperature to form the atoms, which absorb the light from element-specific hollow cathode lamp. Graphite furnace AAS is electrothermal atomization technique in which metal is vaporized using hollow graphite tube. Concentration of the metals in the given samples can be calculated using Lambert-Beer's law.

14.4.10 INDUCTIVE COUPLED PLASMA MASS SPECTROSCOPY (ICP-MS)

ICP-MS is an analytical tool for determination of concentration of various elements in the liquid samples at very low concentrations in the range of ppb and parts per trillion (ppt) levels. In this method, atomic elements in the form of aerosol interact with high-temperature (6000–8000 K) argon plasma where they become ionized. These ions are sorted based on their mass in detector. The advantage of this technology is that multiple elements can be detected only in a single run in the sample.

14.5 CONCLUSION

Requirements of materials are increasing day by day for technological development across the globe. A number of techniques have been applied for production of the material to fulfill the market demand without considering any other environmental impact. However, researchers have also proposed many green alternative routes for the material synthesis. Statistical techniques further help to optimize the process and yield. Characterization tools have their own role for confirmation of structure and comparison of properties of materials with materials prepared via conventional routes. Significant development is made in synthesis of green materials and natural materials, like plant extracts, chitosan, etc., have been utilized for synthesis of novel materials. There are also some challenges related to feasibility of available green routes synthesis to be implemented on industrial scale including limited options of source material, green solvents, and green catalysis for desired products and this area should be further explored in future studies.

REFERENCES

1. Pandit, P., Nadathur, G.T., 2018. Characterization of green and sustainable advanced materials: Processing and characterization, Green and Sustainable Advanced Materials; Wiley Publications: US. 35–66.

2. Ali, M., Kim, B., Belfield, K.D., Norman, D., Brennan, M., Ali, G.S., 2016. Green synthesis and characterization of silver nanoparticles using Artemisia absinthium aqueous extract—A comprehensive study, Material Science and Engineering: C Materials for Biological Applications, 58, 359–65.
3. Geetha, A.R., George, E., Srinivasan, A., Shaik, J., 2013. Optimization of green synthesis of silver nanoparticles from leaf extracts of Pimenta dioica (Allspice), Scientific World Journal, 1–5.
4. Irshad, M.A., Nawaz, R., Rehman, M.Z., Adrees, M., Rizwan, M., Ali, S., Ahmad, S., Tasleem, S., 2021. Synthesis, characterization and advanced sustainable applications of titanium dioxide nanoparticles: A review, Ecotoxicology and Environmental Safety, 212, 111978–111991.
5. Hossein, J., Ensieh, G.L., Roshanak, R.M., Thomas, W., 2018. A review of using green chemistry methods for biomaterials in tissue engineering, International Journal of Nanomedicine, 13, 5953–5969.
6. Chudoba, T., Wojnarowicz, J., 2018. Current trends in the development of microwave reactors for the synthesis of nanomaterials in laboratories and industries: A review, Crystals, 8, 379.
7. Ravichandran, S., Karthikeyan, E., 2011. Microwave synthesis a potential tool for Green chemistry, International Journal of Chemical Technology & Research, 3, 466–470.
8. Lonkar, S.P., Pillai, V., Abdala, A., 2019. Solvent-free synthesis of ZnO-Graphene nanocomposite with superior, Applied Surfactant Science, 465, 1107–1113.
9. Martin, M., Gani, R., Iqbal, M., 2022. Mujtaba, sustainable process synthesis, design, and analysis: Challenges and opportunities, Current Research in Green and Sustainable Chemistry, 30, 686–705.
10. Nayak, S., Kittur, A.A., Nayak, S.K., 2022. Green synthesis of silver-zirconia composite using chitosan biopolymer binder for fabrication of electrode materials in supercapattery application for sustainable energy storage, Separation and Purification Technology, 5, 100292–100301.
11. Dang, X., Yu, Z., Yang, M., Woo, M.W., Song, Y., Wang, X., Zhang, H., 2021. Sustainable electrochemical synthesis of natural starch-based biomass adsorbent with ultrahigh adsorption capacity for Cr(VI) and dyes removal, International Journal of Hydrogen Energy, 46(69), 34264–34275.
12. Kumar, A., Dhameliya, T.M., Sharma, K., Patel, K.A., Hirani, R.V., Bhatt, A.J., 2022. Sustainable approaches towards the synthesis of quinoxalines: An update, Journal of Molecular Structure, 2022, 132732.
13. Carreira, M.A., Oliveira, M.C., Fernandes, A.C., 2022. One-pot sustainable synthesis of valuable nitrogen compounds from biomass resources, Molecular Catalysis, 518, 112094.
14. Kong, D., Dolzhenko, A.V., 2022. Cyrene: A bio-based sustainable solvent for organic synthesis, Sustainable Chemistry and Pharmacy, 25, 100591.
15. Kurmarayuni, C.M., Chandu, B., Yangalasetty, L.P., Gali, S.J., Alam, M.M., Rani, P., Bollikolla, H.B., 2022. Sustainable synthesis of silver decorated graphene nanocomposite with potential antioxidant and antibacterial properties, Materials Letters, 308(A), 131116.
16. Elango, D., Packialakshmi, J.S., Manikandan, V., Jayanthi, P., 2022. Sustainable synthesis of carbon quantum dots from shrimp shell and its emerging applications, Materials Letters, 312, 131667.
17. Kaabipour, S., Hemmati, S., 2022. Green, sustainable, and room-temperature synthesis of silver nanowires using tannic acid—Kinetic and parametric study, Colloids and Surfaces A: Physicochemical and Engineering Aspects, 641, 128495.

Green Materials

18. Chakraborty, G., Mondal, R., Guin, A.K., Paul, N.D., 2021. Nickel catalyzed sustainable synthesis of benzazoles and purines *via* acceptorless dehydrogenative coupling and borrowing hydrogen approach, Organic & Biomolecular Chemistry, 19(33), 7217–7233.
19. Strapasson, G.B., Assis, M., Backes, C.W., Corrêa, S.A., Longo, E., Weibel, D.E., 2021. Microwave assisted synthesis of silver nanoparticles and its application in sustainable photocatalytic hydrogen evolution, Journal of Drug Delivery Science and Technology, 61, 102325.
20. Narayanan, M., Jahierhussain, F.A., Srinivasan, B., Sambantham, M.T., Al-Keridis, L.A., AL-mekhlafi, F.A., 2022. Green synthesizes and characterization of copper-oxide nanoparticles by *Thespesia populnea* against skin-infection causing microbes, Journal of King Saud University—Science, 34(3), 101885.
21. Flores, J.G., Díaz-García, M., Ibarra, I.A., Pliego, J.A., Sánchez-Sánchez, M., 2021. Sustainable M-MOF-74 (M = Cu, Co, Zn) prepared in methanol as heterogeneous catalysts in the synthesis of benzaldehyde from styrene oxidation, Journal of Solid State Chemistry, 298, 122151.
22. Saini, S., Kumar, R., Chawla, J., Kaur, I., 2019. Adsorption of bivalent lead ions from an aqueous phase system: Equilibrium, thermodynamic, kinetics and optimization studies. Water Environment Research, 91(12), 1692–1704.
23. Kaur, I., Gupta, A., Singh, B.P., Kumar, R., Chawla, J., 2019. Defluoridation of water using micelle templated MCM-41: Adsorption and RSM studies. Journal of Water Supply: Research and Technology AQUA, 68(4), 282–294.
24. Kaur, I., Sonal, Kaur, S., Kumar, R., Chawla, J., 2018. Spectrophotometric determination of triclosan based on diazotization reaction: Response surface optimization using Box-Behnken design. Water Science and Technology, 77(9), 2204–2212.
25. Box, G.E.P., Wilson, K.B., 1951. On the experimental attainment of optimum conditions (with discussion), Journal of the Royal Statistical Society, Series B, 13(1), 1–45.
26. Li, Y., Zhang, L., Yuan, Z., Gani, R., 2020. Synthesis and design of sustainable integrated process, water treatment, and power generation networks, Computers & Chemical Engineering, 141, 107041.
27. Khushbu, Jindal, R., 2020. RSM-CCD optimized microwave assisted synthesis of chitosan and sodium alginate based nanocomposite containing inclusion complexes of β-cyclodextrin and amlodipine besylate for sustained drug delivery systems, Catalysis Today, 346, 98–105.
28. Murcia, M.D., Arnaldos, M.S., Requena, S.O., Máximo, F., Bastida, J., Montiel, M.C., 2022. Optimization of a sustainable biocatalytic process for the synthesis of ethylhexyl fatty acids esters, Environmental Technology & Innovation, 25, 102191.
29. Noor, M.H.M., Azli, M.F.Z., Ngadi, N., Inuwa, I.M., Opotu, L.A., Mohamed, M., 2021. Optimization of sonication-assisted synthesis of magnetic *Moringa oleifera* as an efficient coagulant for palm oil wastewater treatment environmental nanotechnology, Monitoring & Management, 16, 100553.
30. Ali, M.E., Abdelmageed, M.O., Maysa, A.E., Naser, G.B., Mohamed, M.H., 2022. Green synthesis of silver nanoparticles using *Aspergillus oryzae* NRRL447 exogenous proteins: Optimization via central composite design, characterization and biological applications, Journal of Drug Delivery Science and Technology, 67, 102976.
31. Vaid, V., Jindal, R., 2021. RSM-CCD optimized in air synthesis of novel kappa-carrageenan/tamarind kernel powder hybrid polymer network incorporated with inclusion complex of (2-hydroxypropyl)-β-cyclodextrin and adenosine for controlled drug delivery, Polymer, 219, 123553.
32. Ghumman, A.S., Shamsuddin, R., Nasef, M.M., Yahya, W.Z., Abbasi, A., 2021. Optimization of synthesis of inverse vulcanized copolymers from rubber seed oil using response surface methodology, Journal of Materials Research and Technology, 15, 2355–2364.

33. Putri, G.E., Rilda, Y., Syukri, S., Labanni, A., Arief, S., 2020. Highly antimicrobial activity of cerium oxide nanoparticles synthesized using *Moringa oleifera* leaf extract by a rapid green precipitation method, Journal of Science: Advanced Materials and Devices, 5, 346–353.
34. Perumal, P., Sivaraj, P., Abhilash, K.P., Soundarya, G.G., Balraju, P., Selvin, P.C., 2022. Green synthesized spinel lithium titanate nano anode material using Aloe Vera extract for potential application to lithium ion batteries, Journal of King Saud University—Science, 34, 101885.
35. Brunauer, S., Emmett, P.H., Teller, E., 1938, Adsorption of gases in multimolecular layers, Journal of the American Chemical Society, 60(2), 309–319.
36. Ghasemlou, M., Daver, F., Ivanova, E.P., Adhikari, B., 2020. Synthesis of green hybrid materials using starch and non-isocyanate polyurethanes, Carbohydrate Polymers, 229, 115535.
37. Henderson, W., McIndoe, J.S., 2005. Mass spectrometry of inorganic, coordination and organometallic compounds: Tools, techniques, tips; John Wiley & Sons: Chichester.
38. Ashcroft, A.E., 1997. Ionization methods in organic mass spectrometry, Royal Society of Chemistry; Royal Society of Chemistry: Cambridge. 97–121
39. Downard, K., 2004. Mass spectrometry: A foundation course; Royal Society of Chemistry: Cambridge.

15 Waste Sewage Sludge Concrete as a Sustainable Building Construction Material

A Review

Jagdeep Singh

CONTENTS

15.1 Introduction 274
15.2 Characterization of Waste Sewage Sludge 275
    15.2.1 Physiochemical Properties 275
    15.2.2 Mineralogical and Morphological
        Properties 275
    15.2.3 Pozzolanic Properties 277
15.3 Properties of Waste Sewage Sludge Concrete 278
    15.3.1 Fresh Properties 278
        15.3.1.1 Workability 278
    15.3.2 Mechanical Properties 279
        15.3.2.1 Compressive Strength 279
        15.3.2.2 Tensile Strength 284
        15.3.2.3 Flexure Strength 284
        15.3.2.4 Modulus of Elasticity 285
    15.3.3 Durability Properties 285
        15.3.3.1 Density 285
        15.3.3.2 Water Absorption 286
        15.3.3.3 USPV 287
        15.3.3.4 Shrinkage 288
        15.3.3.5 Chloride Ion Permeability 289
        15.3.3.6 Sorptivity 289
        15.3.3.7 Electrical Resistivity and Corrosion
                Resistance 290
        15.3.3.8 Thermal Conductivity 290
        15.3.3.9 Frost Resistance 290
        15.3.3.10 Alkali-Silica Reaction (ASR) 291

DOI: 10.1201/9781003297772-15

274   Handbook of Sustainable Materials

15.4   Leachability 291
15.5   Ecological and Economic Analysis 294
15.6   Conclusion 296
15.7   Future Scope 297
References 297

15.1   INTRODUCTION

One of the most hotly debated topics in today's world is waste management. Rapid growth in industrialization and urbanization leads to more consumption of water and production of wastewater. In the treatment of wastewater as a by-product, sludge is produced, which contains various organic and inorganic matter like pathogens, heavy metals, sediments, microorganisms, microplastics, and nutrients [1–3]. This production of sewage sludge is of great concern in the world's second-largest populated country like India. Approximately 120,000 tonnes/day of sludge is produced in India [4]. Urban areas of India produce approximately 72368 MLD of sewage, from this one can imagine the production of sewage sludge [5]. There are various sludge management methods available like landfilling, composting, incineration, pyrolysis, etc. [6–8]. Sometimes, up to 50% of the operational cost of a wastewater treatment plant is associated with the disposal of waste sludge [9]. In the current pandemic situation of COVID-19, it's also very important to properly manage and dispose of sewage water/sludge to lower the risk of virus transmission [10]. So there is a great need for an effective disposal method for such a huge amount of sewage sludge in India and worldwide as well. On the other hand, today's construction industries are facing various challenges related to materials and the environment. As far as materials are concerned, the industry uses various natural materials (nonrenewable) like aggregates, fillers, etc. in the production of concrete. The availability of these materials is limited, which leads to various crises. Similarly, the production of cement causes various environmental issues due to the emission (7–9% globally) of greenhouse gas (CO_2) [11, 12]. So there is a need to look for an alternative construction material before it's too late. Keeping the concept of 3R in mind to achieve sustainable development goals, one can use waste sewage sludge in construction as cementing materials or as an aggregate. This will lead to a reduction in supply chain stress and reduce various environmental impacts like sludge disposal, CO_2 emission, etc.

The major focus of this chapter is on the trends and problems that various researchers have encountered while using waste sewage sludge or sewage sludge ash in concrete. The study conducted an assessment of sewage sludge-related data published in various publications, book chapters, and reports mainly by Elsevier. The chapter presents various properties of concrete incorporating sewage sludge, including environmental and cost-benefit analysis. The information presented in this article shows that sewage sludge can be utilized as a building material to some extent. In terms of sustainable development, this is a viable waste sewage sludge management approach in today's culture.

Waste Sewage Sludge Concrete

15.2 CHARACTERIZATION OF WASTE SEWAGE SLUDGE

15.2.1 PHYSIOCHEMICAL PROPERTIES

The particle size of sewage sludge ash generally remains lower than 125 µm. In most of the studies, it was observed that the particles of sewage sludge ashes were finer or comparable to cement particle size [13]. The specific gravity of raw sewage sludge or sewage sludge ashes ranged between 2.11 2.74. Generally, the moisture content in such ashes was negligible [14]. The density of sewage sludge ashes lies between 1.1 and 2.68 g/cm^3 and these are generally comparable to the density of cement. The porosity of sewage sludge ashes generally remains higher compared to cement or fine aggregates. Most of the ashes represent continuous nonhomogeneous grading as far as particle size distribution is concerned [15]. The median particle size of sewage sludge ashes was 23–100 µm. Brunauer-Emmett-Teller (BET) surface area of sewage sludge ashes was higher because of the uneven structure and porous nature of the ashes. Various physical characteristics of sewage sludge are mentioned in Table 15.1. The colors of such ashes vary with incineration temperature or various treatments applied to them. Generally, the colors of the ashes were black, gray, and reddish. Loss in ignition (LOI) value of unprocessed sewage sludge and ash lies between 52 and 73%, and 0.5 and 5.5%, respectively. Higher LOI of raw sludge is a result of organic matter's presence. LOI is generally inversely proportional to the incineration temperature. In sewage sludge, the main chemical components included are silica, aloxide, ferric oxide, and quicklime. In detail, chemical compositions are mentioned in Figure 15.1. It's generally observed that the silica, alumina, and ferric oxide content in sewage sludge ashes remains lower compared to other types of ashes like fly ash and bagasse ash [13]. It was also observed that P_2O_5 remains higher in sewage sludge/ash. This might be due to the processing of sewage in treatment plants. Higher ferric oxide content was observed as a result of the use of iron-based coagulants like ferric chloride during the wastewater treatment [15].

15.2.2 MINERALOGICAL AND MORPHOLOGICAL PROPERTIES

As per various studies, particles of sewage sludge ash were irregular in shape with variable size, rough surface, highly porous surface, and cavities as shown in Figure 15.2 [13, 15, 21, 26, 31, 33, 40–42]. But if these ashes were properly ground or incinerated at a higher temperature then their particles would become less porous, smoother, and flaky [21, 40, 42] as shown in Figure 15.2. In the case of raw dry sludge, particles look like a smoother surface like siliceous crystals [20, 21]. Sometimes irregular stacking particle shapes were observed in sludges due to the presence of various substances in them [2]. Organic and inorganic components present in sewage sludge include lignin, hemicellulose, and cellulose [20]. Major mineralogical phases observed after XRD analysis were quartz, hematite, magnetite, leucite, and albite [20, 25, 26, 40, 41, 43]. Apatite and fluorapatite were also observed due to the presence of phosphates in sludge [13]. In the case of sewage sludge XRD, large humps represent the presence of organic matter [20]. Amorphous phases level in sewage sludge ashes remains less comparative to other ashes like in fly ash [32].

TABLE 15.1
Physical Characteristics of Sewage Sludge Reported by Various Researchers

Physical Properties

Research By	Sludge Type	Specific Gravity	Density (g/cm³)	Moisture Content (%)	BET Fineness (m²/g)	Specific Surface Area (m²/kg)	Particle Size
Rutkowska et al., 2018 [13]	Ash	-	1.826	-	-	-	34.24% particles are 50–100 µm
Chen et al., 2018 [40]	Ash	2.33–2.74	-	-	-	-	-
Chen et al., 2013 [14]	Ash	-	0.0027	0	-	-	Median size 100 µm
Zhou et al., 2020 [26]	Ash	-	2.740	-	2.866	-	The mean particle sizes 23.77 µm
Hsu and Chen, 2020 [20]	Sludge	2.60	-	3.05	-	-	Mean particle size 9.18 µm
Gu et al., 2021 [27]	Ash	-	-	-	-	421	Mean particle size 24.1 µm
Oliva et al., 2019 [21]	Sludge	-	-	-	22.3	1.1	60% particles are less than 28.6 µm
Chen and Poon, 2017 [32]	Ash	2.74	-	-	17.366	-	Mean diameter 6 µm
Liu et al., 2019 [33]	Ash	-	2.69	-	-	1041	Mean particle size 5.387 µm
Garcés et al., 2008 [35]	Ash	-	2.62	0.5	-	-	Particle size range 1–250 µm
Cyr et al., 2012 [38]	Ash	-	2.43	-	-	-	Mean diameter of 19 µm

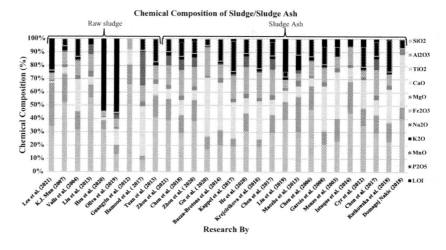

FIGURE 15.1 Chemical characteristics of sewage sludge reported by various researchers [13–39].

Waste Sewage Sludge Concrete

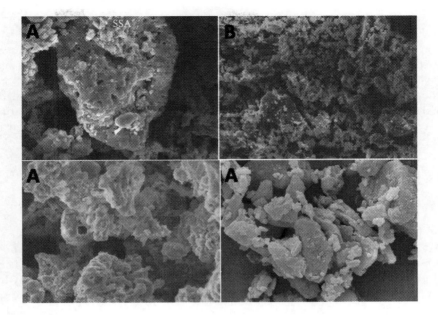

FIGURE 15.2 Morphology images of (A) raw sewage sludge ash and (B) ground sewage sludge ash [2, 15, 32].

15.2.3 Pozzolanic Properties

Pozzolanic materials are those that, when coming in contact with water, start behaving like cementing materials. The reactivity speed of such materials depends on the water-binder ratio, alkali level in cement, the surface area of particles of pozzolan, and temperature. Silica and alumina content generally remain higher in pozzolanic materials. Various test standards exist to judge a material's pozzolanic strength, like strength activity index, saturated lime test, Frattini test, etc. As per Oliva et al., 2019 [21] waste sewage sludge ash is considered as a potential pozzolanic material. During the research, they evaluated the property of the ash with the R3 test procedure. In the study, they evaluated the property of sewage sludge ash, fillers (200 microns), and sewage sludge. As per the results, the pozzolanic strength of sludge ash was 14.4–17.8 times more comparative to filler, and 2.16–2.86 times higher compared to sewage sludge. Incineration of such ashes at a higher temperature also helps in increasing its pozzolanic activity strength. The pozzolanic activity of the ash measured for 24 hours in the form of heat release was reported as 113.05–149.89 J/g. This demonstrates that the ash has sufficient amorphous phases to produce pozzolanic action. [40] In another research, the activity was evaluated with the help of the Frattini test. In this case, samples were prepared with 80% cement and 20% ground/raw sewage sludge ash and fly ash and tested after 8 days. As per the results, all types of ashes qualified for the test and represent good pozzolanic material. It was observed that finer sewage sludge ash shows more pozzolanic activity strength comparative to unprocessed sludge ash. It can be concluded that the grinding process makes these ashes more pozzolanic. Further, the pozzolanic activity of the ashes was evaluated as per the strength activity index. For this, mortar samples of 50 mm cube were prepared and cured in a saturated lime solution. As per the results obtained after testing, the sewage sludge ashes were considered as not a good pozzolanic material. This might be due to the use of more water during the preparation of the samples, which underestimates the property of sewage sludge ashes. As the strength of the sludge ash sample increased over time and keeping the findings of the Frattini test in mind, sewage sludge ash was classified as a material with moderate pozzolanic activity.

15.3 PROPERTIES OF WASTE SEWAGE SLUDGE CONCRETE

15.3.1 Fresh Properties

15.3.1.1 Workability

Workability can be considered as one of the most critical characters of freshly prepared concrete. Generally, the workability of concrete is determined with the help of a standard truncated cone and the test is known as the slump test. In a study by Nakic, 2018 [39] concrete samples were prepared with the incorporation of 10% ash made up of sewage sludge as a substitute for cement. In the mix, the water-binder ratio remained constant at 0.50. No bleeding or segregation was reported in the mix. But lower consistency in the mix was observed when the ash was added. The presence of tiny ash particles, which absorb more water, was the reason for this and reduce the workability of the concrete mix. So, to overcome the problem, a good amount (1% by the weight of binder) of superplasticizer was used in the mix containing sewage sludge ash. As per a report by Rabie et al., 2019 [44] they used two types of sewage sludges, one is in dry form and the other one is wet. Sewage sludges of 5, 10, 15, and 20% were used in the preparation of various concrete mixes. It was reported that the slump value was reduced in both types of sewage sludge concrete samples. Samples containing higher sludge values show a higher reduction in slump value as comparative to standard concrete samples. When the waste sewage sludge is used in the manufacturing of lightweight aggregates, it also affects the workability of concrete. As these aggregates are generally porous. As per a research by Tuan et al., 2013 [24] concrete was made with the use of lightweight aggregates prepared with the help of sewage sludge. In the concrete, 45% of lightweight aggregates were used as compared to the normal concrete. In order to keep proper workability 1.14% of superplasticizer was incorporated into the concrete. This shows a reduction in slump value by using aggregates made from sewage sludge in concrete. A slump value of 260 mm was observed in lightweight aggregate concrete. Along with slump value, flow time (54 seconds) and slump flow value (480 mm) of the sewage sludge lightweight aggregate concrete samples were also reported. The workability of concrete samples can also be correlated with the water content required in making the concrete mix. A rise in water absorption demand caused a downfall in concrete workability. Similar outcomes have been according to Liu et al., 2013 [19] throughout the concrete research made with ceramisite sewage sludge aggregates. In research by Chen et al., 2018 [15] ash made up of sewage sludge ash (10 and 20%)

Waste Sewage Sludge Concrete

was used in the manufacturing of concrete as a cementing material. Application of ash in concrete samples leads to a rise in the water-powder ratio. This implies that the workability of the concrete sample is reduced. As these ash particles are porous and absorb more amount of water and thus reduce the workability of concrete mix. This water absorption of the ash can be reduced with the use of grinding, ash particle size can be reduced. As ground particles have smooth and less porous surfaces. Which aids in the maintenance of proper workability of concrete mix. According to Lee et al., 2021 [16] workability of concrete including lightweight aggregates made from waste sewage sludge rises.

15.3.2 MECHANICAL PROPERTIES

15.3.2.1 Compressive Strength

The most essential character to judge the strength parameter of concrete is compressive strength. The property can be affected due to the presence of internal constituent material of concrete as well. Use of waste sewage sludge also alters the compressive strength of concrete. As per a study by Rutkowska et al., 2018 [13] they have used normal concrete incorporating sewage sludge ash powder as substitute of cement. Maximal cement and water content in the mix was 360.58 kg/m^3 and 166.42 kg/m^3, respectively. Replacement level of sewage sludge ash with cement was 5, 10, and 15% and the strength of the concrete samples were evaluated after 28 and 56 days of moist curing. With the exchange of cement for sewage sludge ash, there is no detrimental impact on the concrete's strength was observed. Even maximum up to 5% variability coefficient was reported in the study which is quite low. Maximum strength in the sample containing 10% waste sewage sludge ash was observed comparative to samples containing 5 and 15% sewage sludge ash. Due to the presence of the ash initially rise in the strength is slow, because of improper binding of cement slurry. But with the passage of time strength starts rising. In a report by Nakic, 2018 [39] waste sewage sludge was used as ash in concrete. The ash was replaced with 10% cement content in the concrete. Average compressive strength of concrete samples were evaluated after 7 and 28 days of water curing. No downfall in the strength of concrete samples were observed with the use of the ash. Concrete samples containing the ash behave much better in comparison to ordinary concrete samples as far as the strength is concerned. This might be due to the use of superplasticizer which enhances workability of concrete samples and leads to better placement of concrete samples in molds. In a study by Chen and Poon, 2017 [32] cylindrical concrete samples of 50 mm diameter and height were prepared for evaluating compressive strength. The concrete was prepared by dry mix method having zero slump value. In this case the water content was not fixed, they have used sufficient quantity of water till the proper cohesiveness was achieved in the mix. In the study glass cullet and sewage sludge were used as substitute of natural aggregates and cement respectively in concrete mix. Replacement level of sewage sludge ash in the study was 10 and 20%. The concrete samples were all assessed after 28 days of proper water curing. Concrete samples in the study show reduction in compressive strength. Between 6 and 13% downfall was noticed in strength of concrete samples containing 20%

sewage sludge ash and 20 and 50% of glass cullet. With the rise in sewage sludge ash content from 10 to 20%, decrease in concrete samples was observed. This was due to slow pozzolanic activity inside the concrete samples due the presence of sewage sludge ashes. It was proposed in the study that with the use of finely grinded sewage sludge ash little rise in compressive strength can be achieved due to better pozzolanic activity in concrete samples. In the same study construction and demolition waste aggregates were used as aggregates along with 20% waste sewage sludge ash as a cementing material in concrete. In this case, 11.6–21.6% concrete's strength is deteriorating. This downfall in the strength was more comparative to the case when glass cullet was used as aggregates in concrete along with sewage sludge ash in concrete. Sewage sludge can also be used in concrete in the form of lightweight aggregates. In another study based on lightweight aggregate (made up of sewage sludge ash), concrete compressive strength reduces (0 to 23.7%) but still remains within permissible limit of ACI, 2004 standards (18–28 MPa) [16]. In a research done by Mun, 2007 [17] lightweight aggregates were prepared with the use of waste sewage sludge. Various types of lightweight aggregates were prepared by varying sewage sludge content (clay-sludge ratio as 1:1, 1:3, and 1:5). Cement and water content in the study remain constant as 320 kg/m^3 and 160 kg/m^3 respectively in preparing concrete mix. It was observed that with the use of higher sewage sludge content in aggregates, higher deterioration in the strength of concrete samples were observed following a 14-day cure period. But the strength achieved by such concrete is still good that the concrete can be used in a nonstructural components. Similarly, in another study by Lau et al., 2018 [45] concrete samples were made using aggregates manufacture from lime treated sewage sludge and palm oil fuel ash. In the case a comparative study was done on two types of concrete samples, out of which cone contains natural aggregates (NWAC) and other contain lightweight aggregates (LWAC) manufactured from sewage sludge and palm oil fuel ash. The concrete mix was designed for target strength of 50 MPa at 28 days, having constant water-cement ratio of 0.47. The compressive strength test was performed on concrete samples of 100 mm cube after 3, 7, 28, 56, 90, and 120 days of curing. In real after 28 days of curing the concrete sample of NWAC and LWAC shows 51.48 and 50.44 MPa compressive strength, which is more than the target strength. With the increase in curing period (3–120 days) the strength of concrete samples were also increases (0–36.7%). So strength of concrete made with aggregates manufactured from waste materials/ashes was comparable to normal concrete. In a study by Tuan et al., 2013 [24] on high-performance concrete manufactured with lightweight aggregates made up of sewage sludge ashes and having constant water-binder ratio of 0.30. The strength achieved by such concrete was 73% as comparative to normal concrete at 28 days of curing period. This also satisfies criteria's of ASTM C 330 and ACI 318 for lightweight concrete. Valls et al., 2004 [18] reported in their research that with the incorporation of waste sewage sludge in concrete compressive strength reduces appreciably. In the study, they have casted cylindrical samples (150 × 300 mm) of concrete containing 0, 2.5, 5, and 10% of sewage sludge as an additive material. Mean compressive strength of concrete samples were evaluated after 7, 28, and 90 days of water curing. Initially (7 days) it was discovered that when sewage sludge was used, the strength of concrete samples reduces drastically. High reduction in the strength occurs with higher addition of

Waste Sewage Sludge Concrete

sewage sludge. But with the passage of time (after 90 days), concrete samples containing sludge recover their strength (58–78%) in comparison to the strength of ordinary concrete. As per an experimental study done by Baezo-Brotons et al., 2014 [28] they have used sewage sludge ash as a cementing material and aggregate replacement material. In concrete mixes 5, 10, 15, and 20% of sewage sludge ash was introduced as cement additive and in one mix 10% of sludge ash was replaced with natural aggregates. All the samples were evaluated for compressive strength after 28 days of curing. Results revealed that using sludge ash as a cementing agent reduction in the strength occurs at all replacement levels except at 5%. More use of sludge ash leads to more reduction (2–20%) in the strength. The reduction in the strength was attributed with high porosity of concrete mixes. However, in the case of sludge ash-containing concrete as aggregates, drastic rise (207.8% relative to control concrete) in the strength was observed. This may be as a result of densification of concrete inner structure as internal voids reduce. As per a report by Rabie et al., 2019 [44] use of waste sewage sludge leads to reduction in concrete's compressive strength. In the study they have used two forms of sewage sludge, one is dry sewage sludge and other one is wet sewage sludge. These sludges were procured from two different firms in the study. Samples of concrete mixes were prepared with 5, 10, 15, and 20% of sewage sludge as additive from cement weight. The strength of concrete samples were evaluated after 7 and 28 days of moist curing. Results show 1.2–63.7% reduction in case of dry sludge concrete and 6.6–73.4% reduction in case of wet sludge concrete samples after 7 days. Maximum reduction in the strength was observed at 20% sludge value. A research study was done by Liu et al., 2013 [19] on concrete containing lightweight aggregates prepared with ceramisite sintering of waste sewage sludge. In the study impact of particle size of aggregates, mineral admixtures, and water-cement ratio was observed on compressive strength of concrete. Results revealed that large-sized aggregates induce a decrease in strength. Whereas the strength of concrete can be increased with the application of mineral admixtures like fly ash or slag. In case of less water-cement ratio more increase in concrete's samples was reported in the study. In general, it was concluded that with the use of appropriate quantity of mineral admixture and low water-cement ratio, higher compressive strength can be achieved in such concrete. This higher compressive strength can be attributed due to dense interface structure between cement slurry and ceramisite (sewage sludge) aggregates in concrete. It's crucial to look into the effects of Na_2O on geopolymer concrete containing sewage sludge. Up to 4% use of Na_2O leads to rise in the strength of concrete samples. This could be owing to a higher alkali concentration which helps in generating higher number of dissolved ions which help in better geopolymerization of concrete [40]. Concrete incorporating waste sewage sludge ash satisfies technical specification of lightweight buildings made up of ash-based concrete. In a study concrete mix was prepared with sewage sludge ash (replacement of 10% with cement and 2% with fine aggregate). The strength of ash-based concrete increases as the curing period increases when sewage sludge is used. But the strength of ash-based concrete was comparatively low with respect to control concrete. The strength achieved by sludge-based concrete after 28 days was 27.1 MPa which was more than 17 MPa, which is a technical specification for lightweight buildings [14]. The impact of the sludge on concrete is summarized in Table 15.2.

TABLE 15.2

Summary of Compressive Strength Property of Sludge-Based Concrete Evaluated by Various Researchers

Study By	Sludge Form	Sludge Used As	Other Materials/ Techniques Used	Sludge Share (%)	Testing Age (days)	Observations
Rutkowska et al., 2018 [13]	Ash	Cement (Replacement)	-	5, 10, 15	28, 56	10% ash-based concrete shows higher strength compared to other replacements. Initially rise in the strength was low, but increases with time.
Nakic, 2018 [39]	Ash	Cement (Replacement)	-	10	7, 28	No reduction in strength was observed.
Chen et al., 2018 [40]	Ash	Cement (Replacement)	Glass cullets (as aggregates)	10, 20	28	The strength reduces (6–13%) with a rise in the ash content.
Lee et al., 2021 [16]	Ash	Coarse aggregates	Lightweight aggregates (LWA)	10, 20, 30, 40	28	Strength reduces (0–23.7%) in LWA concrete but remains higher than the limits as per ACI 2004 standard (18–28 MPa).
Mun, 2007 [17]	Sludge	Coarse aggregates	Lightweight aggregates (LWA)	Clay:sludge (1:1, 1:3, 1:5)	14	Higher sludge content aggregates samples show a greater loss in strength. But the strength is still good so that such material can be utilized in making nonstructural components of buildings.
(Lau et al., 2018 [45]	Sludge	Coarse aggregates	Lightweight aggregates (LWA); palm oil fuel ash (40–60%)	40, 50, 60	3, 7, 28, 56, 90, 120	Comparable strength achieved by LWA concrete with respect to control concrete. No major loss was observed.
Tuan et al., 2013 [24]	Ash	Coarse aggregates	Lightweight aggregates (LWA): glass powder (10–50%)	50, 70, 90	28	73% strength achieved by LWA concrete comparative to normal concrete at 28 days testing.
Valls et al., 2004 [18]	Sludge	Cement (Additive)	-	2.5, 5, 10	7, 28, 90	A larger sludge addition level resulted in a greater drop in strength initially (7 days). With time strength starts rising in all samples.

(Continued)

TABLE 15.2 (*Continued*)

Summary of Compressive Strength Property of Sludge-Based Concrete Evaluated by Various Researchers

Study By	Sludge Form	Sludge Used As	Other Materials/ Techniques Used	Sludge Share (%)	Testing Age (days)	Observations
Baeza-Brotons et al., 2014 [28]	Ash	Cement (Additive)	-	5, 10, 15, 20	28	2–20% decrease in strength observed with the incorporation of the ash in all mixes, except at 5% ash mix.
		Fine aggregates (Replacement)	-	10	28	Drastic rise (105.7%) in the strength observed.
Rabie et al., 2019 [44]	Sludge	Cement (Additive)	Dry sludge; wet sludge	5, 10, 15, 20	7, 28	In the case of wet sludge concrete, there was a greater drop in strength comparative to dry sludge concrete.
Liu et al., 2013 [19]	Sludge	Coarse aggregates	Lightweight aggregates (LWA)	sludge:fly ash:silt—7:3:3, 6:4:3, and 5:5:3 for LWA	7, 14	Strength increases with the use of small-sized aggregates, mineral admixtures like fly ash or slag, and a low water-binder ratio.
Chen et al., 2013 [14]	Ash	Cement and fine aggregates (replacement)	-	10% cement; 2% fine aggregates	7, 28, 90	A rise in the strength of concrete incorporating the ash was less (2.2–9.7%) comparative to normal concrete. Strength increases with time and the concrete fulfills the technical requirements of lightweight building material.

15.3.2.2 Tensile Strength

Tensile strength of hardened concrete is a mechanical attribute. It generally depends on the nature of ingredients used in the production phase of concrete, like the texture of aggregates, the strength of paste and aggregates, and the ITZ character of aggregates and binder. In a study by Lau et al., 2018 [45] on concrete made up of lightweight aggregates, these aggregates were prepared with waste sewage sludge. The concrete sample shows 3.53 MPa tensile strength, which is quite higher than the minimum specified strength (2.0 MPa) as per ASTM-C330, 2017 [46] at 28 days. The strength of sewage sludge-based concrete was 14% less compared to normal concrete.

15.3.2.3 Flexure Strength

Concrete's flexural strength is similar to its tensile strength. It's used to evaluate resistance against bending failure in concrete. It is also known as the modulus of rupture. Concrete's flexure property also gets altered with the incorporation of waste sewage sludge. In a study by Nakic, 2018 [39] on concrete containing 10% waste ash made up of sewage sludge, the study to evaluate flexural strength, three prisms of size $100 \times 100 \times 400$ mm were prepared. All the concrete specimens were tested at 28 days. In the concrete mix to achieve proper workability, 1% (by mass of binder) superplasticizer was used. Test results reveal no major adverse impact of waste sewage sludge ash on concrete. Only a 0.6% reduction took place in the strength of concrete incorporating 10% sludge ash as compared to normal concrete. It was also reported that with time, the strength of ash-based concrete rises. This good strength of sewage sludge ash concrete was evidenced by the better workability of concrete with the use of a superplasticizer. In another experimental study by Mun, 2007 [17], they used waste sewage sludge in the production of light aggregates, and these were used in the manufacturing of concrete. Such aggregates were prepared with varying sewage sludge content (clay-sludge ratio as 1:1, 1:3, and 1:5). Cement and water content in the study remain constant as 320 kg/m^3 and 160 kg/m^3 respectively in preparing the concrete mix. To evaluate the property, prism samples of dimension $60 \times 60 \times 240$ mm were prepared and cured till the testing age of 14 days. Test results revealed that with a hike in the sludge level, little reduction in the strength occurs. But the strength is still good that such concrete can be utilized in nonstructural lightweight elements in construction. Similarly, in another study by Lau et al., 2018 [45] samples of concrete were prepared with aggregates manufactured from waste sludge treated with lime along with ash made up of palm oil fuel. In this case, a comparative study was done on two types of concrete samples, out of which one contains NWAC and the other contains LWAC manufactured from sewage sludge and palm oil fuel ash. No major reduction in the flexural strength was observed with the use of waste sewage sludge. Only 11.9% less flexural strength was achieved by LWAC concrete samples compared to NWAC concrete after 28 days of curing. In research by Valls et al., 2004 [18] on concrete containing waste sewage sludge, a prism of $7.6 \times 7.6 \times 25.4$ cm was cast to evaluate flexural strength. To determine the strength, the average of the three prism's results was evaluated at 28 and 90 days of curing. The dosage of sewage sludge considered in the concrete mixes was 0, 2.5, 5, and 10%.

Waste Sewage Sludge Concrete 285

Results of the study revealed that with a rise in the level of sewage sludge in concrete, flexural strength reduces but rises with the rise in the curing period.

15.3.2.4 Modulus of Elasticity

It represents the measure of resistance to elastic deformation of concrete when stressed. In research by Valls et al., 2004 [18] on concrete containing waste sewage sludge, a modulus of elasticity test was performed on cylindrical concrete samples of size 150×300 mm. The concrete samples were prepared with 0, 2.5, 5, and 10% of sewage sludge and evaluated after 90 days of water curing. In the test, three cycles of loading-unloading were applied to note deformation corresponding to the stress level. The test results of concrete specimens represent reduction a (17.4–36.5%) with the addition of waste sewage sludge. The concrete sample containing the highest sludge content shows the highest deformation, under the same strain. This might be due to the rise in pores inside concrete samples due to sewage sludge. Lau et al., 2018 [45] published research in which concrete samples were prepared with aggregates manufactured from waste sewage sludge subjected to lime treatment and ash made up of palm oil fuel. In this case, a comparative study was done on two types of concrete samples, out of which one contains NWAC and the other contains LWAC. The concrete mix was designed for a target strength of 50 MPa at 4 weeks, having a constant water/cement value, i.e. 0.47. In the study, a reduction (14%) in the modulus value of LWAC concrete was observed compared to NWAC at 28 days of testing. The reduction was marginal and was comparable to other concrete samples prepared with various waste materials.

15.3.3 DURABILITY PROPERTIES

15.3.3.1 Density

Generally, the density of concrete refers to the solidity of concrete. It's a ratio of the weight of concrete and the weight of the same volume of standard liquid. The use of sewage sludge may also change the density of concrete. Let's consider the case in which an experimental study has been conducted by Chen et al., 2018 [40] on concrete containing ash made up of waste sewage sludge as a cement substitute and glass cullets as a replacement of aggregates. In the study, two different forms of waste sludge ash of sewage were used, one was raw sewage sludge ash (10 and 20%) and another one was processed sewage sludge ash (well-ground) (10 and 20%). The density of cylindrical concrete samples was evaluated after 28 days of water curing. It was noticed that with the use of glass cullets, the density of concrete specimens was reduced, but no major impact of sewage sludge ashes was observed in all concrete samples. Whereas the density of concrete samples containing processed sewage sludge ash was a little higher than that of raw sewage sludge concrete samples. This order is also in line with specific gravity as well as the fineness of the ashes. The higher density of processed sewage sludge ash concrete was because of the existence of very small particles of the ash which fill the internal voids of concrete and thus make the concrete denser. In another study done by Mun, 2007 [17] in which they have used sewage sludge in the manufacturing of lightweight aggregates, and then those aggregates were used in concrete. In the manufacturing of lightweight aggregates with the help of sewage sludge, three ratios of clay-sludge content were used as 100:100, 100:300, and 100:500. The density of concrete samples was evaluated after 14 days of water curing. As per the report, a decrease in the density of waste-water sludge-based aggregate concrete was noticed. With a rise in the level of waste sewage sludge inside aggregates, the density of concrete reduces. It was noticed that with the use of such aggregates, the density of concrete can be reduced by 700–800 kg/m³. The density of sewage sludge aggregates concrete samples lies in the range of 1410–1500 kg/m³ and such concrete can be used in the case of light-weight nonstructural elements construction. As per research by Valls et al., 2004 [18] on concrete containing three different doses (2.5, 5, and 10%) of waste sewage sludge, the density of such concrete samples was evaluated after 1, 4, and 12.85 weeks of curing. It was observed that with the use of arid sewage sludge, the density of concrete samples reduces in the same proportion as the content of sewage sludge rises. In simple words, we can say that, with a hike in waste sewage sludge level, the density of concrete samples reduces. This reduction in the density is due to the presence of dry spongy sewage sludge whose density is lower than the density of natural aggregates. It was also observed that at the initial age the density of higher sludge content concrete was low, but with time it also rose. According to experimental data given by Baeza-Brotons et al., 2014 [28] on concrete samples containing ash made up of waste sewage sludge as a cementing material and as a fine aggregate, 5–20% of waste sewage sludge ash was added as cementing material and in one sample, 10% of the ash was replaced with fine aggregates. All the samples were cured in water for 4 weeks and then the density of the specimens was evaluated. A rise (up to 2.1%) in the concrete density was noticed with the rise in the level of waste ash made up of sewage sludge. This is due to the formation of a compact structure as these finer ash particles fill the internal voids of concrete. In the case of ash used as aggregates, the highest density (7.1% more than control concrete) was noticed compared to the samples in which ash was used as cementing material. Whereas contradictory outcomes were seen in a study by Rabie et al., 2019 [44], they have used two different types of sewage sludge and the content of the sludge was 5, 10, 15, and 20%. One of the sludges was dry and the other one was wet. It was noticed that with the use of sludge in concrete density was reduced (2.5–6.8%). Whereas more reduction in concrete samples containing wet sludge (6.8 to 4.1% reduction) was observed compared to the dry sludge samples (6.6 to 2.5% reduction). As per the data, one can say that the density of concrete depends on the nature of the ash used in it.

15.3.3.2 Water Absorption

Water absorption is one of the most prominent durability aspects of concrete. It gives an idea of the total pore volume of concrete. It depends on various aspects like characteristics of ingredients used in concrete, entrained air content, curing duration, presence of microcracks, etc. [47]. To understand this, we can consider a case in which waste sewage sludge ash was used in concrete. In the study, waste sewage sludge ash (one raw sludge ash and other well-ground sludge ash) was used as cementing material, along with glass cullets (aggregate replacement). Cylindrical samples of concrete were prepared with 10 and 20% of the ash and evaluated for the property after 28 days of the curing period. As per the results, the samples incorporating

Waste Sewage Sludge Concrete

glass cullets show less water absorption. But in the case of sewage sludge ash concrete samples, water absorption rises. The highest water absorption was noticed in specimens containing raw sewage sludge ash comparative to the well-ground ash samples. These results are also in line with the strength achieved by these concrete samples [40]. A case discussed by Mun, 2007 [17] about concrete containing waste sewage sludge lightweight aggregates. In the study, they used three different proportions of sewage sludge in the manufacturing of aggregates like clay:sludge ratio as 100:100, 100:300, and 100:500. In the study, no significant difference in water absorption capacity was observed among three different types of concrete samples. Water absorption shown by these samples was 9.6–10.2% [45]. Similarly, in another study related to concrete made up of lightweight aggregates manufactured with the use of waste sludge initially subjected to lime treatment and ash made up of palm oil fuel. A concrete mix was designed for a target strength of 50 MPa and the water-binder ratio in the mix remained constant, i.e. 0.47. Water absorption of the concrete samples was evaluated after 28 days of the curing period. No major difference in the concrete samples containing sewage sludge aggregates or natural aggregates was observed as far as the water absorption was concerned. Only 0.5% higher absorption capacity was observed in sludge aggregate concrete compared to natural aggregate concrete. So it was concluded that the water-permeable pores in both types of concrete were the same. In general, the sewage sludge is porous, so one can expect more water absorption capacity with the use of sludge in concrete at an initial stage. These spongy sludge particles create cavities in the internal structure of concrete initially. At later stages, water absorption capacity starts reducing as concrete achieves good hardness and resistance, thus demanding less water. These outcomes were demonstrated by Valls et al., 2004 [18] during their study, in which they used 2.5 to 10% of sewage sludge in the manufacturing of concrete. All the samples were evaluated after being subjected to moist curing for 7–90 days. All the samples reflect less than 8% of water absorption capacity. As per research by Baeza-Brotons et al., 2014 [28] in which they have used ash made up of waste sewage sludge as cementing stuff (5–20%) and as an aggregate (10%) in concrete. It was noticed that the water absorption of concrete samples reduces (9.2–16%) with an increase in the ash content as cementing material at 28 days. Whereas in the case of the ash used as an aggregate, the lowest water absorption (6.0%) was observed. This reduction was attributed to densification of an internal concrete structure and pozzolanic effect as well inside concrete samples. In the same study, a capillary water absorption test was also performed on concrete samples after 28 days of water curing for 4 hours. As a result, in this circumstance, as in the case of regular water absorption, a similar trend was seen. In the study, statistical analysis was also done to obtain a correlation between normal water absorption and capillary water absorption. With a coefficient of 0.96, they demonstrated a strong link between these absorption capacities. Water permeability increases with the employment of waste ash (10%) formed by burning sewage sludge, according to a study by Nakic, 2018 [39] on concrete.

15.3.3.3   USPV

This test method helps in figuring out the homogeneity and relative nature of concrete. It gives an idea about voids, fissures, and measures the efficacy of crack restorations.

It can also be used to indicate changes in concrete qualities and structural examinations to assess the extent of deterioration. The transit time is measured in this test method, and the distance between the transducers is used to calculate an 'apparent' pulse velocity. Not all types of deterioration affect the material's pulse velocity, but they do affect the trajectory of the pulse from transmitter to receiver. In this way, the internal quality of concrete can be accessed approximately [48]. In general, a high-ultrasonic pulse velocity reading in concrete indicates high-quality concrete. The use of various additional materials in concrete considerably influences its pulse velocity. The pulse velocity of concrete alters when waste sewage sludge lightweight aggregates are used. After 56 days, it was discovered that when such aggregates are used, the pulse velocity value decreases (6.6%) in comparison to the value of control concrete. However, the value of sludge-based concrete is still found in a variety of "good-quality" concrete [24].

15.3.3.4 Shrinkage

It is critical to estimate concrete shrinkage and regulate cracking quantitatively during the design stage to improve the durability of constructions [49]. Concrete drying shrinkage is influenced by two major factors: concrete properties and ambient circumstances. The use of sewage sludge has a significant impact on concrete characteristics. For example, Chen et al., 2018 [40] conducted a study in which they employed two types of waste sewage sludge ash (10 and 20% ash content) along with glass cullets. Sludge ashes were employed as a cementing ingredient, and glass cullets were used as a fine aggregate replacement. The unprocessed sludge ash was one, while the processed sludge (finely ground) ash was the other. In the study, concrete samples of 285 mm length were cast and placed in a humid climate for initial 28 days. After that, the samples were placed in a drying chamber to evaluate the dry shrinkage of the concrete for a further 28 days. The test's results were based on the average of two samples of each concrete mix. It was discovered that when sludge ashes are used, dry shrinkage of concrete increases, however when glass cullets are used, dry shrinkage of concrete decreases. This could be owing to the ashes' porous nature, which allows them to absorb more water. Processed sludge ash has a greater impact because it has smaller particles and a larger surface area, which aids in the drying of water. However, by combining the usage of glass cullets in the aggregates, this problem might be solved because glass cullets can reduce drying shrinkage. Nakic, 2018 [39] published another study in which they employed 10% ash made up of waste sewage sludge as a cementing element in concrete. For 91 days, concrete samples were evaluated for dry shrinkage. As a result, it was observed that the usage of the ash has no negative impact on hardened concrete volume degradation. The total shrinkage of sludge ash-based concrete samples differs slightly from the values found in the control concrete. Chen et al., 2018 [15] researched geopolymer concrete including waste sewage sludge. The impact of Na_2O on the concrete was detected in the research. In the study, specimens were made from various Na_2O concentrations, including 3.5%, 4.0%, and 4.5%. It was discovered that the dry shrinkage of concrete samples increases in the first 7 days, but then decreases as time goes on. It was also discovered that when the Na_2O level is smaller, there is less dry shrinkage. The study found no significant sewage sludge side effects on the concrete dry shrinkage property.

Waste Sewage Sludge Concrete

15.3.3.5 Chloride Ion Permeability

Chloride ion permeability is an important metric to consider when evaluating the durability of concrete. The property generally depends on the nature of the internal structure of concrete and the nature of various materials used in the manufacturing of concrete. So the usage of sewage sludge has an impact on the concrete's properties. Let's take a look at a study by Lau et al., 2018 [45] as an example. They employed posslite aggregates produced of sewage sludge in concrete in the study and compared it to regular concrete made with natural aggregates. When preparing the concrete mix, W/C considered was 0.47, and the maximum cement content was 468 kg/m^3. They employed two standards to assess the chloride ion permeability of concrete in the investigation. The first was RCPT, while the second was salt ponding. To evaluate RCPT value, they have cast a cylindrical concrete sample of size 100 × 200 mm, whereas for salt ponding they used a slab of 280 × 380 × 95 mm size. The RCPT test was performed after a duration of 28 and 90 days, while the salt ponding experiment was done for 90 days. The charge passed in the case of sludge-based aggregate concrete was found to be somewhat higher (0.35% at 28 days and 6.87% at 90 days) than in natural aggregate concrete. With the passage of time, the ability to pass charge decreases (30–34%) in both types of concrete. As per the standard ASTM-C1202, 2019 [50] various forms of concrete used in the study were categorized as having "moderate" chloride penetrability. Thus, the permeability of ions in sludge-based aggregate concrete is quite comparable to natural aggregate concrete. Because of the strong shell of sludge-based aggregates, free admission of chloride-ion via the sludge-based aggregates was limited. Similarly, comparable results were reported in the salt ponding test as in the RCPT. The surface chloride concentration and coefficient of chloride diffusion of Posslite sludge-based aggregate concrete were determined to be comparable to those of natural aggregate concrete based on the findings. Sludge-based aggregate concrete and natural aggregate concrete had average chloride-ion penetration depths of 15.1 mm and 12.5 mm, respectively.

15.3.3.6 Sorptivity

It represents the capability of a material to soak up and transport liquid through capillary action. It is quite important to estimate the nature of the internal microstructure of a material as far as durability is concerned. Nowadays, this property is becoming very common to judge the nature of concrete subjected to harsh circumstances [51]. A good concrete should have a sorptivity rating of less than 0.0130 mm/min0.5. According to a study conducted by Lau et al., 2018 [45] on two distinct types of concrete, one made with natural aggregates and the other with artificial aggregates made with sewage sludge (lime treated) and palm oil fuel ash. The concrete mix was prepared with maximum cement and water content of 468 and 220 kg/m^3, respectively. To evaluate the sorptivity, cylindrical samples of 100 × 200 mm were prepared and cured for 28 days. These samples were cut into a 100 × 50 mm sample piece and then tested for the property at the testing age (28 days). It was observed that the initial sorptivity value of artificial aggregate concrete was 0.0151 mm/s0.5, which was a little higher (0.66%) than the value achieved by the natural aggregate concrete sample. However, because this increase was insufficient, it may be inferred that concrete containing artificial aggregates prepared with sewage sludge is comparable to

290　　　　　　　　　　　　　　　　　　　Handbook of Sustainable Materials

concrete containing natural aggregates. The study found that the absorption capacity of such artificial aggregates is still modest, which is a good thing.

15.3.3.7 Electrical Resistivity and Corrosion Resistance

Electrical resistivity measurement is a nondestructive mechanism used for evaluating concrete internal cracks and determining whether open porosity is expanding or decreasing. The electrical resistivity of concrete is the resistance to ion mobility when an electric field is applied. Generally, it is noticed that if the resistivity value of concrete is more, the longer it will last against corrosion. Electrical resistivity, often known as conductivity, is a material attribute that is affected by pore volume, pore structure, pore solution composition, concrete mixture characteristics, concrete specimen saturation, and temperature [52]. Tuan et al., 2013 [24] published a study on high-performance concrete made up of sludge ash-based light aggregates. In the study, lightweight sewage sludge aggregate concrete was compared with concrete samples prepared with natural aggregates. It was observed that the resistivity value of lightweight sewage sludge aggregates concrete (62.88 kΩ cm) was higher compared to natural aggregate concrete (46.25 kΩ cm) at 56 days of testing. Various research studies have recommended a threshold value of 20 kΩ cm for high-performance concrete resistivity. Thus, in this case, the surface resistivity of lightweight aggregate concrete exceeded the recommended value of 20 kΩ cm, as can be seen. As a result, the lightweight sewage sludge aggregate has high corrosion resistance. As per a study by Song et al., 2021 [53], bio-concrete samples were prepared with granular sludge, prepared from wastewater to use in sewers. It was reported that 1 and 2% use of sludge leads to 17.2 and 42.8% reduction in corrosion rate, respectively, in comparison to standard concrete. In a nutshell, it was reported that granular sludge-based bio-concrete offers a promising option for reducing sewage corrosion.

15.3.3.8 Thermal Conductivity

Thermal conductivity plays a vital role in the evaluation of the quantity of the movement of heat through conduction in concrete. Loss of heat inside buildings through its various elements like walls and roofs can be correlated to the use of energy by the building [54]. In a study conducted by Mun, 2007 [17], discarded sewage sludge was used to make lightweight aggregates. By altering the amount of sewage sludge in the aggregates, different types of lightweight aggregates were created (clay-sludge ratio as 100:100, 100:300, and 100:500). In the investigation, the cement and water content in the concrete mix remained constant at 320 and 160 kg/m^3, respectively. All the samples of concrete were tested after 14 days of the curing period. It was observed that with a rise in the waste sludge content, thermal conductivity value reduces (0.593–0.733 W/mK). This was only because of the calorification of the present organic matter which produced gas. This leads to the highest degrees of porosity internally. Generally, normal concrete has conductivity values of around 1.50–1.60 W/mK. So concrete specimens made up of sewage sludge-based aggregates show a high-insulation effect comparative to normal concrete.

15.3.3.9 Frost Resistance

In areas where temperatures can drop below zero, frost resistance is essential. As this might cause deterioration of concrete. However, just a few studies have looked

Waste Sewage Sludge Concrete 291

at sewage sludge concrete's freeze-thaw resilience so far. According to a study by Rutkowska et al., 2018 [13], concrete prepared with ash made up of waste-water sludge was employed as cement. The mix had a maximum cement and water content of 360.58 and 166.42 kg/m³, respectively. The percentage of sewage sludge ash replaced with cement was 5, 10, and 15%, and the concrete samples were subjected to 150 cycles of freeze-thaw during the testing. In this case, compressive strength and mass loss in concrete samples were determined before and after 150 cycles. As per the findings, the frost resistance of concrete improves with a rise in the level of the ash used. As concrete samples containing 5, 10, and 15% sludge ash show 8.23, 1.84, and 1.80% reduction in compressive strength. Whereas a rise (2.90%) in the compressive strength was observed in the case of normal concrete after 150 cycles. The same pattern was noticed in the mass loss of the samples (sludge content-mass loss: 5%-0.333, 10%-0.209, and 15%-0.085). As per a standard referred to in the study, the reduction in compressive strength and loss in mass should not be greater than 20% and 5%, respectively. In the study, fly ash (5, 10, and 15%) was also used as a substitute material in concrete. In comparison to sewage sludge concrete, fly ash concrete showed a greater fall in compressive strength (53–72%) and mass loss (1.1–58%). From this, one can say that the sewage sludge-based concrete has good frost resistance.

15.3.3.10 Alkali-Silica Reaction (ASR)

An ASR occurs when alkalis like sodium or potassium oxides from cement or other sources mix with certain reactive silica minerals found in coarse or fine aggregates. The chemical reaction produces a gel that is hydrophilic and can cause concrete to expand abnormally and fracture. Low-alkali cement is one approach for avoiding this reaction in concrete, but mineral admixtures are also recommended [55]. The role of ash made up of sewage sludge in the reaction between alkali-silica components of concrete is explored in this section. Bar specimens were made for evaluating ASR during research by Chen et al., 2018 [15]. The dry mix process was used to make the concrete, which had a zero slump value. The water content was not predetermined in this situation; they utilized enough water till the required cohesion was reached in the mix. Glass cullet and sewage sludge were used in the research. Glass cullet and sewage sludge were employed as substitutes for natural aggregates and cement, respectively, in the concrete mix in this investigation. In the study, the replacement level of sewage sludge ash was 10 and 20%. The investigation was done till the age of 14 days immersion in 1 M NaOH solution. As per ASTM-C1567, 2013 [56], the maximum permissible value of expansion is 0.1%. It was noticed that with the use of glass cullets expansion in concrete samples rises, but in the case of waste sewage sludge ash, ASR expansion reduces. The highest reduction level was observed at the 20% level of sewage sludge ash. This was attributed to the positive pozzolanic impact, which hindered ASR expansion.

15.4 LEACHABILITY

A leaching test is commonly performed to determine whether waste has a danger of releasing organic and inorganic pollutants into the environment. As per a study by Nakic, 2018 [39] in which they have used waste ash made up of sewage sludge as cementing stuff in a mortar (10 and 20%) and concrete (10%). In the study, they evaluated leachability as per EN 12457-2 (2002) standard for concrete and tank leaching test for mortar. After 54 days of long study, no major difference in the leaching was noticed in sludge-based mortar and control mortar specimens. From this, one can say that the use of waste ash up to 20% in mortar would not cause any environmental as well as health issues. Similar results were noticed in concrete specimens. Total leaching of 10% sludge ash concrete (0.585 mg/kg) was similar to leaching of control concrete with no sludge ash (0.679 mg/kg). After evaluating the leachability of heavy metals, it was determined that 10% sludge ash concrete behaves similarly to normal concrete. Generally, sewage sludge ashes are considered inert. But in this case, after examination of sewage sludge ash some high concentrations of Pb, Se, Cr, and Mo were observed in it as mentioned in Figure 15.3. As sludge ash is considered nonhazardous, but still there is a chance that the ash will harm the environment (in landfilling). Rutkowska et al., 2018 [13], performed a leachability test on ash made up of sewage sludge and concrete containing 15% ash using European standards, in research. As per the results obtained in the research, they state that the sludge-based ash incorporation in concrete is not harmful to the environment, as far as technical requirements and environmental regulations are concerned. As per the standards used in the research, the maximum limit of the level of heavy metal detected in eluates of the test samples is limited to 10 mg/L. Whereas in the study the concentration shown by sludge ash was 0.587 mg/L and concrete (15% ash) 1.167 mg/L. Although the heavy metal concentration in concrete was higher than in sludge ash, the value was relatively very low in comparison to the standard limit of inert waste. In another study by Chen et al., 2018 [40] leachability of finely ground sewage sludge ash and concrete containing 20% sludge ash (as a cementing material) along with glass cullets (as an aggregate) were evaluated. In the study, they referred to two standards one was EN-13657, 2002 [57] and the other was EPA-Method-1311, 1990 [58]. According to the findings, the concentration of heavy metals present in the eluates of concrete and sludge ash was very low as per the standards concerned. Which defends the utilization of the sludge-based ash in the manufacturing of concrete (cement substitute) safely and sustainably. The amount of heavy metals in concrete was less comparative to the ash, due to the entrapment and paralysis impact of cement hydration products in concrete. Similar results were observed by Chen et al., 2018 [15] during their research on geopolymer concrete containing waste sewage sludge ash. The amount of various heavy metals considered during the study was well below the specified limits. This makes the use of the ash in geopolymer concrete feasible as far as the safety of the environment is concerned. Chen et al., 2013 [14] observed that most of the concentrations of heavy metals were within the permissible limit except Mo and Se in the case of the sludge ash. Whereas in the case of concrete, all the heavy metals amounts were within the specified limits and observed that Mo and Se were stabilized by cement paste. Donatello et al., 2010 [59], found that the sludge-based ash they investigated was not acceptable for dumping in soil because of the presence of good amounts of molybdenum and selenium which are highly soluble. As per research by Lin et al., 2012 [60] on waste-water slurry ash in mortar, after leachate analysis, some heavy metals like Ba, Cr, Pb, and Sr were observed but their concentration was within the permissible limit. But as compared to normal mortar samples,

Waste Sewage Sludge Concrete 293

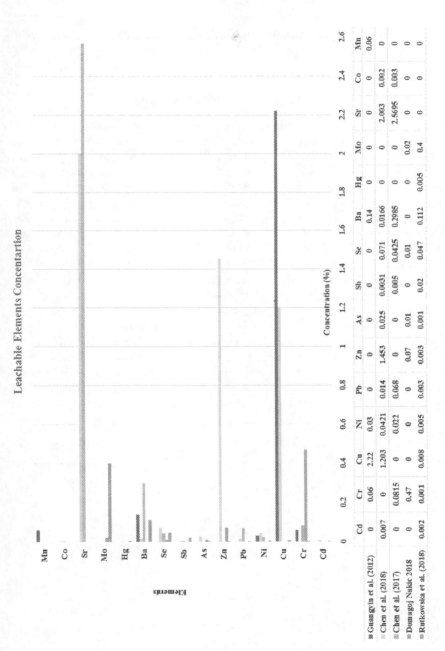

FIGURE 15.3 Various heavy metal compositions were reported in different studies [13, 15, 22, 32, 39].

the concentration of this heavy metal was high. But still, the use of sewage sludge ash was no threat to the environment due to the very low concentration of heavy metals. Similar results were reported by Zhen et al., 2012 [22] during their study. According to a recent study by Wu et al., 2021 [43], no major phases of heavy metal were detected (except Zn but within limit) as leachate after the evaluation on sewage sludge ash-based concrete (5–50% sludge ash).

15.5 ECOLOGICAL AND ECONOMIC ANALYSIS

In today's society, the project's development cost, as well as the project's environmental impacts, are quite important. This necessitates the discovery of alternative construction materials that can cut construction costs while also having a lower environmental impact. In this section, we'll look at the environmental and financial implications of using waste sewage sludge in construction. Baeza-Brotons et al., 2014 [28], incorporated sludge-based ash as a cementing (5–20%) and aggregate (10%) replacement material in an experimental investigation on concrete. In the study to check its cost-benefit analysis, concrete mixes containing 15% sewage sludge ash (cementing material) and 10% sewage sludge ash (fine aggregates) were compared with control concrete (no sludge ash). The main rationale for selecting this 15% ash concrete was because it has identical technical specifications to the control concrete. It was discovered that adding 15% ash to concrete improved a variety of characteristics, which is a good thing. Following this, environmental and economic factors were investigated. The life cycle of concrete is shown in Figure 15.4. According to Figure 15.4, "Materials" is the most important factor that influences the life cycle of concrete built with or without sewage sludge ash. The difference in environmental and economic advantages gained by utilizing sewage sludge ash concrete over regular concrete is due to the "materials." A concept of eco-costs is used to evaluate a product's environmental characteristics and minimize the environmental effect of its

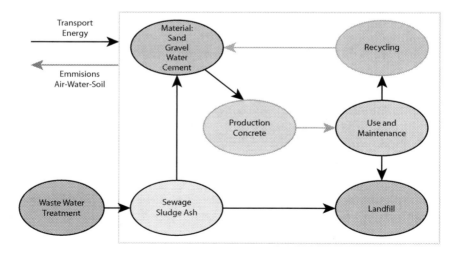

FIGURE 15.4 The life cycle of normal concrete and sewage sludge ash-based concrete [28].

Waste Sewage Sludge Concrete

TABLE 15.3

Price and Eco-Cost Analysis of Concrete Containing Sewage Sludge Ash (Per Ton Concrete Production) [28]

Materials	Concrete – 0% Ash			Concrete – 15% Sludge Ash (As Cement)			Concrete-10% Sludge Ash (As Aggregate)		
	Quantity (kg)	Eco-cost (€)	Price (€)	Quantity (kg)	Eco-cost (€)	Price (€)	Quantity (kg)	Eco-cost (€)	Price (€)
Sand	610.69	1.93	8.05	601.29	1.90	7.93	527.70	1.66	6.96
Gravel	284.19	1.90	1.49	279.82	1.87	1.47	272.93	1.83	1.44
Cement	62.51	10.14	0.64	61.55	9.99	0.63	60.04	9.74	0.61
Water	42.60	0.02	0.02	48.12	0.02	0.02	80.68	0.04	0.04
Additives	-	-	-	9.21	-0.87	0.00	58.65	-5.54	0.00
Total	1000	13.99	10.20	1000	12.91	10.05	1000	7.73	9.04

use. Emissions, energy/transport, and material depletion are all included in the eco-cost. It is a kind of virtual cost related to the environment. In detail, materials cost and eco-cost are mentioned in Table 15.3. In this table, the eco-cost of sewage sludge ash was considered negative because of two facts. Firstly is sewage sludge ash, which is a waste that is not directly related to the production of concrete and so does not add to the eco-cost. Secondly, because sewage sludge ash must be transported and disposed of at a landfill, but if it can be utilized in concrete, the eco-cost of its disposal must be subtracted. The "price" in Table 15.3 represents the market value of the materials. It can be observed from Table 15.3 that 7.7% (per ton) saving occurs in the case of 15% use of the ash in concrete (as cementing material) compared to normal concrete. Whereas in the case when the ash was used as aggregates, more saving was achieved. In nature, these savings are minor, but using sludge ash in concrete instead of natural aggregates decreases landfilling and reduces the usage of natural aggregates. This will be advantageous to the environment. Gu et al., 2021 [27] mentioned in their research that by cement replacement (5%) with the ash made up of sewage sludge, the release of carbon dioxide (4.98%) and monetary expenses (25.42%) were reduced. If all of China's sewage sludge-based ash is recycled in cementing products, annual CO_2 emissions could be reduced by 6.8×10^9 kg CO_2-e, and monetary expenses might be reduced by 16.1 billion RMB yuan. It was also suggested that the findings of the study help in guiding the recycling of the ash made up of sewage sludge in cement-based materials (where the SO_3 level can be as high as 20%). This also helps in assisting garbage disposal and the evolution of environmentally friendly building materials. Hsu and Chen, 2020 [20] observed that with the incorporation of the sludge-based ash in mortar and cement paste, both cost-cutting and sustainability can be achieved. As sludge-based past has high strength and a less costly disposal technique (3.6–30%), compared to landfilling and incineration. Its numerous uses aid in the reduction of land site requirements, CO_2 emissions, and cement usage. The researchers Rabie et al., 2019 [44] utilized two types of sewage sludge in the study: dry sewage sludge and wet sewage sludge. Samples of concrete mixes were made with sewage sludge added at 5, 10, 15, and 20% of the cement weight. After proper cost analysis, it was concluded in the study that with the use of sludge (5%) a reduction in the cost of concrete manufacturing was achieved.

15.6 CONCLUSION

In today's scenario, landfilling is not a very efficient method of sewage sludge disposal. As a result, effective waste management techniques are required, which will benefit both the economy and the environment. With the use of existing research data, the major goal of the work was to demonstrate proof that waste sewage sludge or ash made up of sewage sludge can be employed in the construction field. Only by studying the varied properties of concrete with the addition of such waste can this be shown.

As per the available data related to physiochemical characteristics of sewage sludge/ash, it can be utilized as a potential construction material. The major mineralogical composition available in the sludge or sludge ash was matched with the composition of cement. Waste sewage sludge ash also shows good pozzolanic properties, which give evidence that the material can be used as a pozzolan. With the incorporation of waste sewage sludge, the workability of concrete is reduced. But this can be overcome with the use of superplasticizers. Most of the time, the use of the sludge leads to a reduction in heat liberation in concrete, which is very effective in the case of mass concreting. Sewage sludge also causes a downfall in the mechanical characteristics of concrete. However, various treatments, such as alkali activators, increased sludge incineration temperature, and grinding, can be used to overcome this. Despite the drop in compressive strength, sludge-based concrete still has enough strength to be used in lightweight material construction or for nonstructural parts. The density of concrete also reduces when waste sewage sludge is added. In terms of durability, a variety of contradicting outcomes have been seen. Rise in water absorption, higher dry shrinkage, and low sulfate resistance in most situations when sewage sludge was used. In terms of corrosion resistance, frost resistance, and increased temperature resistance, positive results were reported. Incorporation of sewage sludge/ash in concrete does not exhibit any undesirable levels of certain heavy metals in terms of leachability. This indicates that the usage of sewage sludge has no major adverse impact on the environment or human health. As a result, it can be safely employed in construction as a building material. The use of sewage sludge has both ecological and economic benefits. Because sewage sludge is a waste and, in most situations, readily available, it saves money when employed in the production of concrete. When the ash made up of sludge is incorporated as cementing stuff, less cement is used, resulting in lower CO_2 emissions during the manufacturing process.

It does not follow, however, that if the use of waste material reduces the strength of concrete, the material cannot be utilized as a construction material. Nowadays, economic and environmental impacts are also important along with the strength of the material in the construction field. Various studies have shown that using sludge in concrete saves money while causing no environmental harm. So overall, one can say that waste sewage sludge/ash is a significant building material. In this approach, the use of sewage sludge in concrete can be considered an effective disposal strategy. This aids in the resolution of the most serious challenge that most countries face when it comes to solid waste management.

15.7 FUTURE SCOPE

After the analysis of the available extensive data, it was observed that the study pertaining to sewage sludge usage in concrete is quite less comparative to mortar. Very limited data is available related to the tensile strength of concrete containing sewage sludge or sewage sludge ash. Durability aspects were also not evaluated extensively till the time. So one can go for the evaluation of carbonation, acid resistance, abrasion resistance, sorptivity, sulfate resistance, etc. As sewage sludge contains various organic matters which would be harmful as a construction material. So processing of sludge before using it as a construction material could be a prominent topic of future studies. As the absorption capacity of dry sewage sludge or ash is high, this leads to a loss in workability of concrete. In such a case, superplasticizer is one of the options till date. But one can use various pretreatment techniques to reduce the absorption capacity of sewage sludge before its application in concrete.

REFERENCES

1. Z. Chang, G. Long, J. L. Zhou, and C. Ma, "Valorization of sewage sludge in the fabrication of construction and building materials: A review," *Resour. Conserv. Recycl.*, vol. 154, no. June 2019, p. 104606, 2020, doi: 10.1016/j.resconrec.2019.104606.
2. W. Tu *et al.*, "A novel activation-hydrochar via hydrothermal carbonization and KOH activation of sewage sludge and coconut shell for biomass wastes: Preparation, characterization and adsorption properties," *J. Colloid Interface Sci.*, vol. 593, pp. 390–407, 2021, doi: 10.1016/j.jcis.2021.02.133.
3. N. H. Sabri, A. Muhammad, N. H. Abdul Rahim, A. Roslan, and A. R. Abu Talip, "Feasibility study on co-pyrolyzation of microplastic extraction in conventional sewage sludge for the cementitious application," *Mater. Today Proc.*, vol. 46, pp. 2112–2117, 2021, doi: 10.1016/j.matpr.2021.05.439.
4. World-Economic-Forums, "120,000 tonnes of faecal sludge: Why India needs a market for human waste," 2019.
5. CPCB, "No Title," *Ministry of Environment, Forest and Climate Change, Goverment of India.*, 2021. https://cpcb.nic.in/.
6. A. Taherlou, G. Asadollahfardi, A. M. Salehi, and A. Katebi, "Sustainable use of municipal solid waste incinerator bottom ash and the treated industrial wastewater in self-compacting concrete," *Constr. Build. Mater.*, vol. 297, p. 123814, 2021, doi: 10.1016/j.conbuildmat.2021.123814.
7. R. Chen *et al.*, "Life-cycle assessment of two sewage sludge-to-energy systems based on different sewage sludge characteristics: Energy balance and greenhouse gas-emission footprint analysis," *J. Environ. Sci.*, vol. 111, pp. 380–391, 2022, doi: 10.1016/j.jes.2021.04.012.
8. L. Świerczek, B. M. Cieślik, and P. Konieczka, "Challenges and opportunities related to the use of sewage sludge ash in cement-based building materials—A review," *J. Clean. Prod.*, vol. 287, no. p.125054, 2021, doi: 10.1016/j.jclepro.2020.125054.
9. N. Gao, K. Kamran, C. Quan, and P. T. Williams, "Thermochemical conversion of sewage sludge: A critical review," *Prog. Energy Combust. Sci.*, vol. 79, p. 100843, 2020, doi: 10.1016/j.pecs.2020.100843.
10. S. Ducoli, A. Zacco, and E. Bontempi, "Incineration of sewage sludge and recovery of residue ash as building material: A valuable option as a consequence of the COVID-19 pandemic," *J. Environ. Manage.*, vol. 282, no. December 2020, p. 111966, 2021, doi: 10.1016/j.jenvman.2021.111966.

11. S. Rao Meda, S. Kumar Sharma, and G. D. Tyagi, "Utilization of waste sludge as a construction material—A review," *Mater. Today Proc.*, 46, pp. 4195–4202, 2021, doi: 10.1016/j.matpr.2021.02.762.

12. Y. Liu *et al.*, "The potential use of drinking water sludge ash as supplementary cementitious material in the manufacture of concrete blocks," *Resour. Conserv. Recycl.*, vol. 168, no. November 2020, p. 105291, 2021, doi: 10.1016/j.resconrec.2020.105291.

13. G. Rutkowska, P. Wichowski, J. Fronczyk, M. Franus, and M. Chalecki, "Use of fly ashes from municipal sewage sludge combustion in production of ash concretes," *Constr. Build. Mater.*, vol. 188, pp. 874–883, Nov. 2018, doi: 10.1016/j.conbuildmat.2018.08.167.

14. M. Chen, D. Blanc, M. Gautier, J. Mehu, and R. Gourdon, "Environmental and technical assessments of the potential utilization of sewage sludge ashes (SSAs) as secondary raw materials in construction," *Waste Manag.*, vol. 33, no. 5, pp. 1268–1275, 2013, doi: 10.1016/j.wasman.2013.01.004.

15. Z. Chen, J. S. Li, B. J. Zhan, U. Sharma, and C. S. Poon, "Compressive strength and microstructural properties of dry-mixed geopolymer pastes synthesized from GGBS and sewage sludge ash," *Constr. Build. Mater.*, vol. 182, pp. 597–607, 2018, doi: 10.1016/j.conbuildmat.2018.06.159.

16. K. H. Lee, K. G. Lee, Y. S. Lee, and Y. M. Wie, "Manufacturing and application of artificial lightweight aggregate from water treatment sludge," *J. Clean. Prod.*, vol. 307, no. December 2020, p. 127260, 2021, doi: 10.1016/j.jclepro.2021.127260.

17. K. J. Mun, "Development and tests of lightweight aggregate using sewage sludge for nonstructural concrete," *Constr. Build. Mater.*, vol. 21, no. 7, pp. 1583–1588, Jul. 2007, doi: 10.1016/j.conbuildmat.2005.09.009.

18. S. Valls, A. Yagüe, E. Vázquez, and C. Mariscal, "Physical and mechanical properties of concrete with added dry sludge from a sewage treatment plant," *Cem. Concr. Res.*, vol. 34, no. 12, pp. 2203–2208, 2004, doi: 10.1016/j.cemconres.2004.02.004.

19. J. Liu, M. Ba, Z. He, and Y. Li, "Microstructure and performance of sludge-ceramisite concrete," *Constr. Build. Mater.*, vol. 39, pp. 82–88, 2013, doi: 10.1016/j.conbuildmat.2012.05.004.

20. C.-W. Hsu and C.-T. Chen, "Strength development of cement pastes with alkali-activated dehydrated sewage sludge," *Constr. Build. Mater.*, vol. 255, p. 119243, Sep. 2020, doi: 10.1016/j.conbuildmat.2020.119243.

21. M. Oliva, F. Vargas, and M. Lopez, "Designing the incineration process for improving the cementitious performance of sewage sludge ash in Portland and blended cement systems," *J. Clean. Prod.*, vol. 223, pp. 1029–1041, Jun. 2019, doi: 10.1016/j.jclepro.2019.03.147.

22. G. Zhen, H. Zhou, T. Zhao, and Y. Zhao, "Performance appraisal of controlled low-strength material using sewage sludge and refuse incineration bottom ash," *Chinese J. Chem. Eng.*, vol. 20, no. 1, pp. 80–88, 2012, doi: 10.1016/S1004-9541(12)60366-8.

23. A. Hamood, J. M. Khatib, and C. Williams, "The effectiveness of using Raw Sewage Sludge (RSS) as a water replacement in cement mortar mixes containing Unprocessed Fly Ash (u-FA)," *Constr. Build. Mater.*, vol. 147, pp. 27–34, 2017, doi: 10.1016/j.conbuildmat.2017.04.159.

24. B. L. A. Tuan, C.-L. Hwang, K.-L. Lin, Y.-Y. Chen, and M.-P. Young, "Development of lightweight aggregate from sewage sludge and waste glass powder for concrete," *Constr. Build. Mater.*, vol. 47, pp. 334–339, Oct. 2013, doi: 10.1016/j.conbuildmat.2013.05.039.

25. Y. Zhou, J. Lu, J. Li, C. Cheeseman, and C. S. Poon, "Influence of seawater on the mechanical and microstructural properties of lime-incineration sewage sludge ash pastes," *Constr. Build. Mater.*, vol. 278, p. 122364, 2021, doi: 10.1016/j.conbuildmat.2021.122364.

26. Y. fan Zhou, J. shan Li, J. xin Lu, C. Cheeseman, and C. S. Poon, "Sewage sludge ash: A comparative evaluation with fly ash for potential use as lime-pozzolan binders," *Constr. Build. Mater.*, vol. 242, p. 118160, 2020, doi: 10.1016/j.conbuildmat.2020.118160.
27. C. Gu, Y. Ji, Y. Zhang, Y. Yang, J. Liu, and T. Ni, "Recycling use of sulfate-rich sewage sludge ash (SR-SSA) in cement-based materials: Assessment on the basic properties, volume deformation and microstructure of SR-SSA blended cement pastes," *J. Clean. Prod.*, vol. 282, Feb 1, p. 124511, 2021, doi: 10.1016/j.jclepro.2020.124511.
28. F. Baeza-Brotons, P. Garcés, J. Payá, and J. M. Saval, "Portland cement systems with addition of sewage sludge ash. Application in concretes for the manufacture of blocks," *J. Clean. Prod.*, vol. 82, pp. 112–124, Nov. 2014, doi: 10.1016/j.jclepro.2014.06.072.
29. A. Kappel, L. M. Ottosen, and G. M. Kirkelund, "Colour, compressive strength and workability of mortars with an iron rich sewage sludge ash," *Constr. Build. Mater.*, vol. 157, pp. 1199–1205, Dec. 2017, doi: 10.1016/j.conbuildmat.2017.09.157.
30. P. He, C. S. Poon, I. G. Richardson, and D. C. W. Tsang, "The mechanism of supplementary cementitious materials enhancing the water resistance of magnesium oxychloride cement (MOC): A comparison between pulverized fuel ash and incinerated sewage sludge ash," *Cem. Concr. Compos.*, vol. 109, p. 103562, 2020, doi: 10.1016/j.cemconcomp.2020.103562.
31. B. Krejcirikova, L. M. Ottosen, G. M. Kirkelund, C. Rode, and R. Peuhkuri, "Characterization of sewage sludge ash and its effect on moisture physics of mortar," *J. Build. Eng.*, vol. 21, pp. 396–403, 2019, doi: 10.1016/j.jobe.2018.10.021.
32. Z. Chen and C. S. Poon, "Comparative studies on the effects of sewage sludge ash and fly ash on cement hydration and properties of cement mortars," *Constr. Build. Mater.*, vol. 154, pp. 791–803, 2017, doi: 10.1016/j.conbuildmat.2017.08.003.
33. M. Liu, Y. Zhao, Y. Xiao, and Z. Yu, "Performance of cement pastes containing sewage sludge ash at elevated temperatures," *Constr. Build. Mater.*, vol. 211, pp. 785–795, 2019, doi: 10.1016/j.conbuildmat.2019.03.290.
34. C. H. Chen, I. J. Chiou, and K. S. Wang, "Sintering effect on cement bonded sewage sludge ash," *Cem. Concr. Compos.*, vol. 28, no. 1, pp. 26–32, 2006, doi: 10.1016/j.cemconcomp.2005.09.003.
35. P. Garcés, M. Pérez Carrión, E. García-Alcocel, J. Payá, J. Monzó, and M. V. Borrachero, "Mechanical and physical properties of cement blended with sewage sludge ash," *Waste Manag.*, vol. 28, no. 12, pp. 2495–2502, Dec. 2008, doi: 10.1016/j.wasman.2008.02.019.
36. J. Monzó, J. Payá, M. V. Borrachero, and I. Girbés, "Reuse of sewage sludge ashes (SSA) in cement mixtures: The effect of SSA on the workability of cement mortars," *Waste Manag.*, vol. 23, no. 4, pp. 373–381, Jan. 2003, doi: 10.1016/S0956-053X(03)00034-5.
37. D. B. Istuque *et al.*, "Behaviour of metakaolin-based geopolymers incorporating sewage sludge ash (SSA)," *Mater. Lett.*, vol. 180, pp. 192–195, 2016, doi: 10.1016/j.matlet.2016.05.137.
38. M. Cyr, R. Idir, and G. Escadeillas, "Use of metakaolin to stabilize sewage sludge ash and municipal solid waste incineration fly ash in cement-based materials," *J. Hazard. Mater.*, vol. 243, pp. 193–203, 2012, doi: 10.1016/j.jhazmat.2012.10.019.
39. D. Nakic, "Environmental evaluation of concrete with sewage sludge ash based on LCA," *Sustain. Prod. Consum.*, vol. 16, pp. 193–201, Oct. 2018, doi: 10.1016/j.spc.2018.08.003.
40. Z. Chen, J. S. Li, and C. S. Poon, "Combined use of sewage sludge ash and recycled glass cullet for the production of concrete blocks," *J. Clean. Prod.*, vol. 171, pp. 1447–1459, Jan. 2018, doi: 10.1016/j.jclepro.2017.10.140.
41. Y. fan Zhou, J. xin Lu, J. shan Li, C. Cheeseman, and C. S. Poon, "Hydration, mechanical properties and microstructure of lime-pozzolana pastes by recycling waste sludge ash under marine environment," *J. Clean. Prod.*, vol. 310, no. December 2020, p. 127441, 2021, doi: 10.1016/j.jclepro.2021.127441.

42. S. Chakraborty, B. W. Jo, J. H. Jo, and Z. Baloch, "Effectiveness of sewage sludge ash combined with waste pozzolanic minerals in developing sustainable construction material: An alternative approach for waste management," *J. Clean. Prod.*, vol. 153, pp. 253–263, 2017, doi: 10.1016/j.jclepro.2017.03.059.
43. Z. Wu *et al.*, "The long-term performance of concrete amended with municipal sewage sludge incineration ash," *Environ. Technol. Innov.*, vol. 23, p. 101574, 2021, doi: 10.1016/j.eti.2021.101574.
44. G. M. Rabie, H. A. El-Halim, and E. H. Rozaik, "Influence of using dry and wet waste-water sludge in concrete mix on its physical and mechanical properties," *Ain Shams Eng. J.*, vol. 10, no. 4, pp. 705–712, Dec. 2019, doi: 10.1016/j.asej.2019.07.008.
45. P. C. Lau, D. C. L. Teo, and M. A. Mannan, "Mechanical, durability and microstructure properties of lightweight concrete using aggregate made from lime-treated sewage sludge and palm oil fuel ash," *Constr. Build. Mater.*, vol. 176, pp. 24–34, Jul. 2018, doi: 10.1016/j.conbuildmat.2018.04.179.
46. ASTM-C330, Standard Specification for Lightweight Aggregates for Structural Concrete, ASTM International, West Conshohocken, PA, 2017, www.astm.org.
47. ASTM-D570-98, Standard Test Method for Water Absorption of Plastics, ASTM International, West Conshohocken, PA, 2018, www.astm.org.
48. ASTM-C597, Standard Test Method for Pulse Velocity Through Concrete, ASTM International, West Conshohocken, PA, 2016, www.astm.org.
49. I. Sims, J. Lay, and J. I. Ferrari, Concrete Aggregates, 5th ed. Elsevier Ltd., 2019.
50. ASTM C1202-12, Standard Test Method for Electrical Indication of Concrete's Ability to Resist Chloride Ion Penetration, ASTM International, West Conshohocken, PA, 2012, www.astm.org.
51. H. C. Uzoegbo, "Dry-stack and compressed stabilized earth-block construction," in Nonconventional and Vernacular Construction Materials, Elsevier, 2020, pp. 305–350.
52. ASTM-C1876, Standard Test Method for Bulk Electrical Resistivity or Bulk Conductivity of Concrete, ASTM International, West Conshohocken, PA, 2019, www.astm.org,.
53. Y. Song *et al.*, "A novel granular sludge-based and highly corrosion-resistant bio-concrete in sewers," *Sci. Total Environ.*, vol. 791, p. 148270, 2021, doi: 10.1016/j.scitotenv.2021.148270.
54. I. Asadi, P. Shafigh, Z. F. Bin Abu Hassan, and N. B. Mahyuddin, "Thermal conductivity of concrete—A review," *J. Build. Eng.*, vol. 20, pp. 81–93, Nov. 2018, doi: 10.1016/j.jobe.2018.07.002.
55. B. Singh, "Rice husk ash," in Waste and Supplementary Cementitious Materials in Concrete, Elsevier, 2018, pp. 417–460.
56. ASTM-C1567, Standard Test Method for Determining the Potential Alkali-Silica Reactivity of Combinations of Cementitious Materials and Aggregate (Accelerated Mortar-Bar Method), ASTM International, West Conshohocken, PA, 2013, www.astm.org,.
57. EN-13657, "Characterization of waste. Digestion for subsequent determination of aqua regia soluble portion of elements," 2002.
58. EPA-Method-1311, "Method 1311: Toxicity characteristic leaching procedure, part of test methods for evaluating solid waste," *Physical/Chemical Methods*, 1990 pp. 1–35.
59. S. Donatello, M. Tyrer, and C. R. Cheeseman, "EU landfill waste acceptance criteria and EU Hazardous Waste Directive compliance testing of incinerated sewage sludge ash," *Waste Manag.*, vol. 30, no. 1, pp. 63–71, Jan. 2010, doi: 10.1016/j.wasman.2009.09.028.
60. Y. Lin, S. Zhou, F. Li, and Y. Lin, "Utilization of municipal sewage sludge as additives for the production of eco-cement," *J. Hazard. Mater.*, vol. 213–214, pp. 457–465, 2012, doi: 10.1016/j.jhazmat.2012.02.020.

16 Applications of Sustainable Materials

Rajat Dhawan, Hitendra K. Malik,
and Davoud Dorranian

CONTENTS

16.1 Introduction to Sustainable Materials 301
16.2 Engineering Applications of Sustainable Materials 303
16.3 Applications of Sustainable Materials in Architecture 305
16.4 Applications of Sustainable Materials in Agriculture 307
16.5 Aerospace with Sustainable Materials 310
16.6 Biomedical and Health-Care Applications of Sustainable Materials 312
16.7 Contribution of Sustainable Materials in Green Technology 314
16.8 Conclusion 315
16.9 Future Applications Aspects 316
References 318

16.1 INTRODUCTION TO SUSTAINABLE MATERIALS

In order to bridge the gulf between environment and development, the term "sustainability" is chosen. Initially, this terminology came from groundwater, fisheries, and forestry, where the quantities, like "maximum sustainable pumping rate", "maximum sustainable yield," and "maximum sustainable cut", were dealt with. At the last of pumping period, in order to still have a viable aquifer, how much groundwater can be drawn? At the last of time period, in order to still have an adequate fishery functioning, how many fish can be taken? In order to still have a considerable forest growth, how many trees can be cut? These "maxima" are corresponding to the components of overall ecosystem. The sustainability in the ecosystem might be achieved when these "maxima" are determined.

The lives of all living things on the earth depend on raw materials, energy, water, and food. With the natural environment, there must be a relationship of all living things with the sustainable development. The trade, manufacture products, and raw materials are sourced in the global economic system. A relationship with them should also be made. Our relationship with societies is perhaps most important. In the identical way, sustainable development is not perceived by everyone. The contributions of sustainable development to nurture and protect the natural environment by preserving a thriving biosystem, productive land, pure water, and clean air are judged by an environmentalist. The contributions of sustainable development for sharing and generation of understanding and knowledge are looked by a humanist.

DOI: 10.1201/9781003297772-16

301

As the ultimate metric, the financial health of the corporation is seen by the sustainable development corporative view. Therefore, economic, social, and environmental are three facets of the sustainable developments.

In 1983, the UN General Assembly created a body named World Commission on Environment and Development, which represented a report on the initiation of addressing major issues of sustainable development. After its chairperson Gro Harlem Brundtland, this commission was also known as the Brundtland Commission. The term "without degrading the future generation's abilities, achieving our own needs as sustainable development" was first popularized in the report of Brundtland Commission.

"The prospects for enhancement and maintenance of future living standards should not be impaired by the current decisions" is the core idea of sustainable development. The social-culture, ecological, and economic approaches are three key approaches to sustainable development. The maintenance of cultural and social systems stability comes under the social-cultural approach. The maintenance of physical and biological systems robustness and resilience comes under the ecological approach. While maintaining enhancing stock of capital, the maximization of the income comes under the economic approach [1]. The sustainable social net product is defined as follows [2]:

$$\text{Sustainable Social Net Product(SSNP)}$$
$$= \text{Net National Product(NNP)} - \text{Defensive Expenditures(DE)}$$
$$- \text{Depreciation of Natural Capital(DNC).}$$

For sustainability, a simple rule for savings is defined as follows [1]:

$$\text{Depreciation of natural capital} + \text{Depreciation of human-made capital}$$
$$+ \text{Depreciation of human knowledge}$$
$$\leq \text{savings as percentage of Gross Domestic Product(GDP).}$$

Undesirable consequences, such as increasing inequality, resource depletion, loss of biodiversity and degradation of land, water, and air, are resulted when we neglect the social and environmental impacts of technological developments. The substances, which are responsible for the ozone layer damage, have been banned in the Montreal Protocol. In order to support sustainable development, 27 principles have been set in the Rio Earth Summit. Within the business community, an abrupt acceptance to sustainable development has been given. The corporate practice and thinking now are being integrated with sustainability by several forward-looking businesses and companies. Through three pillars of social progress, ecological balance, and economic growth, sustainable development has been promoted by World Business Council for Sustainable Development.

As a scientific discipline, a direction of development and a socio-philosophical idea can be discussed as the permanent and sustainable development in science. With economic and natural environment, the connections of human life's qualitative aspects are being focused on in scientific essence. The wisdom, the good, and

Applications of Sustainable Materials

the truth are three pillars which must be expressed by scientific essence so that the realization idea of permanent and sustainable development can be supported effectively in actual practice. With the morality and ethics issues, the connections of latest scientific discoveries are also necessary to understand and to attain sustainability of the world in realistic terms. In order to implement the idea of sustainability, the connections of scientific essence with other fields are the major boundary conditions.

16.2 ENGINEERING APPLICATIONS OF SUSTAINABLE MATERIALS

For practical purposes, like operation, manufacturing, and designing of processes and products, the applications of mathematical and scientific principles are termed as Engineering. The sociological, environment, and economic factors are also accounted for in this. Through engineering, we have brought several technical advances [3–10], which shall lead to sustainable development of coating, surface hardening of automobile parts, etching and via holes for semiconductor industries, etc. through plasma processing.

Engineering is linked indirectly to all three components of sustainability, i.e. social sustainability, economic sustainability, and environmental sustainability. In all economic sectors, such as commercial, residential, transportation, industry, etc., the resources are being used by engineering. In engineering, the resources consumed (water, minerals, and fuels) are determined by the environment. The waste obtained from engineering processes, like utilization, storage, transport and production, is also released to the environment. Social and cultural development as well as social stability is often supported, and good living standards are often allowed by the services provided by engineering. In simple language, without depleting the materials and without compromising with the environment, the process of utilizing resources is sustainable engineering. For an increasing world population, the supply of sufficient mobility, shelter, water, and food with a method which diminishes the damage to environment; the waste of one can be utilized as an input for other by designing processes and products; economic considerations, as well as social and environmental constraints incorporation into engineering decisions, are included under the sustainable engineering examples. In all engineering aspects, an interdisciplinary approach is required for sustainable engineering.

In the field of construction engineering, the sustainability concept has been widely adopted nowadays [11, 12]. The rehabilitation and maintenance of infrastructure objects or buildings, construction works, construction planning and management, designing of a structure or building, etc. are involved in the construction engineering, where sustainable and effective decisions regarding all these are sought through alternative solutions. A variety of tools and methodologies are being adopted these days for sustainable construction keeping in mind economic, technological, social, and environmental benefits. These include fundamental decision-making models and methods. Not only this advanced multi-criteria decision-making (MCDM) methods and techniques are being employed for supporting the sustainable decision-making in construction engineering.

In construction engineering, the achievements of fundamental sciences have been published in a large number of research articles. Several review papers have summarized most of these researches [13–15]. Novel applications, MCDM developments, and the fundamental methods have been constantly growing in construction problems. Sustainable decisions are effectively supported by this group of methods. With an increment in the sustainable development interest, MCDM applications have been constantly growing in construction engineering. In the construction problems, the research of sustainable decision-making has been confirmed as of a greater potential. In specific areas of construction building managements and technologies, the applications of MCMD methods have been widely discussed [16–18]. As a result, the method of superiorly modified hybrid multi-criteria decision-making (HMCDM) has acquired a great popularity. The overview of HMCDM methods applications in sustainability problems has also been made. By applying mathematical models, uncertainties, and risk assessments in construction engineering have been dealt with an enhanced attention.

Reduced environmental impact, enhanced efficiency, sustainable processes, sustainable resources and fulfillments of other aspects of sustainability are the key requirements for the engineering sustainability. The buildings, which are consuming equivalent amount of energy to the sum of thermal and electrical energies produced from the renewable energy sources, are designated by net-zero energy buildings. An integrated approach and whole-building systems are required for smart net-zero energy approach. In order to attain health requirements, safety requirements, and indoor environment needs with enhanced efficiency, control and building automation systems must be enhanced, an appropriate designing of building envelopes is requisite and a link must be there between renewable energy components, storage, lightning, and HVAC (Heating, Ventilating, and Air-Conditioning). Figure 16.1 illustrates the aspects requisite in smart net-zero energy communities and buildings.

There are several momentous applications and advantages which engineering sustainability is grabbing from the concept of smart net-zero energy communities

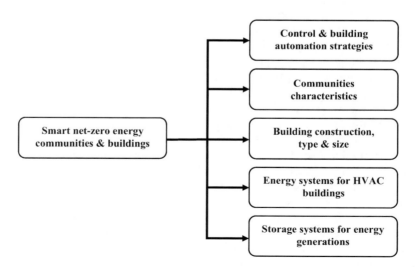

FIGURE 16.1 The aspects requisite in smart net-zero energy communities and buildings.

Applications of Sustainable Materials 305

and buildings. In such systems, the usage of sustainable resources is the first benefit. In their establishment and operation, the involvement of sustainable processes is the second benefit. The environmental benefits are the third advantage whereas enhanced attainable efficiency is the fourth one. Smart net-zero energy communities and buildings also fulfill the other aspects of sustainability. For example, nonsustainable energy resources are not utilized here. Renewable energy resources (geothermal and solar energies) are used to derive the utilization of energy. Therefore, energy sustainability has been significantly contributed by such communities and buildings. During the construction and development of the components and technologies of the systems, several efforts are put forward to enhance the efficiency of the processes included in such communities and buildings. For example, these net-zero energy communities and buildings reduce the environmental impacts through no utilization of non-sustainable energy resources. Engineering sustainability's nontechnical aspects are also contributed by them, such as the energy prices. With the usage of no energy resources, enhancing resource demands on societies is expected to be alleviated by net-zero energy buildings.

In mechanical engineering, the sustainability ranges from reducing energy use and waste to employing recycling and lowering emission [19]. Most of the mechanical engineers believe that all the sustainable design practices are ultimately cost-saving. They always consider that and recommend to others too. The sustainability can be achieved through the designs that reduce material waste in manufacturing and use less energy. To achieve the target of sustainability, more resource- and energy-efficient manufacturing is generally focused. On the other hand, green tribology also contributes to the sustainability in mechanical engineering when it is thought from the point of view of mechanical system [20]. Here we need to consider energy and sustainability, life-cycle assessment, and lubrication aspects as the important facets of the green tribology.

16.3 APPLICATIONS OF SUSTAINABLE MATERIALS IN ARCHITECTURE

New design possibilities and applications will be opened up by the continuing development of hygromorphic shape-changing materials. The opportunities for environmental benefits, material sustainability, and modification in quality of life will be resulted. Within architectural design, the conventional and accepted definitions of sustainable materiality shall find several opportunities to extend by the adoption of multidimensional characteristics of hygromorphic technologies. Within adaptive and environmentally responsive architectures, the opportunities for integrative and varied design applications may be provided. Therefore, a broad range of sustainability considerations can be addressed simultaneously by hygromorphic material systems [21, 22]. The concerns of local relevance, ecological aesthetics, human comfort, and energy efficiency can also be addressed herewith. For building integration and future development of hygromorphic wood composites, one of the major challenges is to exploit and understand the full potentials of the technologies. By doing so, the applications range can be expanded to other sectors also. Within architectural design, a hierarchy of four overlapping typologies has been suggested by literature and current research. These are: functional components or devices (locomotion engines, sensors, micro-generators, actuators, etc.), performance-oriented adaptive systems (energy efficiency, increased occupant comfort, etc.), spatial/aesthetic/formal experience value (dynamic facades increased visual appearance) and location/contextual specific value (local climate and environment's physical representations).

Discrete functional devices for hygromorphic composites are the simplest level of application. Here, the combination and integration of building components and other technologies with the materials is the main design challenge. A substantial increment in the requirement for broader building integration and demand on the technology is resulted when we investigate the impact of hierarchy of applications on aesthetics, formal or spatial experience, human comfort, and building envelope. Highly integrated and complex applications of the same have been focused on by Holstov et al. [23]. For interdisciplinary design approaches and methods, there is an inherent requirement of these applications. In ambient climatic conditions response, self-optimization and adjustments of building properties are enabled by kinetic and adaptive façade systems integration. From an environment barrier to an environment mediator, the shifting in the building envelope's role can be helped by this integration. With passive embedded response for more versatile and simpler adaptive building skins development, an underpinning principle can be served by materials applications with climatic stimuli to intrinsic sensitivity.

The striking visual effects are one of the major engaging and imperative aspects of hygromorphic materials. Through the coupling of adaptive and shape-changing behavior with design integration, we can achieve this striking visual effect. Biomimetic architectural forms are promoted by an environmental-design approach, designated as "Eco-aesthetics". Eco-aesthetics is usually related to hygromorphs biomorphic expressiveness. As ecological footprint or reduced energy consumption are important to sustainable architecture, sensual, aesthetic, and formal values of buildings are also imperative in the same way. A "unique convergence of spatial and environmental experience" is provided by the integration of hygromorphic materials. Dramatic ecologically embedded architecture creation is especially emphasized in this aesthetically driven application. Here, through responsive envelope's silent and subtle movement, the intensification of ever-changing, locally varied, and delicate environmental dynamics has been performed.

A special focus has been given on the development of those buildings which are fully expressive of their location's characteristics as well as in their sites, which are designed to be grounded. Around sustainable architecture in contemporary environmental discourse, emphasis on development of such buildings is one of the key threads. For authenticity, a fundamental concern has been emphasized by this approach. A relation with the concept of place and locality is necessary for truly sustainable buildings. The characteristics of a specific regime impose several possibilities and constraints. Living within and adopting to such possibilities and constraints is the meaning of sustainability here. With specific values of people or place, "universal technological" solutions are generally failed to coincide. Hence, technical designs must be made by the designers for the applications of the hygromorphic materials and such decisions should be relied on the development of place-specific design strategy as well as on the understanding of the site. The benefits of the local

Applications of Sustainable Materials 307

climate as material applications driver and as key constituent are acknowledged by such place-specific design strategies. Specific weather patterns and ecological conditions are uniquely adapted by the hygromorphs continuous response. Based on the prevailing winds and the Sun orientation, timber shingle cladding varies for every building elevation. These existing weathering expressions of patterns' are amplified and emphasized by the utilization of newly added hygromorphic responses.

Hygromorphic wood composites building integrated applications have been firstly introduced by Holstov et al. [23]. In England, the largest man-made woodland and in Northumberland, a large forestry plantation is the Kielder Forest. The sustainable sources of constructional timber are provided by this Kielder Forest. Larch, fir, pine, and spruce are the major species in the forest and it is predominated by conifers. In this specific project, pine cones proliferation at the site has inspired the use of hygromorphs. Additionally, hygromorphic materials' relevance has influenced the use of hygromorphs. Regarding the immediate environment, Forest Park and Kielder Water Trust have educated the visitors. Holstov et al. [23] have investigated the sustainable materialization of responsive architecture in more details.

Origami-based architecture materials have proved their significance in obtaining exceptional properties and enhanced functionalities of thermal-regulatory means in regulating the temperature of the building. Origami folding has potential application in architecture [24]. Recently, BIM-origami-based technique has been efficiently used for energy optimization of such buildings [25]. Here BIM stands for Building Information Modeling. In such a system, an element is modeled which forms a dynamic shell. This is done considering the perpendicular incidence of the Sun into the infrastructure. Through the study, it is understood that BIM is very useful for the projects where the principles of origami are applicable and, based on this, their improved sustainability and habitability are expected.

16.4 APPLICATIONS OF SUSTAINABLE MATERIALS IN AGRICULTURE

Across several sectors of the society, one of the key challenges is to attain a double transition, i.e., digital and economical. This goal has recently been identified by the European Commission. In order to enhance the efficiency, sustainability, quality, and productivity in agriculture sector, several new methodologies, and technologies have been incorporated into this transition process. With multidisciplinary technologies, we can solve various distinctive aspects related to sustainability in agriculture.

In order to attain sustainability in semiarid regions, it is essential to have water use efficiency. Along with the irrigation scheduling, the adjusted water amounts can be determined suitably with the adoption of compact removable weighing lysimeters. The irrigation system efficiency can be improved and the crop water requirements can be determined by the farmers with the help of these devices. Through an isolated soil volume's mass balance, the crop evapotranspiration (ETC) can be measured by a direct method provided by a weighing lysimeter. The waste management in the agriculture is a second important factor related to the sustainability. In terms of economic administration and environmental treatment, a large amount of waste is generated by the agriculture activities, which cannot be managed adequately. In wood-based materials production, it has been assumed that wood residues are socially acceptable economically viable, and environmentally sustainable. However, suitable substitutes are prompted to look in forests because of taking place of deforestation. Through the introduction of mixture of fruit tree pruning, garden tree pruning, whole trees, and forest residues, the usage of wood in particle board might be reduced. In the new materials development, an abrupt increment in research using plant fibers has been made. This increment is because of decreasing supplies of wood-based materials and solid wood. Adopting vegetable fibers, like vine pruning, common cane, jute, bamboo, flax and hemp, tea, coconut, and cane bagasse, particle boards have been manufactured in numerous studies [26–30]. One of the sustainable applications can be the usage of giant reed's waste. The properties of the wood boards can be improved through the addition of giant reed particles. Consequently, environmental and industrial benefits are being resulted [31].

Because of the climate change, excessive use of conventional fertilizers, and an enhancement in the demand for food, a huge number of distinctive challenges are being faced in agriculture sector. The use of bio-stimulant products could be the solution to these problems. The concentration of amino acids (AAs) is high in bio-stimulant products. As complement and/or substituting conventional fertilizers, these substances can be used. Additionally, the impact of climate change on the plants shall be reduced [32]. The other way to attain sustainability in agriculture for optimal natural resource management is the use of computational tools. The resolution and definition of optimization problems are often included in resources and energy dispatch in agricultural systems. In several case studies for solving various resource management aspects, the use of multicriteria problems has been adopted. There is a variation among these aspects. Importantly, the defined system is represented by the deterministic mixed-integer linear problem. In agricultural systems, like photovoltaic systems, greenhouses, etc., this can be applied. For decision-making, i.e., managing storage systems, CO_2 enrichment, schedule irrigation, etc., simulations are being included in these tools [33].

In order to attain sustainability in agriculture, another related applied technology is the use of remote sensing. From satellite imagery, we can obtain the evaluation of the organoleptic, nutritional quality, phenological, and agronomic characteristics of crops. These characteristics are based on Normalized Difference Vegetation Index (NDVI). According to productive capacity, different areas can be classified by this NDVI. The applications of fertilization, irrigation, and management can also be determined adequately by this index [34]. In agriculture sector in terms of sustainability, we can evaluate the modernization impacts of irrigation systems. These impacts are being determined in terms of effectiveness. Hence, water productivity and crop yield are allowed to increase significantly. In order to modify agricultural water management, these modernizations should be implemented by the farmers. The storage volumes can be optimized and crop yields can be increased by this way effectively. Before and after the modernization, the agricultural irrigation and production evaluations and the water productivity assessments are deciding these evaluations [35]. Over time, the irrigation module modernization impacts are also determined by these evaluations [35].

Applications of Sustainable Materials

The concept of sustainability is also related to the concept of food security. Through the combination of demographic evolution, crop yield, and climate modeling under future climatic scenarios, we can study the climate change impact on food security. With these studies, we can also analyze the economic indicators' impacts [36]. In agriculture, another application of water sustainability is irrigation management depending on the soil moisture sensors [37]. The investigation of nutrient management recommendations is another aspect of sustainability. Farmer guidelines and policies are expected to develop by these studies. In the agricultural water environment, the influence of urban development is another imperative consideration. In agricultural water utilization efficiency measurements, robust results can be obtained by stochastic frontier analysis model. For this, several factors, like rice planting ratio, crop planting ratio, balanced urbanization, economic urbanization, and population urbanization, have to be taken into considerations [38]. For sustainable purposes in agriculture, we can use the environmental impact analyses of greenhouses. The time and location of farming mainly influence the cultivation in open fields. Soil characteristics, weather conditions, and quantity and quality of water for irrigation highly influence the time and location of farming.

A good alternative in mixed wood boards manufacturing is the use of common cane. This is because more sustainable materials development is contributed by their use. Usage of common cane in mixed wood boards manufacturing requires lower energy consumption and consequently environmental and industrial benefits are resulted. In agriculture, the environmental sustainability of plastics has been reviewed by Maraveas [39]. The mulching films, plastic covers, polyethylene (PET) and polypropylene (PP) plastic shade nets, polyvinyl chloride, and linear low-density polyethylene (LLDPE) have been focused on in his investigations. The plastics chemical composition influences the biodegradable plastics ecological impacts. Meteorological events, exposure to solar radiation, limited insect infestation, and enhanced nutritional properties and agricultural yields are resulted in farm environments by the plastic shade nets. On plants, the impact of natural elements is limited and the microclimate is altered by the plastic nets. In small-scale as well as large-scale agriculture, it is common to use plastic covers, mulching films, and plastic nets for greenhouses. In modern technology, the plastic shading technology's centrality has been demonstrated by Zhang et al. [40]. The growth of weeds is limited, and temperature and soil moisture content are regulated by the plastic mulching materials. On sustainability and ecological considerations, it is premised to concentrate on biodegradable plastics in agricultural applications. A diminished threat to the environment is resulted from the bio-based plastics, which are obtained from the renewable feedstocks [41]. A considerable amount of CO_2 has been generated in the production process of nonbiodegradable plastics, which are derived from fossil fuels. It is a primary concern for climate change and global warming. It is suitable to adopt biodegradable and bio-based polymers. This is because antimicrobial/antibacterial properties can be posed by customizing the surface properties [42]. In addition to this, these materials have tuneable optical, thermal, radiometric, and mechanical properties.

In "II Iberian Symposium on Horticultural Engineering 2020" held at Portugal, several technological advances in agricultural systems, animal production, viticulture, extensive crops, agricultural construction and mechanization, energy management, and water resources management have been developed. It was organized by Higher Agricultural School – Polytechnic Institute of Viana do Castelo, Portuguese Association of Horticulture and Horticultural Engineering Group of the Spanish Society of Horticultural Sciences. In this event, representatives of companies, universities, and research centers have participated. From the discussion, it was understood that the sustainable cultivation can be benefited by the applications of bio-stimulants comprising amino acids to tomatoes. The efficiency of the water usage, the net assimilation of CO_2, and the growth of the aerial part are contributed by these. Adopting a removable compact weighing lysimeter, the cultivation and the evapotranspiration coefficients of bell pepper can be estimated. For these types of crops, the water requirement determinations can also be performed by these achieved values. From NDVI, organoleptic, nutritional, phenological, and agronomic characteristics of pistachios have been evaluated. For applying fertilization, irrigation, and management, this technology can be useful.

16.5 AEROSPACE WITH SUSTAINABLE MATERIALS

At distinct stages of supply chain, several companies are involved in a long process of aerospace product development. Set-based concurrent engineering commonly steered this aerospace product development. On the specific product features, as decisions are made and as time goes on, a series of requirements and assumptions are adjusted in the beginning of this process. For component and subsystem manufacturers, a challenging situation has been created by long lead time. Before requirements have been signed, there is an urgent need for the initiation of the development process. In addition to improve product performances, reduce cost and weight, and enhanced design robustness, requirements are sought to deal with uncertainty simultaneously. In the design space, instead of focusing on particular solution points, an open series of feasible design solutions are focused in early multidisciplinary design optimization. In early design stages, the implications of decisions made are enabled to understand by engineers through this process. Here, about 20 years of production periods and several decades of operations are featured for the products.

In aerospace product development, the origin of value-driven design (VDD) research field is lying. In VDD, an innovative process is proposed to have best value of a system. For this, traditional design methods are either complemented or replaced. In order to drive a design's multidisciplinary optimization, the VDD methods depend on the adoption of "value function". For design concept trade-off, the requirements of the combination of multiple-valued functions or a single-valued function have been argued by researchers. A specific design solution and monitoring of profits or losses have been often linked with the use of value functions. Surplus-value calculation or a net present value assessment form is taken by the value function in these cases. Instead of providing optimization results, the awareness of decision-makers has been aimed to enhance multidisciplinary trade-off analysis in other situations of VDD models [43].

In order to attain more sustainable solutions, how one can measure sustainability, how one can achieve the sustainability, and what does sustainability actually mean

Applications of Sustainable Materials

are the imperative key factors with which product development team must be familiar. On these ideas, sustainable product development (SPD) is based on. With a back-casting approach, a forecasting approach has been combined by researchers in SPD. The imagination of the success in the future is back-casting. In back-casting, in order to reach that success, numerous ways have been explored and this success definition has been implemented in the current situation. From a back-casting perspective, the definition of prioritized sustainability criteria and from acquisition of raw materials to disposal phase, how environmental and social sustainability are influenced by a design solution are included under such processes [44]. Consequently, comprising indicators and criteria to support SPD, sustainability design space (SDS) is being formulated [44]. The target of prioritization of an aspect for "no hazardous chemicals used" or "no raw material used" is fallen under the term criterion. The level or state of the concerned criterion is indicated by either quantitative or qualitative measurements in an indicator.

Three parts are comprised in SDS, as defined by Hallstedt [44]. On a back-casting perspective basis, based on overarching sustainability principles, the strategic sustainability criteria fall under the first part. For life-cycle phase of every product, the target of ideal long-term sustainability has been included in this first part. Including company-specific and industry-specific expectations and requirements, the criteria of long-term strategic sustainability development are supported by the design guidelines of tactical sustainability in second part of SDS. In relation to the sustainable solutions, up to what degree the performance of a product can be assessed is linked with sustainability compliance index (SCI) which is a qualitative measurement scale. This criterion falls under the third part of SDS. In aerospace product development, with other supporting tools we can use all three parts of SDS in distinct combinations whereas in some situations these parts are being used separately. In the early design stages, the assessment of critical alloys and through the technology method of readiness assessment, the product innovation process of integrating sustainability are some examples of this aerospace product development with SDS. In the innovation process, one of the decisions, which must be taken in early stages, is the material selection. Sustainability risk can be linked with such selections/decisions. This is because of a direct impact on downstream decisions (end-of-life solutions and manufacturing processes selection) and upstream decisions (in rural areas, the extraction activities). Towards strategic sustainable development, opportunities and threats which are because of counteraction to society's transition or an organization's contribution are defined as sustainability risks [45].

For design space exploration, in combining models of SPD and VDD, the common denominators are scarce availability of data, poor maturity, and high heterogeneity. In distinct aerospace development stages, there are various imperative applications of SPD and VDD models. In design space exploration, in order to integrate SPD and VDD models, little effort has been made. For this, product structural simulation's numerical results are being combined with them. For sustainability and value, the use of computational models is made due to three challenges. First challenge is the limited first-hand data availability. This is due to the fact that within the usage environment and product life cycle as well as within the product definition, the populated data of SPD and VDD models is residing. Second challenge is the difficulty in

312

Handbook of Sustainable Materials

identification of the relations among design variables which are affecting sustainability and value [43]. Third challenge is the communication of the results to engineers with effective approaches by coupling sustainability and value models. There should be a close relationship between "natural thinking pattern" of engineers and the form in which such communications are expected to be done. For a new formal method, it is an imperative factor for the acceptance. In current decision situations, in comparison with fluid dynamic, thermal, and mechanical performance of an engine, the weaknesses in understanding and clarifying the sustainability and value implications are responsible for not being addressed these challenges. On a development project, before committing high resources it is necessary to answer some imperative questions such as most valuable component to develop, its sustainability profile and during the entire life cycle, the product's sustainability impact.

16.6 BIOMEDICAL AND HEALTH-CARE APPLICATIONS OF SUSTAINABLE MATERIALS

In several research areas and applications of material science, extreme importance has been given to zinc oxide (ZnO) in recent decades because of its great versatility, low cost, and multifunctional properties. In industries like pharmaceutical products, cosmetics, food packaging, paints, ceramic materials, and rubber, a significant increment in the ZnO market has accompanied their scientific interest. The Food and Drug Administration (FDA) has approved the adoption of ZnO in cosmetic products because of its recognition as a bio-safe material. In nanostructured form, more interesting application of ZnO has been resulted. With specific chemical-physical properties, novel nanodevices and nanomaterials have been enabled to realize [46]. The broad varieties of nanostructures (NStr) such as tetrapods (TPs), nanoribbons (NRBs), nanotubes (NTs), nanosheets (NSs), nanorods (NRs), nanofibers (NFs), nanowires (NWs), and nanoparticles (NPs) can be achieved with various techniques by easy synthesis. Figure 16.2 illustrates the imperative applications of nanostructured ZnO. The main emphasis has been given to sustainability and biomedical applications, which cover diagnostic imaging, flexible electronics, biosensors, tissue engineering scaffolds, antimicrobial materials, drug delivery, wound healing, and fluorescence imaging.

Boosting osteointegration and angiogenesis processes, tissue regeneration as well as cell differentiation, proliferation, and growth can be promoted by the adoption of miniature amounts of ZnO NStr. ZnO antifungal and antibacterial properties further support these processes. With respect to specific cell lines, a selectivity has also been presented by ZnO NStr. Consequently, for killing cancer cells, ZnO NStr makes them potential candidates. For diverse target applications, it is also particularly relevant to have a relationship between biological microenvironments and ZnO-based systems. In order to ensure effective and safe applications, much efforts have been requisite at the manufacturing and designing stages. In order to develop superior-performance hybrid composite materials, numerous methodologies are adopting polymer matrices within ZnO NStr-based novel biomaterials manufacturing processes [47, 48]. With high scalability, ease of processing, and low cost, the most promising and versatile of these techniques is electrospinning. Micro- and nanofibers 2D and 3D scaffolds

Applications of Sustainable Materials

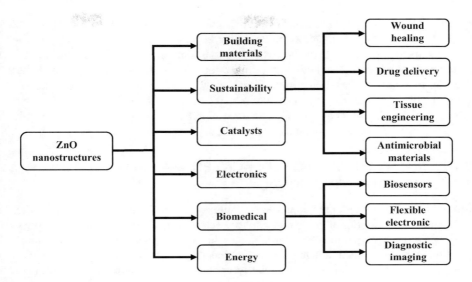

FIGURE 16.2 Sustainability and biomedical applications of ZnO nanostructures.

are allowed to be achieved by this technique [49]. With a large surface-to-volume ratio and with controlled porosity, the materials are allowed to be produced by the electrospinning technique. Due to their mechanical and morphological properties, a suitable interconnected network has been formed for biological applications [50].

Material parameters like homogeneity of filling elements dispersion, topography of the surfaces, interconnection and shape of pores, degree of porosity, and fiber diameter have a strong influence on the biological processes [51]. Through an action on the parameters of electrospinning process, the approximate setting and monitoring of all these aspects may be performed. High-performance materials are possible to be developed with the combination of ease of electrospinning technique fabrication with the nanostructured ZnO properties [52]. ZnO–polymeric nanocomposites and ZnO nanostructures have been shown recently very useful for the sustainability, health, and biomedical applications. In biomedical field, safe, and biocompatible ZnO nanostructures are being made to be synthesized by numerous methods. Ligand-coated ZnO nanocrystals (NCs) are synthesized by latest organometallic self-supporting approach [53]. Through a well-passivated surface and an impermeable shell, a protection to the core has been given by NC–ligand interface characteristics. In comparison with the traditional methods, this method showed low toxicity on human lung cancer cells (A549) and human fetal lung fibroblast cells (MRC-5). For the treatment of tumors, it has been proven that one of the most promising approaches is to use ZnO nanostructures in drug delivery. Here, impact on healthy cells has been minimized and the effect of toxicity is also reduced.

Adopting lipid-coated ZnO NPs, the lymphatic metastatic tumors treatment has been performed by lymphatic targeting drug delivery system. In comparison with non-coated ZnO NPs, lipid-coated ZnO NPs have shown higher biocompatibility, as confirmed by tests on Sprague-Dawley rats. It has been shown that red blood cells (RBCs) did not aggregate in the existence of lipid-coated ZnO NPs. In the situation of lipid-coated ZnO NPs, reversible hepatotoxicity has been shown in the biochemical parameters in the liver, whereas in the situation of ZnO-NPs, nonreversible hepatotoxicity is resulted. In the lungs, kidneys, and intestine, mild-moderate inflammation has been shown in the situation of ZnO-NPs whereas no considerable lesions in the organs have been shown in histopathological analyses for lipid-coated ZnO NPs. During first hours of administration, in the spleen slight congestion is only caused by lipid-coated ZnO NPs.

16.7 CONTRIBUTION OF SUSTAINABLE MATERIALS IN GREEN TECHNOLOGY

The development of green materials is supporting the sustainable development. Unique characteristics of green materials are versatile in terms of chemical and physical properties, economically affordable, less toxic, and abundant in nature. In pollution technology and management and engineering, materials science, construction and infrastructures, building and energy applications, and numerous applications are being realized with the green materials. We can develop the green materials as an energy production source. We can also develop green materials as bioethanol and biodiesel source. For advanced bio-applications, the transformation into advanced functionalized materials can be done by biomass-based materials. Consequently, it can be adopted in tissue engineering, biomaterials, and biomedicine applications. Through the adoption of green materials as a source, an abrupt interest has been gained in the development of pigment materials.

A significant research area has been developed by green technology to answer the important issues of 21st century like efficiency in manufacturing, environment concern, energy consumption, etc. In 21st century manufacturing, two imperative challenges to be considered are deficiency of resources and energy and environmental shift. Sustainable manufacturing, green products, waste reduction, green manufacturing, energy consumption, and energy saving are the major issues of the present time. In order to attain sustainable manufacturing, the implementation of green manufacturing has been increasing rapidly throughout the world. For business practice, such as diminishing supply chain uncertainty and risk, responding to government regulations, and saving energy resources, the most attractive strategy realized by the companies is sustainability. Additionally, the competitiveness can be maintained and it can also respond to the consumer demands. Since future water/food supply systems, defense, dwelling, lifestyle, health/medical, transport, and energy are depending on increasingly precise components and elements, the competitive advantages are the supply of such machine tools which consumed relatively lesser space, material and energy, the manufacturing of alternative energy supply systems, and manufacturing for environmentally and energy-aware consumer for efficient factory operations. For green manufacturing innovations, enhancing resource productivity and effective utilization of resources are the key considerations. The resource effectiveness can be improved by extended life, leveraged resources, ensured low cost of reuse and high reuse yield, reduced footprint of resources, enhanced yield, light-weighting, and avoiding the adoption of a resource in the first place. For an

Applications of Sustainable Materials 315

improvement in product design, manufacturing processes and material conversion, all of these must be implicated for manufacturing.

In order to deal with finite resources and global warming in the aspect of green technology, one of the important issues is the manufacturing process sustainability. Potential for process optimization, part consolidation, recycling, life extension, material saving, and energy saving are the imperative determinations for the manufacturing process sustainability. The manufacturing process operation principle is strongly influencing the manufacturing process sustainability. Without additional processes, it is difficult to fabricate heterogeneous and freeform shapes structures because of the drawbacks in the forming process, subtractive manufacturing (SM) processes, conventional manufacturing (CM) processes, and process principle. During the fabrication of parts, a huge deal of waste material has been created by the SM process.

In computer-aided manufacturing (CAM) and computer-aided design (CAD) environments, through metallic materials layer-by-layer deposition with geometrical complexity, three-dimensional (3D) metallic parts can be fabricated by adopting direct metal additive manufacturing (DMAM) process. In comparison with plastic-based additive manufacturing (AM) process, no secondary process is needed to fabricate metallic parts in DMAM process. Consequently, in the manufacturing of 3D metallic parts, a considerable reduction in the lead time is resulted by the DMAM process. In medical, home appliances, electronics, tooling, automotive, and aerospace industries, a diminishment in the product development's lead time is resulted by rapidly fabrication of the metal parts adopting the DMAM process. In recent years, there is an abrupt expansion of the DMAM applications in the functional metallic parts production such as unified components, lightweight parts, order adaptive parts, and many more [54].

A heterogeneous structure and characteristic shapes have been created easily with a freeform geometry with different materials by the DMAM process layer-by-layer deposition characteristics. Various promising merits of DMAM process are: novel geometries, environmental and performance advantages in application of parts; a modification in the manufacture of metallic parts buy-to-fly ratio; and the designing and manufacturing of the parts with novel geometries [55]. The manufacturing process sustainability can be improved by the promising merits and inherent characteristics of the DMAM [55]. From the viewpoint of an enhancement in the manufacturing process sustainability, a steadily increment in the DMAM processes interest has been seen in research, academy, and industries. Products life extension, remanufacturing and repairing; a modification in the manufacturing process efficiency; and redesigning of the product are three categories in which DMAM process is applicable in terms of sustainability [55]. The social dimension of sustainability, i.e., social responsibility, needs to be addressed at a higher level if we want to shift toward sustainable manufacturing. Numerous methodologies and tools are necessary to be developed for implementing greener production technology at a large scale.

16.8 CONCLUSION

In this chapter, detailed applications of sustainable materials in engineering, architecture, agriculture, aerospace, health care, biomedical, and green technology have been discussed. For sustainability, a simple rule for savings is defined in addition to the sustainable social net product. In order to enhance life quality, sustainability should be incorporated in all engineering fields. Reduced environmental impact, enhanced efficiency, sustainable processes, sustainable resources, and fulfillment of other aspects of sustainability are the key requirements for the engineering sustainability. Within architectural design, the conventional and accepted definitions of sustainable materiality shall find several opportunities to extend by the adoption of multidimensional characteristics of hygromorphic technologies. In agriculture sector in terms of sustainability, we can evaluate the modernization impacts of irrigation systems. These impacts are being determined in terms of effectiveness. Hence, water productivity and crop yield are increased significantly. In order to attain more sustainable solutions, how one can measure sustainability, how one can achieve the sustainability and what does sustainability actually mean are the imperative key factors with which product development team must be familiar with. On these ideas, SPD is based on. This is important to emphasize sustainability and biomedical applications of nanostructured ZnO, which covers diagnostic imaging, flexible electronics, biosensors, tissue engineering scaffolds, antimicrobial materials, drug delivery, wound healing, and fluorescence imaging. In order to attain sustainable manufacturing, the implementation of green manufacturing is also vital and it has been increasing rapidly throughout the world.

16.9 FUTURE APPLICATIONS ASPECTS

Nowadays, the renewable energy conversion systems are being coupled with the sustainable batteries implementations. By doing so, several technological difficulties for electrification like making complete system eco-friendly, enhancing modest technical capabilities, and challenging cost and geographical accessibility can be overcome in remote, poor, and off-grid communities. The emissions of greenhouse gases to the atmosphere will be decreased significantly. For future sustainable batteries, advanced and green electrolyte and electrode materials are being developed at an increasing rate. In order to move toward a zero-carbon future, several researchers are being focused toward sustainable strategies of end-of-life and current second-life battery disposal [56].

In non-agrarian and agrarian sectors, the utilization of pesticides has been increased recently. In different environmental segments, the deposition of their remnants has been increased significantly. A bigger threat to the climate and human well-being is the existence of pesticides in drinkable water sources, groundwater sources, air, soils, lakes, ponds, and rivers. In order to reduce the impact of pesticides on human beings, it is imperative to remove metabolites of pesticides from contaminated water [57, 58]. Therefore, functionalized cellulose and low-cost pristine biomass-based derivatives have been utilized for the removal of pesticides [59]. For this, the behavior of pesticides in the water as well as in the environment is requisite to understand. On percent removal of pesticides, the impact of activated biochar/carbons, cellulose/CdS nanocomposites, clay/cellulose nanocomposites, cellulose derived photo nanocatalyst and magnetite cellulose nanocomposites have also been investigated by several researchers [59–63].

Applications of Sustainable Materials

With an increment in the human population, the sustainable energy production and storage demand is continuously increasing. The development of energy devices requires urgent attention owing to environmental protection and energy utilization demand. Because of superior chemical stability, strong mechanical strength, high surface area, and high electrical and thermal conductivity of 2D structured graphene, they are widely adopted as an excellent energy material. The physicochemical properties of graphene are modified significantly when doped with heteroatoms and results in a promising material in a broad range of applications/fields. Nowadays, in dye-sensitized solar cells, batteries, supercapacitors and energy storage and conversion technologies, the advancement and adoption of heteroatom-doped graphene materials have been reported in several research articles. For heteroatoms doped graphene materials, future challenges, and perspectives have been discussed by Lee et al. [64].

There is an abrupt increment in the demand of eco-friendly products. This is because of enhancing environmental awareness among the societies. Consequently, biopolymers-based eco-friendly materials are being developed significantly owing to ecological concerns and market demand. For a broad sector of applications like biomedical, environmental, and packaging applications, one of the functional materials known to be utilized is marine algae-derived nanocellulose. Adopting hybrid, mechanical, chemical, or microbial methods, researchers have extracted nanocellulose from algae biomass. An enzyme named cellulase is used in the microbial treatment through enzymatic hydrolysis. Microorganisms, such as fungi, yeast, and bacteria, are being used to produce cellulase. Since they are cheaply available, the operating costs have been reduced significantly, which in turn, enhanced the production processes. In numerous applications, the macroalgae utilization as functional and sustainable materials has been highlighted by several researchers [65–69].

There is an abrupt movement toward the usage of energy obtained from sustainable and renewable materials rather than fossil-fuel-driven energy. For advanced materials, a significant attention has been given to the trees. This is because most abundant renewable bioresources are the trees. There are complex components and structures like root, seed, flower, leaf, and truck with which trees are composed. In different fields, like water treatment, biomedical, electronics, and energy, the excellent applications of wood-derived biopolymers have been published by several researchers. In order to attain sustainability, the latest achievements in the structures and materials obtained from distinct parts of trees have been summarized by Liu et al. [70]. The biopolymer-based materials have been emphasized to achieve sustainability in different sectors and these are considered to be explored more and more in the future.

Finally, this would be worth mentioning that the complete implementation of sustainable materials has not been practically possible due to several factors such as materials and technology, construction processes, design processes, and lack of information and cost. The advantages of high-performance buildings are not understood truly due to lack of reliable cost models. This is also attributed to inherently conservative nature of the building industries despite several incentive programs like recognition programs, technical assistance, etc. The affordability of choice of

REFERENCES

1. Rogers P P, Jalal K F and Boyd J A 2012 *An introduction to sustainable development* (Routledge, London, United Kingdom)
2. Daly H E 2014 *Beyond growth: The economics of sustainable development* (Beacon Press, London, United Kingdom)
3. Malik H K and Singh A K 2010 *Engineering physics* (McGraw-Hill Education, India)
4. Dhawan R, Kumar M and Malik H K 2020 Influence of ionization on sheath structure in electropositive warm plasma carrying two-temperature electrons with non-extensive distribution *Phys. Plasmas* **27** 63515
5. Dhawan R and Malik H K 2020 Behaviour of sheath in electronegative warm plasma *J. Theor. Appl. Phys.* **14** 121–8
6. Dhawan R and Malik H K 2020 Sheath formation criterion in collisional electronegative warm plasma *Vacuum* **177** 109354
7. Kumar S, Dhaka V, Singh D K and Malik H K 2021 Analytical approach for the use of different gauges in bubble wakefield acceleration *J. Theor. Appl. Phys.* **15** 1–12
8. Dhawan R and Malik H K 2020 Sheath characteristics in plasma carrying finite mass negative ions and ionization at low frequency *Chinese J. Phys.* **66** 560–72
9. Malik H K and Dhawan R 2020 Sheath structure in electronegative plasma having cold ions: An impact of negative ions' mass *IEEE Trans. Plasma Sci.* **48** 2408–17
10. Dhawan R and Malik H K 2021 Modelling of electronegative collisional warm plasma for plasma-surface interaction process *Plasma Sci. Technol.* **23** 45402
11. Zavadskas E K, Šaparauskas J and Antucheviciene J 2018 Sustainability in construction engineering *Sustainability* **10** 2236
12. Yates J K and Castro-Lacouture D 2018 *Sustainability in engineering design and construction* (CRC Press, Boca Raton, Florida)
13. Siddique N and Adeli H 2017 Nature-inspired chemical reaction optimisation algorithms *Cognit. Comput.* **9** 411–22
14. Yeganeh Fallah A and Taghikhany T 2016 A modified sliding mode fault tolerant control for large-scale civil infrastructure *Comput. Civ. Infrastruct. Eng.* **31** 550–61
15. Ghaedi K, Ibrahim Z, Adeli H and Javanmardi A 2017 Invited review: Recent developments in vibration control of building and bridge structures *J. Vibroengineering* **19** 3564–80
16. Jato-Espino D, Castillo-Lopez E, Rodriguez-Hernandez J and Canteras-Jordana J C 2014 A review of application of multi-criteria decision making methods in construction *Autom. Constr.* **45** 151–62
17. Zavadskas E K, Antuchevičienė J and Kapliński O 2015 Multi-criteria decision making in civil engineering: Part I–applications state-of-the-art survey *Eng. Struct. Technol.* **7** 103–13
18. Zavadskas E K, Antuchevičienė J and Kapliński O 2015 Multi-criteria decision making in civil engineering. Part II–applications *Eng. Struct. Technol.* **7** 151–67
19. Kumar V, Haapala K R, Rivera J L, Hutchins M J, Endres W J, Gershenson J K, Michalek D J and Sutherland J W 2005 Infusing sustainability principles into manufacturing/mechanical engineering curricula *J. Manuf. Syst.* **24** 215–25
20. Anand A, Haq M I U, Vohra K, Raina A and Wani M F 2017 Role of green tribology in sustainability of mechanical systems: A state of the art survey *Mater. Today Proc.* **4** 3659–65
21. Holstov A, Bridgens B and Farmer G 2015 Hygromorphic materials for sustainable responsive architecture *Constr. Build. Mater.* **98** 570–82

Applications of Sustainable Materials

22. Abdelmohsen S, Adriaenssens S, El-Dabaa R, Gabriele S, Olivieri L and Teresi L 2019 A multi-physics approach for modeling hygroscopic behavior in wood low-tech architectural adaptive systems *Comput. Des.* **106** 43–53

23. Holstov A, Farmer G and Bridgens B 2017 Sustainable materialisation of responsive architecture *Sustainability* **9** 435

24. Megahed N A 2017 Origami folding and its potential for architecture students *Des. J.* **20** 279–97

25. Pavón R M, Bazán Á M, Cepa J J, Arcos Álvarez A A, Trigueros J M A, Alberti M G and Tellaeche J R 2022 New use of BIM-Origami-based techniques for energy optimisation of buildings *Appl. Sci.* **12** 1496

26. Bui H, Sebaibi N, Boutouil M and Levacher D 2020 Determination and review of physical and mechanical properties of raw and treated coconut fibers for their recycling in construction materials *Fibers* **8** 37

27. Ferrandez-García M T, Ferrandez-Garcia C E, Garcia-Ortuño T, Ferrandez-Garcia A and Ferrandez-Villena M 2020 Study of waste jute fibre panels (corchorus capsularis L.) agglomerated with Portland cement and starch *Polymers (Basel)* **12** 599

28. Ferrandez-Villena M, Ferrandez-Garcia C E, Garcia-Ortuño T, Ferrandez-Garcia A and Ferrandez-Garcia M T 2020 The influence of processing and particle size on binderless particleboards made from Arundo donax L. rhizome *Polymers (Basel)* **12** 696

29. Ferrandez-García M T, Ferrandez-Garcia A, Garcia-Ortuño T, Ferrandez-Garcia C E and Ferrandez-Villena M 2020 Assessment of the physical, mechanical and acoustic properties of Arundo donax L. biomass in low pressure and temperature particleboards *Polymers (Basel)* **12** 1361

30. Ferrandez-Villena M, Ferrandez-Garcia C E, Garcia-Ortuño T, Ferrandez-Garcia A and Ferrandez-Garcia M T 2020 Analysis of the thermal insulation and fire-resistance capacity of particleboards made from vine (Vitis vinifera l.) prunings *Polymers (Basel)* **12** 1147

31. Ferrandez-Garcia M T, Ferrandez-Garcia C E, Garcia-Ortuño T, Ferrandez-Garcia A and Ferrandez-Villena M 2019 Experimental evaluation of a new giant reed (Arundo Donax L.) composite using citric acid as a natural binder *Agronomy* **9** 882

32. Alfosea-Simón M, Simón-Grao S, Zavala-Gonzalez E A, Cámara-Zapata J M, Simón I, Martínez-Nicolás J J, Lidón V and García-Sánchez F 2021 Physiological, nutritional and metabolomic responses of tomato plants after the foliar application of amino acids Aspartic acid, Glutamic Acid and Alanine *Front. Plant Sci.* **11** 2138

33. Farah A, Hassan H, M Abdelshafy A and M Mohamed A 2020 Optimal scheduling of hybrid multi-carrier system feeding electrical/thermal load based on particle swarm algorithm *Sustainability* **12** 4701

34. Vélez S, Barajas E, Rubio J A, Vacas R and Poblete-Echeverría C 2020 Effect of missing vines on total leaf area determined by NDVI calculated from Sentinel satellite data: Progressive vine removal experiments *Appl. Sci.* **10** 3612

35. Borrego-Marín M M and Berbel J 2019 Cost-benefit analysis of irrigation modernization in Guadalquivir River Basin *Agric. Water Manag.* **212** 416–23

36. Sultan B, Defrance D and Iizumi T 2019 Evidence of crop production losses in West Africa due to historical global warming in two crop models *Sci. Rep.* **9** 1–15

37. Incrocci L, Marzialetti P, Incrocci G, Di Vita A, Balendonck J, Bibbiani C, Spagnol S and Pardossi A 2019 Sensor-based management of container nursery crops irrigated with fresh or saline water *Agric. Water Manag.* **213** 49–61

38. Liu Y and Mao D 2020 Integrated assessment of water quality characteristics and ecological compensation in the Xiangjiang River, south-central China *Ecol. Indic.* **110** 105922

39. Maraveas C 2020 Environmental sustainability of plastic in agriculture *Agriculture* **10** 310

40. Zhang X, You S, Tian Y and Li J 2019 Comparison of plastic film, biodegradable paper and bio-based film mulching for summer tomato production: Soil properties, plant growth, fruit yield and fruit quality *Sci. Hortic. (Amsterdam)* **249** 38–48
41. Cui S, Borgemenke J, Liu Z and Li Y 2019 Recent advances of "soft" bio-polycarbonate plastics from carbon dioxide and renewable bio-feedstocks via straightforward and innovative routes *J. CO_2 Util.* **34** 40–52
42. Zhong Y, Godwin P, Jin Y and Xiao H 2020 Biodegradable polymers and green-based antimicrobial packaging materials: A mini-review *Adv. Ind. Eng. Polym. Res.* **3** 27–35
43. Bertoni M, Bertoni A and Isaksson O 2018 Evoke: A value-driven concept selection method for early system design *J. Syst. Sci. Syst. Eng.* **27** 46–77
44. Hallstedt S I 2017 Sustainability criteria and sustainability compliance index for decision support in product development *J. Clean. Prod.* **140** 251–66
45. Schulte J and Hallstedt S I 2018 Company risk management in light of the sustainability transition *Sustainability* **10** 4137
46. Wang X, Song J and Wang Z L 2007 Nanowire and nanobelt arrays of zinc oxide from synthesis to properties and to novel devices *J. Mater. Chem.* **17** 711–20
47. Sharma B, Malik P and Jain P 2018 Biopolymer reinforced nanocomposites: A comprehensive review *Mater. Today Commun.* **16** 353–63
48. Lin R, Hernandez B V, Ge L and Zhu Z 2018 Metal organic framework based mixed matrix membranes: An overview on filler/polymer interfaces *J. Mater. Chem. A* **6** 293–312
49. Hemamalini T and Dev V R G 2018 Comprehensive review on electrospinning of starch polymer for biomedical applications *Int. J. Biol. Macromol.* **106** 712–8
50. Ding J, Zhang J, Li J, Li D, Xiao C, Xiao H, Yang H, Zhuang X and Chen X 2019 Electrospun polymer biomaterials *Prog. Polym. Sci.* **90** 1–34
51. Han J, Xiong L, Jiang X, Yuan X, Zhao Y and Yang D 2019 Bio-functional electrospun nanomaterials: From topology design to biological applications *Prog. Polym. Sci.* **91** 1–28
52. Aziz A, Tiwale N, Hodge S A, Attwood S J, Divitini G and Welland M E 2018 Core–shell electrospun polycrystalline ZnO nanofibers for ultra-sensitive NO_2 gas sensing *ACS Appl. Mater. Interfaces* **10** 43817–23
53. Wolska-Pietkiewicz M, Tokarska K, Grala A, Wojewódzka A, Chwojnowska E, Grzonka J, Cywiński P J, Kruczała K, Sojka Z and Chudy M 2018 Safe-by-design ligand-coated ZnO nanocrystals engineered by an organometallic approach: Unique physicochemical properties and low toxicity toward lung cells *Chem. Eur. J.* **24** 4033–42
54. Chua C K and Leong K F 2014 *3D Printing and additive manufacturing: Principles and applications (with companion media pack)-of rapid prototyping* (World Scientific Publishing Company, Singapore)
55. Ford S and Despeisse M 2016 Additive manufacturing and sustainability: An exploratory study of the advantages and challenges *J. Clean. Prod.* **137** 1573–87
56. Navalpotro P, Castillo-Martínez E and Carretero-González J 2021 Sustainable materials for off-grid battery applications: Advances, challenges and prospects *Sustain. Energy Fuels* **5** 310–31
57. Sharma S, Dutta V, Raizada P, Hosseini-Bandegharaei A, Thakur V, Nguyen V-H, VanLe Q and Singh P 2021 An overview of heterojunctioned $ZnFe_2O_4$ photocatalyst for enhanced oxidative water purification *J. Environ. Chem. Eng.* **9** 105812
58. Utzeri G, Verissimo L, Murtinho D, Pais A A C C, Perrin F X, Ziarelli F, Iordache T-V, Sarbu A and Valente A J M 2021 Poly (β-cyclodextrin)-activated carbon gel composites for removal of pesticides from water *Molecules* **26** 1426
59. Rana A K, Mishra Y K, Gupta V K and Thakur V K 2021 Sustainable materials in the removal of pesticides from contaminated water: Perspective on macro to nanoscale cellulose *Sci. Total Environ.* **797** 149129

Applications of Sustainable Materials

60. Zhao M, Zhou H, Hao L, Chen H and Zhou X 2021 Natural rosin modified carboxymethyl cellulose delivery system with lowered toxicity for long-term pest control *Carbohydr. Polym.* **259** 117749

61. Zheng L, Yu P, Zhang Y, Wang P, Yan W, Guo B, Huang C and Jiang Q 2021 Evaluating the bio-application of biomacromolecule of lignin-carbohydrate complexes (LCC) from wheat straw in bone metabolism via ROS scavenging *Int. J. Biol. Macromol.* **176** 13–25

62. Zielińska D, Rydzkowski T, Thakur V K and Borysiak S 2021 Enzymatic engineering of nanometric cellulose for sustainable polypropylene nanocomposites *Ind. Crops Prod.* **161** 113188

63. Voicu S I and Thakur V K 2021 Aminopropyltriethoxysilane as a linker for cellulose-based functional materials: New horizons and future challenges *Curr. Opin. Green Sustain. Chem.* **30** 100480

64. Lee S J, Theerthagiri J, Nithyadharseni P, Arunachalam P, Balaji D, Kumar A M, Madhavan J, Mittal V and Choi M Y 2021 Heteroatom-doped graphene-based materials for sustainable energy applications: A review *Renew. Sustain. Energy Rev.* **143** 110849

65. Gola D, Dey P, Bhattacharya A, Mishra A, Malik A, Namburath M and Ahammad S Z 2016 Multiple heavy metal removal using an entomopathogenic fungi Beauveria bassiana *Bioresour. Technol.* **218** 388–96

66. Liyanaarachchi V C, Premaratne M, Ariyadasa T U, Nimarshana P H V and Malik A 2021 Two-stage cultivation of microalgae for production of high-value compounds and biofuels: A review *Algal Res.* **57** 102353

67. Chawla P, Gola D, Dalvi V, Sreekrishnan T R, Ariyadasa T U and Malik A 2022 Design and development of mini-photobioreactor system for strategic high throughput selection of optimum microalgae-wastewater combination *Bioresour. Technol. Reports* **17** 100967

68. Dalvi V, Patil K, Nigam H, Jain R, Pabbi S and Malik A 2021 Environmental resilience and circular agronomy using cyanobacteria grown in wastewater and supplemented with industrial flue gas mitigation *Ecophysiology and Biochemistry of Cyanobacteria* (Springer, Singapore) pp 291–325

69. Mathur M, Kumar A, Ariyadasa T U and Malik A 2021 Yeast assisted algal flocculation for enhancing nutraceutical potential of Chlorella pyrenoidosa *Bioresour. Technol.* **340** 125670

70. Liu C, Luan P, Li Q, Cheng Z, Xiang P, Liu D, Hou Y, Yang Y and Zhu H 2021 Biopolymers derived from trees as sustainable multifunctional materials: A review *Adv. Mater.* **33** 2001654

17 Heusler Alloys
Sustainable Material for Sustainable Magnetic Refrigeration

Chhayabrita Maji

CONTENTS

17.1 Introduction 323
    17.1.1 Sustainable Material and Technology...323
17.2 Other Sections and Work 326
    17.2.1 Heusler Alloys...326
        17.2.1.1 Applications of Heusler Alloys..327
        17.2.1.2 Heusler Alloys with Reversible Martensitic Transition 328
        17.2.1.3 Magnetic Refrigeration Using Heusler Alloys 330
        17.2.1.4 Preparation and Properties of Heusler Alloys for Magnetic Refrigeration 331
17.3 Critical Discussion 334
    17.3.1 Challenges of Heusler Alloys for Magnetic Refrigeration..........334
17.4 Conclusion...337
References..337

17.1 INTRODUCTION

17.1.1 SUSTAINABLE MATERIAL AND TECHNOLOGY

The sustainable environment, economy, and society are mainly governed by science and technology. The environmental changes are the direct consequence of science and technology advancement. The science builds the ladder for technological progress, which in turn affects the productivity that drives the world economic growth [1, 2]. The economic growth directly influences the health of the society, providing an indirect touch of science and technology to the society [3–6]. Thus, a responsible and careful science and technology development is of utmost importance for sustainability of human race on the Earth. Keeping the environmental, economic, and social sustainability as key to our existence, the United Nations defined 17 sustainable development goals (SDGs). Since all 17 SDGs are interlinked to each other or are inseparable, the environment-friendly and sustainable technology encompasses,

DOI: 10.1201/9781003297772-17   **323**

directly or indirectly, all the SDGs. The SDG no. 13, climate action is considered the most important for achieving the sustainability goals that deal with environmental issues. Further, the Paris Agreement COP21 also emphasized the need of limiting global warming within 2°C as compared to preindustrial temperature. Therefore, combating climate change is the common cause of all nations. Thus, environment-friendly technology-related research gained momentum. One such technology is magnetic refrigeration that encompasses the SDG 7 (Affordable and clean energy), SDG 12 (Responsible consumption and production), SDG 13 (Climate action), and SDG 15 (Life on land). It is a very crucial technology to reduce the greenhouse effect and global warming. The magnetic refrigeration technique utilizes solid state material avoiding the chemicals used in prevailing refrigeration or cooling technology. The magnetic refrigerant is a solid material that has zero direct Ozone Depletion and Global Warming Potential. The cooling efficiency up to 60% could be achieved as compared to the 40% efficiency with conventional vapor compression plant. In magnetic refrigerator, water could be used as a secondary fluid. As compared to the vapor compression plants, the energy consumption of magnetic refrigerators is much low due to high efficiency [7, 8]. Thus, it is a crucial technology for reducing greenhouse gases leading to sustainable climate. However, the commercialization of the technology is still not possible because of certain challenges that need to be overcome. The major challenge is the appropriate solid state magnetic refrigeration material. The challenges related to the present magnetic refrigerant material are: limited availability, high cost, and environmental friendliness of the material [9, 10]. All these three challenges lead to the requirement of sustainable material for the sustainability of environment-friendly magnetic refrigeration technology. Thus, development of new magnetic refrigerant material is the need of the hour. The magnetocaloric effect drives the magnetic refrigeration process. The material that exhibits conventional magnetocaloric effect could be cooled by demagnetization process and heated by magnetization process as shown in Figure 17.1(a). On the other hand, there is a class of material that cools down upon magnetization and heats up upon demagnetization exhibiting inverse magnetocaloric effect as shown in Figure 17.1(b).

Both, magnetocaloric effect and inverse magnetocaloric effect, could be used in thermal conditioning, industrial refrigeration, and low-temperature applications.

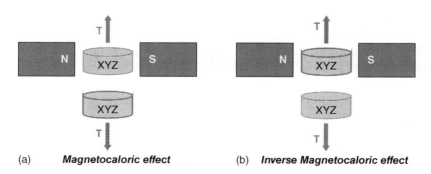

FIGURE 17.1 Process of cooling and heating using magnetic field for (a) magnetic effect and (b) inverse magnetic effect.

Heusler Alloys

Two types of phase transitions are exhibited by magnetocaloric material: first-order phase transition and second-order phase transition. The conventional magneto-caloric effect occurs through both kind of phase transition. The Gd- and Gd-Y-based alloys undergo second-order phase transition and Gd-Si-Ge- and Mn-Fe-P-As-based alloys undergo first-order phase transition giving rise to the conventional magne-tocaloric effect. The Fe-Rh alloys and Ni_2-Mn-In Heusler alloys exhibit first-order transition that gives rise to inverse magnetocaloric effect. The important parameters for magnetic refrigeration, like adiabatic change in temperature (ΔT_{ad}), and refrig-eration capacity (RC) with 2T applied magnetic field, are mentioned in Table 17.1. The ΔT_{ad} with 2T magnetic field is comparable for conventional first-order phase transition and inverse magnetocaloric effect using first-order phase transition. The ΔT_{ad} for Ni-Mn-In-based Heusler alloys is found to be very less as compared to that for above-mentioned alloys. The RC at 2T is very low for Ni-Mn-In-based Heusler alloys as compared to that for other alloys and is shown in Table 17.1.

The Gd-based or Fe-Rh-based alloys exhibit very attractive properties for mag-netic refrigeration. However, the commercial acceptability is still a challenge [9, 10]. The main reason is being the sustainability of material. The sustainability of material depends on the criticality index. The criticality index is defined mainly by Supply Risk and Economic Importance (EI) and (SR) [11]. The EI is defined as

$$EI = \sum_{s} \left(B_s \times V_s \right). SI_{EI} \tag{17.1}$$

B_s = As per NACE Rev. 2 (2-digit level) sector: raw material end use shape;
V_s = As per NACE Rev. 2 (2-digit level): the sector's VA
SI_{EI} = the substitution index of economic alloy important raw material;
s is sector.

TABLE 17.1
The ΔT_{ad}, RC, Criticality Index, and Phase Transition Temperature Comparison for Magnetocaloric Effect (Second-Order and First-Order) and Inverse Magnetocaloric Effect (First-Order) [9]

Material	2T Applied Magnetic Field		Criticality Index	Phase Transition Temperature (K)
	$\Delta T_{ad}(K)$	RC (JKg^{-1})		
Gd	4.7	226.9	5	292
$Gd_{1.97} Y_{0.03}$	4.1	214.5	5	284
$Gd_{1.93} Y_{0.07}$	3.8	206.1	5	278
$Gd_{0.2} Si_{0.08} Ge_{0.08}$	4.9	195.3	6	261.7
$Mn Fe P_{0.55} As_{0.45}$	3.1	88.1	6	315
$Fe_{0.98} Rh_{1.02}$	6.2	153.4	6	316.5
$Ni_{2.01} Mn_{1.4} In_{0.59}$	1.2	23.7	4	301
$Ni_{1.98} Mn_{1.42} In_{0.59}$	1.6	30.5	4	267.7
$Ni_{1.99} Mn_{1.4} In_{0.61}$	0.9	8.5	4	233.6

The SI_{EI} is calculated by the formula:

$$SI_{EI} = \sum_i \sum_a SCP_{i,a} * Sub-share_{i,a} * Share_a \qquad (17.2)$$

Where, i is individual substitute material;
a is individual application of the candidate material;
SCP= cost performance parameter of substitute;
Share = in an end-use application, the share of the raw materials;
Sub-share = within each application, the sub-share of each substitute
    The supply risk is defined as

$$SR = \left[\left(HHI_{WGI,t} \right)_{GS} \cdot \frac{IR}{2} + \left(HHI_{WGI,t} \right)_C \left(1 - \frac{IR}{2} \right) \right] \cdot \left(1 - EoL_{RIR} \right) \cdot SI_{SR} \quad (17.3)$$

Where:
SR= risk of supply;
GS= global supply;
C= actual sourcing of the supply to a country;
HHI= Herfindahl-Hirschman Index;
WGI= scaled World Governance Index;
t= trade parameter adjusting WGI;
IR= import reliance;
EoL_{RIR}= recycling input rate at the end-of-life;
SI_{SR}= related supply risk substitution index.

Other parameters that define the criticality are vulnerability to supply restriction and environmental implications. The assessment of criticality index shows that the Gd-based alloys and Fe-Rh alloys are highly critical material, whereas, Ni-Mn-In alloys are medium in criticality as shown in Table 17.1. Further, the magnetocaloric effect driven by first-order phase transition has lower power consumption than that driven by second-order phase transition [8]. Moreover, the Ni-Mn-based Heusler alloys have high recyclability because constituent elements are almost 100% recyclable [12–14]. Thus, the Ni-Mn-In alloys are commercially attractive materials because they are low-cost, low-toxic, and recyclable. These properties categorize them as sustainable material.

17.2 OTHER SECTIONS AND WORK

17.2.1 HEUSLER ALLOYS

The Heusler alloys emerged as multifunctional material because they exhibit plethora of properties that are attractive for various applications [15]. Figure 17.2 shows the range of Heusler alloy properties that are actively explored for various applications.
    The Heusler alloys are of two types: half-Heusler alloy (XYZ) and full-Heusler alloy (X_2YZ). The full Heusler alloys (X_2YZ) are known to exhibit inverse magnetocaloric effect. The most useful composition consists of X as Ni, Y as Mn and

Z as Sn or In. Interestingly, the stoichiometric Ni$_2$MnZ (Z = Sn or In) does not exhibit magnetocaloric effect, while the off-stoichiometric Ni$_2$Mn$_{1+x}$Z$_{1-x}$ (Z = Sn or In) undergo reversible martensitic phase transition with temperature and magnetic field and exhibits inverse magnetocaloric effect. The high-temperature phase, known as Austenitic phase, has L2$_1$ crystal structure. The L2$_1$ crystal structure has atoms at X(0,0,0), Y$\left(\frac{1}{4},\frac{1}{4},\frac{1}{4}\right)$, X$\left(\frac{1}{2},\frac{1}{2},\frac{1}{2}\right)$, and Z$\left(\frac{3}{4},\frac{3}{4},\frac{3}{4}\right)$ positions. The Heusler alloys are reported to have localized moment with magnetic moment mainly localizing on Mn. The minority spin electrons are excluded from the Mn 3d states that result in localized character. The magnetism is driven by RKKY type indirect exchange interaction. The ferromagnetic behavior is obtained when X-conduction electrons mediate the indirect exchange interaction. The Z-conduction electrons mediated indirect exchange interaction can exhibit ferromagnetism or antiferromagnetism. The fermi-level position in the p-d hybridized states of Mn-Z results in either ferromagnetism or antiferromagnetism [16–20].

17.2.1.1 Applications of Heusler Alloys

The Heusler alloys emerged as attractive multifunctional materials providing boundless possibilities for applications according to the properties [21, 22]. The application-oriented properties are shown in Figure 17.2. The applications are shown in Figure 17.3.

The properties are tunable to the change in composition of the Heusler alloys, providing boundless possibilities for applications. The half-Heusler alloys are most attractive for thermoelectric devices and thermocouples. Many half-Heusler alloys

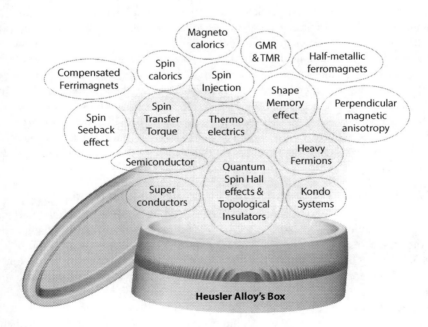

FIGURE 17.2 The properties exhibited by Heusler alloys.

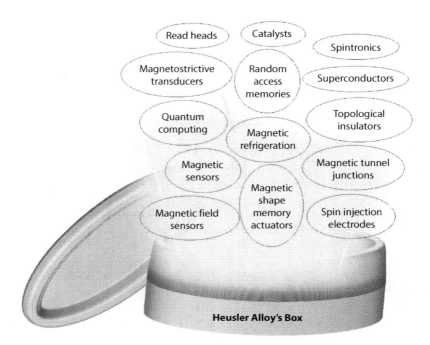

FIGURE 17.3 The range of applications of Heusler alloys.

are identified as topological insulators and have potential applications in quantum computing. The full-Heusler alloys (X_2YZ) with X = Co have high potential in spintronics, random access memory, magnetic tunnel junctions, read head, spin injection electrodes, and magnetic field sensor. The Ni_2-based full-Heusler alloys have potential applications as magnetic shape memory actuators, magnetic sensors, magnetostrictive transducers, and magnetic refrigeration. Some of the full-Heusler alloys are identified as effective catalysts. The Pd_2-based full-Heusler alloys are explored for superconductivity. Recently, the research focused on Ni_2-Mn-Z-based full-Heusler alloys has increased due to their high potential for environmental friendly magnetic refrigeration technique, which is considered to be one of the sustainable solution for climate change.

17.2.1.2 Heusler Alloys with Reversible Martensitic Transition

The Ni_2-Mn-Z (Z = Sn or In)-based full-Heusler alloys undergo reversible martensitic transformation. The cooling of the alloy leads to martensitic phase accompanied by the change in crystal structural phase and magnetic phase as shown in Figure 17.4. The strong magneto-structural coupling is observed during martensitic transition.

As discussed, the high-temperature Austenite phase has $L2_1$ crystal structure. The martensitic phase exhibits either tetragonal, orthorhombic, or modulated orthorhombic structure. The magnetic phase depends on the composition of the Heusler alloy [23–25]. As discussed, the most interesting composition for magnetic refrigeration is off-stoichiometric $Ni_2Mn_{1+x}Sn_{1-x}$ and $Ni_2Mn_{1+x}In_{1-x}$. In this chapter, we focus

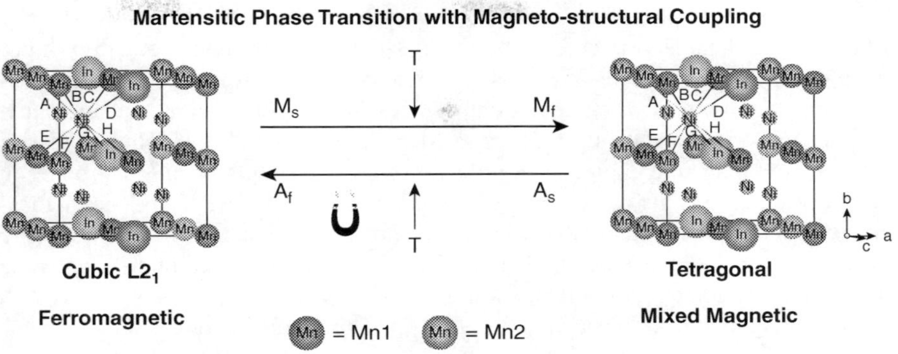

Heusler Alloys

Martensitic Phase Transition with Magneto-structural Coupling

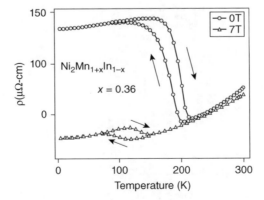

FIGURE 17.4 The temperature and magnetic field induced reversible martensitic transition in $Ni_2Mn_{1+x}Z_{1-x}$ (Z = Sn or In) Heusler alloys.

on the $Ni_2Mn_{1+x}In_{1-x}$ and Co-doped $Ni_2Mn_{1+x}In_{1-x}$, Heusler alloys. The $Ni_2Mn_{1+x}In_{1-x}$ alloys undergo magnetic phase transformation from ferromagnetic to mixed magnetic phase (ferro + anti − ferro). Upon heating of the alloys, the crystal structure reverses back to $L2_1$ crystal structure along with recovery of magnetic phase. The martensitic phase transition occurs over a range of temperature. The onset of magneto-structural martensitic transition, during cooling, is known as martensitic start (M_s). The completion of the martensitic transition is known as martensitic finish (M_s). While heating for reverse martensitic phase transition, the onset of transition is known as Austenitic start (A_s) and completion of transition to Austenitic phase is known as Austenitic finish (A_f). The martensitic phase transition is first-order phase transition, and hysteresis is observed during cooling and heating as shown in Figure 17.5 [26].

FIGURE 17.5 With or without magnetic field, the resistivity vs temperature behavior of $Ni_2Mn_{1+x}In_{1-x}$ (x = 0.36). The first-order martensitic transition shows hysteresis during cooling and heating (marked by arrows). The martensitic transition shifts to lower temperature with magnetic field [26].

The martensitic phase structural stability depends upon the Ni-Mn-In Heusler alloy composition. The stoichiometric composition Ni$_2$MnIn does not exhibit martensitic transition. However, the off-stoichiometric Ni$_2$Mn$_{1+x}$In$_{1-x}$ undergo martensitic transition in a narrow range of composition. The original Mn at Mn site (Mn1) couples antiferromagnetically with extra Mn at the In site (Mn2). This gives rise to the mixed magnetic phase. The antiferromagnetic coupling becomes stronger in martensitic phase. This narrow range of composition gives rise to the magnetocaloric effect for useful magnetic refrigeration application. The reversible martensitic transformation could also be obtained by applying magnetic field. Interestingly, the shift in martensitic transition temperature toward low temperature is observed with applying magnetic field as shown in Figure 17.5. This phenomenon results in inverse magnetocaloric effect in Ni$_2$Mn$_{1+x}$In$_{1-x}$ and Co-doped Ni$_2$Mn$_{1+x}$In$_{1-x}$, which is utilized for magnetic refrigeration [27–29].

17.2.1.3 Magnetic Refrigeration Using Heusler Alloys

The Ni$_2$Mn$_{1+x}$In$_{1-x}$ and Co-doped Ni$_2$Mn$_{1+x}$In$_{1-x}$ Heusler alloys exhibit inverse magnetocaloric effect. According to Figure 17.1(b), the applied magnetic field cools down the material that in turn cools the surrounding. The inverse magnetocaloric effect is influenced by the total entropy change of the material with applied magnetic field. The variation of total entropy as a function of temperature, with or without applied magnetic field, for inverse magnetocaloric effect in these Heusler alloys is shown in Figure 17.6. The solid line in Figure 17.6 depicts the change in total entropy upon martensitic transition without magnetic field. The entropy decreases in martensitic phase. The hysteresis is observed in entropy during reverse cycle. The solid line (without applied magnetic field) shifts to lower temperature as shown by dashed line (with applied magnetic field). This shifting of martensitic transition toward low temperature with magnetic field increases the entropy of material (ΔS_T) isothermally

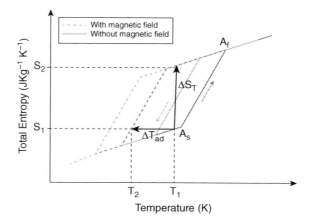

FIGURE 17.6 The adiabatic cooling and isothermal entropy change with applied magnetic field for Heusler alloys that exhibit inverse magnetocaloric effect and in martensitic transition temperature shift with applied magnetic field.

Heusler Alloys

and cools the material (ΔT_{ad}) adiabatically. Thus, if the martensitic phase of the material is around room temperature, then application of magnetic field to the material, at room temperature, will lower the temperature. This property could be used to cool the surrounding material. Thus, $Ni_2Mn_{1+x}In_{1-x}$ and Co-doped $Ni_2Mn_{1+x}In_{1-x}$ Heusler alloys are potential candidate for magnetic refrigeration. The ΔT_{ad} depends on the ΔS_T following the formula [9]:

$$\Delta T_{ad} = \frac{-TVS_T}{C_p} \qquad (17.4)$$

Where C_p is the specific heat of the material. Another important property is refrigeration capacity (RC) that defines the practical application of the material for magnetic refrigeration. The RC also depends on the ΔS_T using the following equation [27]:

$$RC = \int_{T_1}^{T_2} [\Delta S_T]_{\Delta B} \, dT \qquad (17.5)$$

Where, ΔB is change in magnetic field. Thus, the ΔS_T is a crucial factor that is responsible for magnetic refrigeration.

If the Curie temperature is above martensitic transition temperature, then the martensitic transition is from high magnetization to low magnetization state.

This increases the entropy of the material following the Clausius-Clapeyron relation:

$$\Delta S = \frac{\Delta M}{\Delta T / \Delta B} \qquad (17.6)$$

where, ΔM and ΔS are the magnetization and entropy change between low and high temperature phases. Also, the occurrence of martensitic transition temperature and Curie temperature at same temperature is not desirable. At Curie temperature, the $\Delta M > 0$ and is second-order phase transition that gives rise to negative ΔS, i.e. conventional magnetocaloric effect. At martensitic transition, $\Delta M < 0$ and is first-order magneto-structural phase transition that gives rise to positive ΔS, i.e. inverse magnetocaloric effect [25].

17.2.1.4 Preparation and Properties of Heusler Alloys for Magnetic Refrigeration

The $Ni_2Mn_{1+x}In_{1-x}$ and Co-doped $Ni_2Mn_{1+x}In_{1-x}$ Heusler alloys are prepared using arc melting technique, which is a common technique for recycling steel and aluminum. The preparation process utilizes the already existing technique, which makes also the process sustainable. The off-stoichiometric Heusler alloys of $Ni_2Mn_{1+x}In_{1-x}$ and $Ni_{1.8}Co_{0.2}Mn_{1+x}In_{1-x}$ were prepared using high purity ($\geq$99.99%) elements of appropriate amount within $\pm$0.1 mg accuracy. The polycrystalline ingot was prepared using arc-melting furnace in the presence of argon atmosphere and water-cooled Cu crucible. To ensure homogeneity of alloy composition, the remelting was done 6–8 times. After which, the alloys were vacuum sealed (3–4×10^{-6} mbar) with

TABLE 17.2
The Intended and Actual Composition Using EDAX [27, 28]

Intended Composition	Actual Composition
$Ni_2Mn_{1.36}In_{0.64}$	$Ni_{2.01}Mn_{1.36}In_{0.63}$
$Ni_{1.8}Co_{0.2}Mn_{1.46}In_{0.54}$	$Ni_{1.81}Co_{0.22}Mn_{1.45}In_{0.52}$ (NCMI2)

Mo foil wrapping. The annealing of vacuum-sealed ingots was done at 1173 K for 24 hours followed by ice-water quenching [30]. For various characterization and measurements, the alloys were cut into appropriate sizes.

The energy dispersive X-ray analysis (EDAX) technique is usually used for deducing chemical composition. Table 17.2 shows the actual and intended composition of the Heusler alloys of our interest [27, 28].

The intended composition is within ±1% of actual composition. The homogeneity at various parts of the alloy is within ±3%. The characteristic transition temperatures are, in general, deduced from the analysis of differential scanning calorimetry (DSC) as shown in Figure 17.7 [27, 28].

Table 17.3 mentions the characteristic transition temperatures for $Ni_2Mn_{1+x}In_{1-x}$ and Co-doped $Ni_2Mn_{1+x}In_{1-x}$.

The Co doping decreases the martensitic transition temperature to around room temperature and increases the Curie temperature for $Ni_{1.81}Co_{0.22}Mn_{1.45}In_{0.52}$, whereas, the $Ni_2Mn_{1.36}In_{0.64}$ has martensitic transition temperature much below room temperature. Since the martensitic transition temperature is above T_c^A, the $Ni_2Mn_{1.42}In_{0.58}$ is in paramagnetic phase at room temperature. Thus, the Co-doped $Ni_2Mn_{1+x}In_{1-x}$ is more attractive for magnetic refrigeration. To ensure the formation of martensitic twin planes, the scanning electron microscopy (SEM) is done at room temperature and shown in Figure 17.8 for $Ni_2Mn_{1.42}In_{0.58}$ [31].

The width of the twin plane is within 2–4 μm. The SEM analysis confirms the absence of impurity phase in the material. The X-ray diffraction (XRD) analysis is undertaken for investigation of alloy crystal structure. As mentioned earlier, the Austenitic phase has $L2_1$ crystal structure. Figure 17.9 shows Rietveld refined crystal structure analysis of $Ni_2Mn_{1.36}In_{0.64}$ and NCMI2. The XRD is carried out at room temperature with CuK_α radiation. The lattice constant for $Ni_2Mn_{1.36}In_{0.64}$ is found

TABLE 17.3
The Characteristic Transition Temperatures for $Ni_2Mn_{1+x}In_{1-x}$ and Co-Doped $Ni_2Mn_{1+x}In_{1-x}$ As Deduced Using DSC [28]

Composition	M_S (K)	M_f (K)	A_S (K)	A_f (K)	T_c^A (K)
$Ni_2Mn_{1.36}In_{0.64}$	258	201	215	266	314
$Ni_2Mn_{1.42}In_{0.58}$	391	369	374	394	
$Ni_{1.81}Co_{0.22}Mn_{1.45}In_{0.52}$	335	292	312	349	397

Heusler Alloys

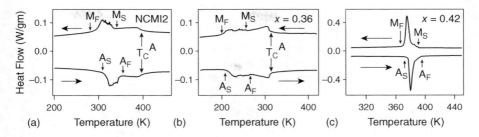

FIGURE 17.7 The characteristic transition temperatures (M_S, M_f, A_S, A_f, and T_c^A) for (a) $Ni_{1.81}Co_{0.22}Mn_{1.45}In_{0.52}$ [27] (b) & (c) $Ni_2Mn_{1+x}In_{1-x}$ [28].

to be 6.001 Å with cubic $L2_1$ structure [28]. However, there are also few martensitic peaks (marked as M in Figure 17.9(a)) observed. The grinding process induces residual stress. The process can shift the temperature of martensitic phase transition toward higher temperature. Thus, further annealing under vacuum is required to eliminate the stress-induced martensitic phase. Figure 17.9(b) shows the martensitic phase crystal structure of NCMI2 [27]. The characteristic martensitic transition temperature (Table 17.3) shows that at room temperature the NCMI2 has martensitic phase. Hence, XRD pattern shows the martensitic phase crystal structure. The crystal structure is found to be tetragonal a = b = 5.6 Å, c = 6.8 Å with Fmmm space group [27]. The alloys have expected characteristic transition temperatures, twin structure in martensitic phase good chemical composition and single-phase crystal structure. Thus, the quality of alloy allows it to be explored for magnetocaloric effect.

The change in entropy ΔS_m (due to applied magnetic field) as a function of temperature is obtained using magnetization curves and Maxwell relation [27]:

$$\Delta S_m(P, T, H)_{P,T,\Delta H} = \int_0^H \left(\frac{\partial M(P,T,H)}{\partial T}\right)_{P,H} dH \tag{17.7}$$

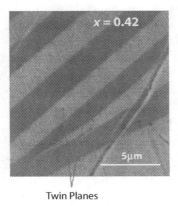

Twin Planes

FIGURE 17.8 The twin planes in martensitic phase of $Ni_2Mn_{1.42}In_{0.58}$ at room temperature using SEM [31].

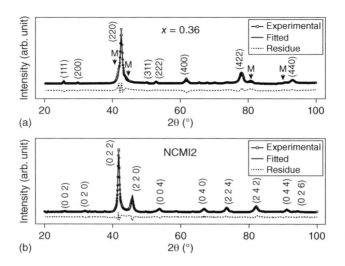

FIGURE 17.9 The room temperature XRD pattern with Rietveld refinement of (a) Ni$_2$Mn$_{1+x}$In$_{1-x}$ (x = 0.36) [31] and (b) NCMI2 (Ni$_{1.81}$Co$_{0.22}$Mn$_{1.45}$In$_{0.52}$) [27].

Where ΔT_{ad} and RC are calculated using equations (17.4) and (17.5), respectively. Also, the temperature difference at half maximum of the ΔS_m curve can be used to deduce ΔT_{ad} [27]. As observed from Table 17.1, the inverse magnetocaloric effect is exhibited by Ni-Mn-In below room temperature except Ni$_2$Mn$_{1.4}$In$_{0.56}$. Also, Ni-Mn-In Heusler alloys have very less ΔT_{ad} and RC as compared to other magnetic refrigeration alloys. The substitution of Co at Ni site increases the Curie transition temperature and martensitic transition temperature is around room temperature as evident from Table 17.3. Thus, Ni-Co-Mn-In is explored for possible application in magnetic refrigeration. Table 17.4 reports the various compositional combinations of Co-doped Ni$_2$Mn$_{1+x}$In$_{1-x}$ with their respective ΔS_m, ΔT_{ad} and RC values.

In general, the ΔT_{ad} of Co-doped Ni$_2$Mn$_{1+x}$In$_{1-x}$ is either better than the Gd-based alloys or comparable to Gd-based alloys at 2T. However, the RC at 2T is less than that of Gd-based alloys. The most attractive compositions are Ni$_{1.81}$Co$_{0.22}$Mn$_{1.45}$In$_{0.52}$ and Ni$_{1.81}$Co$_{0.22}$Mn$_{1.46}$In$_{0.51}$ that show ΔT_{ad} and RC comparable to Gd-based alloys. Also, the martensitic phase exists around room temperature for these two compositions with T_c^A higher than M_s. The Co doping decreases the e/a ratio causing decrease in M_S. Thus, Co-doped Ni$_2$Mn$_{1+x}$In$_{1-x}$ alloys are attractive and sustainable for the environment-friendly magnetic refrigeration technology.

17.3 CRITICAL DISCUSSION

17.3.1 Challenges of Heusler Alloys for Magnetic Refrigeration

The foremost challenge for sustainability of the magnetic refrigeration technique is material. The Co-doped Ni$_2$Mn$_{1+x}$In$_{1-x}$ is identified as environmental friendly, low cost and sustainable material for magnetic refrigeration. However, the scientific and

Heusler Alloys

TABLE 17.4

The Various Magnetocaloric Parameters (ΔS_m, ΔT_{ad}, and RC) Related to Magnetic Refrigeration Application Using Co-Doped $Ni_2Mn_{1+x}In_{1-x}$ Heusler Alloys

Composition	ΔS_m (JKg^{-1}K^{-1})	ΔT_{ad} (K)	RC (JKg^{-1})
$Ni_{1.8}Mn_{1.47}In_{0.52}Co_{0.2}$ [32]		−6.2 (2T)	
$Ni_{1.83}Mn_{1.46}In_{0.54}Co_{0.16}$ [32]		−8 (2T)	
$Ni_{1.81}Co_{0.2}Mn_{1.44}In_{0.54}$ [32]	16.7 (5T)		237 (5T)
$Ni_{1.8}Co_{0.2}Mn_{1.46}In_{0.53}$ [32]			198 (5T)
$Ni_{1.68}Co_{0.32}Mn_{1.5}In_{0.49}$ [33]	14.30 (6T)		549 (6T)
			57.3 (1T)
$Ni_{1.96}Mn_{1.56}Co_{0.04}In_{0.44}$ [34]	−1.09 (1.8T)		27.4 (1.8T)
$Ni_{1.8}Co_{0.2}Mn_{1.5}In_{0.5}$ [35]	30 (7T)	14 (7T)	267 (7T)
$Ni_{1.8}Co_{0.2}Mn_{1.5}In_{0.5}$ [36]	43 (5T)	−6.4 (5T)	
$Ni_{1.8}Co_{0.2}Mn_{1.46}In_{0.54}$ [37]	22 (5T)		
$Ni_{1.8}Co_{0.2}Mn_{1.47}In_{0.53}$ [38]	7 (5T)		
$Ni_{1.83}Mn_{1.46}In_{0.54}Co_{0.17}$ [39]	13.8 (1.95T)	−8 (1.95T)	
$Ni_{1.8}Co_{0.2}Mn_{1.48}In_{0.52}$ [40]	31 (5T)		317 (5T)
$Ni_{1.8}Co_{0.2}Mn_{1.46}In_{10.54}$ [40]	30 (5T)		488 (5T)
$Ni_{1.92}Co_{0.08}Mn_{1.4}In_{0.6}$ [41]	20.5 (5T)	−3.7 (5T)	268 (5T)
$Ni_{1.8}Mn_{1.47}In_{0.53}Co_{0.2}$ [38]	7 (4T)		
$Ni_{1.8}Co_{0.2}Mn_{1.46}In_{0.54}$ [38]	14.7 (1.3T)		
$Ni_{1.81}Co_{0.22}Mn_{1.46}In_{0.51}$ [27]	9.7 (1.5T)	8 (1.5T)	71.2 (1.5T)
$Ni_{1.81}Co_{0.22}Mn_{1.45}In_{0.52}$ [27]	5.4 (1.5T)	17 (1.5T)	66.1 (1.5T)

technological challenges are still remaining to be solved. The two important challenges related to material are:

i. To obtain martensitic transition temperature around room temperature for cooling below room temperature with applied magnetic field.
ii. To obtain cooling with sub-tesla magnetic field. The low magnetic field will enable the use of low-cost permanent magnets instead of high-priced superconducting magnets. This will contribute toward the sustainability of the technology.

The martensitic transition temperature around room temperature is highly dependent on the composition of the Heusler alloy. The Co doping at Ni site increases the Curie temperature through ferromagnetic coupling [42]. The martensitic transition temperature of $Ni_2Mn_{1.48}In_{0.52}$ and $Ni_{1.81}Co_{0.22}Mn_{1.45}In_{0.52}$ is around 457 K and 335 K, respectively. The Curie temperature of $Ni_2Mn_{1.48}In_{0.52}$ is around 314 K, while that of $Ni_{1.81}Co_{0.22}Mn_{1.45}In_{0.52}$ is around 397 K [27, 28]. The $Ni_2Mn_{1.48}In_{0.52}$ does not show magnetocaloric property, while $Ni_{1.81}Co_{0.22}Mn_{1.45}In_{0.52}$ exhibits inverse magnetocaloric property at room temperature. This is possible because the martensitic transition temperature (335 K) is below Curie temperature (397 K) for $Ni_{1.81}Co_{0.22}Mn_{1.45}In_{0.52}$ alloy composition, whereas the martensitic transition temperature (457 K) of $Ni_2Mn_{1.48}In_{0.52}$ alloy is above Curie temperature (314 K). Thus, the alloy composition should have martensitic transition temperature below Curie temperature. However, the large difference between Curie temperature and martensitic transition temperature results in decrease of inverse magnetocaloric effect [33]. Thus, difference between martensitic transition temperature and Curie temperature needs to be less. The characteristic transition temperatures also depend on the Mn and In concentration [43]. The increase in Mn2 increases the martensitic transition temperature. Therefore, the alloy composition needs to be explored for balance between the two above-mentioned criteria.

Another important challenge is to obtain the operation of magnetic refrigeration at sub-tesla magnetic field. The total entropy change is

$$\Delta S_T = \Delta S_{el} + \Delta S_m + \Delta S_l \tag{17.8}$$

The ΔS_{el} is electronic contribution, which is negligible. The ΔS_l is lattice contribution, which has positive magnitude. The ΔS_T for inverse magnetocaloric effect has positive magnitude. Thus, the ΔS_l plays the dominant role in increasing the inverse magnetocaloric effect. Therefore, the long range structural ordering is necessary for low magnetic field operation. The ΔS_m is magnetic entropy change involving two types of contributions that compete with each other:

$$\Delta S_{\mu B} + \Delta S_{sd} \tag{17.9}$$

The $\Delta S_{\mu B}$ is entropy change due to magnetic moment. The increase in magnetic moment increases positive entropy change. Thus, it is important to increase the magnetic moment. The increase in Mn2 decreases the net magnetic moment by inducing antiferromagnetism. Thus, the composition of the alloy should have less Mn2. The ΔS_{sd} contribution is due to spin disorder resulting in negative entropy change. Thus, spin disorder needs to be decreased in order to increase the net positive ΔS_m. The spin disorder might increase if short-range antiferromagnetic coupling exists. The short range antiferromagnetic coupling arises due to atomic site disorder. Thus, the structural ordering is very important to eliminate the possibility of short-range antiferromagnetic coupling. The change in entropy in these alloys shown in Figure 17.6 is, also, dependent on the shift of martensitic transition temperature with applied magnetic field. The increase in shift of martensitic transition with applied magnetic field increases the change in entropy. In order to increase the shift of martensitic transition with low magnetic field, the magnetic disorder needs to be reduced. This could, again, be obtained by reducing the structural disorder since the magnetic disorder is influenced by structural phase transition through magneto-structural coupling. The magnetocaloric properties influenced by magnetic and structural disorder will determine the extent of magnetic field requirement. Thus, understanding and controlling of disorder are very important to achieve low magnetic field application. The GE labs have already come up with Ni-Mn-based magnetic refrigerator prototype [44]. However, the

Heusler Alloys

size of the refrigerator needs to be reduced. This in turn is, again, dependent on the material property.

17.4 CONCLUSION

The magnetic refrigeration is an environment-friendly technique that is necessary for sustainable climate. However, it requires sustainable material for its commercial usefulness. The Heusler alloys that exhibit reversible martensitic transition along with the shift of martensitic transition temperature to lower temperature with applied magnetic field are attractive materials for magnetic refrigeration technology. They are sustainable materials since their criticality index is medium. However, the material research needs further input in order to optimize the material properties for room temperature and sub-tesla magnetic field operation. Presently, the attractive magnetocaloric properties of Co-doped $Ni_2Mn_{1+x}SIn_{1-x}$ are obtained at 1.5 T magnetic field around room temperature. The composition and structural and magnetic disorder need an in-depth research consideration for reducing the magnetic field requirement. Also, the composition influences the room temperature operation. Thus, a delicate balance of composition needs to be explored for attractive magnetocaloric properties at room temperature and with sub-tesla magnetic field. The sustainability of magnetic refrigeration depends on the attractive properties of Co-doped $Ni_2Mn_{1+x}SIn_{1-x}$, which is a sustainable material for industry use.

REFERENCES

1. Mokyr J. 2018. Building Taller Ladders. Finance Dev. 55: 32–35.
2. Pakes A and Sokoloff K.L. 1996. Science, Technology, and Economic Growth. Proc. Natl. Acad. Sci. USA 93: 12655–12657.
3. Martinez W. 2018. How Science and Technology Developments Impact Employment and Education. PNAS 115: 12624–12629.
4. Burke J., Bergman J. and Asimov I. 1985. The Impact of Science on Society. NASA SP-482.
5. Naikoo A.A., Thakur S.S., Guroo T.A. and Lone A.A. 2018. Development of Society under the Modern Technology—A Review. Scholedge Int. J. Bus. Policy Gov. 5: 1–8.
6. Giacomelli G. and Giacomelli R. 2005. Science, Technology and Society. Non-Accelerator Astroparticle Physics. Edited by R. A. Carrigan, Jr, G. Giacomelli and N. Paver, 277–284.
7. Botoc D., Siroux M. and Salceanu A. 2021. Magnetic Refrigeration: Emerging Technology for Sustainable Refrigeration. E3S Web Conf. 294: 03001.
8. Aprea C., Greco A. and Maiorino A. 2014. Magnetic Refrigeration: A Promising New Technology for Energy Saving. Int. J. Ambient Energy 37: 294–313.
9. Gottschall T., Skokov K.P., Fries M., Taubel A., Radulov I., Scheibel F., Dimitri Benke D., Riegg S. and Oliver Gutfleisch O. 2019. Making a Cool Choice: The Materials Library of Magnetic Refrigeration. Adv. Energy Mater. 9: 1901322.
10. Communication on the review of the list of critical raw materials for the EU and the implementation of the Raw Materials Initiative 2020. (https://ec.europa.eu/growth/sectors/raw-materials/specific-interest/critical_en).
11. European Commission 2017. Methodology for establishing the EU list of critical raw materials. ISBN 978-92-79-68051-9.

12. https://nickelinstitute.org/policy/nickel-life-cycle-management/nickel-recycling/
13. Chan K.H., Anawati J., Malik M. and Azimi G. 2021. Closed-Loop Recycling of Lithium, Cobalt, Nickel, and Manganese from Waste Lithium-Ion Batteries of Electric Vehicles. ACS Sustainable Chem. Eng. 12: 4398.
14. Hagelstein K. 2009. Globally Sustainable Manganese Metal Production and Use. J. Environ. Manage. 90: 3736.
15. Graf T., Felser C. and Parkin S.S.P. 2011. Simple Rules for the Understanding of Heusler Compounds. Prog. Solid State Chem. 39: 1–50.
16. Heusler F. 1903. Verhandl. Deut. Physik. Ges. 5: 219.
17. Oxley D.P., Tebble R.S. and Williams K.C. 1963. Heusler Alloys. J. Appl. Phys. 34: 1362.
18. Dubowik J., Gościańska I., Szlaferek A. and Kudryavtsev Y.V. 2007. Films of Heusler Alloys. Mater. Sci. (Poland) 25: 583.
19. Hamzic A., Asomoza R. and Campbell I.A. 1981. The Transport Properties of Heusler Alloys: 'Ideal' Local Moment Ferromagnets. J. Phys. F: Metal Phys. 11: 1441.
20. Kübler J., Williams A.R. and Sommers C.B. 1983. Formation and Coupling of Magnetic Moments in Heusler Alloys. Phys. Rev. B. 28: 1745.
21. Graf T., Parkin S.S.P and Felser C. 2011. Heusler Compounds—A Material Class with Exceptional Properties. IEEE Trans. Magn. 47: 367–373.
22. Kojima T., Kameoka S. and Tsai A.P. 2019. The Emergence of Heusler Alloy Catalysts. Sci. Technol. Adv. Mate. 20: 445–455.
23. Sutou Y., Imano Y., Koeda N., Omori T., Kainuma R., Ishida K. and Oikawa K. 2004. Magnetic and Martensitic Transformations of $NiMnX(X=In,Sn,Sb)$ Ferromagnetic Shape Memory Alloys. Appl. Phys. Lett. 85: 4358.
24. Brown P.J., Gandy A.P., Ishida K., Kainuma R., Kanomata T., Neumann K.-U., Oikawa K., Ouladdiaf B. and. Ziebeck K.R.A. 2006. The Magnetic and Structural Properties of the Magnetic Shape Memory Compound $Ni_2Mn_{1.44}Sn_{0.56}$. J. Phys.: Condens. Matter 18: 2249.
25. Planes A., Mañosa L. and Acet M. 2009. Magnetocaloric Effect and Its Relation to Shape-memory Properties in Ferromagnetic Heusler Alloys. J. Phys.: Condens. Matter 21: 233201.
26. Singh S., Glavatskyy I. and Biswas C. 2014. Field-cooled and Zero-field Cooled Magnetoresistance Behavior of $Ni_2Mn_{1+x}In_{1-x}$ Alloys. J. Alloys Compd. 615: 994.
27. Singh S., Glavatskyy I. and Biswas C. 2014. The Influence of Quench Atomic Disorder on the Magnetocaloric Properties of Ni–Co–Mn–In Alloys. J. Alloys Compd. 601: 108.
28. Singh S, Pal S and Maji C. 2021. The Resistivity Upturn in Ni–Mn–In Alloys Exhibiting Magnetic Field Induced Shift in Martensitic Transition with and without Co Doping. J. Alloys Compd. 856: 157405.
29. Maji C. 2017. Properties of Magnetic Shape Memory Alloys in Martensitic Phase. Curr. Sci. 112: 1390.
30. Singh S., Pal S and Biswas C. 2014. Disorder Induced Resistivity Anomaly in $Ni_2Mn_{1-x}Sn_{1+x}$. J. Alloys Compd. 616: 110–115.
31. Singh, S. 2015. Properties of Ni-Mn based heusler alloys with martensitic transition. PhD Thesis, Calcutta University. chrome-extension://efaidnbmnnnib-pcajpcglclefindmkaj/viewer.html?pdfurl=https%3A%2F%2Fwww.bose.res.in%2Flinked-objects%2Facademic-programmes%2FPhD%2520Thesies%2F2015%2FSandeep%2520Singh.pdf&clen=11218949&chunk=true
32. Zhenzhuang L., Zongbin L., Bo Y., Xiang Z. and Liang Z. 2018. Giant Low-field Magnetocaloric Effect in a Textured $Ni_{45.3}Co_{5.1}Mn_{36.1}In_{13.5}$ Alloy. Scripta Materialia 151: 61–65.

Heusler Alloys **339**

33. Cheng F., Gao L., Wang Y., Wang J., Liao X. and Yang S. 2019. Large Refrigeration Capacity in a $Ni_{42}Co_8Mn_{37.7}In_{12.3}$ Magnetocaloric Alloy. J. Magn. Magn. Mater. 478: 234–238.
34. Modak R. and Srinivasan A. 2019. Enhanced Room Temperature Magneto-caloric Effect in Ni-Mn-In Films with Fe/Co Substitution. J. Appl. Phys. 125: 085302.
35. Bourgault D., Tillier J., Courtois P., Maillard D. and Chaud X. 2010. Large Inverse Magnetocaloric Effect in $Ni_{45}Co_5Mn_{37.5}In_{12.5}$ Single Crystal above 300 K. Appl. Phys. Lett. 96: 132501.
36. Guillou F, Courtois P, Porcar L, Plaindoux P, Bourgault D and Hardy V. 2012. Calorimetric Investigation of the Magnetocaloric Effect in $Ni_{45}Co_5Mn_{37.5}In_{12.5}$. J. Phys. D: Appl. Phys. 45: 255001.
37. Chen L., Hu F.X., Wang J., Shen J., Zhang J., Sun J.R., Shen B.G., Yen J.H. and Pan L.Q. 2010. Magnetoresistance and Magnetocaloric Effect in Metamagnetic Alloys $Ni_{45}Co_5Mn_{36.5}In_{13.5}$. J. Appl. Phys. 107: 09A940.
38. Recarte V., Pérez-Landazábal J.I., Kustov S. and Cesari E. 2010. Entropy Change Linked to the Magnetic Field Induced Martensitic Transformation in a Ni–Mn–In–Co Shape Memory Alloy. J. Appl. Phys. 107: 053501.
39. Gottschall T., Skokov K.P., Frincu B. and Gutfleisch O. 2015. Large Reversible Magnetocaloric Effect in Ni-Mn-In-Co. Appl. Phys. Lett. 106: 021901.
40. Ghahremani M., Aslani A., Hosseinnia M. and Bennett L.H. 2020. Hysteresis Loss Reduction and Magnetocaloric Effect Improvement in the Ni-Co-Mn-In Alloys. AIP Advances 10: 015227.
41. Pathak A.K., Dubenko I., Xiong Y., Adams P.W., Stadler S. and Ali N. 2011. Effect of Partial Substitution of Ni by Co on the Magnetic and Magnetocaloric Properties of $Ni_{50}Mn_{35}In_{15}$ Heusler Alloy. J. Appl. Phys. 109: 07A916.
42. Ollefs K., Schöppner C., Titov I., Meckenstock R., Wilhelm F., Rogalev A., Liu J., Gutfleisch O., Farle M., Wende H. and Acet M. 2015. Magnetic Ordering in Magnetic Shape Memory Alloy Ni-Mn-In-Co. Phys. Rev. B 92: 224429.
43. Dubenko I., Samanta T., Pathak A.K., Kazakov A., Prudnikov V., Stadler S., Granovsky A., Zhukov A. and Ali N. 2012. Magnetocaloric Effect and Multifunctional Properties of Ni–Mn-based Heusler Alloys. J. Mag. Mag. Mater. 324: 3530.
44. https://www.ge.com/news/reports/not-your-average-fridge-magnet

18 Chemical Synthesis or Phytofabrication of Metal/Metal Oxide Nanoparticles

Amulya Giridasappa, Ismail Shareef M., and Gopinath S. M.

CONTENTS

18.1 Introduction 342
    18.1.1 Dimensions of Nanoparticles 342
    18.1.2 Major Forms of Nanoparticles 343
        18.1.2.1 Tin Dioxide Nanoparticles 343
        18.1.2.2 Properties of SnO_2 Nanoparticles 343
        18.1.2.3 Aluminum Oxide Nanoparticles 344
        18.1.2.4 Properties of Al_2O_3 Nanoparticles 344
        18.1.2.5 Silver Nanoparticles 344
        18.1.2.6 Properties of Ag Nanoparticles 344
        18.1.2.7 Cupric Oxide Nanoparticles 345
        18.1.2.8 Properties of CuO Nanoparticles 345
    18.1.3 Synthesis of Nanomaterials 345
        18.1.3.1 Chemical Synthesis of Nanoparticles 346
        18.1.3.2 Green Synthesis of Nanoparticles 346
18.2 Methodology 347
    18.2.1 Chemical Synthesis of SnO_2 and Al_2O_3 NPs 347
    18.2.2 Plant Materials 347
        18.2.2.1 Plant Extract Mediated Green Synthesis of Ag and CuO NPs 349
    18.2.3 Characterization of Synthesized Nanoparticles 350
18.3 Results and Discussion 352
    18.3.1 UV-Vis Spectroscopic Analysis 352
    18.3.2 FTIR Spectrum 353
    18.3.3 XRD Analysis 355
    18.3.4 SEM and EDX Analysis 356
    18.3.5 TEM Analysis 357
    18.3.6 BET Analysis 358
18.4 Conclusion 360
References 360

DOI: 10.1201/9781003297772-18

18.1 INTRODUCTION

Nanotechnology is fast emerging as state-of-the-art technology in current epoch that gives a discourse of one billionth of a meter in measurement (i.e., 1 nm = 10^{-9} m) and consociates to the fabrication of nanostructures felicitous in science and modern technology of which nanoparticles serve as the central fundamental element with size of 1–100 nm [1, 2]. The word 'nano' was introduced using the Greek term 'nanos' that is synonym with tiny, and further-on particles possess remarkable and altered physical and chemical features as consequence of their distinct electronic configuration; ample reactive accessible surface area; along with quantum of magnitude properties [3]. History related to nanomaterials dates back as early as 4500 years before, when nanofibers in the form of asbestos were utilized, then by Egyptians who synthesized lead sulfide nanoparticles for dying and also developed synthetic pigment known as 'Egyptians blue', which was used by many countries for decorative reasons [4]. During 1950s, the ideas and concepts got popularized from the talk 'There's Plenty of Room at the Bottom' by eminent physicist Richard Feynman at 'American Physical Society' of 'California Institute of Technology' and the word was coined by Norio Taniguchi who was a Japanese scientist at Tokyo University of Science.

18.1.1 DIMENSIONS OF NANOPARTICLES

Nanoparticles are specified as 'zero-dimensional' nanomaterials and their rationale elucidate their existences, due to the matter of absoluteness by which their dimensions were within the nano range in contrast to one-dimension materials that possess an increased dimension beyond the nano range, example includes nanowires and nanotubes, and also to the two-dimensional materials that retain twofold enhancements of nano size, as observed in case of self-assembled monolayer films [5]. The investigators realized that these materials were valuable not until they got to know that their size could affect the physiochemical characteristics of material, like for instances their optical features, when we consider the 20–30 nm gold, platinum, silver, and palladium nanoparticles that acquire a typical aspect of hue of claret, yellowish gray, black, and dark black that get changed with differences in size and shape and these type of applications were well-utilized in bioimaging [6].

Like all the various domains of science, nature offers mankind with reservoir of ideas in furtherance of manufacturing progressive nanomaterials, indicating selective options for the researchers [7, 8]. Nanoparticles were primarily employed for the advancement of wide spectrum of prepared materials like carbon nanotube (CNT), quantum dots (QD), polyepoxides-sheathed CNTs, polymer-layered materials, super magnetic iron oxide nanoparticles (SPION), lipid-based nanoparticles, mesoporous ceramics, catalysts, cascade molecules, nanolayers, nanofibers, and enhanced nanocomposites [9]. The nanoparticles themselves are complex molecules and consequently incorporate three layers viz., (i) the outer surface layer that can be improvised using assortment of diminutive molecules, metal ions, super active materials, polymers, etc., (ii) the middle shell layer that are distinct substances in contrast to the core layer in every matter, and (iii) the inner core layer that necessarily occupies the central portion of the nanoparticles and are generally addressed as nanoparticles themselves.

18.1.2 Major Forms of Nanoparticles

Excluding crystalline forms and dimensions, nanoparticles were further subdivided into carbon-related, organic, and inorganic nanoparticles. Carbon-based nanoparticles, as the name suggests, are mainly composed of carbon and exist in forms like CNT, graphene, carbon nanofibers, fullerenes, and carbon black [10]. They are synthesized basically by the methods like laser ablation, arc discharge, and chemical vapor deposition (CVD). Organic nanoparticles are known due to their biocompatible and nontoxic composition also they are linked by non-covalent bonds. They might be present in the form of liposomes, micelles, or dendrimers that have a hollow center which is filled by drug of required interest [11]. This is a renowned application of drug delivery and is sensitive to thermal and electromagnetic radiations. The nanoparticles which are not composed of carbon are known as inorganic nanoparticles [12]. They can be either synthesized as metal or metal oxide nanoparticles, the former includes nanoparticles like gold, silver, palladium, copper, nickel, etc., and the latter includes nanoparticles like zinc oxide, magnesium oxide, calcium oxide, titanium oxides, etc. There is also the presence of composite-based nanoparticles which are integration of any of these two varieties of nanoparticles amongst the three types. Physicochemical properties will be distinct based on the type of nanoparticle synthesized. The characteristics, like optical, mechanical, suspension, diffusion, thermal, magnetic, and electrical, and other such are defined under physical properties, while the chemical properties comprised of reactivity, stability, and sensitivity that sheerly depend on external environment [13].

18.1.2.1 Tin Dioxide Nanoparticles

Tin is present both as divalent (Sn^{2+}) and tetravalent (Sn^{4+}) forms and its ore generally occurs as a type of tin oxide. In semiconductor metal-based nanoparticles, tin dioxide (SnO_2) having 3.6 eV of the broad, band gap energy at room temperature, and their intermixture with similar band gap energy were quite affirmatively considered for the reason of their exceptional optoelectronic, photocatalytic, sensing, and biological utilities [14]. They exist as n-type semiconductors which are known to possess intrinsic weakness by the means of available oxygen vacancies along with interstitial atoms that correspond to n-type carriers.

18.1.2.2 Properties of SnO_2 Nanoparticles

SnO_2 NPs are prepared using varied procedures such as sol-gel, solvothermal, precipitation, electrochemical, solid-state reaction, green synthesis, chemical vapor, or pulsed laser deposition (CVD/PLD), microwave irradiation, etc. [15]. The characteristics like elevated level of transparency within visible spectrum, immense physiochemical reactions between adsorbed ions, less temperature requirement, immense thermal stability until 500°C induce SnO_2 NPs as a favorable material and their numerous utilizations are ascribed to bulk structure, surface, and interface features [16]. The surface area of SnO_2 is advantageous and is present because of large number of surface active groups. Moreover, SnO_2 NPs are extensively implemented in the fields like sensors, pollution control, solar cells, ceramic coatings, photocatalysts, detectors, transparent electrodes, antimicrobials, etc. [17].

18.1.2.3 Aluminum Oxide Nanoparticles

Aluminum oxide (Al_2O_3) is an inorganic mineral compound which is amphoteric in nature and is also called by the name, alumina. Alumina is very well used in industrial supplies because of its unique features namely high strength, melting point, corrosion resistance, phase stability, reduced thermal conductivity, and good electrical insulation. Alumina exists in seven transition forms which is differentiated based on phases and classified using the Greek denominations like alpha (α), chi (χ), kappa (κ), gamma (γ), theta (θ), delta (δ), and eta (η) [18]. These forms are further recognized depending on the arrangement of oxygen anions into face-centered cubic phase which consists of γ, θ (monoclinic), δ (tetragonal or orthorhombic), η (cubic), and into hexagonal close-packed (hcp) phase that contains α (trigonal), κ (orthorhombic), and χ (hexagonal) [19].

18.1.2.4 Properties of Al₂O₃ Nanoparticles

Similar to SnO_2, Al_2O_3 NPs can be synthesized by a number of methods yet solgel and coprecipitation are generally used to prepare these nanoparticles. Alumina sustains the large surface areas and also holds appropriate surface acidic-basic characteristics which usually give rise to beneficial improvements in catalytic activities [20]. During neutral pH, Al_2O_3 NPs will have positive charge on outer surface and this reacts electrostatically with bacterial surface using hydrophobicity and polymer binding force thereby permitting strong adhesion between them [21] and this accounts for their antimicrobial potential.

18.1.2.5 Silver Nanoparticles

On considering the valuable nanoparticles in association with noble metal nanoparticles, silver nanoparticles (Ag NPs) have captured enormous consideration in science and technology by the reason of acquiring interesting characteristics like good catalysis, conductivity, and particularly as excellent medicinal and pharmaceutical agents [22]. This ancient metal is renowned across various countries worldwide for its therapeutic usage. Silver is noble but rare metal which is considerably harder, ductile, and malleable than gold and comes with great electrical and thermal conductivity and minimal contact resistance [23]. There are four oxidation states of silver and among them Ag^0 and Ag^+ are relatively stable. Since, World War I, silver is being utilized in the form of antiseptic to treat wound and other injuries; this motivated the scientists in discovering the functions of Ag NPs as antimicrobial operatives. During the 18th century, silver was employed to heal acne, ulcers, epilepsy, venereal, gonococcal, and ophthalmic infections [24, 25].

18.1.2.6 Properties of Ag Nanoparticles

By assimilation with macro silver correspondent, the Ag NPs demonstrate varied physical and chemical features and are principally assignable to their minute size, ensuing a remarkable surface area to volume ratio [26]. Ag NPs are prepared by different physical, chemical, and biological routes using defined reducing or stabilizing agents and fabricated into diverse shapes like spherical, rod, triangle, hexagonal, octagonal, flower-like, etc. [27]. Nanotechnology has inspired the development of Ag NPs featured as nonpoisonous to mankind and holds greater bacteriostatic potential,

Chemical Synthesis or Phytofabrication of Metal/Metal Oxide Nanoparticles 345

drug delivery, diagnostic and imaging, catalyst, analytical, antitumor, and therapeutics [28, 29]. Its scope is ever demanding in industries like food, textiles, chemicals, environment protection, diagnostics, forensics, and medicine. The process of action of Ag NPs in addition to interaction with biological entities such as microbes and cancer cells will be discussed in detail in subsequent chapters.

18.1.2.7 Cupric Oxide Nanoparticles

From the time of coinage, copper the noble metal which is readily available was famous for its ductility, malleability, thermal and electrical conductivity, and is less priced in comparison to the other noble metals [30]. Copper was the first ever metal known to humans and was also there essential form of diet. In recent period, copper is extensively used in smart devices as electric circuits in computers, mobiles, remotes, etc., and is immensely produced using photolithography technique as this procedure is convenient to produce in large-scale and high-quality integrated circuits [31]. Copper in the structure of nanowires is operated in nanoelectronics being implemented in magnetic devices, nanosensors, electron emitters, semiconducting antiferromagnetic materials, etc. [32].

18.1.2.8 Properties of CuO Nanoparticles

Copper is reactive toward oxygen at specified temperature and thus converts itself to copper oxide. The copper oxide produced will either be cupric oxide (CuO) or cuprous oxide (Cu_2O) and is classified as monoclinic structure. CuO NPs are also synthesized using different chemical, physical, and conservative biological procedures. A significant amount of atoms will exist on the surface because of the minute particle size of CuO NPs and this modifies the features comprising electronic energy and transitions, electron affinity, magnetism, solar energy conservation, superconductors, phase transformation, melting point, polymer relativity, and other organic compounds [33–35]. Furthermore, CuO NPs being a p-type semiconductor presents exorbitant capability of 1.85 eV band gap by their cost effectiveness, nontoxic, photocatalysis, and optical and biological propensities [36, 37]. Among the various theories for CuO NPs activity on microorganisms, usually comprises reaction between the formation of oxidative stress and the potential ability exhibited by them through cupric oxidation states [38] and are generally observed, and such related mechanisms are described in further sections.

18.1.3 Synthesis of Nanomaterials

In the present era, the distinct characteristics along with magnificent implementation of synthesized nanomaterials from transition metal oxides have drawn the eye of researchers for the reason of usage in health care and treatment [39]. On this note, the fabrication methods used to prepare these nanomaterials become crucial to understand. Synthesis involves constructing nanomaterials of defined size, shape, structure, and applications which are apart from their classifications. Synthesis is of either obtaining through large molecules also known as 'bottom-up' process by increment of particles via fine molecules gathering or through small molecules also known as 'top-down' process wherein the particles fragments into tiny powders [40].

The bottom-up approach involves methods like sol-gel, CVD, spinning, spray pyrolysis, etc., while top-down approaches involve laser ablation, nanolithography, sputtering, thermal decomposition, ball milling, etc.

The nanomaterials are perhaps synthesized by virtue of regulating basic characteristics, like thermal conductivity, resistivity, melting point, resistance against corrosion, density, charge capacity, etc., with no alteration in their chemical composition, in addition to taking advantage of such capabilities they can be conducive to superior energy efficient materials and inventions that didn't happen earlier [41]. From past many years, there is obvious approachability of many types of synthesis methods which were accessible to prepare them and based on their physical examination, there was an increment in functioning in contrast to their bulk nature for various utilities that is enhanced by their physical parameters by sustainable means.

18.1.3.1 Chemical Synthesis of Nanoparticles

The chemical method is the general technique used to prepare metal and also its derived nanoparticles that arised by reduction of metal precursors using most suitable reducing agents which are either organic or inorganic substances and are usually stabilized after reaction, due to maintenance of parameters viz., pH, temperature, duration, and concentration. Weak and strong reducing agents, like sodium citrate, ascorbate, sodium borohydride, urea, citric acid, etc., are used as stabilizing or capping agents during reaction [42]. These chemicals will cause reduction of precursor to metal/metal oxide nanoparticles by donating electrons to metal ions to form atoms that later get united. Furthermore, the nanoparticles avoid formation of agglomerates by using certain chemicals known as surfactants that consist of functional groups which prevent them from coming close to each other and thereby ensure the stability of synthesized nanomaterials.

18.1.3.2 Green Synthesis of Nanoparticles

Recently, there exists an increasing urge to establish an environmentally sustainable approach so as to eliminate the usage of chemicals which impose lethal and precarious effects on humans and for this purpose, a potential method has to be chosen from the available physical, chemical, and biological route of synthesis for nanomaterials [43]. Meanwhile, the reprehensible consequence of chemical synthesis has created the conditions for biological synthesis of nanoparticles based on their persistence, nonpolluting, and energy saving in addition to owning advantages toward numeric organic reactivity [44]. Thereby, the several noxious types of synthesis are displaced by the green nanotechnology for enormous applications [45]. The nanotechnology is steadily improving its prominence of synthesizing nanoparticles through biological routes, and researchers have advanced their study due to their phenomenal effects and distinct performance in the field of public health services and medication, the properties executed by the defined nanoparticles in conjunction with their mechanism of action have made themselves the most suitable prospects for medicinal remedies.

We can convert metal ions into nanoparticles by utilizing the organic macromolecules existing in medicinal plants which consist of numerous phytochemicals (like phenols, flavonoids, saponins, coumarins, alkaloids, lipids, legnins, etc.) that

Chemical Synthesis or Phytofabrication of Metal/Metal Oxide Nanoparticles **347**

are easily dissolve in water [46]. Exclusion of maintenance of mammalian cell culture and similar tedious procedures, effortless steps to scale for large production during synthesis and low cost are some of the various conveniences and advantages observed while employing green plants to synthesize nanoparticles [47]. β-D glucose was utilized for the first time to synthesize Ag NPs along with starch that behaved as capping agent, while the former functioned as reducing agent, and this experimentation was conducted by Raveendran and coworkers by the method involving green synthesis [48]. To enable advancement in biocompatible nanoparticles, multiple biomimetic procedures and techniques are being investigated. Certainly, the explorations on utilization in therapeutics are enhanced by broad gamut and this demands the requirement of naturally favorable nanotechnology.

18.2 METHODOLOGY

18.2.1 Chemical Synthesis of SnO_2 and Al_2O_3 NPs

For synthesis of SnO_2 NPs, the procedure explained by Dong et al. [49] was followed that involves the preparation of NPs by combustion method. Combustion is performed as per the propellant chemistry concepts; by calculating the total valency of oxidizing and reducing agents from their molecular weight and they ignite themselves with this composition at specified temperature. This typical process of synthesis of SnO_2 NPs includes combining the precursor, i.e., 1 mM of stannous chloride with 6.2 mM of nitric acid, drop by drop in a beaker containing 30 mL of deionized water. They were dissolved completely in the solution and were accompanied by addition of 8.1 mg of reducing agent like urea. This solution was magnetically stirred at a uniform speed for about 10–20 minutes to allow full dissolution of all chemicals present in them. Later, this mixture was filtered using Whatman grade 1 filter paper placed on the glass funnel and the filtrate collected in conical flask was poured to 100 mL silica crucible kept in the muffle furnace, set to 600°C temperature. Finally, the dark brown color powder got was the synthesized SnO_2 NPs and was further converted into smooth powder using the pestle and mortar.

Next, Al_2O_3 NPs were also prepared using the same process in accordance with propellant chemistry, but the precursor molecule here is aluminum nitrate. The method observed by Branquinho et al. [50] was used wherein 10 mM of precursor was dissolved until no residues were found in 30 mL of deionized water. It was continued with addition of 25 mM of urea to the solution and then allowed to mix thoroughly on magnetic stirrer for 10–20 minutes. The temperature of muffle furnace was set to 1200°C, while the mixture was filtered. Then, as usual, the crucible was kept in furnace for flaming to occur and within a few minutes, we got the white powder of Al_2O_3 NPs, which was further smoothened using mortar and pestle.

18.2.2 Plant Materials

In the present research work, the choice amongst the medicinal plants was made depending on their therapeutic value and habitual knowledge from reputed plant resource centers and institutes mainly focused on anticancer ability contents existing in them. We have short listed two ethno-pharmaceutical plants notably known by their binomial nomenclature as *Simarouba glauca* and *Celastrus paniculatus*. They were recognized, gathered, and got validated by the herbarium reports specified with the voucher number 120199 and 120198 from Foundation for Rehabilitation of Local Health Tradition (FRLHT), Bengaluru, Karnataka. Various known phytochemicals, abundant distribution, and potency to fight against diseases and disorders are the reasons which make these two trees very well known in several classifications of medicinal systems.

S. glauca DC, belonging to the class of angiosperms and commonly called by the name, like Paradise tree, Lakshmi taru, or bitterwood, is domestic to Central-South America and their surroundings. Their leaves were plucked on 27.04.2018 from FRLHT, Jarakabande Kaval of Bengaluru district at an altitude of 920 msl from ethno-pharmaceutical garden (Figure 18.1). The tree, grows up to 15–20 meters height and shade tolerant, has well-developed root system, produces edible fleshy fruits, and also checks soil erosion. Traditionally they were used for treatment against malaria, inflammation, parasites, diarrhea, digestive problems, hemorrhages, fever, protozoans, viruses, etc. [51, 52]. They incorporate secondary metabolites such as alkaloids, polypeptides, flavonoids, steroids, coumarins, anthraquinones, triterpene like quassinoids, and other constituents [53].

C. paniculatus Willd. are the woody climbers indigenous to Australia and many countries in Asia including India and have vernacular names like Jyothishmati, black oil plant, or intellect tree. The aerial parts of this medicinal herb (Figure 18.2) were

FIGURE 18.1 Photograph showing the *Simarouba glauca* tree and inset is the twig with fruits.

FIGURE 18.2 Photograph of *Celastrus paniculatus* climbers and inset are the images of aerial parts (left) and the plant found creeping on other trees for support (right).

collected from same place, garden of FRLHT, Jarakabande Kaval, Bengaluru, at an altitude of 920 msl on same day, i.e., 27.04.2018. It grows up to 20 meters with thin brown bark, oblong-elliptic leaves, yellow or greenish-white flowers bearing subglobose capsules dehiscing into three valves with three to six seeds present in them. This plant is used as a neuro tonic, memory enhancer, to cure inflammation, rheumatism, dysentery, facial paralysis, hemiplegia, gout, spasm including a few fungal growths [54]. It contains various secondary metabolites like triglycerides under which stearodiolein, triolein, palmito-oleopalmitin, palmito-oleostearin, palmitodiolein, palmito-oleolinolein sesquiterpene alkaloids, polyalcohols, phenolic triterpenoids, and quinone-methide are seen [55, 56].

18.2.2.1 Plant Extract Mediated Green Synthesis of Ag and CuO NPs

The leaves of SG and aerial parts of CP were rinsed thoroughly in water for several times and once with deionized water. The water content on them was reduced using clean cotton cloth and later they were spread on white cardboard sheets for drying in shade for 7–12 days. After drying they were ground into coarse powder using grinder and stored in separate air-tight containers named with labels for more than a month and used when in need. For preparation of extraction using water, 200 mL of deionized water was held in sterile beaker and allowed to heat until boiled. To this, 20 g of each plant material was added distinctively to separate beakers and let to boil for around half an hour. It was followed by cooling of prepared extracts and filtering using Whatman paper to eliminate undissolved residues. Later, the filtered extracts were stored in sterile screw cap bottles at 4°C and used within 2–3 days.

FIGURE 18.3 Filtration of leaf extract of *Simarouba glauca* (left) and aerial extract of *Celastrus paniculatus* (right) and inset are the pictures of respective residual extracts.

In order to synthesize Ag NPs from silver nitrate (AgNO$_3$), a beaker containing 100 mL of deionized water with 5 mM of AgNO$_3$ dissolved in it was taken, and to this 200 mL of the prepared extracts was added slowly with constant stirring at 70–80°C. It can be noted the conversion of AgNO$_3$ solution to Ag NPs upon addition of extracts specified by their color change to brown and later the resultant solution was centrifuged to remove impurities. The procedure as described by Saha et al. [43] was followed to synthesize Ag NPs. After centrifugation, they were allowed to dry to get powder form which was further smoothened using pestle and mortar to obtain fine nanoparticles.

In similar manner, the CuO NPs were synthesized by using leaf extract of SG and aerial extract of CP as reducing agents of copper nitrate trihydrate (Cu(NO$_3$)$_2$.3H$_2$O). To achieve this, the procedure detailed by Saif et al. [57] with some minor changes was employed. A beaker containing 100 mL of 10 mM of Cu(NO$_3$)$_2$.3H$_2$O solution to which 200 mL of the mentioned plant extracts were added slowly. The solution mixtures were agitated steadily in conjunction with boiling at 70–80°C for more than an hour and on succeeding, the formation of nanoparticles was evidenced by visualization of color change (blue to green); the resultant solution was cooled to room temperature. The so formed reddish brown CuO NPs were centrifuged and pellets collected were kept in muffle furnace for 5 minutes maintained at 400 ± 10°C and then converted to fine particles utilizing mortar-pestle and stored separately in air-tight containers.

18.2.3 Characterization of Synthesized Nanoparticles

To comprehend the crystallinity, chemical adsorption of functional groups, size, and surface morphology along with elemental composition of prepared nanoparticles, their characterization was executed which also helped in gaining information about

Chemical Synthesis or Phytofabrication of Metal/Metal Oxide Nanoparticles 351

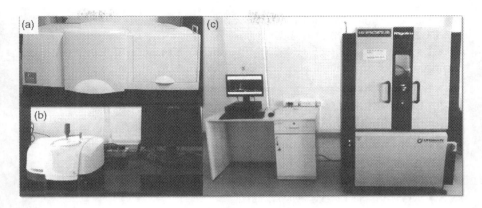

FIGURE 18.4 Instruments used for characterization (a) UV-Vis spectrophotometry, (b) FT-IR, and (c) XRD analyzers.

their biological effects. The generation of nanoparticles was initially verified using UV-visible spectroscopy which is of PerkinElmer Lambda-750 make and was operated at defined wavelength (i.e., 400–4000 cm^{-1}) and resolution. XRD of 2θ range from 10° to 70° was performed using Rigaku Ultima IV having Cu-Kα of 0.15406 nm. Next, PerkinElmer UATR Spectrum two model FTIR was employed for recognizing the functional groups, then the surface morphological analysis was carried out utilizing Hitachi SU1510 model SEM and additionally Ametek-EDAX was utilized to know the percentage of elements in prepared nanoparticles. To obtain

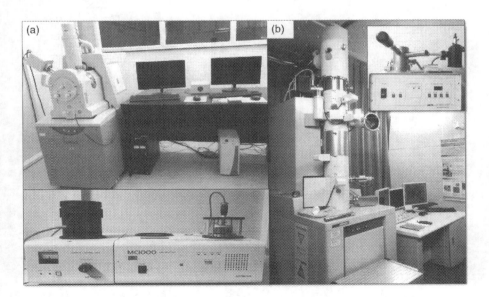

FIGURE 18.5 Instruments used for morphological characterization (a) SEM-EDX in the top with sputtering unit below and (b) TEM analyzer with drying pump inset.

better magnification of about 100,000× and using LaB$_6$ filament with ZrO$_2$/W (100) field emission gun, TEM (Jeol/JEM 2100F) images were ascertained. BELSORP-mini II was accessed to find out the specific surface area as in BET analysis where the nanoparticles were at temperature of 200°C before the measurement was taken.

18.3 RESULTS AND DISCUSSION

Based on characterization results, we convince the effectiveness and use of synthesis of desired nanoparticles, whether they meet the required physicochemical parameters and how they are relatable to utilizations for which they were fabricated.

18.3.1 UV-Vis Spectroscopic Analysis

The Al$_2$O$_3$ NPs spectra in Figure 18.6a indicate the adsorption edge at 220 nm; during the same time SnO$_2$ NPs were also checked for the maximum adsorption which was evident at 300 nm. The specified peaks noticed at 402 and 412 nm for Ag-SG and Ag-CP NPs (Figure 18.6b) were prominent because of observed change in color

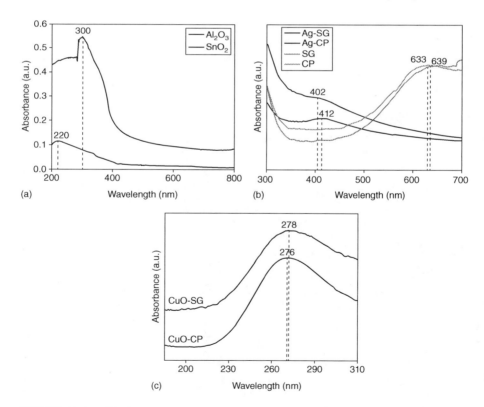

FIGURE 18.6 UV-Vis analysis of synthesized (a) Al$_2$O$_3$ and SnO$_2$ NPs, (b) Ag-SG, Ag-CP NPs, extracts of *Simarouba glauca* and extracts of *Celastrus Paniculatus*, and (c) CuO-SG and CuO-CP NPs.

Chemical Synthesis or Phytofabrication of Metal/Metal Oxide Nanoparticles   353

during reaction, this was consistent with, as reported by Priyadharshini et al. [58]. The visual hue change is the first indication of conversion of Ag⁺ to Ag⁰, also as in when the incubation period incremented, the color got much deeper. The leaf extract of SG and aerial extract of CP also showed adsorption maximum at 639 and 633 nm. With reference to green synthesized CuO NPs in the reaction mixture, the peaks at 276 nm and 278 nm for CuO CP and CuO SG NPs in UV-visible spectrum as seen in Figure 18.6c, and these results were in accord with other reports [59, 60].

18.3.2   FTIR SPECTRUM

The characteristic frequency peaks obtained for synthesized nanoparticles are represented in Figure 18.7, of which Al_2O_3 and SnO_2 NPs functional groups are evidenced in Figure 18.7a. For Al_2O_3 NPs, the transmittance exerted was found to be at 3435 cm⁻¹ and from 1000 to 400 cm⁻¹ range. The peak around 3435 cm⁻¹ is because of adsorption of O-H elongation and expansion demonstrated by alcohol respectively. In connection with metal oxides, we can observe bands from 1000 to 400 cm⁻¹ portion that

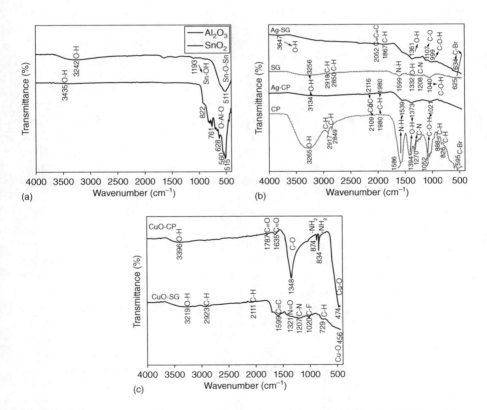

FIGURE 18.7   FTIR spectra of prepared (a) Al_2O_3, SnO_2, (b) Ag-SG, Ag-CP NPs, extract of *Simarouba glauca*, extract of *Celastrus paniculatus*, and (c) CuO-SG and CuO-CP NPs.

is attributed to stretching of O-Al-O group present in them [61]. Further considering the SnO_2 NPs spectrum, we can notice prominent peaks in regions of 3242, 1193, and 511 cm^{-1} amongst which 3242 cm^{-1} was consigned to the stretching of water or alcohol. The small peak in 1193 cm^{-1} region is ascertained because of oscillations caused by Sn–OH bending and similarly, the symmetric Sn–O–Sn vibrations of metal oxides were providently seen at 511 cm^{-1} wavenumber accordingly [62].

Proceeding with the examination of secondary metabolites that enacted, like stabilizers and reducers, and readily converted Ag^+ ions during synthesis was found by the FTIR spectrum (Figure 18.7b) involving transmittance. The notable peak at 3647, 3265, 3256, and 3134 cm^{-1} corresponds to vibrations encountered because of hydroxyl (-OH) group [63]. Next, due to stretching produced by alkenes (C-H) resulted in transmittance in 2918, 2917, 2850, and 2849 cm^{-1} and due to Ag NPs synthesis in comparison to the extracts they were retarded and found missing. This might be due to reason of stretching produced by phytochemicals like triterpenoid saponins that got vanished after synthesized to nanoparticles.

The minor peaks observed in the regions, like 2116 and 2109 cm^{-1}, were attributable to C≡C, while 2052 cm^{-1} is linked to the presence of phytoconstituents of extracts defined for C=C=C vibrations [64]. The other peaks found by aromatic C-H stretching were marked in regions like 1980 and 1867 cm^{-1}. Then corresponding to vibrations produced by N-H extension mode pertaining to amides, specifically amide II caused the peak appearance in 1500 cm^{-1}. Furthermore, the intense peaks seen in 1394, 1373, 1361, and 1332 cm^{-1} regions are retable to O-H bending of phenolic or aromatic amines [65]. The small peaks noticed in regions of 1270 and 1208 cm^{-1} were by the virtue of vibrations of amine (C-N). The ether group causing the stretching of C–O is appeared in the form of band at 1107 cm^{-1}. There was also transfer of C–OH bending in plant extracts' peaks observed in 1052 and 1040 cm^{-1} moved toward lower wavenumber 1021 and 999 cm^{-1} in synthesized nanoparticles [66] and this is mainly by the act of aldehydes. There was disappearance of peaks found in plant extracts at 888 and 826 cm^{-1} in amongst synthesized nanoparticles. At the end, the peaks notable in 634, 625, and 595 cm^{-1} regions are attributable to metal-O bond of alkyl halides that verify formation of silver ion in reaction mixture [67].

These functional groups of CuO NPs synthesized by extracts of *S. glauca* and *C. paniculatus* are identified utilizing FTIR spectroscopy (Figure 18.7c). Broad transmittance noticed in 3219 and 3396 cm^{-1} were connected with hydroxyl (-OH) groups, indicating existence of phenolics or amino acids. There was an asymmetrical bending of alkane and alkyne groups that appeared in 2923 and 2111 cm^{-1} [68]. Then the vibration stretching of carbonyl (C=O) belonging to amide group was observed in regions of 1787 and 1635 cm^{-1} whereas C=C reverberation of aromatic ring resulted in beam at 1599 cm^{-1}. Peaks at 1348, 1321, 1207, 1020, 874, 834, and 729 cm^{-1} had primarily appeared due to bending of several functional groups of phytoconstituents like polysaccharides or alcohol leading to C-O, nitro resulting in N-O, and other C-N, C-halo, etc. [69]. Finally, the metal vibrations encountered in 474 and 456 cm^{-1} were also prominent.

18.3.3 XRD ANALYSIS

The XRD pattern of Al_2O_3 and SnO_2 NPs are as demonstrated in Figure 18.8 (a & b) which signifies their crystallographic nature. The Al_2O_3 NPs indicated strong intensity peaks at 25.49°, 35.06°, 43.25°, and 57.41° respectively which are attributed to (012), (014), (113), and (116) miller indices [70]. This data was matching with the Joint Committee on Powder Diffraction Standards (JCPDS) number 00-010-0173, displayed rhombohedral crystallography and corresponding space group being R3c(167) in addition [71] and by using Scherrer's equation $D = K\lambda/\beta cos\theta$, the particle size was known to be 34.04 nm. Meantime, we also performed the XRD analysis of SnO_2 NPs which showed the cassiterite particularity along with $P4_2/mnm$ space group [72]. The diffraction peaks were comparable with the JCPDS card number 88-0287 that revealed highest intensities at 26.54°, 33.90°, and 51.80° respectively that was consistent with their miller planes like (110), (101), and (211) [73], and particle size was found out using the same Scherrer's formula that gave result as 24.57 nm.

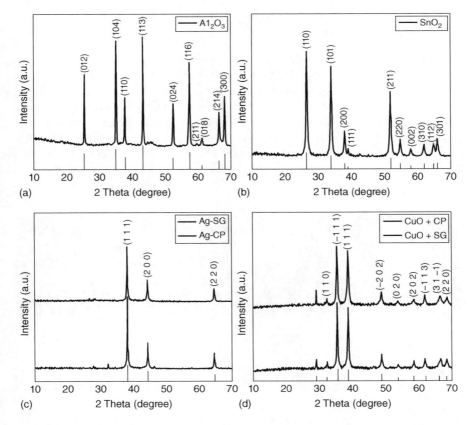

FIGURE 18.8 XRD plots of prepared (a) Al_2O_3, (b) SnO_2, (c) Ag SG and Ag CP, and (d) CuO SG and CuO CP NPs.

There is existence of precise ascribable beams in XRD pattern diffractogram amongst the synthesized Ag NPs prepared utilizing extracts of medicinal plants at 2θ = 38.21°, 44.21°, and 64.46° possessing the miller indices at (111), (200), and (220) planes and were defined as face-centered cubic form of crystallite structure (Figure 18.8c). The information of JCPDS card No. 87-0717 was in good accord with our findings and can be correlated with the spherical-shaped appearance [74]. Furthermore, we gathered details which indicated that it belongs to space group of 225: Fm-3m respectively and could find the particle size as 42.52 and 37.88 nm for Ag-SG and Ag-CP NPs.

Moving to peak produced by CuO NPs prepared from extracts of SG and CP indicated as 2θ = 32.50°, 35.50°, 38.68°, and 48.81° for CuO-SG NPs and 32.54°, 35.54°, 38.73°, and 48.66° for CuO-CP NPs that relates to (011), (11-1), (111), and (20-2) miller indices (Figure 18.8d). The sharp intensity and in-depth wide pattern of peak affirm the formation of crystal structure and this resembles with JCPDS card (No. 48-1548) [75, 76]. The pattern represents the defined monoclinic framework of CuO NPs along with space group 15: C12/c1, unique-b, cell-1 from which the crystalline size was found out to be 35 nm.

18.3.4 SEM AND EDX ANALYSIS

The information relating to surface morphology of Al_2O_3 and SnO_2 NPs was obtained by the images acquired using SEM (Figure 18.9 a1 & a2). The images of both the prepared nanoparticles were captured at different dimensions as indicated in figure for better comprehension. The Al_2O_3 NPs possessed partial depictions of spherical shapes observed at majority of portions of surface [77]. Their elemental mapping showed high intensity at 1.5 keV for 'Al' including the next higher intensity at 0.5 keV for oxygen. Subsequently reviewing the images of SnO_2 NPs (Figure 18.9 b1 & b2) we can observe immense filling of spherical formations of nanoparticles

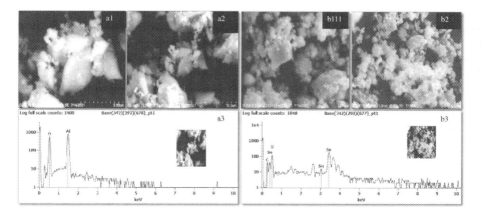

FIGURE 18.9 Different resolution images captured using SEM for Al_2O_3 (a1 & a2) and SnO_2 (b1 & b2) NPs and their elemental mapping by EDX (a3 & b3).

Chemical Synthesis or Phytofabrication of Metal/Metal Oxide Nanoparticles 357

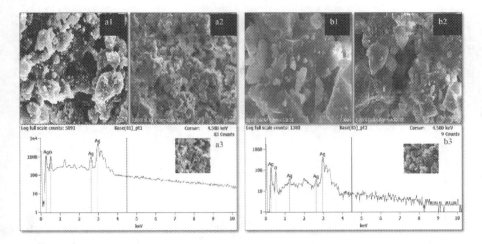

FIGURE 18.10 Different resolution images captured using SEM for Ag SG (a1 & a2) and Ag CP (b1 & b2) NPs and their elemental mapping by EDX (a3 & b3).

in almost every corner with some overlapping denoting aggregations [78]. Likewise their EDX spectra had high and more intensities of 'Sn' element at 3.4, 3.0, and 0.3 keV in addition to oxygen at 0.5 keV and carbon at 0.2 keV respectively, carbon was present due to the film material.

From the images captured using SEM revealed that there was scatter disposition of spherical particles, while this scattering was permissible due to their polarity and linking by electrostatic attraction existing between silver ions and extracts of medicinal plants. The images we have obtained using SEM were similar to as seen in [79]. The elemental composition of synthesized Ag NPs was scrutinized using EDX mapping (Figure 18.10 a3 & b3) by which it was possible to examine the biosynthesized NPs with their prominent indication of Ag element and also generous mapping of O element due to their participation during the reduction into NPs by the phytochemicals. The highest peak was observed for Ag element marked at 3 keV.

The SEM images of CuO NPs depicted spherical structure as shown in Figure 18.11 (a1, a2, and b1, b2). The outline of EDX mapping revealed elemental peaks of C, Cu, and O only. The mapping of copper element was observed in 1, 8, and 9 keV, while oxygen was noted at 0.5 keV and there was also the presence of elemental carbon. Such indications verify the qualitative and quantitative phytoconstituents of synthesized CuO NPs and these results were in resemblance to the research performed by Nasrollahzadeh et al. [80].

18.3.5 TEM Analysis

TEM analysis helps to acquire knowledge pertaining to the particles' average size, their distribution, shape, and crystal structure of nano range. The TEM images of chemically synthesized Al_2O_3 NPs denote that they were touching each other at their

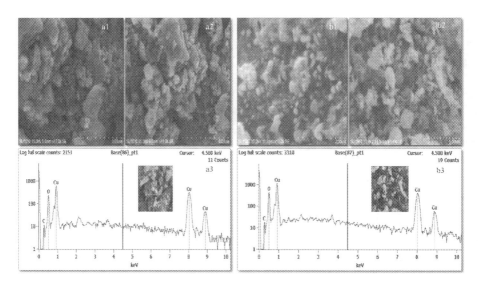

FIGURE 18.11 Different resolution images captured using SEM for CuO SG (a1 & a2) and CuO CP (b1 & b2) NPs and their elemental mapping by EDX (a3 & b3).

corners, this was in line with the observation made by Yalamaç et al. [81]. Moreover, on analyzing the measurement of particle size, it resulted about 30 nm in approximate. Thereafter, considering the shape morphology of SnO_2 NPs, it appeared to be spherical with not much overlapping of nanoparticles and this helps in defining their shape with 20 nm size [82]. Looking into the images captured we can conclude that most of nanoparticles were of spherical shaped along with smooth edges (Figure 18.12 c & d), although some were seen marginally ellipsoid and these findings go in hand with the explanation for monodispersed nature of Ag-SG and Ag-CP NPs. Meanwhile, for CuO NPs, there was agglomeration observed because of submergence in filaments, this is in good accord with findings of Sivaraj et al. [83].

18.3.6 BET Analysis

The BET plot indicating valuation of relative pressure on x-axis with respect to BET formula on y-axis is used to determine the surface area of only chemically synthesized Al_2O_3 and SnO_2 NPs and is depicted in Figure 18.13. They are usually obtained in terms of nitrogen adsorption and from BET theory values indicating if their value is high then surface area also increases and this is consequently related to scale-down size of nanoparticles [84, 85]. The values relating to BET analysis are revealed, surface area of Al_2O_3 NPs was found to be 57.30 m²/g, whereas for SnO_2 NPs was 106.46 m²/g respectively. Additionally, it is also evident from the data that SnO_2 NPs are more porous and have better surface area distribution when correlated to Al_2O_3 NPs.

Chemical Synthesis or Phytofabrication of Metal/Metal Oxide Nanoparticles 359

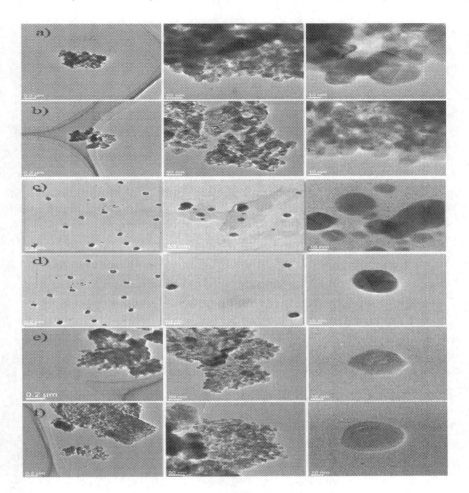

FIGURE 18.12 TEM images of synthesized (a) Al$_2$O$_3$, (b) SnO$_2$, (c) Ag SG, (d) Ag CP, (e) CuO SG, and (f) CuO CP NPs.

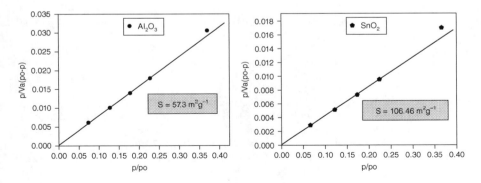

FIGURE 18.13 Surface area analysis of synthesized Al$_2$O$_3$ and SnO$_2$ NPs.

18.4 CONCLUSION

We favorably conducted the synthesis of nanoparticles both by chemical means to gather Al_2O_3 and SnO_2 NPs and by green approach to obtain Ag and CuO NPs employing leaf extract of *S. glauca* and aerial extract of *C. paniculatus*. Following synthesis, the physicochemical properties of materials were studied by characterization techniques. The crystal structure was identified to be rhombohedral for Al_2O_3 NPs, cassiterite for SnO_2 NPs, face-centered cubic pattern for green synthesized Ag NPs, and monoclinic for green synthesized CuO NPs. Al_2O_3 NPs demonstrated the size of 34.04 nm and SnO_2 NPs with 23.27 nm in addition to uniform shape visualizations, although the nanoparticles were slight in touch with each other. Morphological confirmation was obtained using SEM and TEM, while EDX affirmed their elemental constitutions. SnO_2 NPs possessed better specific surface area than Al_2O_3 NPs which display their high porosity. Ultimately, validating the evidential proof of requisite characteristic parameters of synthesized nanoparticles.

REFERENCES

1. Goutam SP, Avinashi SK, Yadav M, et al. 2018. Green Synthesis and Characterization of Aluminium Oxide Nanoparticles Using Leaf Extract of Rosa. Advanced Science, Engineering and Medicine 10:719–722. https://doi.org/10.1166/asem.2018.2236
2. Hu X, Zhang Y, Ding T, et al. 2020. Multifunctional Gold Nanoparticles: A Novel Nanomaterial for Various Medical Applications and Biological Activities. Frontiers in Bioengineering and Biotechnology 8:1–17. https://doi.org/10.3389/fbioe.2020.00990
3. Malik B, Pirzadah TB, Kumar M, Rehman RU. 2017. Biosynthesis of Nanoparticles and Their Application in Pharmaceutical Industry. In: Metabolic Engineering for Bioactive Compounds. Springer Singapore, Singapore, pp 331–349.
4. Jeevanandam J, Barhoum A, Chan YS, et al. 2018. Review on Nanoparticles and Nanostructured Materials: History, Sources, Toxicity and Regulations. Beilstein Journal of Nanotechnology 9:1050–1074. https://doi.org/10.3762/bjnano.9.98
5. Murthy Shashi K. 2007. Nanoparticles in Modern Medicine: State of the Art and Future Challenges. International Journal of Nanomedicine 2:129–141.
6. Khan I, Saeed K, Khan I. 2019. Nanoparticles: Properties, Applications and Toxicities. Arabian Journal of Chemistry 12:908–931. https://doi.org/10.1016/j.arabjc.2017.05.011
7. Somashekharappa KK, Rajendrachari S. 2022. Sustainable Development Information Management of Carbon Nanomaterial-based Sensors. In: Carbon Nanomaterials-Based Sensors. Elsevier, pp 3–12.
8. Shetty M, Manickavasakam K, Kulal N, et al. 2021. Bismuth Oxycarbonate Nanoplates@α-Ni(OH)$_2$ Nanosheets 2D Plate-on-sheet Heterostructure as Electrode for High-performance Supercapacitor. Journal of Alloys and Compounds 860:1–12. https://doi.org/10.1016/j.jallcom.2020.158495
9. Mousavi SM, Hashemi SA, Ghasemi Y, Atapour A, Amani AM, Dashtaki AS, Babapoor A, Arjmand O. 2018. Green Synthesis of Silver Nanoparticles toward Bio and Medical Applications: Review Study. Artificial Cells, Nanomedicine, and Biotechnology 46:S855–S872. https://doi.org/10.1080/21691401.2018.1517769
10. Patel KD, Singh RK, Kim H-W. 2019. Carbon-based Nanomaterials as an Emerging Platform for Theranostics. Materials Horizons 6:434–469. https://doi.org/10.1039/C8MH00966J

Chemical Synthesis or Phytofabrication of Metal/Metal Oxide Nanoparticles 361

11. Wakaskar RR. 2018. General Overview of Lipid–Polymer Hybrid Nanoparticles, Dendrimers, Micelles, Liposomes, Spongosomes and Cubosomes. Journal of Drug Targeting 26:311–318. https://doi.org/10.1080/1061186X.2017.1367006

12. Parijat P, Mandeep D. 2016. A Brief Review on Inorganic Nanoparticles. Journal of Critical Reviews 3:18–26.

13. Halligudra G, Paramesh CC, Mudike R, et al. 2021. Pd II on Guanidine-Functionalized Fe_3O_4 Nanoparticles as an Efficient Heterogeneous Catalyst for Suzuki–Miyaura Cross-Coupling and Reduction of Nitroarenes in Aqueous Media. ACS Omega 6: 34416–34428. https://doi.org/10.1021/acsomega.1c04528

14. Sudhaparimala S, Vaishnavi M. 2016. Biological Synthesis of Nano Composite SnO_2-ZnO – Screening for Efficient Photocatalytic Degradation and Antimicrobial Activity. Materials Today: Proceedings 3:2373–2380. https://doi.org/10.1016/j.matpr.2016.04.150

15. Merlin M, Chitra S., Nalini NJ. 2018. Synthesis and Characterization of Tin Oxide Nanoparticles Using Plant Extract. Der Pharma Chemica 10:20–23.

16. Kamaraj P, Vennila R, Arthanareeswari M, Devikala S. 2014. Biological Activities of Tin Oxide Nanoparticles Synthesized Using Plant Extract. World Journal of Pharmacy and Pharmaceutical Sciences 3:382–388.

17. Roopan SM, Kumar SHS, Madhumitha G, Suthindhiran K. 2015. Biogenic-Production of SnO_2 Nanoparticles and Its Cytotoxic Effect against Hepatocellular Carcinoma Cell Line (HepG2). Applied Biochemistry and Biotechnology 175:1567–1575. https://doi.org/10.1007/s12010-014-1381-5

18. Jbara AS, Othaman Z, Ati AA, Saeed MA. 2017. Characterization of γ-Al_2O_3 Nanopowders Synthesized by Co-precipitation Method. Materials Chemistry and Physics 188:24–29. https://doi.org/10.1016/j.matchemphys.2016.12.015

19. Rajaeiyan A, Bagheri-Mohagheghi MM. 2013. Comparison of Sol-Gel and Co-precipitation Methods on the Structural Properties and Phase Transformation of γ and α-Al_2O_3 Nanoparticles. Advances in Manufacturing 1:176–182. https://doi.org/10.1007/s40436-013-0018-1

20. Farahmandjou M, Golabiyan N. 2019. Synthesis and Characterisation of Al_2O_3 Nanoparticles as Catalyst Prepared by Polymer Co-precipitation Method. Materials Engineering Research 1:40–44. https://doi.org/10.25082/MER.2019.02.002

21. Mukherjee A, Sadiq M, Chandrasekaran N. 2011. Antimicrobial Activity of Aluminium Oxide Nanoparticles for Potential Clinical Applications. Science against Microbial Pathogens: Communicating Current Research and Technological Advances 2011:245–251.

22. He Y, Wei F, Ma Z, Zhang H, Yang Q, Yao B, Huang Z, Li J, Zenga C, Zhang Q. 2017. Green Synthesis of Silver Nanoparticles Using Seed Extract of Alpinia katsumadai, and Their Antioxidant, Cytotoxicity, and Antibacterial Activities. RSC Advances 7:39842–39851. https://doi.org/10.1039/C7RA05286C

23. Nouri A, Wen C. 2021. Noble Metal Alloys for Load-bearing Implant Applications. In: Structural Biomaterials. Elsevier, pp 127–156.

24. Politano AD, Campbell KT, Rosenberger LH, Sawyer RG. 2013. Use of Silver in the Prevention and Treatment of Infections: Silver Review. Surgical Infections 14:8–20. https://doi.org/10.1089/sur.2011.097

25. Konop M, Damps T, Misicka A, Rudnicka L. 2016. Certain Aspects of Silver and Silver Nanoparticles in Wound Care: A Minireview. Journal of Nanomaterials 2016:1–10. https://doi.org/10.1155/2016/7614753

26. Calderón-Jiménez B, Johnson ME, Montoro Bustos AR, et al. 2017. Silver Nanoparticles: Technological Advances, Societal Impacts, and Metrological Challenges. Frontiers in Chemistry 5:1–26. https://doi.org/10.3389/fchem.2017.00006

27. Zhang X-F, Liu Z-G, Shen W, Gurunathan S. 2016. Silver Nanoparticles: Synthesis, Characterization, Properties, Applications, and Therapeutic Approaches. International Journal of Molecular Sciences 17:1–34. https://doi.org/10.3390/ijms17091534
28. Hamouda RA, Hussein MH, Abo-elmagd RA, Bawazir SS. 2019. Synthesis and Biological Characterization of Silver Nanoparticles Derived from the Cyanobacterium Oscillatoria Limnetica. Scientific Reports 9:1–19. https://doi.org/10.1038/s41598-019-49444-y
29. Siddiqi KS, Husen A, Rao RAK. 2018. A Review on Biosynthesis of Silver Nanoparticles and Their Biocidal Properties. Journal of Nanobiotechnology 16:1–28. https://doi.org/10.1186/s12951-018-0334-5
30. Seehra MS, Bristow AD. 2018. Introductory Chapter: Overview of the Properties and Applications of Noble and Precious Metals. In: Noble and Precious Metals – Properties, Nanoscale Effects and Applications. InTech, pp 1–10.
31. Kamikoriyama Y, Imamura H, Muramatsu A, Kanie K. 2019. Ambient Aqueous-Phase Synthesis of Copper Nanoparticles and Nanopastes with Low-Temperature Sintering and Ultra-High Bonding Abilities. Scientific Reports 9:1–10. https://doi.org/10.1038/s41598-018-38422-5
32. Chandra S, Kumar A, Tomar PK. 2014. Synthesis and Characterization of Copper Nanoparticles by Reducing Agent. Journal of Saudi Chemical Society 18:149–153. https://doi.org/10.1016/j.jscs.2011.06.009
33. Din MI, Rehan R. 2017. Synthesis, Characterization, and Applications of Copper Nanoparticles. Analytical Letters 50:50–62. https://doi.org/10.1080/00032719.2016.1172081
34. Ren G, Hu D, Cheng EWC, et al. 2009. Characterisation of Copper Oxide Nanoparticles for Antimicrobial Applications. International Journal of Antimicrobial Agents 33:587–590. https://doi.org/10.1016/j.ijantimicag.2008.12.004
35. Akintelu SA, Folorunso AS, Folorunso FA, Oyebamiji AK. 2020. Green Synthesis of Copper Oxide Nanoparticles for Biomedical Application and Environmental Remediation. Heliyon 6:1–12. https://doi.org/10.1016/j.heliyon.2020.e04508
36. Muthuvel A, Jothibas M, Manoharan C. 2020. Synthesis of Copper Oxide Nanoparticles by Chemical and Biogenic Methods: Photocatalytic Degradation and in vitro Antioxidant Activity. Nanotechnology for Environmental Engineering 5:1–19. https://doi.org/10.1007/s41204-020-00078-w
37. Rabiee N, Bagherzadeh M, Kiani M, et al. 2020. Biosynthesis of Copper Oxide Nanoparticles with Potential Biomedical Applications. International Journal of Nanomedicine 15:3983–3999. https://doi.org/10.2147/IJN.S255398
38. Bezza FA, Tichapondwa SM, Chirwa EMN. 2020. Fabrication of Monodispersed Copper Oxide Nanoparticles with Potential Application as Antimicrobial Agents. Scientific Reports 10:1–18. https://doi.org/10.1038/s41598-020-73497-z
39. Christy AJ, Nehru LC, Umadevi M. 2013. A Novel Combustion Method to Prepare CuO Nanorods and Its Antimicrobial and Photocatalytic Activities. Powder Technology 235:783–786. https://doi.org/10.1016/j.powtec.2012.11.045
40. Singh J, Dutta T, Kim K-H, et al. 2018. 'Green' Synthesis of Metals and Their Oxide Nanoparticles: Applications for Environmental Remediation. Journal of Nanobiotechnology 16:1–24. https://doi.org/10.1186/s12951-018-0408-4
41. Rane AV, Kanny K, Abitha VK, Thomas S. 2018. Methods for Synthesis of Nanoparticles and Fabrication of Nanocomposites. In: Synthesis of Inorganic Nanomaterials. Elsevier, pp 121–139.
42. Iravani S, Korbekandi H, Mirmohammadi SV, Zolfaghari B. 2014. Synthesis of Silver Nanoparticles: Chemical, Physical and Biological Methods. Research in Pharmaceutical Sciences 9:385–406.

Chemical Synthesis or Phytofabrication of Metal/Metal Oxide Nanoparticles 363

43. Saha J, Begum A, Mukherjee A, Kumar S. 2017. A Novel Green Synthesis of Silver Nanoparticles and Their Catalytic Action in Reduction of Methylene Blue Dye. Sustainable Environment Research 27:245–250. https://doi.org/10.1016/j.serj.2017.04.003

44. Mohan AN, B. M. 2019. Biowaste Derived Graphene Quantum Dots Interlaced with SnO_2 Nanoparticles – A Dynamic Disinfection Agent against Pseudomonas aeruginosa. New Journal of Chemistry 43:13681–13689. https://doi.org/10.1039/C9NJ00379G

45. Abbasi BA, Iqbal J, Mahmood T, et al. 2019. Biofabrication of Iron Oxide Nanoparticles by Leaf Extract of Rhamnus virgata : Characterization and Evaluation of Cytotoxic, Antimicrobial and Antioxidant Potentials. Applied Organometallic Chemistry 33:1–15. https://doi.org/10.1002/aoc.4947

46. Kajani AA, Bordbar A-K, Zarkesh Esfahani SH, et al. 2014. Green Synthesis of Anisotropic Silver Nanoparticles with Potent Anticancer Activity Using Taxus baccata Extract. RSC Adv 4:61394–61403. https://doi.org/10.1039/C4RA08758E

47. Rasheed T, Bilal M, Iqbal HMN, Li C. 2017. Green Biosynthesis of Silver Nanoparticles Using Leaves Extract of Artemisia Vulgaris and Their Potential Biomedical Applications. Colloids and Surfaces B: Biointerfaces 158:408–415. https://doi.org/10.1016/j.colsurfb.2017.07.020

48. Raveendran P, Fu J, Wallen SL. 2003. Completely "Green" Synthesis and Stabilization of Metal Nanoparticles. Journal of the American Chemical Society 125:13940–13941. https://doi.org/10.1021/ja029267j

49. Dong C, Liu X, Xiao X, et al. 2014. Combustion Synthesis of Porous Pt-Functionalized SnO_2 Sheets for Isopropanol Gas Detection with a Significant Enhancement in Response. J Mater Chem A 2:20089–20095. https://doi.org/10.1039/C4TA04251D

50. Branquinho R, Salgueiro D, Santos L, et al. 2014. Aqueous Combustion Synthesis of Aluminum Oxide Thin Films and Application as Gate Dielectric in GZTO Solution-based TFTs. ACS Applied Materials & Interfaces 6:19592–19599. https://doi.org/10.1021/am503872t

51. Richon VM, Sandhoff TW, Rifkind RA, Marks PA. 2000. Histone Deacetylase Inhibitor Selectively Induces p21WAF1 Expression and Gene-associated Histone Acetylation. Proceedings of the National Academy of Sciences 97:10014–10019. https://doi.org/10.1073/pnas.180316197

52. Hussain MS, Hussain M, Khan MD. 2021. Pharmacological Uses of Simarouba glauca: A Review. Plant Archives 21:648–655. https://doi.org/10.51470/PLANTARCHIVES.2021.v21.no1.090

53. Vikas B, Akhil BS, Suja SR, Sujathan K. 2017. An Exploration of Phytochemicals from Simaroubaceae. Asian Pacific Journal of Cancer Prevention : APJCP 18:1765–1767. https://doi.org/10.22034/APJCP.2017.18.7.1765

54. Kulkarni Y, Agarwal S, Garud M. 2015. Effect of Jyotishmati (Celastrus paniculatus) Seeds in Mouse Models of Pain and Inflammation. Journal of Ayurveda and Integrative Medicine 6:82–88. https://doi.org/10.4103/0975-9476.146540

55. Saroya AS, Singh J. 2018. Neuropharmacology of Celastrus Paniculatus Willd. In: Pharmacotherapeutic Potential of Natural Products in Neurological Disorders. Springer Singapore, Singapore, pp 135–139.

56. Alves IABS, Miranda HM, Soares LAL, Randau KP. 2014. Simaroubaceae Family: Botany, Chemical Composition and Biological Activities. Revista Brasileira de Farmacognosia 24:481–501. https://doi.org/10.1016/j.bjp.2014.07.021

57. Saif S, Tahir A, Asim T, Chen Y. 2016. Plant Mediated Green Synthesis of CuO Nanoparticles: Comparison of Toxicity of Engineered and Plant Mediated CuO Nanoparticles towards Daphnia Magna. Nanomaterials 6:1–15. https://doi.org/10.3390/nano6110205

58. Priyadharshini Raman R, Parthiban S, Srinithya B, et al. 2015. Biogenic Silver Nanoparticles Synthesis Using the Extract of the Medicinal Plant Clerodendron serratum and Its in-vitro Antiproliferative Activity. Materials Letters 160:400–403. https://doi.org/10.1016/j.matlet.2015.08.009

59. Meghana S, Kabra P, Chakraborty S, Padmavathy N. 2015. Understanding the Pathway of Antibacterial Activity of Copper Oxide Nanoparticles. RSC Advances 5: 12293–12299. https://doi.org/10.1039/C4RA12163E

60. Gopinath V, Priyadarshini S, Al-Maleki AR, et al. 2016. In vitro Toxicity, Apoptosis and Antimicrobial Effects of Phyto-mediated Copper Oxide Nanoparticles. RSC Advances 6:110986–110995. https://doi.org/10.1039/C6RA13871C

61. Dhawale VP, Datta J, Late SDK. 2019. Synthesis, Characterization of α-Al_2O_3 Nanoparticles and Its Application in Decolorization of Methyl Orange Azo Dye in the Presence of UV Light. Journal of Nanoscience and Technology 5:580–583. https://doi.org/10.30799/jnst.192.19050101

62. Wan W, Li Y, Ren X, et al. 2018. 2D SnO_2 Nanosheets: Synthesis, Characterization, Structures, and Excellent Sensing Performance to Ethylene Glycol. Nanomaterials 8:1–20. https://doi.org/10.3390/nano8020112

63. Mariychuk R, Porubská J, Ostafin M, et al. 2020. Green Synthesis of Stable Nanocolloids of Monodisperse Silver and Gold Nanoparticles Using Natural Polyphenols from Fruits of Sambucus nigra L. Applied Nanoscience 10:4545–4558. https://doi.org/10.1007/s13204-020-01324-y

64. Ahmed S, Saifullah, Ahmad M, et al. 2016. Green Synthesis of Silver Nanoparticles Using Azadirachta indica aqueous Leaf Extract. Journal of Radiation Research and Applied Sciences 9:1–7. https://doi.org/10.1016/j.jrras.2015.06.006

65. Lopes CRB, Courrol LC. 2018. Green Synthesis of Silver Nanoparticles with Extract of Mimusops coriacea and Light. Journal of Luminescence 199:183–187. https://doi.org/10.1016/j.jlumin.2018.03.030

66. Samari F, Salehipoor H, Eftekhar E, Yousefinejad S. 2018. Low-temperature Biosynthesis of Silver Nanoparticles Using Mango Leaf Extract: Catalytic Effect, Antioxidant Properties, Anticancer Activity and Application for Colorimetric Sensing. New Journal of Chemistry 42:15905–15916. https://doi.org/10.1039/C8NJ03156H

67. Bhuvaneswari R, John Xavier R, Arumugam M. 2015. Biofabrication and Its in vitro Toxicity Mechanism of Silver Nanoparticles Using Bruguiera cylindrica Leaf Extract. Karbala International Journal of Modern Science 1:129–134. https://doi.org/10.1016/j.kijoms.2015.08.003

68. Arya A, Gupta K, Chundawat TS, Vaya D. 2018. Biogenic Synthesis of Copper and Silver Nanoparticles Using Green Alga Botryococcus braunii and Its Antimicrobial Activity. Bioinorganic Chemistry and Applications 2018:1–9. https://doi.org/10.1155/2018/7879403

69. Ramamurthy NKS. 2007. Fourier Transform Infrared Spectroscopic Analysis of a Plant (Calotropis gigantea Linn) from an Industrial Village, Cuddalore dt, Tamil Nadu, India. Rom J Biophys 17:269–276.

70. Kim HS, Park N-K, Lee TJ, Kang M. 2014. Effect of AlF_3 Seed Concentrations and Calcination Temperatures on the Crystal Growth of Hexagonally Shaped α-alumina Powders. Ceramics International 40:3813–3818. https://doi.org/10.1016/j.ceramint.2013.08.033

71. Komeili S, Ravanchi MT, Taeb A. 2015. The Influence of Alumina Phases on the Performance of the Pd–Ag/Al_2O_3 Catalyst in Tail-end Selective Hydrogenation of Acetylene. Applied Catalysis A: General 502:287–296. https://doi.org/10.1016/j.apcata.2015.06.013

72. Singh D, Kundu VS, Maan AS. 2016. Structural, Morphological and Gas Sensing Study of Zinc Doped Tin Oxide Nanoparticles Synthesized via Hydrothermal Technique. Journal of Molecular Structure 1115:250–257. https://doi.org/10.1016/j.molstruc.2016.02.091

73. Selvi NSS, DK. 2014. Interfacial Effect on the Structural and Optical Properties of Pure SnO_2 and Dual Shells (ZnO; SiO_2) Coated SnO_2 Core-shell Nanospheres for Optoelectronic Applications. Superlattices Microstruct 76:277–287.

74. Pattanayak S, Mollick MMR, Maity D, et al. 2017. Butea Monosperma Bark Extract Mediated Green Synthesis of Silver Nanoparticles: Characterization and Biomedical Applications. Journal of Saudi Chemical Society 21:673–684. https://doi.org/10.1016/j.jscs.2015.11.004

75. Ethiraj AS, Kang DJ. 2012. Synthesis and Characterization of CuO Nanowires by a Simple Wet Chemical Method. Nanoscale Research Letters 7:1–5. https://doi.org/10.1186/1556-276X-7-70

76. Hwa KY, Karuppaiah P, Gowthaman NSK, et al. 2019. Ultrasonic Synthesis of CuO Nanoflakes: A Robust Electrochemical Scaffold for the Sensitive Detection of Phenolic Hazard in Water and Pharmaceutical Samples. Ultrasonics Sonochemistry 58:1–9. https://doi.org/10.1016/j.ultsonch.2019.104649

77. Han Q, Setchi R, Evans SL. 2016. Synthesis and Characterisation of Advanced Ball-milled Al-Al_2O_3 Nanocomposites for Selective Laser Melting. Powder Technology 297:183–192. https://doi.org/10.1016/j.powtec.2016.04.015

78. Sun X. 2020. Morphosynthesis of SnO_2 Nanocrystal Networks as High-capacity Anodes for Lithium Ion Batteries. Ionics 26:3841–3851. https://doi.org/10.1007/s11581-020-03552-2

79. Sudha A, Jeyakanthan J, Srinivasan P. 2017. Green Synthesis of Silver Nanoparticles Using Lippia nodiflora Aerial Extract and Evaluation of Their Antioxidant, Antibacterial and Cytotoxic Effects. Resource-Efficient Technologies 3:506–515. https://doi.org/10.1016/j.reffit.2017.07.002

80. Nasrollahzadeh M, Sajadi SM, Maham M. 2015. Tamarix gallica Leaf Extract Mediated Novel Route for Green Synthesis of CuO Nanoparticles and Their Application for N-Arylation of Nitrogen-containing Heterocycles under Ligand-free Conditions. RSC Advances 5:40628–40635. https://doi.org/10.1039/C5RA04012D

81. Yalamaç E, Trapani A, Akkurt S. 2014. Sintering and Microstructural Investigation of Gamma–alpha Alumina Powders. Engineering Science and Technology: An International Journal 17:1–6. https://doi.org/10.1016/j.jestch.2014.02.001

82. Mueller F, Bresser D, Chakravadhanula VSK, Passerini S. 2015. Fe-doped SnO_2 Nanoparticles as New High Capacity Anode Material for Secondary Lithium-ion Batteries. Journal of Power Sources 299:398–402. https://doi.org/10.1016/j.jpowsour.2015.08.018

83. Sivaraj R, Rahman PKSM, Rajiv P, et al. 2014. Biogenic Copper Oxide Nanoparticles Synthesis Using Tabernaemontana Divaricate Leaf Extract and Its Antibacterial Activity against Urinary Tract Pathogen. Spectrochimica Acta Part A: Molecular and Biomolecular Spectroscopy 133:178–181. https://doi.org/10.1016/j.saa.2014.05.048

84. Rashad MM, Ibrahim IA, Osama I, Shalan AE. 2014. Distinction between SnO_2 Nanoparticles Synthesized Using Co-precipitation and Solvothermal Methods for the Photovoltaic Efficiency of Dye-sensitized Solar Cells. Bulletin of Materials Science 37:903–909. https://doi.org/10.1007/s12034-014-0024-3

85. Kenchappa Somashekharappa K, Lokesh SV. 2021. Hydrothermal Synthesis of K 2 Ti 6 O 13 Nanotubes/Nanoparticles: A Photodegradation Study on Methylene Blue and Rhodamine B Dyes. ACS Omega 6:7248–7256. https://doi.org/10.1021/acsomega.0c02087

19 Antioxidant, Bactericidal, Antihemolytic, and Anticancer Assessment Activities of Al_2O_3, SnO_2, and Green Synthesized Ag and CuO NPs

Amulya Giridasappa, Ismail Shareef M., and Gopinath S. M.

CONTENTS

19.1 Introduction 368
    19.1.1 Significance of Metal-based Nanoparticles
        as Antioxidant Agents 370
    19.1.2 Significance of Metal-based Nanoparticles
        as Antimicrobial Agents 371
    19.1.3 Significance of Metal-based Nanoparticles
        as Anticancer Entities 371
19.2 Methodology 372
    19.2.1 In Vitro Evaluation of Antioxidant Activities 372
    19.2.2 Antibacterial Activity by Diffusion Method 373
    19.2.3 In Vitro Anticancer Efficacy Employing Cell Lines 373
        19.2.3.1 Apoptosis Evaluation by Annexin V-FITC/PI
               Staining.. 374
        19.2.3.2 Assessment of Cell Cycle Analysis 374
        19.2.3.3 Caspase-3 Expression Study 375
        19.2.3.4 Fluorescence Microscopic Analysis Using
               AO/EB Staining.. 375
    19.2.4 Hemolysis Activity of Synthesized Nanoparticles 375
        19.2.4.1 Effect of Green Synthesized Nanoparticles
               on RBCs Morphology... 376
    19.2.5 In Vivo Study for Anticancer Evaluation 376
        19.2.5.1 Experimental Design of Subacute Oral Toxicity......... 376
        19.2.5.2 Evaluation of Survivability... 377

DOI: 10.1201/9781003297772-19

		19.2.5.3	Transformation in Hematological Profile and Oxidative Stress Measurement	377

19.2.5.3 Transformation in Hematological Profile and
Oxidative Stress Measurement 377
19.2.5.4 Diagnosis of Organ Histology 378
19.3 Results and Discussion 378
    19.3.1 Measurement of Antioxidant and Antibacterial
Capability 378
        19.3.1.1 Decolorization Activity of ABTS Radical Cation 379
        19.3.1.2 Analysis by DPPH Radical Scavenging 380
        19.3.1.3 Measurement of NO Radical-Scavenged Activity 380
    19.3.2 Estimation of Antibacterial Efficacy 380
    19.3.3 Effectiveness of Antiproliferation on Cancer and
Non-cancer Cell lines 381
        19.3.3.1 Apoptosis by FITC Annexin V-PI 383
        19.3.3.2 Propidium Iodide-based Cell Cycle Analysis 383
        19.3.3.3 Expression of Caspase-3 in MCF-7 Cell Line 384
        19.3.3.4 Dual Stain Analysis of MCF-7 Cell Line 386
    19.3.4 Prevention of Hemolysis by Prepared Nanoparticles 387
        19.3.4.1 Effectiveness of Nanoparticles on RBC
Morphology 387
    19.3.5 In Vivo Anticancer Threshold 388
        19.3.5.1 Survivability and Growth Performance 389
        19.3.5.2 Hematological Indices after Treatment 391
        19.3.5.3 Accomplishment of Tissue Antioxidant Activity 391
        19.3.5.4 Efficiency of Organ Histopathology 392
19.4 Conclusion 394
References 395

19.1 INTRODUCTION

The inadequate bioavailability of phytochemical formulations on the site-specific regions of living entities is the evident drawback observed in case of utilization of plant-based medicinal remedies; to overcome such problem, the convincible access is obtained through nanoscience. Due to this reason, nanotechnology is highly credible to be utilized in pharmaceutics and therapeutics as it is believed to bring drastic improvement in terms of medical examination and remedy [1]. The fundamental units of biology, like DNA, antibodies, enzymes, proteins, and similar entities, have measurements in the range of nano and therefore the surface science unveiled by nanomaterials is to be given a paramount importance. The favorable features of using these structures comprised of betterment in solubility, ease of bioavailability, defense against toxicity, advancement in drug-induced delivery and stability, even administration of tissue macrophages, allowing persistent interaction and aegis against physical and chemical deterioration, and so forth. Consequently, the inherent ability of nano-based drug delivery in improving the process of reaction augmented with ethno-pharmaceutical properties to surmount the issues related to herbal formulations was observed [2, 3]. With all these advantages, the domain of nanobiotechnology during recent times has advanced imposingly which is attributable to synthesis of significant number of nanostructures holding spectacular utilizations, thus an attempt to study their applications is made here.

Metal and their related compound nanoparticles have acquired the feature of localized surface plasmon resonance that has gained distinctive optoelectrical characteristics in comparison with its specified bulk nature. The size-reliant nanoparticles show varied physicochemical properties which conduce specified electrical, mechanical, optical, and imaging features and these are majorly desirable in applications like pharmaceutical, commercial equipment, ecological conservatives, etc. [4]. The advancement in nanotechnology-related drugs formulated by various pharmaceutical companies during some recent years has obtained authorization through the US Food and Drug Administration (FDA) [5]. Having such instances in mind, it is required to consider applications of metal and metal compound nanoparticles.

If the nanomaterials prepared are exactly acting on specified region at right dosage then the total drug requirement and adverse impacts can be remarkably reduced and with this the tissue engineering that deals with restoration and revival of worn-out tissues could be favorably achieved and will be an efficient alternative to conventional artificial implants, organ transplants, etc. There are several concepts supporting nanotechnology in medicine, such as the usage of polyethylene oxide and polylactic acid nanoparticles for treatment through intravenous systems, polymers

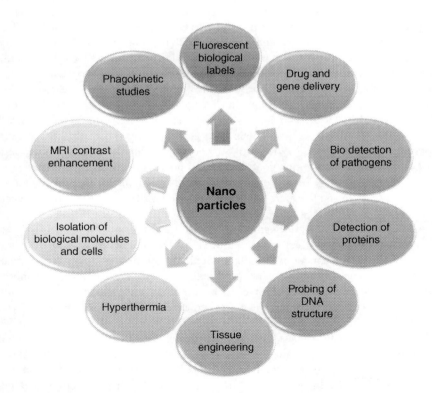

FIGURE 19.1 Multiple applications of nanostructures in biotechnology.

370 Handbook of Sustainable Materials

that can administer medicine to specified region and in recent times, ideal drug carriers in the form of liposomes that possess the drawbacks, like inherent health problems, squat encapsulation capability, sudden water outflow within the affinity of blood components, etc., are being replaced by polymeric nanoparticles due to organized drug delivery [4].

19.1.1 Significance of Metal-based Nanoparticles as Antioxidant Agents

The oxidative stress-related conditions could be overcome by remedial abilities acquired by phytonutrients of diverse selection obtained employing plant-based restorative entities. Oxidative damage, which is found mainly due to disparity observed during the generation and elimination of free radicals under oxidative stress conditions, is provoking concerned effect related to ill health of either acute or chronic circumstances. The general determinants for oxidative stress include environmental pollution, ingestion of toxic drug, consumption of high-fat diet, moldy food, oxidized oil, etc. To surmount such complications, the involvement of antioxidants is considered to be adequate to fight against them.

Antioxidants can combine with free radicals and prevent harmful reaction by modifying them to benign by liberation of an electron. Each antioxidant has potent biological properties for instance the antioxidant like polyphenols derived from green plants is well-known in pharmaceutics due to their medicinal values and are also believed to actively eliminate free radicals and control lipid peroxidation. Thus, these green plants behave as a major ethanomedicinal and remedial compound for treatment of various disorders and similar antioxidants have developed these unique applications relating to therapeutics. Frequently we utilize organic sources of secondary metabolites, like curcumin, resveratrol, catechin, quercetin, etc., obtained from medicinal plants as promising controller of reactive species generation but the pharmaceutical effect of such entities are confined because of their poor bioavailability as they are less soluble, stable, and absorbed weakly in alimentary canal along with some adverse carryovers [6]. Although every human being has got natural endogenous antioxidant protection and healing power to encounter injuries caused during oxidation yet by reason of their insufficient availability, they are unable to monitor the ravage and therefore additional supplements from other sources are necessary for normal functioning of cells [7].

At present, innovations through nanotechnology have created functional goals to enhance drug delivery effectively in required site by making them target-specific, which leads to advancement in potentiality of phyto-bioactive compounds and innate antioxidants when synthesized along with nanoparticles. Development in nanoscience has utilized various nanoparticles of other organic or inorganic sources; organic derivations, like melanin nanostructures, reduced glutathione nanoparticles, etc., are notable among such groups while inorganic materials include metal nanoparticles, transition metal oxides, etc. in order to evaluate the antioxidant potential. Nanocarriers, nanoencapsulated materials, nanoemulsions, and several other nanomaterials are some of the recent advancements of biotechnology invented to address good bioavailability, increased shelf-life, effective solubility, targeting specified site, controlled release of drug, and reduced after effects, such improvements were

Antioxidant, Bactericidal, Antihemolytic, and Anticancer Assessment Activities **371**

experimentally proved using animal models [6]. Overall, they provide magnified therapeutic replacements to ward off free radicals.

19.1.2 Significance of Metal-based Nanoparticles as Antimicrobial Agents

Over the past few decades, nanoparticles have indicated immense influence on controlling the growth of microorganisms, which are infective and may lead to advancement of grievous disease. Nanoparticles will provide easy destruction as they permit effortless penetration into microbial cells that result in healing of infections and protection against disease [8]. Also, there are incidences of resistance acquired against multidrug by certain dreadful microorganisms which is a very serious issue and needs to be resolved as early as possible and due to the reason as it would affect the entire population. Such situations were handled by the range of antibiotics, which enacted accordingly based on mechanism of action and objective to be met in order to destroy fatal microbes. Example includes the development of penicillin, beta-lactam antibiotic to address the synthesis of peptidoglycan that is mostly observed in case of Gram-positive bacteria and their crucial regulator, involvement in cell division and similarly, Rifampicin which can aim either cell wall or nucleic acid synthesis to restrict the growth of microorganisms [9].

Although there are various mechanisms explained for the destruction of microorganisms by nanoparticles yet the mode of action needs to be studied more intensely to altogether understand the way by which the nanoparticles hinder the growth of the microbes. One such possibility is that the nanoparticles would bring about damage to cell membrane by establishing pits, distortion, deformation, etc., which accomplish the failure of membrane integrity and finally induce cellular death [10]. Another such reason is that the size, shape, structure, and morphology of nanoparticles that get attached to cell membrane of the microbes, result in physiological changes, and the generation of oxidative stress as they sustain dehydrogenation in cellular respiration in the same region [11, 12]. It impacts the direct linkage between the size and morphology of nanoparticles to the antimicrobial potential. Oxidative stress created because of production of free radicals like hydroxyls, carbon monoxide, nitrogen species, hydrogen peroxides, singlet oxygen, etc., would cause increased permeability in cells and also disintegrate the intracellular substance matter of microbial cells.

19.1.3 Significance of Metal-based Nanoparticles as Anticancer Entities

The comprehensive ability of nanoparticle-related drug delivery systems for anticancer has allowed novel renovation of medications to be highly successful, noninvasive and precisely innocuous as the nanoparticles are having the enormous benefits in comparison with traditional treatment practices [13]. Nanoscience is multidisciplinary subject area evolved by merging of advanced technology with material science and chemistry; they are known for their excessive utilization in tumor science for instance like identification of target site, exploration of biomarkers on tumor cells, and innovation of suitable medication.

To administer the anticancer remedies, we have now encountered nanodevices loaded with drug that precisely reach the defined target site, one such example is liposomes employed as nanostructures which are used as 'first generation' artifices. Another such illustration is photodynamic cancer therapy that kills tumor cells when irradiated under laser atomic oxygen that works based on their selectivity of tumors over normal cells; nonetheless, the dye left out gets transferred to the skin and eyes of person who has undergone such treatment will become vulnerable to sunlight that may proceed for one to two months [14]. Henceforth, to overcome this problem, the porous nanoparticles that confine the dye compound of hydrophobic type such that they remain within the ormosil nanoparticles and do not get liberated to surrounding body regions while concurrently the oxygen production occurs with no repercussions having 1 nm pore size for exemption of oxygen molecule [15]. Similarly, with these developments, various tools have been generated to treat cancer that needs to be assessed justly.

19.2 METHODOLOGY

19.2.1 In Vitro Evaluation of Antioxidant Activities

Antioxidant activities of prepared nanoparticles were evaluated to know their capability to extinguish the production of excess-free radicals in laboratory conditions. There were numerous methods available to determine the antioxidant capacity and, based on reproducibility, effectivity, and less chemical consumption, we have chosen estimation by 2,2'-azino-bisethylbenzo thiazoline-6-sulphonic acid (ABTS), 2,2-diphenyl-1-pikryl-hydrazyl (DPPH), and nitric oxide (NO) radical scavenging methods. We accomplished these assays using extracts of medicinal plants, Ag and CuO NPs synthesized using same plant extracts, along with SnO_2 and Al_2O_3 NPs, which were synthesized by chemical method. The activity was conducted with these test substances of varying concentration from 62.5, 125, 250, 500 to 1000 µg/mL, with ascorbic acid as standard. The optical density emitted by these samples was recorded using ELISA plate reader (Spectra Fluor plus, Tecan, US) and the percentage of inhibition of free radicals was computed by the following mentioned formula:

$$\text{Inhibition of free radicals}(\%) = \frac{(\text{ODcontrol} - \text{ODsample})}{\text{ODcontrol}} \times 100$$

The ABTS-free radical scavenging assay was proceeded as per technique adopted by Adebiyi et al. [16] with small modifications, the stock solution comprising of 7 mM ABTS was mixed with equal quantity of 2.4 mM of potassium persulfate and left overnight at room temperature. Now to initiate the activity, 0.4 mL of different dilutions of the samples were taken, including standard, each was allowed to react with 0.32 mL of stock solution of ABTS and was diluted using Phosphate Buffer Saline (PBS) to make up volume to reach the total of 3 mL. The specified reaction mixtures were transferred into microtiter plate to observe optical density at 734 nm using ELISA plate reader, and the activity of each sample was determined by the formula mentioned above. The conventional procedure explained by Hemlata et al. [17] was followed to measure the antioxidant capacity by scavenging the stable DPPH radicals formed during the assay along with few alterations. The stock solution was

Antioxidant, Bactericidal, Antihemolytic, and Anticancer Assessment Activities   **373**

made by mixing 100 μM of DPPH in methanol. For the reaction, 300 μL of methanolic DPPH was combined thoroughly with 15 μL of prepared concentrations samples including standard in microtiter plate, incubated at room temperature for 20 minutes and the absorbance was read using plate reader at 517 nm. To evaluate the NO radical scavenging capacity of samples, we were consistent with the procedure detailed by Asemani and Anarjan [18]. Briefly, 50 μL of the aforementioned samples were mixed with 200 μL of 10 mM sodium nitroprusside and 50 μL of PBS. The procedure followed was same for standard nitrite of 100 μM concentration. They were allowed to react for 150 minutes at 25°C with dimethyl sulfoxide (DMSO) as control. Then 100 μL of mentioned mixtures were taken from each sample and placed in another microtiter plate and were allowed to interact with 200 μL of Griess reagent for about 30 minutes. This diazotization resulted in the formation of a pink-color chromophore that was read at 540 nm.

19.2.2 ANTIBACTERIAL ACTIVITY BY DIFFUSION METHOD

For present investigation, Ag and CuO NPs prepared using *Simarouba glauca* and *Celastrus paniculatus* along with Al_2O_3 and SnO_2 NPs were scrutinized for antibacterial potentiality by means of well diffusion assay similar to method detailed by Kumar et al. [19]. Antibacterial activity was performed against both strains viz., Gram-negative bacteria like *Escherichia coli* and Gram-positive bacteria like *Streptococcus pyogenes* (for green synthesized NPs) and *Pseudomonas aeruginosa* (for SnO_2 and Al_2O_3 NPs), tested in comparison with the commercially available antibiotic. Microorganisms used for evaluation were procured from American Type Culture Collection (ATCC), Manassas, Virginia, and stored in −20°C until next use. Bacteria were revived individually using nutrient broth medium by inoculation of each strain and then subjected to incubation at 37°C for 24 hours. When the cell mass reaches 0.5 McFarland absorbance, it indicates that it is prepared to participate in experiment.

To begin with the experiment, the sterile Petri plate containing approximately 15 mL of nutrient agar medium was taken under laminar airflow and after solidification, bacterial inoculum of about 100 μL was spread uniformly on agar surface with the help of an L glass spreader. Then the plates were punched to create wells (3 mm in diameter) through a sterile cork borer in order to load the samples used for testing. Later, as described by Choudhary et al. [20], wells were filled with samples of varied concentrations along with standard (30 μg) in each plate for easy correlation. The inoculated plates were placed in an incubator maintained with the same temperature (i.e., 37°C) for 24 hours in order to accomplish the zone of inhibition.

19.2.3 IN VITRO ANTICANCER EFFICACY EMPLOYING CELL LINES

The cell lines were procured from ATCC (Mediatech, Manassas, VA) and used as monolayer in Dulbecco's Modified Eagle's Medium enriched by 10% fetal bovine serum, minimal inorganic supplements (0.1%), L-glutamine (200 mM), penicillin-streptomycin (1%), with 5% CO_2, and humidified 95% air at 37°C. Cells were cultivated in 75 cm^2 tissue culture flask and retained until the number of cells reached

374 Handbook of Sustainable Materials

a confluence of 80%. To evaluate cytotoxicity of synthesized SnO_2, Al_2O_3, green synthesized Ag, CuO NPs, and plant extracts on cancer cells, conventional MTT calorimetric assay was adopted based on their mitochondrial dehydrogenase content. The nanoparticles were diluted to concentrations of 10, 20, 40, 80, 160, and 320 µg/mL using DMSO.

The analysis was initiated by employing MCF-7, HT 29 along with MCF-10A cell lines separately after trypsinization and passaging, later they were inoculated in 96 multi-well plate at the density of 3×10^4 cells/100 µL/well. The cells were favored to attach and mature in 5% CO_2 incubator for 24 hours, 37°C within designated wells. Afterward, they were administered with varying dilutions of synthesized NPs and plant extracts. Following incubation for 24 hours at 37°C in 5% CO_2 atmosphere, the media was taken out and washed with freshly prepared PBS twice and replenished using 100 µL of MTT (0.5 mg/mL) and again maintained under same conditions for 4 hours in dark. Then the media was casted out without disturbing formazan crystals and they were supplied with 100 µL of DMSO. The antiproliferation activities were evaluated by optical density values of each sample at 570 nm to obtain the concentration at which 50% of cell growth was hindered (IC_{50}). This optimum dose was calculated using the following formula:

$$\text{Inhibition of growth of cells } (\%) = 1 - \frac{\text{OD at 570 nm (treated)}}{\text{OD at 570 nm (control)}} \times 100$$

19.2.3.1 Apoptosis Evaluation by Annexin V-FITC/PI Staining

To know the cytomorphological conversion in cancer cells due to treatment of nanoparticles, the apoptosis study was executed incorporating Annexin V-fluorescein isothiocyanate (FITC)/propidium iodide (PI) staining process, taking one of the best IC_{50} value exhibited i.e., CuO-CP NPs dosed on MCF-7 cell line. Briefly, 3×10^5 cells/2 mL medium was placed in 6-well plate and then incubated at 37°C in 5% CO_2 for 24 hours. Later, the cells were administered with known half-maximal inhibitory concentration of CuO-CP NPs and incubated, and then cells were harvested into a sterile polystyrene tube. The cells were recovered by trypsin – EDTA, centrifuged at 300× g. Using apoptosis Detection Kit (BD Pharmingen) and pellets were suspended in 5 µL of Annexin V-FITC, incubated for about 15 minutes and fixed. Subsequently, the above resultant was treated with RNase at 37°C and stained by PI for 15 minutes. 1× Annexin-V binding buffer was used for washing and then analyzed from flow cytometry (FACS Calibur, BD Biosciences, USA). The laser was excited at 488 nm wavelength sourcing 585 nm, PI fluorochrome and about 10,000 gated events were fed.

19.2.3.2 Assessment of Cell Cycle Analysis

The MCF-7 cell line at the density of 3×10^5 cells/2 mL was cultured in 6-well plate and same steps as in apoptosis were followed until centrifugation. The cells were rinsed using buffer and BD Cycletest, Plus DNA kit was employed to analyze cell cycle phase. Then the cells were fixed using 0.7 mL of 70% absolute ethanol placed

Antioxidant, Bactericidal, Antihemolytic, and Anticancer Assessment Activities **375**

on ice for 30 minutes, centrifuged at 2000 rpm and pellet was gently vortexed to break open. The cells were washed again with PBS and suspended in Ribonuclease A (50 µg/mL). Lastly stained using PI for 15 minutes, and cell cycle was evaluated by utilizing flow cytometry. Results were analyzed using 10,000 gated cell singlets in PI histogram where cells were distributed in Sub G_0/G_1, G_0/G_1, S, and G_2/M phases existing in samples.

19.2.3.3 Caspase-3 Expression Study

The expression of Caspase-3 protein was accomplished to identify the protease enzyme that gets activated at early phase of apoptosis in tumor cells. The activity was determined, as per Giridasappa et al. [21], by the specific cleavage of synthetic tetra-peptide Ac-DEVD-AMC fluorogenic substrate. The experiment till fixation, as followed in cell cycle analysis, was performed and then incubated with 500 µL of cytofix solution for 10 minutes. Then they were washed employing 0.5% bovine serum albumin, later mixed completely in 20 µL of anti-Caspase-3 antibody, followed by incubation for 30 minutes in dark. The samples were washed with PBS, 0.1% sodium azide, and cleavage were forthwith detected at 390 nm excitation and emission at 460 nm using flow cytometry.

19.2.3.4 Fluorescence Microscopic Analysis Using AO/EB Staining

The morphological variations that the cancer cells undergo in the process of apoptosis could be visualized using fluorescence microscope (Olympus, CKX41, Japan). About 2×10^5 number of cells of MCF-7 in growth medium were cultured on the coverslips, which were previously smeared with cell adhesive. The synthesized NPs at which their IC_{50} value occurred were noted, while the control remains untreated. The process was resumed by incubation for 24 hours and was accompanied by staining the cells for 2–3 minutes using 10 µL of AO and EB, followed by addition of mounting medium. Finally, images were observed under microscope at 40× magnification with filter cube of excitation 560/40 nm, emission of 645/75 nm for EB, and with excitation 470/40 nm, emission of 525/50 nm for AO, they were connected to digital imaging system.

19.2.4 HEMOLYSIS ACTIVITY OF SYNTHESIZED NANOPARTICLES

The proficient anticancer capability of prepared nanoparticles was intended to be assessed for their compatibility with blood, especially red blood corpuscles (RBCs), so as to verify their benign characteristics. The process can be determined based on enhanced permeability to complete lysis of RBCs and results in release of hemoglobin into blood. The oxidative disruption caused because of elevated levels of lipid molecules, oxygen concentration, hemoglobin, etc. that act as inducers were detected readily by RBCs. To consider the safer efficacy of green synthesized nanoparticles on RBCs, the test procedure adopted by Kalita et al. [22] was followed without any modifications. Blood pellet was washed for 3–4 times using a defined PBS at 7.2 pH. PBS represented the negative control and water acted as positive control. They were incubated at 37°C for 60 minutes and centrifuged, supernatant from the resultants were used to note down the optical density at 540 nm from which hemolysis percentage was enumerated using the following formula:

$$\text{Percentage of hemolysis} = \frac{\left(\text{OD of sample} - \text{OD of negative control}\right)}{\left(\text{OD of positive control} - \text{OD of negative control}\right)} \times 100$$

19.2.4.1 Effect of Green Synthesized Nanoparticles on RBCs Morphology

The morphology of RBCs contributes immensely to quantify the habitual blood count; an accomplished observation of proper smear of blood comprises the extremely precious individual method in hematological research. The RBCs of human beings are discoid or bi-concave in shape with 6–8 µm in size at their center; it has a middle area of pallor, which is considerably hemoglobinized and the outer red cell diameters without any inclusions [23]. Healthy RBCs separated from complete blood were rinsed with PBS at pH 7.2 for couple of times until the red pigment gets completely settled at the bottom. The settled pellet was reconstituted with same PBS in 1:10 ratio and to this solution mixture, each green synthesized nanoparticles, their respective plant extracts at 1000 µg/mL were added to incubate at 37°C for about 60 minutes. It was followed by fixing with 5% formaldehyde, dehydration using dilute ethanol and finally drying under vacuum.

19.2.5 IN VIVO STUDY FOR ANTICANCER EVALUATION

A sum of 48 healthy C-57 mice of body weight (BW) measuring about 28–32 g were engaged in the study and distributed into six groups (n = 8) in equal fractions. Animals (5–7 weeks old) were allowed to adapt to lab environment with temperature controlled at 22 ± 3°C and were kept in polyacrylic cages with sawdust or husk bedding, preserved at animal house facility. The animal house was maintained with 12 hours light/12 hours dark phase, 60 ± 5°C humidity and was fed with standard mice pellets and good source of water. Overall, the animals were handled with extreme care and compassion, and the protocol adhered was certified by Institutional Animal Ethics Committee with registration no: 997/c/06/CPCSEA. After acclimatization, the Organization for Economic Co-operation and Development (OECD) test guideline 425 was used to determine the acute oral toxicity [24] on C57 mice. For the induction of tumor, inoculum containing EAC cell suspensions, which were procured from Amala Cancer Research Centre, Trissur, India, were sustained by transplantation in mice.

19.2.5.1 Experimental Design of Subacute Oral Toxicity

For the repeated dose studies, animals in each group except for Group I consisting of negative control were implanted with EAC cells of concentration, 2×10^5 cells per 0.2 mL volume. Group II–VI denotes the mice bearing EAC cells. Group II represented the control without any treatment acquired; while Groups III and VI illustrated the low (50 mg/kg BW) and high (100 mg/kg BW) dose treatment with CuO-CP NPs. Subsequently, Group V was dosed using dietary flavanones called Hesperetin (30 mg/kg BW) and Group VI obtained the medication used in chemotherapy called

Antioxidant, Bactericidal, Antihemolytic, and Anticancer Assessment Activities 377

5-Fluorouracil (20 mg/kg BW). The day when this event took place was aforethought to be as Day 0 and sequentially treatment with as mentioned was initiated from Day 1 and proceeded until 21 days. After which, four mice from specified group were famished for 18 hours and anesthetized slightly to collect blood samples by means of retro-orbital plexus [25]. Internal organs were excised and preserved at −20°C for biochemical and histopathological investigations.

19.2.5.2 Evaluation of Survivability

It was necessary to know the number of days the animals survived after EAC inoculation in each group in order to correlate with the duration of life of normal mice. The comparison was made in the following mentioned criteria between control and treated groups employing the following formulae:

$$\text{Gain in body weight}(BW)$$

$$= \left(\frac{\text{Weight of mice on } 21^{st} \text{day} - \text{Weight of mice on } 0^{th} \text{day}}{\text{Weight of mice on } 0^{th} \text{day}} \right) \times 100$$

$$\text{Mean Survival Time}(MST)$$

$$= \frac{\sum \text{Time of survivability of each mice in a group}}{\text{Total number of mice in same group}} (\text{days})$$

$$\text{Increase in life span}(ILS)$$

$$= \left(\frac{\text{MST of treated group} - \text{MST of control group}}{\text{MST of control group}} \right) \times 100$$

These equations were also witnessed in studies conducted by Islam et al. [26] and Sunil et al. [27], which further provide evidences for the research conducted in this section.

19.2.5.3 Transformation in Hematological Profile and Oxidative Stress Measurement

The collected blood samples in test tubes containing EDTA were used immediately to examine the hematological profile of mice in individual group. The oxidative stress, which can cause the potential cellular damage, was measured to estimate tissue antioxidant activity in specified biological samples taken for analysis. Here, oxidative stress caused due to presence of ROS was determined by evaluating the levels of LPO, MPO, and GSH in liver homogenate prepared with PBS, pH 7.4 of treated and control groups. The antioxidant capacity on EAC harboring mice was tested utilizing liver tissue homogenate (10% w/v), which was placed in ice-cold PBS, then immersed in 10% KCl for a few seconds, centrifuged at 1500 rpm for 14 minutes at 4°C.

Malondialdehyde (MDA), a product of LPO when combined with other oxidation process, can cause severe injuries and hence their quantification was performed from method described by Ohkawa et al. [28]. MPO, the highly oxidative enzyme which is involved in many diseases, as they affect adversely was assessed by the procedure followed as described by Mullane et al. [29] with some modifications and for the GSH evaluation, Ellman's reagent was used in particular [30]. To the 10% w/v homogenate of 0.2 mL; 0.2 mL of 8% SDS, 1.5 mL of 20% acetic acid, and 1.5 mL of 0.8% TBA were mixed together. This mixture was made up to 5 mL using double deionized water and later heated in oil bath for 60 minutes at 95°C. It was followed by centrifugation at 4000 rpm for 12 minutes in order to separate the organic layer and absorbance was noted at 532 nm. From the standard compound analyzed, yielding a graph, the levels of MDA and hence LPO were determined in tissue sample.

$$\text{Concentration of MDA}(\text{nmol / mL}) = \frac{\text{Absorbance at 532 nm} - \text{intercept of line graph}}{\text{Gradient of line graph}}$$

19.2.5.4 Diagnosis of Organ Histology

The organs like kidney, liver, and spleen were brought down to habitual temperature and placed in containers consisting of 10% neutral formalin as fixatives. After 24 hours, they were hydrated consecutively using water, and to make them dehydrated, ethanol was used. The specimens were refined from ethanol to promote molten paraffin wax penetration in order to get embedded within blocks and sectioned precisely to 5 μm thickness employing Leica Rotary Microtome (RM 2125, China). Earlier to staining, the slides were cooled and coatings of paraffin were removed by xylene. Later the slides were stained using hematoxylin and rinsed with 1% acid alcohol to withdraw leftover stain and also to set them apart. This completes H and E staining, to dehydrate them in ascending levels, ethanol was used and was mounted by DPX and covered with slips.

19.3 RESULTS AND DISCUSSION

19.3.1 Measurement of Antioxidant and Antibacterial Capability

The small molecules especially nanoparticles are known to possess superior ingress to the radical site; for this reason, the steric receptiveness of the free radicals is the primary governing factor in the biological process of antioxidant determination. The synthetic antioxidants like butylated hydroxytoluene and butylated hydroxyanisole are known to stimulate consequences particularly relating to hepatic impairment and are also presumed for causing mutagenesis and neurotoxicity [31] and accordingly we have synthesized metal oxides using traditional plant sources. Furthermore, there are features ascertained in various experimental trials specifying the relation between progress of cancer and endogenous antioxidant entities and therefore the estimation of antioxidant capacity of synthesized nanoparticles was necessary. It can be easily identified that antioxidant activities of all the materials investigated were dose-dependent and their IC_{50} values are indicated in Figure 19.2 for simple correlation of antioxidant efficacy.

The zone of inhibition is simple, convenient, and expeditious technique to perceive the antibacterial efficacy of substances that can prevent the growth of pathogenic

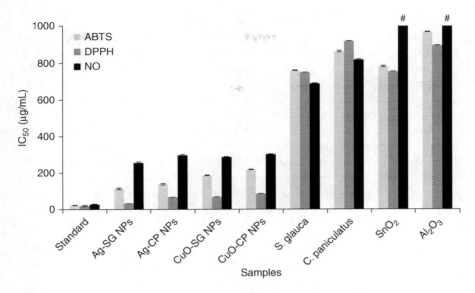

FIGURE 19.2 Minimal inhibition percentage of free radicals expressed by extracts, nanoparticles, and standard evaluated using scavenging activity of ABTS, DPPH, and NO radicals.

microorganisms and is marked by the formation of increasing clear zone around the material used for evaluation [32]. It is generally understood that secondary metabolites present in these medicinal plants aid in synthesizing the aforementioned nanoparticles will bring about an increment in therapeutic index. Therefore, employing such materials for health concern will be an added advantage.

19.3.1.1 Decolorization Activity of ABTS Radical Cation

The decolorization of blue-green ABTS•+ from the electron acquired by antioxidant will produce colorless neutral ABTS form and this process was quantified by over through of representative far wave (i.e., 734 nm) optical density [33]. It can be observed from Figure 19.2 that ascorbic acid had the capacity to inhibit free radicals at 24.10 ± 0.06 µg/mL concentration and this was the highest among the materials used for evaluation. Considering the materials we had prepared, Ag-SG NPs exhibited the maximal 112.56 ± 3.47 µg/mL of antioxidant activity compared to the remaining nanoparticles and was nearly fourfold less than the standard. Effective inhibition was noticed in the case of Ag-CP NPs and then by CuO-SG and CuO-CP NPs. Meanwhile, the medicinal plant extracts revealed the activity of 757.21 ± 2.90 µg/mL and 861.03 ± 2.98 µg/mL respectively, which was less than six- to sevenfold than the synthesized nanoparticles using the same extracts. This clearly signifies that by the synergistic effect of therapeutic extracts and nanoparticles, the antioxidant activity could be readily elevated. The chemically prepared SnO_2 indicated 778.64 ± 0.68 µg/mL scavenging ability while Al_2O_3 showed 966.83 ± 0.99 µg/mL.

19.3.1.2 Analysis by DPPH Radical Scavenging

The DPPH-free radical, when correlated to rest of free radicals, is distinguished on the account of being a stable radical due to its property of delocalization over the entire molecule as a spare electron with an intention to prevent dimerization [34]. The spare electron has absorption maxima at 517 nm and without scavenging, intense violet is commonly seen, but by the antioxidant activity the color starts decreasing due to pairing of electrons. Formally the standard ascorbic acid illustrated the superior antioxidant activity at 20.80 ± 2.60 µg/mL concentration. Meanwhile, Ag-CP NPs also demonstrated potent antioxidant ability at 67.99 ± 0.11 µg/mL dosage and CuO-SG NPs indicated similar results. The extracts of these prepared nanoparticles i.e., *S. glauca* and *C. paniculatus* revealed their prevention of DPPH radical formation at 745.61 ± 0.46 µg/mL and 916.42 ± 0.69 µg/mL concentration. Relating to chemically synthesized nanoparticles, SnO_2 was effective in inhibiting the oxidants at 751.66 ± 1.68 µg/mL, whereas Al_2O_3 expressed activity at 894.90 ± 0.38 µg/mL concentration.

19.3.1.3 Measurement of NO Radical-Scavenged Activity

Nitric oxide radical emits high intensity of pink color at 540 nm produced by nitrite with the help of Griess reagent. Much like the other two assays, the standard ascorbic acid here again exceeded high in scavenging the free radicals by 28.21 ± 0.73 µg/mL concentration. Nevertheless, Ag-SG NPs proved efficient antioxidant capacity at the concentration of 254.43 ± 3.76 µg/mL. This accessibility of antioxidants was immediately followed by Ag-CP, CuO-SG, and CuO-CP NPs and was certainly realized by their IC_{50} values. For comparison, we have measured the activity expressed by extracts of SG and CP and found their half inhibition (IC_{50}) percentage at 685.70 ± 1.45 µg/mL and 816.69 ± 2.38 µg/mL dosages. The chemically synthesized nanoparticles did not possess any antioxidant ability by this method.

19.3.2 ESTIMATION OF ANTIBACTERIAL EFFICACY

The bacterial growth inhibition zones displayed by concentration-dependent Ag-SG, Ag-CP, CuO-SG, CuO-CP NPs along with chemically synthesized SnO_2 and Al_2O_3 NPs are mentioned in Table 19.1. To determine the efficiency of antibacterial activity, the correlation was made with the standard cefpodoxime that is considered as positive control.

The silver nanoparticles synthesized using extract of *C. paniculatus* had exhibited the highest antibacterial efficacy against both the strains, with 18 mm and 15 mm of inhibition zones at 5 µg concentration against *S. pyogenes* and *E. coli*, while the standard had shown 24 mm and 19 mm clear zones at 30 µg concentration against the same organisms. For chemically synthesized nanoparticles, the bactericidal activity against *E. coli* by Al_2O_3 NPs had exhibited 12 mm ZOI for 10 µg, while SnO_2 NPs and ciprofloxacin both revealed 14 mm zone for 10 µg which imply that the synthesized NPs had acquired potent antibacterial activity similar to standard ciprofloxacin (21 mm) against the tested microbe. However, in case of bactericidal analysis against *P. aeruginosa*, either of prepared NPs unveiled 9 mm ZOI for 10 µg, and on the other hand, ciprofloxacin (14 mm) evidenced highly significant control on their growth.

TABLE 19.1
The Dose-dependent Antibacterial Ability of Standard and Green Synthesized Ag and CuO NPs against *S. pyogenes* and *E. coli* organisms

Sl. No	Nano-particles	*S. pyogenes* Zone of Inhibition in Diameter (mm)					*E. coli* Zone of Inhibition in Diameter (mm)				
		2.5 µg	2.5 µg	5 µg	5 µg	Standard (30 µg)	2.5 µg	2.5µg	5 µg	5 µg	Standard (30 µg)
1	AG-SG	9	8	11	12	22	11	10	14	14	18
2	AG-CP	12	12	18	18	24	5	6	15	15	19
3	CuO-SG	5	4	12	13	22	2	4	15	14	21
4	CuO-CP	3	5	8	7	24	5	6	9	9	24

19.3.3 EFFECTIVENESS OF ANTIPROLIFERATION ON CANCER AND NON-CANCER CELL LINES

Succeeding the powerful antioxidant activity exhibited by the green synthesized nanoparticles from previous examinations, although SnO_2 and Al_2O_3 NPs did not possess sufficient antioxidant capacity. For the remaining nanoparticles, the next predominant action to be considered for the anticancer applications is by knowing its proficiency in hindering the growth of cancer cells. Henceforth, the antiproliferative effect of these nanoparticles along with their plant extracts was conducted by MTT method, using cancerous cell lines like MCF-7 and HT-29.

Almost all samples used for analysis indicated efficient IC_{50} value that substantiates the potential antitumor activity on cancer cell lines except for SnO_2 and Al_2O_3 NPs as they could not inhibit cancer cells. Also, green synthesized nanoparticles showed superior cytotoxic activity than crude plant extracts. Ag-SG NPs displayed exceptionally good inhibition of both MCF-7 and HT-29 cell lines with IC_{50} values of 70.84 ± 0.67 and 158.24 ± 0.89 µg/mL in correlation to other nanoparticles. Next CuO-CP NPs indicated better activity with 97.39 ± 0.48 and 205.11 ± 0.39 µg/mL followed by CuO-SG NPs and Ag-CP NPs showed efficacious anticancer activity.

TABLE 19.2
ZOI Method to Determine the Bactericidal Activity of Al_2O_3 and SnO_2 NPs

Sl. No	Sample Name	Concentration (µg)	ZOI (mm)	
			E. coli	*P. aeruginosa*
1	Al_2O_3 NPs	10	12	9
2	SnO_2 NPs	10	14	9
3	Ciprofloxacin	10	14	21

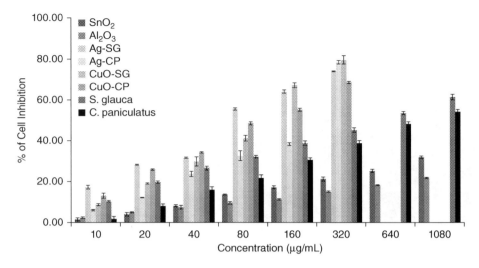

FIGURE 19.3 Histogram revealing the dose-dependent anticancer activities of chemically synthesized NPs, green synthesized NPs, and their extracts on MCF-7 cell line.

Since the green synthesized Ag and CuO NPs were effective on cancer cells they had to be checked for their noncytotoxic ability on normal cells. The ability not to hinder the growth of normal cells could be evidenced in Figure 19.5 by utilizing MCF-10A cell line. All the green synthesized nanoparticles that were capable of inhibiting cancer growth showed concentration above 1000 µg/mL for growth inhibition of normal cells. To further comprehend the cancer-killing ability at cellular level, CuO-CP NPs were selected among the eight treatments in order to validate anticancer analysis by apoptosis, cell cycle evaluation, and Caspase-3 expression.

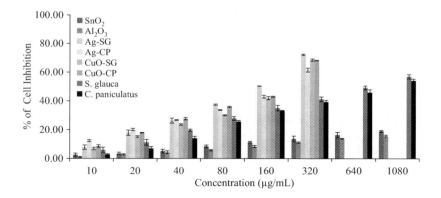

FIGURE 19.4 Histogram revealing the dose-dependent anticancer activities of chemically synthesized NPs, green synthesized NPs, and their extracts on HT-29 cell line.

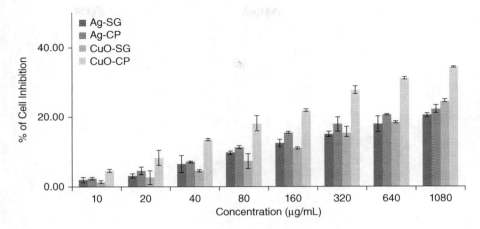

FIGURE 19.5 Histogram revealing the dose-dependent anticancer activities of green synthesized nanoparticles on normal MCF-10A cell line.

19.3.3.1 Apoptosis by FITC Annexin V-PI

The apoptotic mechanism is very much essential for the elimination of worn-out and damaged cells present within the defined tissue. Due to the fact that cancer cells will be having trouble to undergo apoptosis, it is important to treat them using anticancer agents that would allow them to experience cell death. Measurable analysis of FITC – Annexin V was made apparent by obstructed cell membrane. In apoptosis, phospholipid will be exposed outward, gets attached by Annexin V, and will be scrutinized by PI. The inhibition of growth of cancerous cells at lesser value is a standard distinctive of probable anticancer therapy. The efficient IC_{50}-valued CuO-CP NPs (97.39 ± 0.48 µg/mL) administered on cancerous cells had majority of cells in late apoptotic region (71.29%) and then in early apoptotic and necrotic region. The control had almost all of the cells in viable region, while the cisplatin (standard) treated sample indicated more cells in necrotic region (39.52%) followed by distribution in late apoptotic (35.84%) and viable region (23.11%). Hence, it can be emphasized that the anticancer activity of CuO-CP NPs was more enhanced than the standard.

19.3.3.2 Propidium Iodide-based Cell Cycle Analysis

Quantitative identification and cell cycle specificity analysis of apoptosis are very much essential for examination of the molecular mechanism of apoptosis and cell cycle progression. The PI mapping of singlet cells of cell cycle analysis in MCF-7 cell line discovered with CuO-CP NPs at 97.39 ± 0.48 µg/mL concentration for 24 hours was characterized in Figure 19.7. Cancer cells indicated DNA fragmentation as observed in Figure 19.7(e) with increment in percentage of cells i.e., 75.28% in sub G_0/G_1 (apoptotic) phase in contrast to control which had majority of cell count in G_0/G_1 (growth) phase (65.75%). This implies that CuO-CP NPs administered within MCF-7 cell line were not qualified to go course of G_2/M (metaphase) and hence G_2/M passageway was considered to be collapsed. PI was also discolored against side scatter (SSC) using the same cells, was found that morphology was impaired

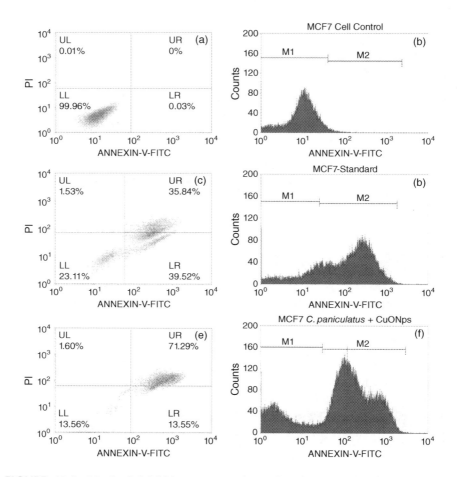

FIGURE 19.6 Maximal inhibition concentrations of control, standard (cisplatin), and CuO-CP NPs (a, b, & c) against MCF-7 cell line determined for entire types of death in specified quadrants, and DNA demethylation in cancer cells toward positive Annexin V site in respective samples (d, e, & f).

in control cells, and was aligned when treated with CuO-CP NPs and/or standard. This event was marked by the movement of cells toward negative side of PI in Figure 19.7(b, d, & f), which further supports our cell cycle assessment.

19.3.3.3 Expression of Caspase-3 in MCF-7 Cell Line

Caspases are a class of genes prominent for enduring homeostasis through regulating cell death and inflammation. Besides cell cycle analysis, the advanced substantiation recommends that proteases of caspase family render a crucial task in functioning of apoptosis. Caspase-3 activity as an effector caspase is found to be connected with late apoptosis phase, as there was an increment in Caspase-3 expression with elevation in late apoptotic cells percentage [35]. To examine that apoptosis had occurred due to Caspase-3 in cancer cells, cell

Antioxidant, Bactericidal, Antihemolytic, and Anticancer Assessment Activities 385

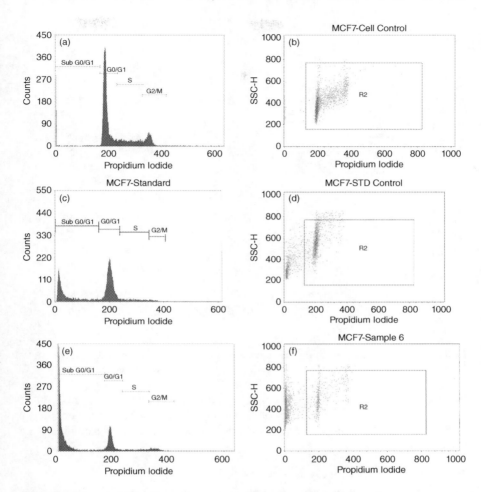

FIGURE 19.7 FACS Calibur indication of 10,000 singlet cells selected for cell cycle interpretation of control, standard (cisplatin), and CuO-CP NPs against MCF-7 cells (b, d, & f), PI mapping of gated cell singlets distinguishes cells at their respective cycle phases (a, c, & e).

culture dosed using CuO-CP NPs of known IC_{50} value were scrutinized for this test. The relative expression of mean fluorescence intensity emitted by MCF-7 cell line administered using CuO-CP NPs and control for 24 hours is indicated in Figure 19.8. It could be perceived that the Caspase-3 specific substrate, i.e., fluorogenic Ac-DEVD-AMC, had increment in proteolytic activity due to their treatment.

Therefore, the intensity of Caspase-3 expression was found to be high in MCF-7 cell line dosed by CuO-CP NPs showing 53.81 Mean Fluorescence Unit (MFU) when compared to 11.22 MFU of control. Additionally, during differentiation between negative (M1) and positive expression (M2) of Caspase-3 FITC-influenced apoptosis, control cells were found to be present predominately with

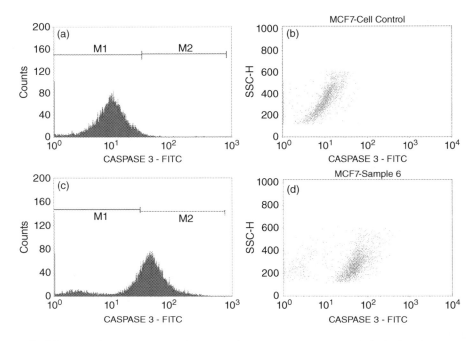

FIGURE 19.8 Caspase-3 expression indication of (a & c) control and CuO-CP NPs treatments in MCF-7 cell line, (b & d) their scattering marked in flow cytometry.

98.55% in M1 expression signifying the living cells. Nonetheless, the MCF-7 treated CuO-CP NPs had majority of cells (i.e., 68.90%) in M2 region compared to M1 with 31.43%, which promotes death of cells which further confirms the Caspase-3 activity.

19.3.3.4 Dual Stain Analysis of MCF-7 Cell Line

The process of cell death stimulated by CuO-CP NPs and Ag-SG NPs was analyzed using AO/EB dual staining method. The most distinguished morphological changes in cancer cells, when treated with nanoparticles, noticed during apoptosis after staining were cytoplasmic condensation, contraction of cell membrane, generation of several cell surface protuberances at the plasma membrane, and accumulation of nuclear chromatin into dense masses, beneath nuclear membrane [36]. The fluorescent AO/EB dye that binds DNA illustrated distinct stages in life and death and those cells that took up green color exhibit viable cells, bright green depicts early apoptotic stage, red staining indicates nonviable or necrotic cells while reddish or orange color describes the late apoptotic cells. To analyze such processes, the fluorescent images of untreated and nanoparticles administered cancer cells were represented in Figure 19.9. The control cells were indicated in the first row, whereas CuO-CP NPs and Ag-SG NPs treated cells were presented in the second and the third rows. Further, CuO-CP NPs with their IC_{50} value (97 µg/mL) treated cells could selectively take up EB dye as their membrane integrity was lost.

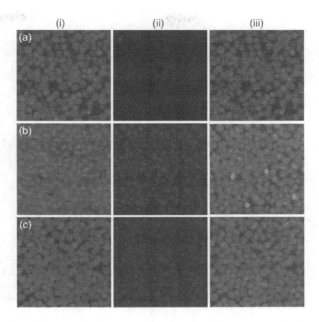

FIGURE 19.9 Images obtained by fluorescent microscope at 40× magnifications with (i) AO, (ii) EB, and (iii) overlay of both stains of MCF-7 cell line treated with (a) control, (b) CuO CP NPs, and (c) Ag SG NPs.

19.3.4 Prevention of Hemolysis by Prepared Nanoparticles

The hemolysis assay is the method that is regularly employed by therapeutic-related clinical laboratories and research laboratories concerned with in complement-based examination [37]. Entry of any foreign particles into blood could be figured out using this assay, as they are sensitive, reliable upon exposure, and can be regarded as preliminary examination for animal studies.

As per the Standard Test Method for Analysis of Hemolytic Properties of Nanoparticles (E2524-08), if the substance used to test compatibility results in more than 5% of hemolysis, then it is called hemolytic. Percentage of hemolysis exhibited by above-mentioned samples are illustrated in Figure 19.10, and it is clearly perceived that Ag-SG NPs indicated the lowest activity i.e., 0.34 ± 0.23% and hence it is considered as nonhemolytic. Moreover, Ag NPs showed slightly better nonhemolytic activity when correlated with CuO NPs. Whereas the extracts that were used for synthesis revealed slightly hemolytic activity as their percentage of hemolysis were between 2–5%.

19.3.4.1 Effectiveness of Nanoparticles on RBC Morphology

The above experimental observations could be more evidenced by looking into the morphological changes in RBCs using microscopic analysis and for this reason we have used SEM for better magnifications. These foreign particles can provoke the pore generation on membrane of RBCs and eventually proceed to osmotic lysis [38].

388 Handbook of Sustainable Materials

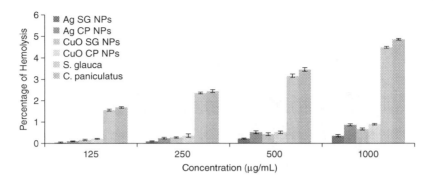

FIGURE 19.10 Indication of hemolysis activity exhibited by (a) Ag-SG NPs, (b) Ag-CP NPs, (c) CuO-SG NPs, (d) CuO-CP NPs, (e) extracts of *S. glauca*, and (f) extracts of *C. paniculatus*.

It can be witnessed from Figure 19.11 that Ag and CuO NPs synthesized using medicinal plants when treated with RBCs resulted in deposition of nanoparticles on their outer membrane but there was no alteration evidenced and hence proving their compatibility. When the extracts of SG and CP at same concentration of NPs were incorporated, we could notice more deposition of them on membrane of RBCs, yet the changes in their cytology were not so prominent. We could only observe the change in morphology, when RBCs treated with water were making the cells to bulge and might lead to bursting out their organelles.

19.3.5 In Vivo Anticancer Threshold

As a continuation of research, the prepared CuO-CP NPs had to be evaluated at *in vivo* environmental circumstances in order to determine their efficacy in live

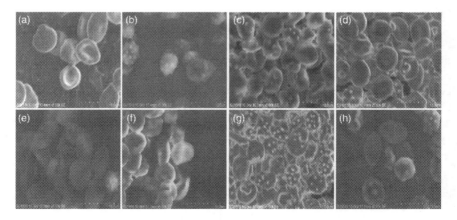

FIGURE 19.11 Examining the morphological evidence of RBCs treated with (a) Ag-SG NPs, (b) Ag-CP NPs, (c) CuO-SG NPs, (d) CuO-NPs, (e) PBS, and (f) Deionized water.

Antioxidant, Bactericidal, Antihemolytic, and Anticancer Assessment Activities **389**

organisms. Laboratory mice model-bearing tumors are the incidences with subject where the biochemical and molecular component specificities and histopathological indications can be designed as they are fundamental of most cancer chemotherapy screening test [39]. There are evidences with data that tumor requires oxygen from the body fluid, considering salubrious admired nourishment to evolve as giant explosion by once a malignant tumor builds to a certain proportion and hence tumor addresses the chemical-signals that stimulate advancement of new blood vessels, which bring the body fluid toward themselves [40]. The first toxicity analysis executed on formulation was the assessment of acute toxicity evaluated by the administration of single exposure. The median lethal dose of CuO-CP NPs was demonstrated to be 2000 mg/kg BW and the animals with this concentration exhibited some changes in behavior, ingestion, sluggish movement, and loss of hair at end of their life. The experiment performed by repeated dosage of CuO-CP NPs with one-twentieth and one-fortieth concentration of LD_{50} i.e., 50 mg/kg and 100 mg/kg BW were delivered orally for 21 days without their routine diet being affected.

19.3.5.1 Survivability and Growth Performance

Prolongation of longevity has been considered as amongst the influential determinant to demonstrate the anticancer remedy of a synthesized medication. Based on the fundamental physical analysis performed, the gain in BW, MST, and %ILS measurements are enumerated in Table 19.3. The body mass of all animals increased as in days passed although the EAC-bearing mice had acquired more weight due to existence of fluid buildup because of tumor within them. In contradictory, the

---

TABLE 19.3

Variation in Growth and Survivability Due to the Dosage of CuO-CP NPs and Standards in Laboratory Mice

Groups	Group I	Group II	Group III	Group IV	Group V	Group VI
Gain in body weight (g)	$11.13 \pm 0.27^*$	$20.61 \pm 1.38^{**}$	$16.31 \pm 1.41^{**}$	$12.82 \pm 0.5^{**}$	$14.65 \pm 0.75^{**}$	$12.32 \pm 0.85^{**}$
Percentage decrease in BW compared with EAC control	-	-	20.87	37.71	28.83	40.14
MST (days)	$26.98 \pm 2.05^*$	$13.37 \pm 0.94^{**}$	$20.62 \pm 1.17^{**}$	$23.89 \pm 1.23^{**}$	$21.12 \pm 1.18^{**}$	$26.87 \pm 0.63^{**}$
Percentage ILS	-	-	54.17	78.48	57.91	100

Values were determined as mean $\pm$ S.E.M (n = 8), of which * indicates p < 0.01 and ** indicates p < 0.001.

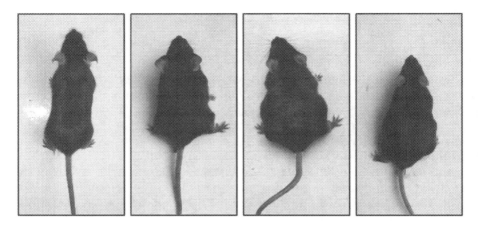

FIGURE 19.12 External appearance of C57 mice present in groups on the zeroth day i.e., before the initiation of subacute toxicity studies.

animal group dosed with CuO-CP NPs in EAC-bearing mice indicated a significant (p < 0.001) decrement in BW correlated to EAC control group. It is evident from the values showed, the mice of Group IV treated with CuO-CP NPs (100 mg/kg BW) and Group VI treated with 5-FU (20 mg/kg BW) had equivalent weights, correlated to the animals in Group I i.e., normal mice group. Likewise, survivability of animals in each group was estimated using MST and % ILS assessments. It can be specifically observed from Table 19.3 that animals treated with CuO CP NPs (100 mg/kg) had significant (p < 0.001) increment of up to 23.88 ± 1.23 days.

This was comparatively high toward MST of EAC control mice of Group II i.e., with 13.38 ± 0.94 days indicating anticancer potential of CuO-CP NPs. Treating EAC-bearing mice with CuO-CP NPs and standards had elevated their lifespan to

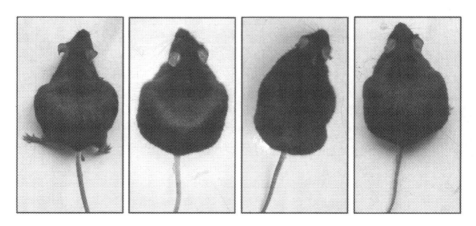

FIGURE 19.13 External view of C57 mice existent in groups on the twenty-first day, i.e., after the subacute toxicity studies.

Antioxidant, Bactericidal, Antihemolytic, and Anticancer Assessment Activities   391

TABLE 19.4
Differences in Hematological Profile of Laboratory Mice Observed in Each Group of Study

Groups	Group I	Group II	Group III	Group IV	Group V	Group VI
WBC ($10^3/\mu L$)	4.73 ± 0.12**	13.33 ± 0.29*	8.73 ± 0.27**	6.00 ± 0.30**	7.70 ± 0.50**	5.74 ± 0.09**
RBC ($10^6/\mu L$)	6.06 ± 0.05**	2.81 ± 0.21*	4.06 ± 0.18**	5.41 ± 0.19**	4.45 ± 0.26**	5.90 ± 0.01**
Hb (g/dL)	12.68 ± 0.05**	6.823 ± 0.38*	9.570 ± 0.28**	12.04 ± 0.56**	11.26 ± 0.48**	12.52 ± 0.04**

Values were determined as mean ± S.E.M (n = 8), of which * indicates p < 0.01 and ** indicates p < 0.001.

more than 50%. Group IV or the animals dosed with CuO-CP NPs (100 mg/kg) presented 78.47% rise in the span of life, nevertheless 5-FU had a cent percent enhancement juxtaposed to normal control group. Therefore, it is apparent by the results of this analysis that, with increase in concentration of CuO-CP NPs administered to mice induced with EAC, their anticancer activity was improved and animals treated with CuO-CP NPs (100 mg/kg) had similar values to that of commercially available standard 5-FU. The study performed for this activity was in agreement with work conducted by Uddandrao et al. [41].

19.3.5.2   Hematological Indices after Treatment

The consequential divergence succeeding the single and repeated oral delivery in EAC harboring mice contradictory to normal mice group were generally perceived more in WBC and hemoglobin count. These variations were more evident after subacute administration, suggesting the capability for summative antitumor effects of CuO-CP NPs. By estimating the hematological parameters, it is certain that the RBC and Hb level in tumor-bearing mice, substantially reduced to 2.81 ± 0.21 $10^6/\mu L$ and 6.82 ± 0.38 g/dL and also the WBC level was elevated to 13.33 ± 0.29 $10^3/\mu L$. These drastic differences of hematological values, in contrast to normal mice group, were found to be reverted back to accustomed range when treated with CuO-CP NPs and 5-FU.

19.3.5.3   Accomplishment of Tissue Antioxidant Activity

Antioxidant enzymes are competent of inhibiting or delaying oxidation by donating an electron to unstable free radicals to prevent cellular deterioration. In the process of routine metabolic activity, extremely reactive species called free radicals are produced in the body; although, they can further be brought in by the environment and these compounds are intrinsically unstable because they have lone pair of electrons and therefore get converted to immensely reactive [42]. It is evident from the results that treating the mice with CuO-CP NPs of 50 mg/kg BW did not signify effective advancement in GSH antioxidant capacity within the cancer-induced mice; however, with the dose elevation to 100 mg/kg BW, there was a remarkable increment of

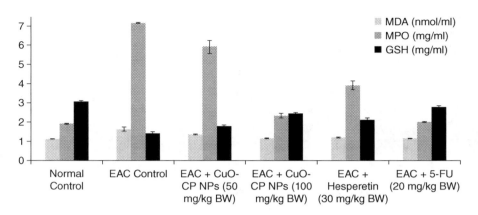

FIGURE 19.14 Measurement of oxidative stress in C57 mice of experimental groups, estimated using MDA, MPO, and GSH methods.

2.44 ± 0.06 mg/mL in GSH content indicating the potent antioxidant activity. They were proportional to the activity exhibited by the standard, 5-FU with 2.78 ± 0.07 mg/mL, although hesperetin showed 2.13 ± 0.09 mg/mL of endogenous nonenzymatic GSH content. The amount of reduced GSH was decreased in EAC-bearing mice and it might be a consequence of their usage by exorbitant content of free radicals [43].

There was a consequential increment in the values of cellular MDA and MPO in untreated Group II with 1.635 ± 0.11 nmol/mL and 7.167 ± 0.17 mg/mL, as lipid peroxidation in liver tissue was considerably raised in this group in contrast to the normal mice group. Our investigation also showed that delivering 100 mg/kg BW of CuO-CP NPs to mice injected with tumor could proficiently decrease the levels of MDA and MPO to 1.178 ± 0.02 nmol/mL and 2.347 ± 0.12 mg/mL, which were equivalent to the values obtained by using 5-FU. Thus, demonstrates the potentiality of CuO-CP NPs like inhibitors of cancer-initiated intracellular oxidative stress, depicting their consortium correlation amongst antioxidant and anti-inflammatory activity to antitumor capability.

19.3.5.4 Efficiency of Organ Histopathology

By the excision of organs, like kidney, liver, and spleen, the gross examinations, like infiltration, degeneration, inflammation, restoration, etc., were evidenced in experimental mice groups. Free radicals induce toxicity in multiple organs and cause failure in their functioning that could be observed from histology. The scrutiny performed on mice over 21 days was utilized to interpret the consequences of utilizing CuO-CP NPs as antitumor agents. Histopathology of EAC control mice group disclosed superior deformity in each and every organ examined under study yet degeneration was restored in mice group treated with CuO-CP NPs and 5-FU. The histological section of kidney belonging to normal control had typical composition of glomerulus-bearing tuft of capillaries, thinly covered by epithelial cells placed within Bowman's capsule (arrow, as shown in Figure 19.15). The tumor-bearing mice had glomeruli

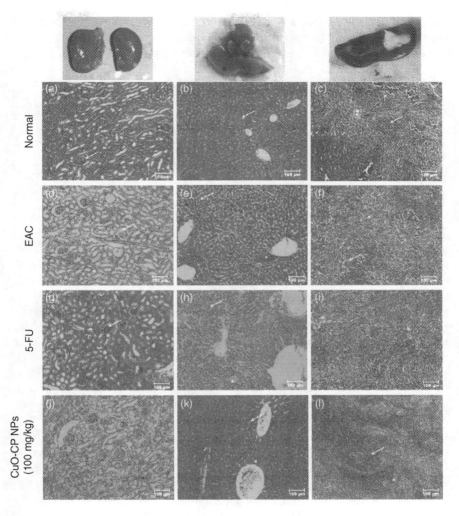

FIGURE 19.15 Pictorial representation of H & E staining concerned with (A, D, G, & J) kidney, (B, E, H, & K) liver, and (C, F, I, & L) spleen exhibiting reversible conditions of (J, K, & L) CuO-CP NPs compared to (G, H, & I) standard, (D, E, & F) EAC induced and (A, B, & C) normal laboratory mice.

hypercellularity with vacuoles which might have occurred because of more number of intrinsic cells or from inflation of leukocytes in capillary lumina; tubular necrosis, renal lesions, and hyaline casts in proximal and distal convoluted tubules with fluid accumulation were prominent (asterisk, as shown in Figure 19.15). Congestions near glomerulus might have partitioned them from Bowman's capsule (arrow, as shown in Figure 19.15).

The mice from EAC harboring group in the liver section showed conspicuous nephrotoxicity due to cellular and glomerular infiltration (asterisk, as shown in

Figure 19.15), focal angiectasis with disrupted cords, sporadic zones of coagulation, condensed nuclei (arrow, as shown in Figure 19.15) with portal inflammation and nodule formation. The microscopic examination of spleen exhibited usual white pulp with T-dependent lymphocytes (arrow, as shown in Figure 19.15) and red pulp along with healthy sinusoids and proper filtration to impart red pigment (asterisk, as shown in Figure 19.15). EAC harboring group indicated loss of splenic architecture, dominant pathology, cellular proliferation, and increased megakaryocyte with irregular lymphocytes (arrow, as shown in Figure 19.15) and red blood cells (asterisk, as shown in Figure 19.15).

19.4 CONCLUSION

Despite Al_2O_3 and SnO_2 NPs showed slight antioxidant capacity by ABTS and DPPH methods yet their activity was not so effective. Nevertheless, the herbal extracts (*S. glauca* and *C. paniculatus*) indicated better antioxidant ability and further when they were used to synthesize Ag and CuO NPs, the activity almost increased to four times, and this could be noted by the IC_{50} value of Ag SG NPs that showed 34.01 ± 0.64 µg/mL of scavenging ability by DPPH method. The functionalization of green synthesized and chemically prepared nanoparticles along with standard inhibited the growth of Gram-negative and Gram-positive organisms in concentration-dependent manner. It's prevalent that green synthesized nanoparticles had better antibacterial activity than chemically synthesized nanoparticles from ZOI test method.

The differentiating peculiarity of noncytotoxic effect of Ag and CuO NPs synthesized from medicinal plants on virulent cells against normal cells was adequately accomplished. The chemically synthesized nanoparticles, i.e., SnO_2 and Al_2O_3 NPs, could not control the growth of cancer cells. The medicinal plant extracts showed inhibition of cancer cells yet their activities were further enhanced four- to fivefold by synthesizing them with efficient nanoparticles which could destroy cancer cells. Likewise, to understand the anticancer effect at cellular level, one best antiproliferative nanoparticles was chosen i.e., CuO-CP NPs for further analysis. Apoptosis resulted in finding the majority of cells i.e., about 71.29% were in late apoptotic phase. Then cell cycle analysis indicated that the cancer cells were arrested in sub G_0/G_1 phase, which would help to block the growth stages of cancer. Furthermore, the green synthesized NPs signified the antihemolytic and biocompatible nature and hence we could proceed with animal testing using them.

The *in vivo* studies performed using C-57 mice were successful due to restoring ability and delay in mortality exhibited by treating repeated doses of CuO-CP NPs. Using subacute concentration, it was possible to decrease the outgrown mass and increase the duration of laboratory mice. Hematological profile could be restored using nanoparticles and antioxidant activities determined using the liver tissue homogenate showed effective clearance of free radicals observed by increment in GSH content with concomitant decrement of LPO and MPO levels. This ascitic tumor model was efficiently used for verification of antineoplastic properties exhibited by CuO-CP NPs in C-57 mice, as they proficiently inhibited the tumor growth and therefore this might provide the optimistic cancer therapy.

REFERENCES

1. de Jong. 2008. Drug delivery and nanoparticles: Applications and hazards. International Journal of Nanomedicine 3:133–149. https://doi.org/10.2147/IJN.S596
2. Ansari S, Sameem M, Islam F. 2012. Influence of nanotechnology on herbal drugs: A review. Journal of Advanced Pharmaceutical Technology & Research 3:142–146. https://doi.org/10.4103/2231-4040.101006
3. Bisht S, Feldmann G, Soni S, et al. 2007. Polymeric nanoparticle-encapsulated curcumin ('nanocurcumin'): A novel strategy for human cancer therapy. Journal of Nanobiotechnology 5:1–18. https://doi.org/10.1186/1477-3155-5-3
4. Khan I, Saeed K, Khan I. 2019. Nanoparticles: Properties, applications and toxicities. Arabian Journal of Chemistry 12:908–931. https://doi.org/10.1016/j.arabjc.2017.05.011
5. BarathManiKanth S, Kalishwaralal K, Sriram M, et al. 2010. Anti-oxidant effect of gold nanoparticles restrains hyperglycemic conditions in diabetic mice. Journal of Nanobiotechnology 8:1–15. https://doi.org/10.1186/1477-3155-8-16
6. Vaiserman A, Koliada A, Zayachkivska A, Lushchak O. 2020. Nanodelivery of natural antioxidants: An anti-aging perspective. Frontiers in Bioengineering and Biotechnology 7:1–19. https://doi.org/10.3389/fbioe.2019.00447
7. Khalil I, Yehye WA, Etxeberria AE, et al. 2019. Nanoantioxidants: Recent trends in antioxidant delivery applications. Antioxidants 9:1–30. https://doi.org/10.3390/antiox9010024
8. Giridasappa A, Ismail SM, Rangappa D, et al. 2021. Antioxidant, antiproliferative and antihemolytic properties of phytofabricated silver nanoparticles using Simarouba glauca and Celastrus paniculatus extracts. Applied Nanoscience 11:2561–2576. https://doi.org/10.1007/s13204-021-02084-z
9. Sharif MS, Aqeel M, Haider A, et al. 2021. Photocatalytic, bactericidal and molecular docking analysis of annealed tin oxide nanostructures. Nanoscale Research Letters 16:1–16. https://doi.org/10.1186/s11671-021-03495-1
10. Kamaraj P, Vennila R, Arthanareeswari M, Devikala S. 2014. Biological activities of tin oxide nanoparticles synthesized using plant extract. World Journal of Pharmacy and Pharmaceutical Sciences 3:382–388.
11. Sudhaparimala S, Vaishnavi M. 2016. Biological synthesis of nano composite SnO_2-ZnO – Screening for efficient photocatalytic degradation and antimicrobial activity. Materials Today: Proceedings 3:2373–2380. https://doi.org/10.1016/j.matpr.2016.04.150
12. Nasir Z, Shakir M, Wahab R, et al. 2017. Co-precipitation synthesis and characterization of Co doped SnO_2 NPs, HSA interaction via various spectroscopic techniques and their antimicrobial and photocatalytic activities. International Journal of Biological Macromolecules 94:554–565. https://doi.org/10.1016/j.ijbiomac.2016.10.057
13. Srinivasan M, Rajabi M, Mousa S. 2015. Multifunctional nanomaterials and their applications in drug delivery and cancer therapy. Nanomaterials 5:1690–1703. https://doi.org/10.3390/nano5041690
14. Salata OV. 2016. Applications of nanoparticles in biology and medicine. Journal of Nanobiotechnology 2:1–6. https://doi.org/10.1186/1477-3155-2-3
15. Roy I, Ohulchanskyy TY, Pudavar HE, et al. 2003. Ceramic-based nanoparticles entrapping water-insoluble photosensitizing anticancer drugs: A novel drug–carrier system for photodynamic therapy. Journal of the American Chemical Society 125:7860–7865. https://doi.org/10.1021/ja0343095
16. Adebiyi OE, Olayemi FO, Ning-Hua T, Guang-Zhi Z. 2017. In vitro antioxidant activity, total phenolic and flavonoid contents of ethanol extract of stem and leaf of Grewia carpinifolia. Beni-Suef University Journal of Basic and Applied Sciences 6:10–14. https://doi.org/10.1016/j.bjbas.2016.12.003

17. Hemlata, Meena PR, Singh AP, Tejavath KK. 2020. Biosynthesis of silver nanoparticles using cucumis prophetarum aqueous leaf extract and their antibacterial and antiproliferative activity against cancer cell lines. ACS Omega 5:5520–5528. https://doi.org/10.1021/acsomega.0c00155
18. Asemani M, Anarjan N. 2019. Green synthesis of copper oxide nanoparticles using Juglans regia leaf extract and assessment of their physico-chemical and biological properties. Green Processing and Synthesis 8:557–567. https://doi.org/10.1515/gps-2019-0025
19. Kumar V, Gundampati RK, Singh DK, et al. 2016. Photo-induced rapid biosynthesis of silver nanoparticle using aqueous extract of Xanthium strumarium and its antibacterial and antileishmanial activity. Journal of Industrial and Engineering Chemistry 37:224–236. https://doi.org/10.1016/j.jiec.2016.03.032
20. Choudhary MK, Kataria J, Cameotra SS, Singh J. 2016. A facile biomimetic preparation of highly stabilized silver nanoparticles derived from seed extract of Vigna radiata and evaluation of their antibacterial activity. Applied Nanoscience 6:105–111. https://doi.org/10.1007/s13204-015-0418-6
21. Giridasappa A, Rangappa D, Shanubhoganahalli Maheswarappa G, et al. 2021. Phytofabrication of cupric oxide nanoparticles using Simarouba glauca and Celastrus paniculatus extracts and their enhanced apoptotic inducing and anticancer effects. Applied Nanoscience 11:1393–1409. https://doi.org/10.1007/s13204-021-01753-3
22. Kalita S, Kandimalla R, Devi B, et al. 2017. Dual delivery of chloramphenicol and essential oil by poly-ε-caprolactone–Pluronic nanocapsules to treat MRSA-Candida co-infected chronic burn wounds. RSC Advances 7:1749–1758. https://doi.org/10.1039/C6RA26561H
23. Samson Adewoyin A, Adeyemi O, Omolola Davies N, Abiola Ogbenna A. 2019. Erythrocyte Morphology and Its Disorders. In: Erythrocyte. IntechOpen, pp 1–10. https://doi.org/10.5772/intechopen.86112
24. OECD Test Guideline 425. 2008. Test No. 425: Acute Oral Toxicity: Up-and-Down Procedure. OECD.
25. Prasanna R, Ashraf EA, Essam MA. 2017. Chamomile and oregano extracts synergistically exhibit antihyperglycemic, antihyperlipidemic, and renal protective effects in alloxan-induced diabetic rats. Canadian Journal of Physiology and Pharmacology 95:84–92. https://doi.org/10.1139/cjpp-2016-0189
26. Islam F, Ghosh S, Khanam JA. 2014. Antiproliferative and hepatoprotective activity of metabolites from Corynebacterium xerosis against Ehrlich ascites carcinoma cells. Asian Pacific Journal of Tropical Biomedicine 4:284–292. https://doi.org/10.12980/APJTB.4.2014C1283
27. Sunil D, Isloor AM, Shetty P, et al. 2013. In vivo anticancer and histopathology studies of Schiff bases on Ehrlich ascitic carcinoma cells. Arabian Journal of Chemistry 6:25–33. https://doi.org/10.1016/j.arabjc.2010.12.016
28. Ohkawa H, Ohishi N, Yagi K. 1979. Assay for lipid peroxides in animal tissues by thiobarbituric acid reaction. Analytical Biochemistry 95:351–358. https://doi.org/10.1016/0003-2697(79)90738-3
29. Mullane KM, Kraemer R, Smith B. 1985. Myeloperoxidase activity as a quantitative assessment of neutrophil infiltration into ischemie myocardium. Journal of Pharmacological Methods 14:157–167. https://doi.org/10.1016/0160-5402(85)90029-4
30. Ellman GL. 1959. Tissue sulfhydryl groups. Archives of Biochemistry and Biophysics 82:70–77. https://doi.org/10.1016/0003-9861(59)90090-6
31. Chaouche TM, Haddouchi F, Ksouri R, Atik-Bekkara F. 2014. Evaluation of antioxidant activity of hydromethanolic extracts of some medicinal species from South Algeria. Journal of the Chinese Medical Association 77:302–307. https://doi.org/10.1016/j.jcma.2014.01.009

32. Abdel-Aziz MS, Shaheen MS, El-Nekeety AA, Abdel-Wahhab MA. 2014. Antioxidant and antibacterial activity of silver nanoparticles biosynthesized using Chenopodium murale leaf extract. Journal of Saudi Chemical Society 18:356–363. https://doi.org/10.1016/j.jscs.2013.09.011

33. Konan KV, Le Tien C, Mateescu MA. 2016. Electrolysis-induced fast activation of the ABTS reagent for an antioxidant capacity assay. Analytical Methods 8:5638–5644. https://doi.org/10.1039/C6AY01088A

34. Alam MN, Bristi NJ, Rafiquzzaman M. 2013. Review on in vivo and in vitro methods evaluation of antioxidant activity. Saudi Pharmaceutical Journal 21:143–152. https://doi.org/10.1016/j.jsps.2012.05.002

35. Sari LM, Subita GP, Auerkari EI. 2019. Areca nut extract demonstrated apoptosis-inducing mechanism by increased caspase-3 activities on oral squamous cell carcinoma. F1000Research 7:1–37. https://doi.org/10.12688/f1000research.14856.5

36. Namvar F, Mohammad R, Baharara J, et al. 2014. Cytotoxic effect of magnetic iron oxide nanoparticles synthesized via seaweed aqueous extract. International Journal of Nanomedicine 9:2479–2488. https://doi.org/10.2147/IJN.S59661

37. Pham CTN, Thomas DG, Beiser J, et al. 2014. Application of a hemolysis assay for analysis of complement activation by perfluorocarbon nanoparticles. Nanomedicine: Nanotechnology, Biology and Medicine 10:651–660. https://doi.org/10.1016/j.nano.2013.10.012

38. Huang H, Lai W, Cui M, et al. 2016. An evaluation of blood compatibility of silver nanoparticles. Scientific Reports 6:1–15. https://doi.org/10.1038/srep25518

39. Hashem MA, Mahmoud EA, Abd-Allah NA. 2020. Alterations in hematological and biochemical parameters and DNA status in mice bearing Ehrlich ascites carcinoma cells and treated with cisplatin and cyclophosphamide. Comparative Clinical Pathology 29:517–524. https://doi.org/10.1007/s00580-019-03089-5

40. Khan SA, Kanwal S, Rizwan K, Shahid S. 2018. Enhanced antimicrobial, antioxidant, in vivo antitumor and in vitro anticancer effects against breast cancer cell line by green synthesized un-doped SnO_2 and Co-doped SnO_2 nanoparticles from Clerodendrum inerme. Microbial Pathogenesis 125:366–384. https://doi.org/10.1016/j.micpath.2018.09.041

41. Uddandrao VVS, Parim B, Nivedha PR, et al. 2019 Anticancer activity of pomegranate extract: Effect on hematological and antioxidant profile against Ehrlich-ascites-carcinoma in Swiss albino mice. Oriental Pharmacy and Experimental Medicine 19:243–250. https://doi.org/10.1007/s13596-018-0348-4

42. Krishnamurthy P, Wadhwani A. 2012. Antioxidant Enzymes and Human Health. In: Antioxidant Enzyme. InTech, pp 3–18. https://doi.org/10.5772/48109

43. Karmakar I, Dolai N, Suresh Kumar RB, et al. 2013. Antitumor activity and antioxidant property of Curcuma caesia against Ehrlich's ascites carcinoma bearing mice. Pharmaceutical Biology 51:753–759. https://doi.org/10.3109/13880209.2013.764538

20 Sustainability of Wind Turbine Blade

Instantaneous Real-Time Prediction of Its Failure using Machine Learning and Solution based on Materials and Design

Lohit Malik, Gurtej Singh Saini, Mayand Malik, and Abhishek Tevatia

CONTENTS

20.1 Introduction 400
    20.1.1 Wind Turbine Blade and Its Principle.......................................400
    20.1.2 Historical Evolution of Wind Turbine Blade 401
    20.1.3 Developments in Wind Turbine Blades: Designing
        and Materials 402
    20.1.4 Failure of Wind Turbine Blade 402
        20.1.4.1 Strain Measurement Detection 403
        20.1.4.2 Acoustic Emission 403
        20.1.4.3 Ultrasound Detection Methods.....................................403
        20.1.4.4 Machine Vision Detection Methods..............................404
20.2 Core Analysis of Wind Turbine Blade 404
    20.2.1 Design of the Wind Turbine Blade 404
    20.2.2 Computational Fluid Dynamics (CFD)
        of the WTB 405
        20.2.2.1 Meshing 406
        20.2.2.2 Numerical Solution..407
        20.2.2.3 Outcome of the Analysis 407
    20.2.3 Finite Element Analysis (FEA) of the WTB 410
        20.2.3.1 Numerical Solution Strategy 410
        20.2.3.2 Plate Theory 410
        20.2.3.3 Shell Theory 411
    20.2.4 Building Algebraic Equations..411
    20.2.5 Meshing 412
    20.2.6 Importing Pressure Load from CFD to FEA................................413

DOI: 10.1201/9781003297772-20

400 Handbook of Sustainable Materials

20.3	Machine Learning Model		415
	20.3.1	Overview	416
	20.3.2	Tools Used	417
	20.3.3	Fault Detection	417
	20.3.4	Dataset Acquisition	417
	20.3.5	Model Usage	420
		20.3.5.1 Logistic Regression	420
		20.3.5.2 Support Vector Machine (SVM)	421
		20.3.5.3 K-nearest Neighbors (KNN)	421
		20.3.5.4 Decision Tree	422
		20.3.5.5 Random Forest	423
		20.3.5.6 Neural Network	424
20.4	Conclusion, Prospects, and Challenges		428
References			429

20.1 INTRODUCTION

World demand for electrical energy is increasing day by day. In such scenarios, a permanent shift to renewable sources of energy becomes a requisite. Using Wind Energy to generate electricity via Wind Turbine Blade (WTB) promises a fruitful future with new materials and designs evolving every now and then to improve the efficiency and the power output [1]. However, these structures are very susceptible to damage from various sources controlled by the geometry, strength of composite, strength of adhesive, stress distribution, and, most importantly, the environmental conditions such as the temperature, humidity, UV effects, etc. [2, 3]. Hence, sustainable materials are important for their design and work. Various techniques to detect the damage including strain measurement, vibration, and machine vision have been proposed [4, 5]. But due to humongous costs concerned with the development, manufacturing, and setting up the wind turbines, the topic that entices the researchers is the failure of wind turbines.

20.1.1 WIND TURBINE BLADE AND ITS PRINCIPLE

The wind turbine rotates when wind is blown over the blades that cause the rotor to spin. The functioning principle of the rotor blades is same as that of the aircraft wings. Due to the difference in the curvature of the blades, the wind flows faster along the longer (curved) edge that develops a pressure difference on either side. For balancing this pressure difference, the blades are pushed by the air causing the blades to rotate.

The majority of wind turbines consist of more than two blades mounted very high on a tower. The high-rise turbines exploit the high-wind speed. With the flow of the wind, a low-pressure air zone is developed, and blade is pulled toward it; hence turning the rotor. This mechanism is termed as lift. The force of the drag is much weaker than the lift force. The small rotations of the turbine are enhanced by 100 times by employing a series of gears system, as shown in Figure 20.1. The key components of turbine mainly comprising of the rotor, gears, and generator are enclosed in a housing called nacelle.

Sustainability of Wind Turbine Blade

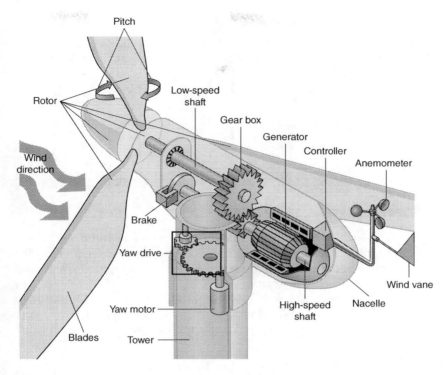

FIGURE 20.1 Interior components of a Wind Turbine Blade.

20.1.2 Historical Evolution of Wind Turbine Blade

The two main categories for blade design include the flat blade and the curved blade types. Flat blades being the oldest were used for thousands of years on windmills. These blades generate power in opposition to the power output as they rotate back on the up stroke, resulting in very slow rotation. They require less design and construction skills as they are the easiest to understand, but their efficiency is very low. Curved blade design is very close to that of an airplane wing (also known as an airfoil) having a curved top surface. The lift force being normal to the blade's upper surface, they rotate around the central hub. This rotational speed and the lift produced are directly dependent on the incoming wind speed. In comparison to the flat blades, lift forces on the curved blades allow its blade tips to move faster than the wind; hence generating relatively higher power and efficiencies. In a nutshell, the axis of wind turbines can be horizontal or vertical, the former being both commonplaces. On the other hand, vertical designs are not that common and produce lower power. These designs are shown in Figure 20.2 and are described below.

- Big three-bladed wind turbines with horizontal-axis (HAWT) are very common and produce almost all the wind energy in the today's world. The axis of most of the horizontal axis turbines is upwind of the supporting tower.

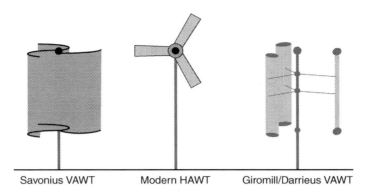

FIGURE 20.2 Types of Wind Turbine Blades.

- Wind turbines with vertical axis (VAWT) have the main rotor shaft along the vertical. This technique innates an advantage that the turbine is not required to be pointed toward the wind, which is a merit typically where the wind direction is not constant.
- Darrieus turbines are known for their good efficiency along with producing large torque ripple that leads to poor reliability. As the starting torque is very low, external power source is commonly a requisite.

20.1.3 Developments in Wind Turbine Blades: Designing and Materials

WTBs are being fabricated with longer lengths to improve their efficiency, which requires them to be strong and light at the same time. A commonly used method to improve the WTB efficiency is to increase the rotor diameter. Presently used WTB materials include glass and carbon fibers known for their higher stiffness and lower density, hybrid reinforcements to reduce weight, and nano-engineered polymers and composites that help increase the lifetime by 1500%. With all these developments, we have seen a drastic improvement in the efficiency as well as in the performance characteristics of the WTBs, which has made them 50 times more powerful than the ones 20 years ago. The development in the size and power output from 1990 to 2016 is in Figure 20.3.

20.1.4 Failure of Wind Turbine Blade

Reasons associated with failure of a WTB are as follows:

- Separation of whole blade from the hub at the joint (known as failure at root joint)
- Load buckling of extreme magnitudes
- Damages due to lightning
- Manufacturing defects leading to weakening of the bonding agents employed at the interface between different structural elements
- Blades striking the tower at high speed of rotation

Sustainability of Wind Turbine Blade

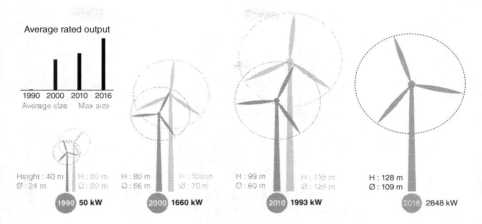

FIGURE 20.3 Development in the WTB technology.

- Errors during scheduled maintenance leading to damage
- Poor manufacturing quality control leading to delamination

WTBs suffer damage due to lightning and birds' impacts, apart from the ones mentioned above. Generally, repair tasks are carried out by workers hanging from the blades using ropes. At present, alternative systems, such as drones, have opened new ideas to prevent operators from climbing up to the turbines [6–12].

For the prevention of WTB from failure, several maintenance methods can be used. Preventative maintenance and the corrective maintenance are the two types of maintenance methods. As the name suggests, preventative maintenance is comprised of timely inspections. On the other hand, corrective maintenance focuses on the repair of the components to correct the damage. However, the techniques that make sure that the blade is in good condition and is properly functioning are Damage Detection Techniques. Blades can be prevented from lightning strikes with the use of lightning detectors. Some of the detection techniques are discussed here.

20.1.4.1 Strain Measurement Detection

This technique uses strain sensors to detect minute deformations of the blade due to load application. Strains classified as direct and shear strains are utilized in WTB inspection.

20.1.4.2 Acoustic Emission

This method helps detect the electrical signals translated from the generated elastic waves as a result of release of energy from damage, cracks, and deformation. This technique can also monitor structural changes at micro level in WTBs.

20.1.4.3 Ultrasound Detection Methods

Ultrasound can easily detect the waves reflected from the damage. Specific wave patterns are generated based on the material or structure in use, which makes it easier to estimate the location and other characteristics of the damage.

404 Handbook of Sustainable Materials

20.1.4.4 Machine Vision Detection Methods

Based on the principle of stereoscopic view of human vision, machine vision uses sequences images from different areas. Based on the algorithms, this method can improve the detection accuracy and efficiency.

The aim of the present chapter is to discuss the development of a Machine Learning (ML) model that will predict this failure for any WTB based on the forecast. This will help us take the necessary actions required to avoid the failure beforehand. To validate the ML model results, the core analysis is done using Finite Element Analysis (FEA) and Computational Fluid Dynamics (CFD) of the WTB. At the end, a comparative study of the results is made based on which a reliability test for the ML model can be conducted.

20.2 CORE ANALYSIS OF WIND TURBINE BLADE

Due to the impact of wind approaching the wind turbine perpendicular to its cross-section, a pair of Lift and Drag forces is generated that are responsible for rotating the three blades of the WTB. Indirectly, wind power (wind speed) is used to generate electricity. The generated power is controlled by factors related to both the wind turbine and the fluid medium. Concerning the failure of the WTB, the main parameter is the incoming wind speed. After a point, the blades deform to an extent leading to fracture and ultimately the wind turbine will fail. This threshold value of wind speed is dependent on the weight, material, and, most importantly, the design parameters of the blade.

The approach of the core analysis is to initially design the WTB followed by its CFD with realistic boundary conditions to evaluate the pressure distribution on a blade due to the incoming air. This distribution responsible for the deformation of the WTB is then taken as input to carry out the FEA.

20.2.1 Design of the Wind Turbine Blade

The WTB is designed in SolidWorks. The most important parameter in the design controlling the lift force, and hence the efficiency of the turbine is the air-foil shape apart from the length of the blade. For the present case, three different well-known air-foil shapes are taken starting from airfoil S818 at the root, airfoil S825 at the middle, and airfoil S826 at the tip while the length of the blade is 40 *m* (Figure 20.4.),

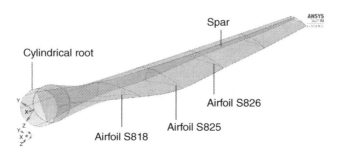

FIGURE 20.4 Wind Turbine Blade design comprising of three different air-foils.

Sustainability of Wind Turbine Blade

TABLE 20.1
Material Properties

Density (kg/m^3)	Poisson's Ratio XY	Poisson's Ratio YZ	Poisson's Ratio ZX	Young's Modulus X (Pa)	Young's Modulus Y (Pa)	Young's Modulus Z (Pa)	Shear Modulus XY (Pa)	Shear Modulus YZ (Pa)	Shear Modulus ZX (Pa)
1550	0.30	0.35	0.33	1.14×10^{11}	7.6×10^{9}	7.6×10^{9}	5.4×10^{9}	3×10^{9}	3×10^{9}

which is similar to a GE 1.5XLE turbine [13]. These air-foils are plotted on three different parallel planes at a fixed distance from each other. Finally, the lofted boss/base feature of SolidWorks is used to build the three-dimensional model with a thickness of 0.07 m. On observing the projection of the 3D model on the YZ plane, it is clear that the air-foils are rotated clockwise as we move from the root to the tip, generating a net pitch angle of four degrees at the tip of the blade. A blade also has a spar inside that is required for structural rigidity. The material chosen for this blade is an orthotropic composite (different properties in three mutually perpendicular directions) whose properties have been detailed in Table. 20.1. These properties have been approximated from the work of Ji et al. [14].

20.2.2 COMPUTATIONAL FLUID DYNAMICS (CFD) OF THE WTB

This study is carried out using ANSYS Fluent 20.2 version. The fluid medium is taken as air and its properties are introduced. The incoming wind speed is taken as 12 m/s approaching the WTB in the negative Z-direction that rotates the turbine at 2.2 rad/s about the Z-axis in the clockwise direction. Considering the height of the wind turbine, the ambient temperature, fluid density, fluid viscosity, and pressure are taken as $15°C$, $1.225 \frac{kg}{m^3}$, $1.79\times10^{-5} \frac{kg}{ms}$, and $1.013\times10^{-5} \frac{N}{m^2}$, respectively.

The said problem is solved using the Navier-Stokes equations comprising of the mass conservation equation, momentum equation, and energy equation. As the speeds associated with this problem are not huge (less than Mach 0.3), the energy equation can be neglected as in this case the fluid is incompressible. Note that these equations have been written with respect to a frame of reference rotating with the blade. This leads to additional terms for the Coriolis force ($2\vec{\omega}\times\vec{u}$) and the centripetal acceleration ($\vec{\omega}\times\vec{\omega}\times\vec{r}$) in the momentum equation [15]. At the same time, it helps reduce the computation time as the mesh of the blade doesn't rotate anymore. Hence, the equations governing this problem include the mass conservation and the momentum equation as shown below:

$$\frac{D\rho}{Dt}+\nabla\cdot\rho\vec{u} \tag{20.1}$$

$$\nabla\cdot\left(\rho\vec{u}\times\vec{u}\right)+\rho\left(2\vec{\omega}\times\vec{u}+\vec{\omega}\times\vec{\omega}\times\vec{r}\right)=-\nabla\rho+\nabla\cdot\vec{\tau} \tag{20.2}$$

where, $\vec{u}$ is the velocity relative to the rotating frame of reference in $\frac{m}{s}$, ρ is the density of the fluid medium in $\frac{kg}{m^3}$, p is the mechanical pressure in $\frac{N}{m^2}$, $\vec{\tau}$ is the torque

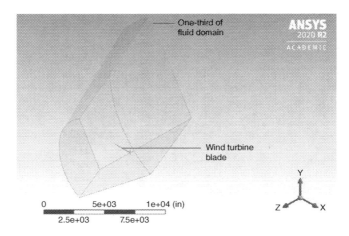

FIGURE 20.5 Fluid domain with the Wind Turbine Blade.

generated in Nm, and $\bar{\omega}$ is the angular velocity in $\frac{rad}{s}$. The study is limited to a rotation of 120 degrees (one-third of the total swept volume) for a single blade considering the limits of our software (Figure 20.5.). These results can be later on extrapolated to compute the simulation results for the whole fluid volume. The turbulent intensity (measure of unsteadiness in the flow) is taken as 5% which is typically found in the regions with wind turbines while the turbulent viscosity ratio is taken as 10. One of the primary boundary conditions includes the no-slip condition (no relative movement between the fluid layer and the structure boundary). The SST K-omega turbulence model (a model widely used for many aerodynamic applications) is used to close these equations.

20.2.2.1 Meshing

One of the most important aspects in FEA, meshing, refers to breaking down the geometry into a very small finite number of elements. A mesh with higher quality will lead to more accurate results. After importing the geometry to ANSYS, the mesh is created. Mainly the model comprised of tetrahedral elements, but the mesh has been refined in the regions near the WTB by using face sizing and body sizing (Figure 20.6) for better accuracy of the results. Match control settings are

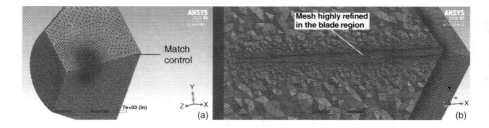

FIGURE 20.6 (A) Complete mesh; (B) Refined mesh near blade.

Sustainability of Wind Turbine Blade 407

implemented to make sure that the two nonadjacent faces share the same nodes on the intersection line. In total, the model comprised of 71643 nodes and 356628 elements. To judge the quality of the mesh, skewness and the orthogonal quality are the main factors; the former being even more important for CFD [16, 17]. Generally, skewness less than 0.5 and orthogonal quality of greater than 0.7 is considered very well. In the present case, 90% of the elements have skewness below 0.5 and orthogonal quality above 0.7.

20.2.2.2 Numerical Solution

The rotating frame of reference rotating at 2.22 rad/s in the clockwise direction is selected by editing in the Cell zone conditions. The boundary conditions mentioned above are input into the model. To translate the idea to carry out the CFD for one-third of the fluid domain, a new mesh interface of rotational type is created with an offset angle of 120 degrees. Once Fluent is told what we want to solve using the geometry and the boundary conditions, minor changes are done under solution section in the setup in order to make that it is on quickest path to the most accurate results. The value of residuals is changed to as small as 10^{-6} and set up 1500 iterations which ascertain that the calculations don't stop until 1500 iterations are completed.

20.2.2.3 Outcome of the Analysis

For visualizing the velocity vectors on every point on the blade, vectors icon is selected in the results section in CFD-Post and velocity is chosen as variable in stationary frame in order to see the velocity with respect to the ground and not with respect to the frame on the turbine. The velocity rises as we move away from the axis of rotation (Figure 20.7).

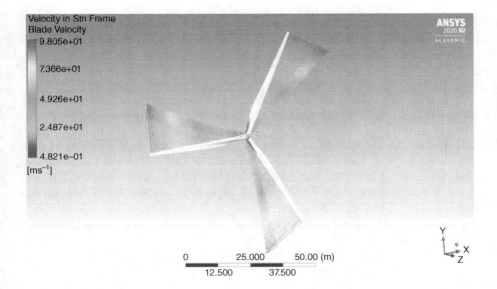

FIGURE 20.7 Blade velocity vectors.

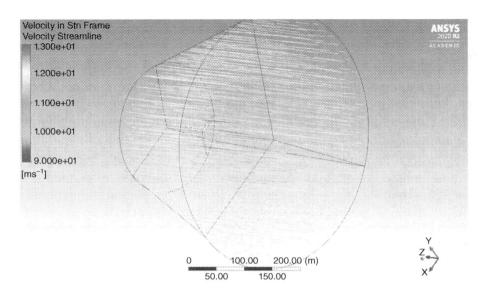

FIGURE 20.8 Velocity streamlines.

To obtain the velocity streamlines and how they behave while passing through the rotating wind turbine, streamline icon is chosen in CFD-Post and inlet is set as the starting point for the streamline. To clearly notice the changes in the streamline, a range of velocities is chosen from 9 *m/s* to 13 *m/s* that corresponds to the blue and the red colors in the color bar (Figure 20.8). This is clear from the yellow color of the streamline that the velocity at the inlet is 12 *m/s* at this position. A drop in velocity is noticed behind the turbine as depicted by the blue color of streamlines at this position. This behavior is correct as it shows a wake behind the turbine. The orange-colored streamline clearly shows an acceleration of the flow around the wake that is expected by mass and momentum balance.

The most important observation with regard to the pressure contours on the three blades is that the pressure on the front surface is higher (~ −400 *Pa*) than the pressure on the back surface (~ −600 *Pa*) which is clear from the color gradient in Figure 20.9. The blue regions on the back surface are under much higher pressure in terms of magnitude (~ −4000 *Pa*). This pressure difference generates the lift force whose one component is in the direction of rotation (in the XY plane) and the other component is in the Z-direction that is responsible for the blade deflection.

In order to observe what is happening around the air-foil, the pressure contours and the velocity vectors are plotted on the YZ plane passing through the blade. Figure 20.10(A) shows the relative velocity vectors passing around the blade which is the vector sum of the wind velocity and the velocity of blade at that point. The corresponding pressure contours are shown in Figure 20.10(B) whose value varies between −800 *Pa* to 420 *Pa*. This pressure difference generates the lift which is by

Sustainability of Wind Turbine Blade

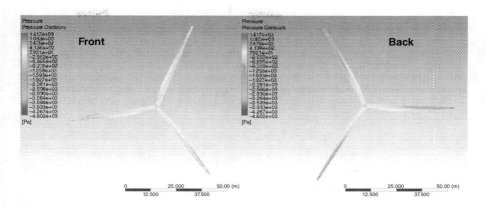

FIGURE 20.9 Pressure contours.

definition perpendicular to the direction of the free stream; hence rotating the blades clockwise.

For a better clarity of results, we create a line passing through the center of the wind turbine (X, Y = 0) and study the static pressure variation along this line. As shown in Figure 20.11, an overall pressure drop is observed as we cross the plane of rotation of the turbine. The lift forces hence generated are ascribed to this pressure difference. This line can be moved anywhere in the space to see how the pressure varies as the fluid crosses the turbine.

The net torque responsible for rotating a single blade comes out to be −140902.98 Nm out of which −158530.43 Nm is due to the pressure difference while 17627.454 Nm is due to the viscous forces of the fluid. The net negative magnitude denotes that the

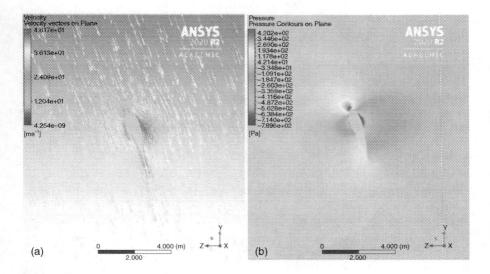

FIGURE 20.10 (A) Velocity vectors on YZ plane; (B) Pressure contours on YZ plane.

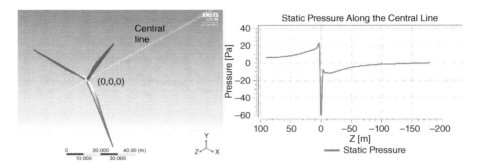

FIGURE 20.11 Static pressure variation along the central line.

torque is applied in the clockwise direction. Using this, we can calculate the output power coefficient of the turbine using

$$C_P = \frac{P_{rotor}}{P_{wind}} = \frac{N\tau_{blade}\omega}{0.5\rho A V_{incoming}^3}$$

where, N is the total number of blades, τ_{blade} is the torque generated on a single blade, A is the area swept by the rotating blades. On substituting the corresponding values, we get $Cp = 0.18$. Hence, 18% of the wind power is converted into electric power or the efficiency of the wind turbine is 18%.

20.2.3 Finite Element Analysis (FEA) of the WTB

The pressure distribution/load is transferred to ANSYS Mechanical for carrying out the FEA to determine the stresses generated and the deformations produced due to this and the centripetal force. At first, let us discuss the numerical solution strategy followed by the detailed FEA of the blade.

20.2.3.1 Numerical Solution Strategy

The mathematical model that is solved behind the scenes is based on the shell theory. As the mathematics involved in the WTB problem is comprised of humongous equations and colossal matrices, we just take an overview of what ANSYS will solve during the computation. We start with an overview of the plate theory on which the shell theory is based.

20.2.3.2 Plate Theory

The plate theory focuses on the mid-surface of the component under consideration. Instead of assuming that the cross-section rotates as a whole it rather assumes that the normal on the mid surface rotates together. For a lucid explanation of the phenomena, we take the example of a cantilever beam fixed at one end and force is applied at the other end. If we are able to calculate the displacement of any point on the normal due to the load applied, we can extrapolate the same for all the points on the beam. Depending on the load vector, a particular point may move along all the

Sustainability of Wind Turbine Blade

three directions along the X, Y, and Z axes and also rotate about these three axes. Hence, if we obtain the displacements u_x, u_y, and u_z and the rotations ϕ_z, ϕ_y, and ϕ_x for one point, we can back calculate the same for all the points on the mid-surface. Using these we can obtain the strains and the stresses based on the material properties and ultimately the potential energy (π) that has to be minimized. This shows that the properties of the material used are important ingredients.

20.2.3.3 Shell Theory

The shell theory translates the idea of plate theory to curved surfaces, and hence generalizes the concept. In this case, the three axes change for ever point making it a curvilinear coordinate system. One of the axes (X-axis) is along the curved mid surface while the Z-axis is perpendicular to this axis. We basically create local coordinate systems for each point considered and later on transfer the results to the global coordinate system. The mathematics involved is very complex, but the algorithms involved in ANSYS Mechanical take care of the same.

20.2.4 BUILDING ALGEBRAIC EQUATIONS

Figure 20.12 shows the beam in its initial and deformed states (due to load application in the vertical direction at the free end) along with the nodes at which the displacement-related parameters are to be calculated. At each node, we have to estimate the values of u_x, u_y, u_z, ϕ_z, ϕ_y, and ϕ_x i.e., 6 parameters. To do so, we obtain all the components of the strain, and hence the stress from which the potential energy expression (π) is formulated in terms of the displacement parameters. This potential energy function is minimized by differentiating it with respect to these parameters (such as $\frac{\partial \pi}{\partial u_{z2}} = 0$) to generate algebraic equations which are solved by ANSYS. In the mid-plane shown in Figure 20.12, we take 9 nodes, and hence we have to calculate $9 \times 6 = 54$ parameters. From the values of parameters obtained at nodes 1, 2, 8, and 9, we can conclude these values for the central node as well (node 10).

As the pressure distribution on the blades has been taken as input load for carrying out the FEA and the blade can be considered stationary, we start with choosing the static structural analysis in ANSYS Mechanical. All the effects that would have occurred as a result of the rotation of the blade and the impacting wind have been already coded in the pressure load; hence going for the static structural analysis is the correct approach. After entering the material properties in the engineering data

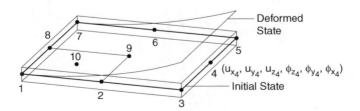

FIGURE 20.12 The deformed and initial states of mid surface of the beam.

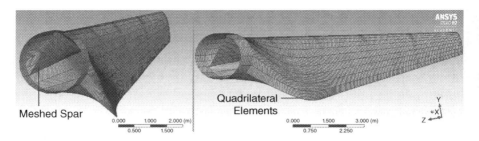

FIGURE 20.13 Meshing of WTB for FEA.

section as listed in Table 20.1 that are close to the ones in actual turbines, we transfer the geometry of the WTB from the earlier CFD portion to the FEA portion.

20.2.5 Meshing

Our geometry majorly consists of the complete blade surface and the spar as shown in Figure 20.13. The mapped face meshing and face sizing properties are understood as the perfect approach for the perfect mesh. As the spar has a relatively simple geometry, we apply the face meshing feature only to the blade surface which generates quadrilateral elements that is suitable for FEA. The face sizing feature is applied to the complete geometry with an element size of 0.2 meters. The final mesh for our blade consists of 5082 nodes and 5241 elements and is shown in Figure 20.13. As discussed earlier, here also more than 99.6% of the elements have orthogonal quality above 0.9 and skewness factor below 0.5, as plotted in Figure 20.14 that translates to an excellent mesh.

We create a new coordinate system that is a copy of the global coordinate system. This is important as we are using shell elements based on the shell theory in which the X-axis of each element changes as we move from point to point. By creating a new coordinate system, the global X-axis is projected onto the face of each of the elements. The thickness of the spar and the blade surface rises linearly from 0.035 m to 0.1 m and from 0.0045 m to 0.1 m, respectively, as we move from the tip to the root. This variation is depicted in Figure 20.15, where the red color corresponds to the thickest portion of 0.1 m in the blade and the spar. To translate our idea that the rotating blade is fixed at the center we can either add a fixed support at the root or add a remote point with zero remote displacement to which all the nodes on the

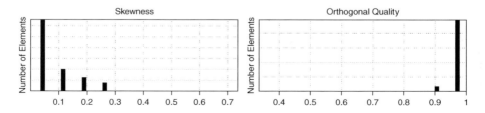

FIGURE 20.14 Skewness and orthogonal quality of the mesh generated.

Sustainability of Wind Turbine Blade 413

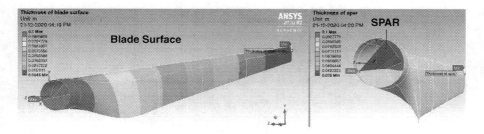

FIGURE 20.15 Thickness variation of the blade and the spar.

outermost root edge are connected rigidly. By doing so, we will also be able to study the bending moments and the torsional reactions as well. To include the effect of the centripetal force, we have to input the rotational velocity of the blade i.e., -2.22 *rad/s* about the Z-axis so that ANSYS can evaluate the expression $\delta\omega^2 r$ for each of the elements and sum it up at the end. Note that the effect of wind speed on rotating the turbine is already coded in the pressure distribution.

20.2.6 Importing Pressure Load from CFD to FEA

In order to procure the pressure distribution obtained from CFD for loading the blades to carry out the FEA, we transfer the solution cell from the CFD to the setup cell of the FEA. We need to make sure that the geometry selection for the pressure imported is the blade surface as the pressure is experienced by the outer blade surface and a mismatch between the results of CFD and loading in FEA can cause errors. It is also equally important to select the interpolation type as CFD results interpolator to obtain the pressure load in the form of vectors.

The pressure load ranges from approximately 2 *Pa* to 4100 *Pa* corresponding to the blue and the red color in the color bar, respectively. The vectors represent the direction and the magnitude of the load at the point from which they originate. A greater load is noticed on the right-hand side of Figure 20.16 that creates a net pressure difference between the two sides of the blade, and the blade is expected to bend in the direction of smaller load. The wireframe in Figure 20.17 shows the undeformed model while the colored blade depicts its deformed state. The deformation value

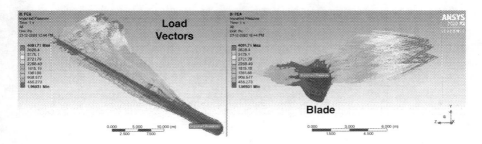

FIGURE 20.16 Pressure load imported from CFD to FEA.

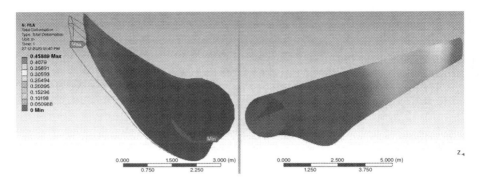

FIGURE 20.17 Total deformation of the blade.

averages at 0.10666 *m* while the maximum deformation of 0.45889 *m* is observed at the tip of the blade and as expected, the root remains undeformed.

It can be seen from Figure 20.18 that there are some regions of high stresses (red colored) whose magnitude gradually decreases as we move away from it radially. The location of these localized stresses is dependent on the WTB model we are studying. The design parameters and the properties of the material decide the same. The average stress value is ~9 *MPa* while the maximum value is 3.5 times this value i.e., 32 *MPa*.

The temporal evolution of equivalent stress is shown in Figure 20.19. It is interesting to note that the stresses emerge from a point and gradually rise in magnitude as depicted by the color variation from light blue ($t = 0.28$ second) to red ($t = 1$ second). Note that this time difference of one second is an ANSYS feature and can be altered for a better visualization. One can conclude that the red-colored regions are always the areas of highest stress, and hence material properties and design aspects can be modified specifically at these locations to improve the response and the role of sustainable materials comes here.

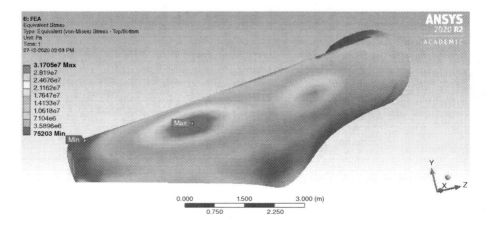

FIGURE 20.18 Von-Mises stress (equivalent stress) distribution on the blade.

Sustainability of Wind Turbine Blade 415

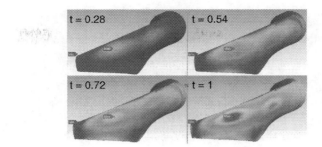

FIGURE 20.19 Time dependence of stress distribution.

On calculating, the force reaction vector comes out to be $\vec{F}(N) = 1.5824 \times 106 \hat{i} + 14223 \hat{j} + 71935 \hat{k}$, which is a vector of magnitude 1.5841×10^6 N directed almost along the positive X-axis. This makes sense as this direction is opposite to the centripetal force generated due to the blade rotation. We can confirm the same by analytically calculating the centripetal force $F_r = -m\omega^2 r$, where m is the mass of a single blade, ω is the rotational velocity, and r is the distance of the blade's center of mass from the point of rotation. Using ANSYS, we obtain these values as 22473 kg, −2.22 *rad/s*, and −14.232 *m*, respectively. Hence,

$$F_r = -(22473) \times (-2.22)^2 \times (-14.232) = 1.5763 \times 10^6 N$$

This value is very close to the reaction force magnitude, and hence the results are correct. The moment reaction is obtained as $\vec{M}$ (*N-m*) $= 2 \times 10^4 \hat{i} + 2.5 \times 10^6 \hat{j} + 2 \times 10^5 \hat{k}$, which is a vector almost in the YZ plane and of magnitude 2.51×10^6 *N-m* inclined at an angle of 4.57° from the Y-axis. Considering the loads, the blades' experience, and how they deform, this is also correct.

The above results confirm the successful completion of the journey that started with the designing and modeling of a WTB followed by its computational fluid analysis and finally importing the pressure distribution to study the deformation and the stresses generated due to this and the centripetal force via FEA. The same process is repeated for different wind speeds and environmental parameters to know the threshold value of wind speed at which the failure might occur, and these results are compared with the ones we obtain from the ML model. The computational study deduces that the failure occurs at a wind speed of 48−54 m/s while other environmental parameters including the air density, viscosity, ambient pressure, and temperature are the same as taken before.

20.3 MACHINE LEARNING MODEL

The main motivation for building an ML model is that sometimes due to adverse climate conditions and effects of dynamic loading in turbine components like blades and also electric fatigue causes some components in the wind turbine to fail and lead to faults in the whole system.

20.3.1 OVERVIEW

ML makes a computer learn from a dataset using some defined algorithm. When the ML model is being deployed, since the machine (computer) has learnt from the dataset, it gives a virtual feeling to the user that the machine is smart. In the crux, ML techniques are used wherever sufficient data (records, historical data, etc.) is present, which can be used to train an ML model. The following are the most basic and crucial steps in solving any ML problem:

Acquiring the data – Mostly, the data in any ML model is acquired in two ways. First is by generating data points in a working mechanical, electrical, physical, or software system. The other idea is to use historical data to make future predictions.

Data preprocessing – In most cases, the data is just records put together. The type of problem and the model parameters are the deciding factors that demand cleaning and processing of the data. Removing null values, redundant values, removing outliers, etc. are the common steps that are carried out in this step.

Data visualization – Once the data has been processed, identifying trends in the dataset can be very much beneficial for selecting best features and creating insights from the data with respect to the problem that is to be solved.

Feature engineering – Feature engineering is a very important step in building an ML model. Selecting best features that will lead to a robust model is a big task in ML model building. Techniques, like Principal Component Analysis (PCA) and Linear Discriminant Analysis (LDA), are used for dimensionality reduction that helps in building a robust ML model.

Identifying the problem – A given ML problem can be of multiple types. Broadly ML techniques are classified as follows:

Supervised Learning – In this type of ML, the dataset used has both labeled data and unlabeled data. The model is made to learn to generate what type of output given a set of input features. Supervised learning is further divided as follows:

- *Regression* – A regression problem is when given some input features, one or more output parameters are needed to be predicted. For example – Predicting house prices from historical data of house prices in an area is a standard regression problem. The common regression algorithms used in ML are linear regression, Support Vector Regression (SVR), Decision Tree Regression, Random Forest Regression, etc.

- *Classification* – A classification problem is when a model is trained to predict a category out of some categories. For example – Predicting gender of a person using his/her handwriting is a classification problem. The common classification algorithms used in ML are Logistic Regression, Support Vector Machine (SVM), Decision Tree Classification, Random Forest Classification, Gaussian naïve Bayes, etc.

Sustainability of Wind Turbine Blade

Unsupervised learning – In this type of ML technique, only labeled dataset is provided. The most common type of unsupervised learning is clustering, which is a technique in which formation of separate clusters is done using some defined algorithmic paradigm. Clustering is often used to find hidden trends in the data. K-means clustering is the most popular clustering algorithm.

Reinforcement learning – This type of learning category is commonly used in games. Basically, it consists of labeled data only like unsupervised learning. But from a state, reward is provided to every substate that after some time makes the machine learn. In human vs computer chess, this learning is used at the backend to make a computer move. Reinforcement learning is still not that used, unlike supervised and unsupervised learning.

20.3.2 Tools Used

Python is an expressive, object-oriented, and very high-level programming language that originated in 1991 by Guido Van Rossum (.py is its extension). It has various notable advantages:

- Supported by multitude of platforms (Raspberry Pi, Windows, Mac, etc.).
- Python has syntax similar to that of English and helps in rapid prototyping.
- Manipulation of data, analysis of data, and scripting is enhanced by it.
- Indentation is a vital part of python and adds to the readability of code.
- Interpreted language with vast libraries for usage in artificial intelligence.

Windows, macOS, Linux, android, and iOS support various python GUI like kivy, PyQT, tkinter, etc. Environment of Python is free and open source thus rendering ease of access and data manipulation. It has two modes namely Interactive and Script modes. Script mode is more advantageous when dealing with large chunks of code as compared to interactive mode. In the present study Python version 3.9 was used along with Anaconda Navigator version 1.9.12.

20.3.3 Fault Detection

Multiple wind turbines are coupled together to form a wind field that is the biggest source of generation of nonrenewable sources of energy in areas like Australia, and in India, wind energy is being used to produce electricity in parts of Rajasthan and Thar desert.

20.3.4 Dataset Acquisition

A timestamp dataset known as **SCADA dataset** is chosen as the adequate data for the intended solution for training the model to be made. SCADA stands for 'Supervisory Control and Data Acquisition'. It is basically an information system that keeps on appending and acquiring data on a computer software that is widely used in areas

1	DateTime	34	Sys 2 inverter 3 cabinet temp.
2	Time	35	Sys 2 inverter 4 cabinet temp.
3	Error	36	Sys 2 inverter 5 cabinet temp.
4	WEC: ava. windspeed	37	Sys 2 inverter 6 cabinet temp.
5	WEC: max. windspeed	38	Sys 2 inverter 7 cabinet temp.
6	WEC: min. windspeed	39	Spinner temp.
7	WEC: ava. Rotation	40	Front bearing temp.
8	WEC: max. Rotation	41	Rear bearing temp.
9	WEC: min. Rotation	42	Pitch cabinet blade A temp.
10	WEC: ava. Power	43	Pitch cabinet blade B temp.
11	WEC: max. Power	44	Pitch cabinet blade C temp.
12	WEC: min. Power	45	Blade A temp.
13	WEC: ava. Nacel position including cable twisting	46	Blade B temp.
14	WEC: Operating Hours	47	Blade C temp.
15	WEC: Production kWh	48	Rotor temp. 1
16	WEC: Production minutes	49	Rotor temp. 2
17	WEC: ava. reactive Power	50	Stator temp. 1
18	WEC: max. reactive Power	51	Stator temp. 2
19	WEC: min. reactive Power	52	Nacelle ambient temp. 1
20	WEC: ava. available P from wind	53	Nacelle ambient temp. 2
21	WEC: ava. available P technical reasons	54	Nacelle temp.
22	WEC: ava. Available P force majeure reasons	55	Nacelle cabinet temp.
23	WEC: ava. Available P force external reasons	56	Main carrier temp.
24	WEC: ava. blade angle A	57	Rectifier cabinet temp.
25	Sys 1 inverter 1 cabinet temp.	58	Yaw inverter cabinet temp.
26	Sys 1 inverter 2 cabinet temp.	59	Fan inverter cabinet temp.
27	Sys 1 inverter 3 cabinet temp.	60	Ambient temp.
28	Sys 1 inverter 4 cabinet temp.	61	Tower temp.
29	Sys 1 inverter 5 cabinet temp.	62	Control cabinet temp.
30	Sys 1 inverter 6 cabinet temp.	63	Transformer temp.
31	Sys 1 inverter 7 cabinet temp.	64	RTU: ava. Setpoint 1
32	Sys 2 inverter 1 cabinet temp.	65	Inverter averages
33	Sys 2 inverter 2 cabinet temp.	66	Inverter std dev

FIGURE 20.20 SCADA dataset features (before preprocessing).

where wind parks are made. The administrator can keep monitoring wind turbines remotely using TCP/IP protocols. The dataset consists of timestamp data in an interval of 10 minutes of different parameters related to components in a wind turbine i.e., Stator and rotor temp, wind speed, power output, bearing temp, etc. Another part of the dataset consists of fault data along with the same timestamp ID as in the SCADA data that is recorded whenever some fault occurs in the turbine. Five types of faults are being recorded namely AF: Air-cooling fault, GF: Generator heating fault, FF: Feeding fault, EF: Excitation fault, MF: Main's failure fault.

Since the data is a timestamp data of records, it needs to be cleaned and processed before training it into the model. The SCADA dataset consisted of the columns is shown in Figure 20.20.

We merge the 'SCADA dataset' and the 'fault dataset' based on timestamps which result in getting a single dataset. As we want to solve the problem of predicting faults given some input parameters, we remove some columns that are of no use with respect to the problem of classification of fault as what type of fault it is. Thus, we have dropped some of the columns and finally, we are left with the columns as shown in Figure 20.21.

On analyzing the data, we observe that since failure/fault in the turbine occurs rarely, we have a 'class-imbalance' problem in the dataset as depicted in Figure 20.22. It means that 'No fault' is the majority class and the fault classes constitute the minority class.

To solve this problem, a technique known as Synthetic Minority Over-sampling Technique (SMOTE) was used to oversample the minority classes. In SMOTE, a

Sustainability of Wind Turbine Blade

```
1    DateTime
2    Time
3    Error
4    WEC: ava. windspeed
5    WEC: ava. Rotation
6    WEC: ava. Power
7    WEC: ava. Nacel position including cable twisting
8    WEC: Operating Hours
9    WEC: ava. blade angle A
10   Spinner temp.
11   Front bearing temp.
12   Rear bearing temp.
13   Blade B temp.
14   Rotor temp. 1
15   Stator temp. 1
16   Nacelle temp.
17   Yaw inverter cabinet temp.
18   Fan inverter cabinet temp.
19   Ambient temp.
20   Tower temp.
21   Control cabinet temp.
22   Transformer temp.
```

FIGURE 20.21 SCADA dataset features (after preprocessing).

random sample of the minority class is chosen, and k of its nearest neighbors is chosen and out of these, randomly a neighbor is generated as a new synthetic data point. The SMOTE oversamples the minority classes such that their number becomes equal to the majority class number. Figure 20.23 shows the results so obtained. A comparison shown below in Figure 20.24 depicts the increase in the data points using SMOTE.

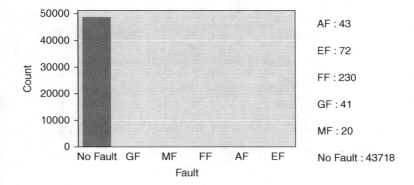

FIGURE 20.22 Class imbalance in the SCADA dataset and depiction of excess data points for 'No fault' in comparison with fault classes.

```
{'AF': 34980,
 'EF': 34980,
 'FF': 34980,
 'GF': 34980,
 'MF': 34980,
 'No Fault': 34980}
```

FIGURE 20.23 Equivalent sample generation using SMOTE.

Since we have 19 features that may lead to overfitting of the model, feature selection is carried out for reducing the dimensions of the dataset to eight features, as shown in Figure 20.25.

20.3.5 Model Usage

First of all, the data generated after SMOTE is divided into 'training data' that is to be used for training the model and 'testing data' to validate and analyze the accuracies of the model used. A train-test split of 0.2 was used, meaning that 80% data was used for training and 20% data was used for validation.

20.3.5.1 Logistic Regression

In logistic regression, the whole dataset is trained in such a way that it fits into a logistic regression function that is used to calculate the probabilities of an input belongs to which class. Since our input data had varying input values, first of all scaling is done using 'Standard Scaler' that scales the data between 0 and 1 for homogeneity in the data analysis. We use multinomial logistic regression that classified a given set of conditions whether fault will occur or not. The confusion matrix and classification report of Logistic Regression are shown in Figure 20.26. The model was pretty much inclined toward classifying the data point as 'No Fault' and it was

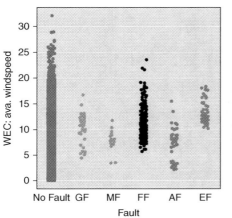
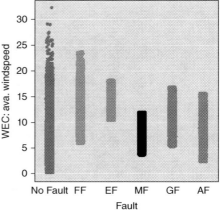

FIGURE 20.24 Increase in data points of minority classes (before and after using SMOTE).

Sustainability of Wind Turbine Blade

```
cols = selector.get_support(indices=True)
cols
```

```
array([ 0,  2,  4,  7,  8, 10, 17, 18], dtype=int64)
```

```
x_train_selected = final_x_train_res.iloc[:,cols]
x_test_selected = X_testing_features.iloc[:,cols]
```

```
x_test_selected.columns
```

```
Index(['WEC: ava. windspeed', 'WEC: ava. Power', 'WEC: Operating Hours',
       'Front bearing temp.', 'Rear bearing temp.', 'Rotor temp. 1',
       'Control cabinet temp.', 'Transformer temp.'],
      dtype='object')
```

FIGURE 20.25 Selected columns in the dataset after feature selection.

leading to 'underfitting'. Clearly, the accuracy for 'No Fault' is good, but, apart from this, the model failed in detecting faults.

20.3.5.2 Support Vector Machine (SVM)

SVM is a technique that first plots all the data points in an N-dimensional space and then iterates over to find the best possible 'Hyperplane' that is able to segregate the data points into separable classes. In SVM too, the data was first scaled using 'Standard Scaler' and was then fed into an SVM model to train the model. The confusion matrix and classification report of SVM are shown in Figure 20.27. It performed slightly better than *Logistic Regression* as it was able to separate data points but the model was still biased toward the majority class, i.e., 'No Fault'. The accuracy got improved than *Logistic Regression*, but the Fault was still not identifiable using this model.

20.3.5.3 K-nearest Neighbors (KNN)

KNN algorithm uses the technique of finding K-nearest neighbors of some data points and identifying to which class should it belong to. For the present problem, k = 5 was selected and confusion matrix and classification report are shown in

col_0	AF	EF	FF	GF	MF	No Fault		precision	recall	f1-score	support
Fault							AF	0.01	0.60	0.01	10
							EF	0.03	0.86	0.05	14
AF	6	0	0	0	2	2	FF	0.06	0.74	0.11	54
EF	0	12	2	0	0	0	GF	0.41	1.00	0.58	7
FF	1	9	40	0	0	4	MF	0.00	0.50	0.00	2
GF	0	0	0	7	0	0	No Fault	1.00	0.67	0.80	8738
MF	0	0	0	0	1	1					
No Fault	1177	451	613	10	655	5832	accuracy			0.67	8825
							macro avg	0.25	0.73	0.26	8825
							weighted avg	0.99	0.67	0.79	8825

FIGURE 20.26 Confusion matrix and classification report when Logistic Regression is used.

col_0	AF	EF	FF	GF	MF	No Fault
Fault						
AF	7	0	0	0	2	1
EF	0	9	4	0	0	1
FF	0	4	50	0	0	0
GF	0	0	0	7	0	0
MF	2	0	0	0	0	0
No Fault	199	99	153	2	69	8216

	precision	recall	f1-score	support
AF	0.03	0.70	0.06	10
EF	0.08	0.64	0.14	14
FF	0.24	0.93	0.38	54
GF	0.78	1.00	0.88	7
MF	0.00	0.00	0.00	2
No Fault	1.00	0.94	0.97	8738
accuracy			0.94	8825
macro avg	0.36	0.70	0.41	8825
weighted avg	0.99	0.94	0.96	8825

FIGURE 20.27 Confusion matrix and classification report when SVM is used.

Figure 20.28. This algorithm also performed similar to SVM, but it was still not able to classify Faults correctly such that it can be accepted as the final model. The use of Logistic Regression, SVM, and KNN suggested that using a decision-based algorithm would yield much better results as the data does not fit into some curve.

20.3.5.4 Decision Tree

A decision tree is an algorithm that uses 'Tree' data structure used in computer science to build a classifier. The tree is built recursively in a top-down fashion from root-leaf. The root node is the start and the leaf nodes are the final decision constituting that whether the data point should sit in which class. Whenever a decision is taken from a node in the tree, there is a cost related to that split. Once the tree has been built recursively, individual costs of every branch are combined to compute that whether a data point should be put into which class. A 'Gini Score' is used to determine the feasibility of a split, as

$$G = \sum \left(p_k \left(1 - p_k \right) \right)$$

where, p_k is the proportion of class inputs present in that group. The tree is often prone to 'overfitting' in the case of large datasets like ours. Thus, we limited the max-depth of the tree. The confusion matrix and classification report are shown in Figure 20.29. The model performed better than previously used models, but hyperparameter tuning is needed to improve the results.

col_0	AF	EF	FF	GF	MF	No Fault
Fault						
AF	5	0	1	0	3	1
EF	0	5	4	0	0	5
FF	0	3	41	0	0	10
GF	0	0	0	7	0	0
MF	2	0	0	0	0	0
No Fault	45	17	63	3	15	8595

	precision	recall	f1-score	support
AF	0.10	0.50	0.16	10
EF	0.20	0.36	0.26	14
FF	0.38	0.76	0.50	54
GF	0.70	1.00	0.82	7
MF	0.00	0.00	0.00	2
No Fault	1.00	0.98	0.99	8738
accuracy			0.98	8825
macro avg	0.40	0.60	0.46	8825
weighted avg	0.99	0.98	0.99	8825

FIGURE 20.28 Confusion matrix and classification report when KNN is used.

Sustainability of Wind Turbine Blade

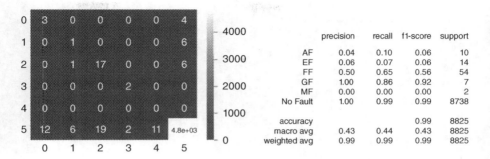

FIGURE 20.29 Confusion matrix and classification report when Decision Tree is used.

20.3.5.5 Random Forest

A Random Forest is also a decision-based algorithm like 'Decision Tree'. A group of trees is called a forest. This algorithm builds multiple trees and combines their individual results as the final result. The advantage that Random Forest has over decision tree is that it uses subset of randomly generated trees each time when it combines the results. Thus, generally Random Forest is not prone to overfitting. The confusion matrix and classification report are shown in Figure 20.30.

Out of all the ML models used so far, Random Forest classifier was able to give good results. So, further hyperparameter tuning was done to tune the parameters to give best possible accuracy.

20.3.5.5.1 Random Forest after Hyperparameter Tuning

For tuning the Random Forest classifier, 'GridSearchCV' was used, which is a function of 'model_selection' package present in 'scikit-learn' library. 'GridSearchCV' makes use of a 'param_grid' in which user-defined parameters are passed with an estimator. The model is trained using every possible combination in the 'param_grid' to get the best accuracy. Finally, the best parameters are chosen for the estimator (Figure 20.31).

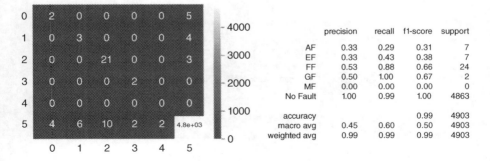

FIGURE 20.30 Confusion matrix and classification report when Random Forest is used.

424   Handbook of Sustainable Materials

```
In [106]:  from sklearn.model_selection import GridSearchCV
           from sklearn.ensemble import RandomForestClassifier

In [107]:  rf2 = RandomForestClassifier()

In [118]:  param_grid = [
               {'n_estimators': [10, 25], 'max_features': [5, 8],
                'max_depth': [10, 50, None], 'bootstrap': [True, False]}
               ]

In [119]:  grid_search_forest = GridSearchCV(rf2, param_grid, cv=4,verbose=2)

In [120]:  grid_result = grid_search_forest.fit(x_train_selected, final_y_train_res)
```

FIGURE 20.31   GridSearchCV for hyperparameter tuning in Random Forest Classifier.

20.3.5.5.2 *Parameters in Random Forest*

The parameters were used as, n_estimators: The number of trees in the forest, max_features: The number of features for best possible split, max_depth: The maximum depth of the tree, bootstrap: The samples used for training. After, the hyperparameter tuning was done, the results obtained are shown in Figure 20.32.

The results so obtained for best accuracy are summarized as: n_estimators – 25, max_features – 5, max_depth – 50, bootstrap – False (the whole dataset to be used for tree construction). The confusion matrix and classification report for the same are shown in Figure 20.33.

In addition, the metrics analysis reads: Precision – The model had an average precision of 46%. This meant that the model did not classify a 'No Fault' as 'Fault' with a probability of 0.46; Recall – The model had an average recall of 62%. This meant that the model was able to classify 'Fault' into 'Fault' with a probability of 0.62; F1-score – The model had a weighted f1_score of 0.99. This meant that the model was able to classify 99% of 'Fault' predictions correctly.

20.3.5.6 Neural Network

Neural network is a technique that belongs to *deep learning*. It has been developed from the idea of how a human neuron works. When a stimulus is generated from our

```
grid_result.best_estimator_

RandomForestClassifier(bootstrap=False, max_depth=50, max_features=5,
                       n_estimators=25)

rf3=RandomForestClassifier(bootstrap=False, max_depth=50, max_features=5,
                           n_estimators=25)

rf3.fit(x_train_selected,final_y_train_res)

RandomForestClassifier(bootstrap=False, max_depth=50, max_features=5,
                       n_estimators=25)
```

FIGURE 20.32   Best parameters for Random Forest Classifier.

Sustainability of Wind Turbine Blade

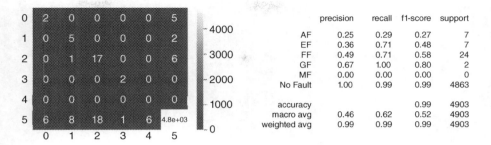

FIGURE 20.33 Confusion matrix and classification report of the tuned Random Forest Classifier.

sense organs, the signal is carried to our neurons and the neurons generate electrical signals so that we can act as per the stimuli. The following are the steps involved in training of any neural network:

Assign weights to the neurons in each layer when going forward in the neural network => Back-propagate after calculating the losses => Adjust the weights again.

All the above steps are carried out for some iterations known as 'epochs'. For the present problem, a sequential neural network was trained using the training data after using the SMOTE to oversample the data. The neural network that was built consisted of the following layers:

- The input layer of 24 nodes.
- A dropout of 20% results.
- A hidden layer of 15 nodes.
- A dropout of 20% results.
- A hidden layer of 12 nodes.
- An output layer of 6 nodes (each node corresponding to the 6 classes that we need to predict).

Figure 20.34 depicts the model layers when it was made using Keras Sequential API.

The NN was then trained onto 100 epochs. The analysis of the loss and the accuracy is shown in Figure 20.35. The plots here show that the neural network is getting underfitted. Thus, hyperparameter tuning of the neural network was carried out to get the optimal parameters for training the neural network.

20.3.5.6.1 Neural Network after Hyperparameter Tuning

Hyperparameter tuning of the neural network was carried out to identify the number of layers and batch_size of the dataset to be fed into the network (Figure 20.36).

After performing the hyperparameter tuning, the following results were obtained: Batch_size – 64; Layers – (20, 16). The neural network was trained with the parameters obtained through tuning, as shown in Figure 20.37.

The model was trained for 100 epochs and the results on the loss and accuracy are shown in Figure 20.38. The graphs show an increase in accuracy and decline in loss.

```
Model: "sequential_5"
_____
Layer (type)                 Output Shape              Param #
=================================================================
dense_16 (Dense)             (None, 24)                480
_____
dropout_9 (Dropout)          (None, 24)                0
_____
dense_17 (Dense)             (None, 15)                375
_____
dropout_10 (Dropout)         (None, 15)                0
_____
dense_18 (Dense)             (None, 12)                192
_____
dense_19 (Dense)             (None, 6)                 78
=================================================================
Total params: 1,125
Trainable params: 1,125
Non-trainable params: 0
_____
None
```

FIGURE 20.34 Neural network model layers.

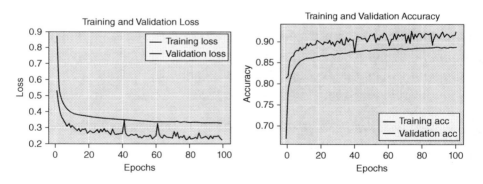

FIGURE 20.35 Graphs showing Loss and Accuracy curves of the neural network.

```
layers = [(20,), (20, 16), (32, 16, 16)]
activations = ['relu']
param_grid = dict(layers=layers, activation=activations, batch_size = (64, 128), epochs=[30])
grid = GridSearchCV(estimator=model, param_grid=param_grid,cv=3)

grid_result = grid.fit(X_train, y_train)
```

FIGURE 20.36 Hyperparameter Tuning using GridSearchCV in the neural network.

Sustainability of Wind Turbine Blade

```
model = Sequential()
model.add(Dense(24,input_dim=6, activation='relu'))
#model.add(Dropout(0.2))

model.add(Dense(20, activation='relu'))
#model.add(Dropout(0.2))

model.add(Dense(16, activation='relu'))
#model.add(Dropout(0.2))

model.add(Dense(6, activation='softmax'))

model.compile(optimizer = 'rmsprop',loss = 'categorical_crossentropy', metrics = ['accuracy'])
print(model.summary())
```

Model: "sequential_22"

Layer (type)	Output Shape	Param #
dense_64 (Dense)	(None, 24)	216
dense_65 (Dense)	(None, 20)	500
dense_66 (Dense)	(None, 16)	336
dense_67 (Dense)	(None, 6)	102

Total params: 1,154
Trainable params: 1,154
Non-trainable params: 0

None

FIGURE 20.37 Tuned neural network.

Also, the validation and training metrics converge, thus depicting that the neural network has been trained without underfitting. The confusion matrix and classification report are shown in Figures 20.39 and 20.40, respectively.

The metrics analysis shows: Precision – The model had an average precision of 95%. This meant that the model did not classify a 'No Fault' as 'Fault' with a probability of 0.95. Recall – The model had an average recall of 95%. This meant that the model was able to classify 'Fault' into 'Fault' with a probability of 0.95. F1-score – The model had a weighted f1_score of 0.95. This meant that the model was able to classify 95% of 'Fault' predictions correctly. Finally, it was found that the

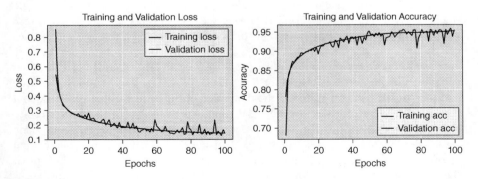

FIGURE 20.38 Graphs showing Loss and Accuracy curves of the tuned neural network.

```
[[9538    0   22    0   58   74]
 [   0 9576   39    0    0    0]
 [  11  858 8905    0    2   84]
 [   0    0    4 9641    0    0]
 [ 433    0    3    0 9310   32]
 [ 545  165  300   14  134 8550]]
```

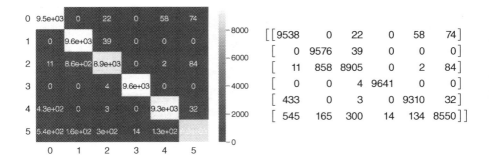

FIGURE 20.39 Confusion matrix of the tuned neural network.

	precision	recall	f1-score	support
0	0.91	0.98	0.94	9692
1	0.90	1.00	0.95	9615
2	0.96	0.90	0.93	9860
3	1.00	1.00	1.00	9645
4	0.98	0.95	0.97	9778
5	0.98	0.88	0.93	9708
accuracy			0.95	58298
macro avg	0.95	0.95	0.95	58298
weighted avg	0.95	0.95	0.95	58298

```
Accuracy: 0.952348
Precision: 0.952348
Recall: 0.952348
F1 score: 0.952348
```

FIGURE 20.40 Classification report of the tuned neural network.

WTB is failing at velocity of 50.4 – 56.5 *m/s* from the NN model under the features in which the model was trained.

20.4 CONCLUSION, PROSPECTS, AND CHALLENGES

In the nutshell, after a brief introduction about the WTBs and their evolution, the most enticing topics on the failure of the WTB comprising of its reasons, damage detection, and maintenance techniques were discussed. For the computational study, a complete three-dimensional blade was designed in SolidWorks made up of a single spar and an orthotropic material. Based on the CFD results, interpretations were made, and the possible reasons were discussed for the obtained velocity streamlines and the pressure distribution on the blades. Along with this, a focus on the pressure variation at the cross-section of a blade was also made. The obtained pressure distribution was imported as a load for the blades' FEA. The deformations produced and the stresses generated due to the combined effect of these pressure loads and the centrifugal force were studied. The time-dependence of Von-Mises stresses was also uncovered. For the ML approach, numerous models including Logistic Regression, SVMs, K-nearest Neighbors, Decision Tree, Random Forest, and Neural Network were trained based on the data obtained and cleaned. It was deduced that the Random Forest classifier and Neural Network stand out as the two most accurate models for classification of a fault in the turbine. Among the two, the Neural

Sustainability of Wind Turbine Blade

Network at an accuracy of 95% was found out to be the best classifier for fault detection in the wind turbine system ascribed to the much lower precision and recall of the Random Forest classifier. It was concluded that the threshold value of wind speed at normal environmental conditions for the failure of the turbine lies in the range of 50.4–56.5 *m/s*. This range is very close to the one obtained from the computational study with an average percentage difference of only 4.6% translating the proposed ML model as a reliable one.

Sustainable materials can play a vital role in solving the issue of failure of wind turbine. Multi-criteria decision method can be implemented for the selection of sustainable material for the development of turbines, namely HAWT, VAWT, and Darrieus turbines. However, it is worthwhile to note that it turns out to be challenging to virtually design the WTBs ascribed to their complex geometric constraints, specifically the thin skin characteristics of the blades which make the failure analysis sensitive. Additionally, this topic includes the transfer of CFD data to FEA platform that itself is a rare problem and requires synchronization between the two. This means that the development of pressure on the whole body of the blades needs to be explored with high-performance computing; the same should be done for the stresses generated and the deformations produced due to uneven pressure and the centripetal force. The temporal evolution of the stress can provide the precise determination and prediction of possible wind turbine failure. As understood in the present study, the ML can be explored by training the Random Forest Classifier and Neural Network with sufficient data along with the inclusion of appropriate material with desired properties.

REFERENCES

1. Battisti, L., Benini, E., Brighenti, A., Dell'Anna, S. and Castelli, M.R., 2018. Small wind turbine effectiveness in the urban environment. *Renewable Energy, 129*, pp. 102–113.
2. Garolera, A.C., Madsen, S.F., Nissim, M., Myers, J.D. and Holboell, J., 2014. Lightning damage to wind turbine blades from wind farms in the US. *IEEE Transactions on Power Delivery, 31*, pp. 1043–1049.
3. Mishnaevsky Jr, L., 2019. Repair of wind turbine blades: Review of methods and related computational mechanics problems. *Renewable Energy, 140*, pp. 828–839.
4. Du, Y., Zhou, S., Jing, X., Peng, Y., Wu, H. and Kwok, N., 2020. Damage detection techniques for wind turbine blades: A review. *Mechanical Systems and Signal Processing, 141*, p. 106445.
5. Ciang, C.C., Lee, J.R. and Bang, H.J., 2008. Structural health monitoring for a wind turbine system: A review of damage detection methods. *Measurement Science and Technology, 19*(12), p. 122001.
6. Shihavuddin, A.S.M., Chen, X., Fedorov, V., Nymark Christensen, A., Andre Brogaard Riis, N., Branner, K., Bjorholm Dahl, A. and Reinhold Paulsen, R., 2019. Wind turbine surface damage detection by deep learning aided drone inspection analysis. *Energies, 12*(4), p. 676.
7. Wang, Y., Yoshihashi, R., Kawakami, R., You, S., Harano, T., Ito, M., Komagome, K., Iida, M. and Naemura, T., 2019. Unsupervised anomaly detection with compact deep features for wind turbine blade images taken by a drone. *IPSJ Transactions on Computer Vision and Applications, 11*(1), pp. 1–7.

8. Montanyà, J., López, J.A., Fontanes, P., Urbani, M., Van Der Velde, O. and Romero, D., 2018, September. Using tethered drones to investigate ESD in wind turbine blades during fair and thunderstorm weather. In *2018 34th International Conference on Lightning Protection (ICLP)* (pp. 1–4). IEEE.
9. Khadka, A., Afshar, A., Zadeh, M. and Baqersad, J., 2021. Strain monitoring of wind turbines using a semi-autonomous drone. *Wind Engineering*, p. 0309524X211027814.
10. Chen, X., Shihavuddin, A.S.M., Madsen, S.H., Thomsen, K., Rasmussen, S. and Branner, K., 2021. AQUADA: Automated quantification of damages in composite wind turbine blades for LCOE reduction. *Wind Energy*, *24*(6), pp. 535–548.
11. Fontanes, P., Montanyà, J., Arcanjo, M., Guerra-Garcia, C. and Tobella, G., 2022. Experimental investigation of the electrification of wind turbine blades in fair-weather and artificial charge-compensation to mitigate the effects. *Journal of Electrostatics*, *115*, p. 103669.
12. Suurnäkki, P., Jokela, T. and Tiihonen, M., 2022. Applying an Icing Wind Tunnel for Drone Propeller Research, Validation of New Measurement Instrument. In *New Developments and Environmental Applications of Drones* (pp. 31–49). Springer, Cham.
13. Elsherif, D.M., Abd El-Wahab, A.A. and Abdellatif, M.H., 2019, September. Factors affecting stress distribution in wind turbine blade. *IOP Conference Series: Materials Science and Engineering*, *610*, p. 012020.
14. Ji, B., Zhong, K., Xiong, Q., Qiu, P., Zhang, X. and Wang, L., 2022. CFD simulations of aerodynamic characteristics for the three-blade NREL phase VI wind turbine model. *Energy*, p. 123670.
15. Zhou, Y., Xiao, Q., Liu, Y., Incecik, A., Peyrard, C., Wan, D., Pan, G. and Li, S., 2022. Exploring inflow wind condition on floating offshore wind turbine aerodynamic characterisation and platform motion prediction using blade resolved CFD simulation. *Renewable Energy*, *182*, pp. 1060–1079.
16. Malik, L. and Tevatia, A., 2021. Comparative analysis of aerodynamic characteristics of F16 and F22 combat aircraft using computational fluid dynamics. *Defence Science Journal*, *71*(2), pp. 137–145.
17. Malik, L., Rawat, S., Kumar, M. and Tevatia, A., 2021. Simulation studies on aerodynamic features of Eurofighter Typhoon and Dassault Rafale combat aircraft. *Materials Today: Proceedings*, *38*, pp. 191–197.

21 A Self-sustained Machine Learning Model to Predict the In-flight Mechanical Properties of a Rocket Nozzle by Inputting Material Properties and Environmental Conditions

Lohit Malik, Gurtej Singh Saini, and Abhishek Tevatia

CONTENTS

21.1 Introduction 432
    21.1.1 Functioning of a Rocket Nozzle 432
    21.1.2 Two Major Disasters in the History of Space
        Industry 433
        21.1.2.1 Space Shuttle Challenger Disaster 433
        21.1.2.2 Space Shuttle Columbia Disaster 434
21.2 A Glance at the Recent Work on Nozzle 434
21.3 Core Analysis of the Rocket Nozzle: Rocketdyne F-1 Engine 435
    21.3.1 Design of the Rocket Nozzle 436
    21.3.2 Finite Element Analysis (FEA) of the Rocket Nozzle 437
        21.3.2.1 Loading Types and Boundary Conditions 437
        21.3.2.2 Meshing 437
        21.3.2.3 Model Setup 438
    21.3.3 Outcome of the Core Analysis 441
        21.3.3.1 Force Reaction in the Axial (Z) Direction 441
        21.3.3.2 Circumferential Stress 442
        21.3.3.3 Total Deformation 442
        21.3.3.4 Stress 443
        21.3.3.5 Gap Generated between the Lower and the
                Middle Parts of the Nozzle 444

DOI: 10.1201/9781003297772-21

	21.3.4	Motivation for Employing Machine Learning............................445
		21.3.4.1 Design of Experiments 445
21.4	Machine Learning Model...446
	21.4.1	Tools Used..447
		21.4.1.1 IDE Used: Google Colab..448
	21.4.2	Model to Predict Mechanical Properties of Rocket Nozzle In-flight 448
		21.4.2.1 Dataset Acquisition..449
		21.4.2.2 Data Cleaning and Visualization..............................449
	21.4.3	Model Usage 450
		21.4.3.1 Decision Tree Model 451
		21.4.3.2 Random Forest Model 451
		21.4.3.3 XGBoost Model.. 452
	21.4.4	Some Interesting Results.. 452
21.5	Conclusion, Prospects, and Challenges..454
References..454

21.1 INTRODUCTION

A rocket engine nozzle is a propelling nozzle used in a rocket engine to expand and accelerate the hot gases from combustion so as to produce thrust. Saturn V rocket sent the first men to the moon. The engine used in Saturn V was the 'Rocketdyne F1'. The Saturn V rocket measured 110 m high and weighed 2840 tons at lift-off, and it still remains the largest and most powerful launch vehicle till date. The power of this rocket was due to the five 'Rocketdyne F1' engines used in the rocket that were located at the first propulsion stage with a power output of 60 Gigawatts. They burned more than two-thirds of the volume of an Olympic swimming pool full of water. The fuel mixture consisted of a mixture of oxygen and RP1 propellant that accelerated the rocket from 0 to 9660 km/hr in just two minutes!

21.1.1 FUNCTIONING OF A ROCKET NOZZLE

The huge flow rate of oxygen and the propellant required by each engine was provided by a turbo pump that was situated at the upper part of the engine, as shown in Figure 21.1. The turbo pump consisted of two high-pressure pumps in a turbine

FIGURE 21.1 Turbo pump in the engine.

A Self-sustained Machine Learning Model 433

FIGURE 21.2 High-pressure helium system used to pressurize the propellant.

mounted on the same shaft. The turbine connected thermal energy to mechanical energy used for propelling the rocket. The mechanical energy was used to rotate the shaft, which in turn drove the pumps. To ensure proper suction at the pump inlets, a dedicated high-pressure helium system (Figure 21.2.) was used to pressurize the propellant tank and high-pressurized oxygen tank to pressurize the oxygen tank. Once these gases were pressurized, they were sent to the mixing chamber where they were used to combine RP1 propellant and oxygen, as shown in Figure 21.3. The mixture was then sprayed into the combustion chamber below the injector plate. This mixture was subsequently burned, giving rise to large energy. Till this burning mixture reached the throat of the rocket, they would have reached the speed of sound. Then this mixture entered the divergence part of the nozzle, where they would have reached supersonic speeds due to increase in nozzle area. Then finally, from the thruster extension, this mixture was released into the atmosphere, giving rise to large energy, this large that this engine remains the most powerful rocket built till date.

21.1.2 Two Major Disasters in the History of Space Industry

21.1.2.1 Space Shuttle Challenger Disaster

This space shuttle disaster happened on 28 January 1986. It is one of the most devastating incidents recorded in the history of space exploration. Just after a minute the

FIGURE 21.3 Combustion chamber.

space shuttle lifted off, spacecraft's O-rings started failing, the rubber seals started separating its rocket boosters and caused a fire to start which destabilized the boosters. The shuttle was moving faster than the speed of sound and quickly began to break apart. The worst aspect of the disaster was that it led to the deaths of all astronauts on board.

21.1.2.2 Space Shuttle Columbia Disaster

Another tragic incident was the disintegration of the space shuttle Columbia on 1 February 2003. The root cause for the accident was during lift-off by the breaking off of a piece of foam that was intended to absorb and insulate the fuel tank of the shuttle from heat and to stop ice from forming but created a hole. Knowing that the foam regularly had fallen off of previous shuttles and had not caused critical damage, NASA officials believed there was nothing to worry about. But when the Columbia attempted re-entry after its mission was complete, gases and smoke entered the left wing through the hole and caused the wing to break off, leading to the disintegration of the rest of the shuttle seven minutes from landing. The entire crew of six American astronauts and the first Israeli astronaut in space died in the accident.

21.2 A GLANCE AT THE RECENT WORK ON NOZZLE

Rocket nozzles are subject to various kinds of loadings such as the pressure due to hot exhaust gas, forces due to high-speed flow of the propellant, thermal strains due to the high temperatures experienced [1], and bolt preloads [2]. All these activities lead to several interesting phenomena. As experiments turn out to be an expensive option for testing the nozzles, engineers majorly rely on the well-established computational and numerical methods/simulations [3-7]. The working environment/ extreme conditions are virtually created, and the computer-aided designed (CAD) models are imported for finally optimizing the rocket nozzles. Lijo et al. [8] carried out numerical simulation of flows in a thrust-optimized rocket nozzle and found the results including the features on the nozzle wall to be in close agreement with the experimental ones. Marchi et al. [1] developed a one-dimensional analytical model to study the flows in a nozzle with regenerative channels using which the thrust can also be readily calculated. Peake et al. [9] updated on the sustainable carbonized rayon for the rocket motor nozzles (solid). Ali and Nived [10] provided a review of the composites materials for sustainable space industry. The use of lunar materials and CMC materials in rocket propulsion was discussed by Valero Sánchez [11] and Olufsen and Orbekk [12]. There are studies on the 3D printing of materials [13–16] and fuel [17] for the rocket design and propulsion. Malik et al. [6, 7] have recently proposed a three-coil setup for the controlled divergence of magnetic nozzle and gave a concept for plasma detachment, which will help precise launching of the spacecraft [18, 19]. Mehta et al. [20] and Zhou et al. [21] have worked toward characterization and cooling of the rocket jet for its control. Numerical studies of water spray on launchpad have been conducted by Lu et al. [22].

Computational fluid dynamics (CFD) has been used for the solid propellent rocket simulation [23] and designing of a converging-diverging nozzle for Kappa-DX experimental rocket engines [24]. Recently, machine learning (ML) has been incorporated for optimizing the rocket engine performance and design characteristics. Waxenegger-Wilfing et al. [25] have employed ML for determining noise in a rocket-thrust chamber by making a classification model for predictive detection of engine characteristic anomalies. This model could be used to improve thrust chamber characteristics for better performance. Using neural networks, the mixing ratios of propellant and air mixture have been optimized [26]. Apart from this, Min et al. [27] uncovered the role of ML and artificial intelligence in improving the aircraft design in general.

The aim of this chapter is to move a step forward by building an ML model that will be trained via in-house generated data for predicting the mechanical properties of the rocket nozzle based on the material properties and working conditions inputted. The data is generated using the reliable computational approach. The major advantage of shifting to ML is the drastic reduction in the high computational time associated with simulations, and hence the computational costs. The quick results generated not only help us save time toward a project but also support the systems in the rocket to warn beforehand of any catastrophic activity that might happen during operation.

21.3 CORE ANALYSIS OF THE ROCKET NOZZLE: ROCKETDYNE F-1 ENGINE

We start with the computational analysis (Finite Element Analysis, FEA) of the rocket nozzle using ANSYS. This is the first major step before delving into the ML portion of the work. Once this is successfully done, the computational step is repeated 200 times to generate good enough data to train the ML model. This model will be able to determine mechanical properties such as deformation, stresses, etc. when the rocket is in operation. For the present study, the Rocketdyne F-1 is chosen, as shown in Figure 21.4. This particular engine consists of a bolted flange joint that connects the mid and lower portions of the engine.

FIGURE 21.4 Rocketdyne F-1 engine.

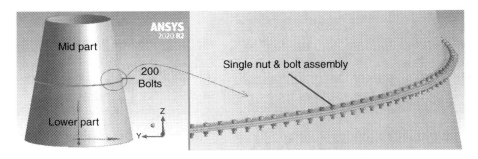

FIGURE 21.5 Rocket nozzle design comprising of lower and middle parts.

21.3.1 Design of the Rocket Nozzle

The rocket nozzle has been designed in Design Modeler. In reality, the nozzle is bell-shaped but to simplify the design for obtaining results quickly and to be able to easily compare them with the analytical calculations, we assume a conical shape. The overall geometry is approximated to two frusta of cones (the middle and the lower nozzles) stapled together by a series of nuts and bolts. In total, there are 200 bolts and nuts joining the two parts. The overall height of the nozzle is 4 *m*, top radius is 1 *m*, bottom radius is 1.77 *m*, wall thickness is 1.27 *cm*, and the bolt diameter is 0.8 *cm*. The side view and a closer look at the CAD model have been depicted in Figure 21.5.

It is important to note that simulating the whole arrangement consisting of 200 bolts will be cumbersome, and hence as the geometry is symmetric, we can reduce the computation to only studying half a bolt and nut, the middle nozzle, and the lower nozzle without the loss of generality. This means that we will analyze 1/400th of the model or a 0.9° slice cut from the complete model and later on extrapolate the results to the whole nozzle. The 1/400th of the model that we will be focusing on is shown in Figure 21.6.

The materials chosen for the nut and bolt and the nozzle parts are different and such that their properties do not change much with temperature variation. The nuts and bolts are made up of A-286 steel, a material that finds use in applications requiring high strength, paired with corrosion resistance up to 700°C [28]. Whereas the middle and lower nozzle parts are made up of the 300 series stainless steel [29]. The related material properties are tabulated in Table. 21.1 which are inputted in the

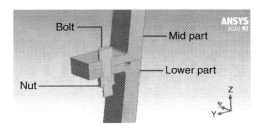

FIGURE 21.6 1/400th of the CAD model consisting of four major parts.

A Self-sustained Machine Learning Model

TABLE 21.1
Material Properties

Component	Material	Young's Modulus (*Pa*)	Poisson's Ratio (μ)	Coefficient of Thermal Expansion (per °C) (α)
Bolt and Nut	A-286 Steel	2×10^{11}	0.31	1.71×10^{-5}/°C
Middle and Lower Nozzles	300 Series Stainless Steel	2×10^{11}	0.27	1.8×10^{-5}/°C

engineering data section of ANSYS to complete the set of mathematical equations being solved behind the scenes.

21.3.2 FINITE ELEMENT ANALYSIS (FEA) OF THE ROCKET NOZZLE

The FEA is carried out in ANSYS 20.2 student version. We analyze the bolted joint which is prone to failure and talk about the gaps that might develop between the two parts of the nozzle due to loadings. The governing equations for the problem are derived from the 3D elasticity model that consists of 3D differential equations of equilibrium, 3D stress-strain relations based on Hooke's law, and finally the strain-displacement relations. This ends up in 15 equations and 15 unknowns for each node which are solved to get the displacements in order to obtain the final results. As discussed above, the loads are of various types.

21.3.2.1 Loading Types and Boundary Conditions

Forces are generated due to the tightening of the bolt and defined by the extent to which it is done, termed as the bolt preload. We take the bolt preload equal to 50% of the ultimate tensile strength of the material from which it is made of [2]. The breaking strength of A-286 steel is 620 *MPa*; hence the equivalent breaking force on half a bolt equals (620 *MPa*) × (*Cross–sectional area of half a bolt*) = 15345 *N* which means that the bolt preload is 7672.5 *N*.

Thermal strains are produced due to temperature change from room temperature of 20°C to ~ 400°C, which is achieved when the hot exhaust gas is initially thrown out of the nozzle. The regeneration channels could not be designed in the CAD model in order to simplify the design as much as possible. But to observe this effect, an equivalent force of 4450 *N* is applied at both the middle and the lower parts of the nozzle that is responsible for pulling apart these two [1]. The pressure variation due to the exhaust gas is calculated using one-dimensional gas dynamics and decreases linearly from 330 kPa to 84 kPa as we move from the top to the bottom part of the nozzle (i.e., along the negative Z-axis) [1]. The excerpt about the loading types is given in Table 21.2.

21.3.2.2 Meshing

After importing the geometry to ANSYS, the mesh is created. The model majorly comprised of hexahedral elements, but the mesh has been refined in the regions near the nut and the bolt by using body sizing feature, as shown in Figure 21.7 for better

TABLE 21.2
Loading Types and Related Details

Loading Type	Details
Thermal strain	Due to temperature change from 20°C to 400°C
Pressure variation	Linear variation from 330 kPa to 85 kPa along the negative Z-axis
Forces from regeneration channels	4450 N at both the middle and the lower parts of the nozzle
Bolt preload	50% of breaking strength that equals to 7672.5 N

accuracy of the results. The element size for the nozzle is 0.762 cm while for the nut and bolt it is even smaller at 0.2 cm. For keeping the total number of elements under the student version limit of ANSYS, method feature is used, and the mesh is made Hex dominant. In total, the model comprised of 26089 nodes and 4225 elements.

To judge the quality of our mesh, there are two most important factors namely the skewness and the orthogonal quality, the former being even more important for FEA. Generally, skewness less than 0.5 and orthogonal quality of greater than 0.7 is considered very well. In the present case, 95% of the elements have skewness below 0.5 and orthogonal quality above 0.7. The plot for the number of elements vs element metrics is also shown in Figure 21.7. As clearly seen, more than 85% of the elements have element quality above 0.9 out of which one translates to a very good mesh. This is an initial mesh that can be refined later but, as of now, it is the most optimized one.

21.3.2.3 Model Setup

In this section, we apply the loads in three steps. For translating this to ANSYS, we need to change the load steps in the ANSYS static structural analysis settings to three from the default one. In the first load step, the bolts are tightened (bolt preload) which gives rise to some stresses. The pressure variation and the separation forces from regeneration channels are turned on in the second load step. Finally, in the last load step, the thermal effect is also incorporated. This series of operations is done to imitate the real launch conditions as much as possible. At the end, all the loading types function together, and the results so obtained are analyzed.

We select the thermal condition from the load section and choose all four bodies. The room temperature is modified to 20°C. To add the thermal data, tabular data

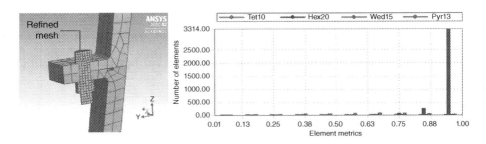

FIGURE 21.7 Complete mesh and mesh metrics.

A Self-sustained Machine Learning Model

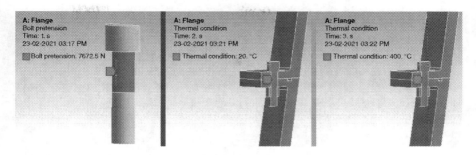

FIGURE 21.8 Preload and thermal conditions.

type is selected and a temperature of 20°C is added in the first two steps while 400°C is inputted in the third step, as shown in Figure 21.8. For applying the bolt preload, we choose bolt pretension from the load section and select the area between the nut and the bolt head for the force application. To add the preload data, tabular data type is selected and a force of 7672.5 N is entered at the first load step, as depicted in Figure 21.8. This is continued for the remaining two load steps as well.

Apart from this, we need to make some additions to make sure that the symmetry condition is fulfilled and that the top portion of the nozzle (top surface of the middle nozzle) is fixed as it is connected to the upper nozzle (not modeled) and cannot move relatively. The two symmetry regions, as shown in Figure 21.9, make sure that these surfaces are free to move parallel to themselves, but no perpendicular displacement is allowed. If this condition is not considered, it would mean that the half-a-bolt system is free to penetrate into its symmetric counterpart. This way we would have only modeled half a bolt system without caring about the whole geometry. Apart from this crucial condition, we need to add frictionless support at the top surface which would restrict any displacement in the normal and tangential directions of the surface. Both these essential boundary conditions are shown in Figure 21.9.

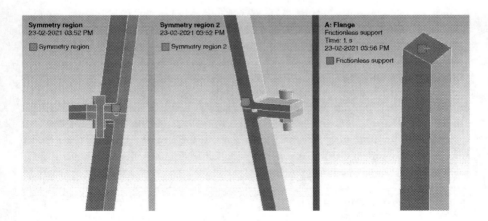

FIGURE 21.9 Symmetry and frictionless support.

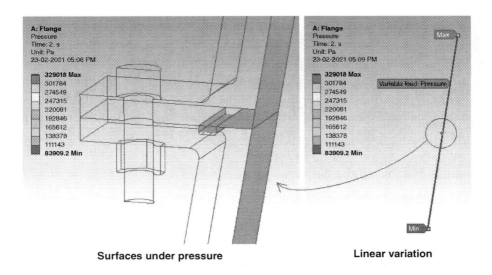

FIGURE 21.10 Pressure load and its linear variation.

The pressure due to the hot gas is applied by first selecting the surfaces experiencing it. Such surfaces are not only the inner walls of the nozzle but also some hidden surfaces between the middle and the lower parts of the nozzle, as shown by green-colored regions in Figure 21.10. The pressure data points are entered in a tabular form with 84 kPa at the exit and 330 kPa at the top surface of the nozzle, respectively, and the variation between these two values is taken linear along the Z-axis, as depicted in Figure 21.10. As, this loading type is turned on in the second step, the necessary changes are done in the model.

The equivalent forces due to the regeneration channels are inputted by selecting force under the load section and then choosing the surfaces experiencing it under the geometry section. A force of magnitude ~4450 N is entered in a tabulated manner so that this loading type is completely turned on after the second loading step. In Figure 21.11, the vectors at the probes A and B represent this force acting on the two red-colored surfaces in opposite directions, responsible for pulling them apart. As there are four major bodies under consideration in this problem, we need to specify the contact surfaces carefully. The contact between the nut and the bolt is taken as bonded so that they act like a single unit and the nut simply shrinks the upper part of the bolt when preload is applied. The contacts between the lower and the middle parts of the nozzle, middle nozzle and the bolt, lower nozzle and the nut, and the bolt and the hidden surfaces of the two parts of the nozzle are taken as frictionless. This type of contact makes sure that none of these surfaces penetrate into each other, which is done by introducing an equivalent traction in direction opposite to the penetration direction. With this, our model is ready, and we can move on to solving the model.

A Self-sustained Machine Learning Model 441

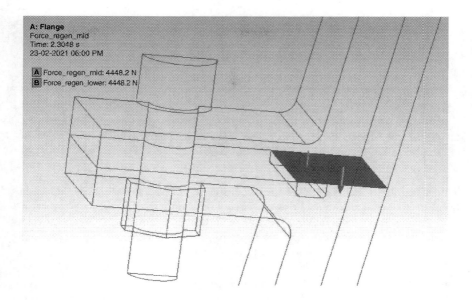

FIGURE 21.11 Forces from regeneration channels.

21.3.3 Outcome of the Core Analysis

The simulation run took around 30 minutes to solve. All the results have been successfully obtained which include the total deformation, stresses, and gaps developed at three different load steps in a cumulative manner. Apart from these, we also obtain the force reaction and the circumferential stress at the bottom of the nozzle. In this chapter, we only show the last two results as these can be verified by some simple analytical calculations, as shown in the following sections.

21.3.3.1 Force Reaction in the Axial (Z) Direction

At the end of the third load step, the reaction force comes out to be as $\vec{F_R} = 0.18984\hat{i} + 0.58357\hat{j} - 3041.2\hat{k}$ which is a vector with a magnitude of 3041.2 N inclined almost completely toward the negative Z-axis. Note that this is the reaction by only 1/400th of the model or half a bolt. Let's try computing the same analytically as follows.

We know that this reaction is against the pressure that is experienced by the inner walls of the nozzle due to the hot exhaust gas. The axial reaction should exactly cancel out the pressure due to the hot gas. The force due to the applied pressure is given by $\vec{F_R} = \left(P_{avg} \times (\text{projected area})\right)$. As shown in Figure 21.12, the projected area is the area of the concentric circle that is formed at the base of the nozzle with an inner radius of 1 m and an outer radius of 1.77 m while the P_{avg} is taken as the average of pressure at the entry and the exit of the nozzle. On calculating these and substituting them in the equation, we get $\vec{F_R} = 1280800$ N which is the reaction generated by the complete model. Hence, the reaction by 1/400th of the model is $\frac{1280800}{400} = 3202$ N which is close to the results from simulation with a difference of only 5.3%.

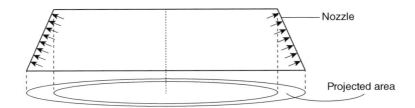

FIGURE 21.12 Hand calculation model for the axial reaction.

21.3.3.2 Circumferential Stress

This is the stress that is found in the circumferential direction and is also termed as hoop stress. We first calculate this by hand at the exit of the nozzle. The hoop stress is given by $\sigma_\theta = \frac{pr}{t}$, where p is the pressure at the exit, r is the radius at the exit, and t is the wall thickness of the nozzle. Inputting these values as 84 kPa, 1.77 m, and 1.27 cm, respectively, we obtain $\sigma_\theta = 1.166 \times 10^7$ Pa.

For verifying this numerically, we obtain the Normal stress along the X-axis as this is the value of the stress that comes out from the cross-section along the circumference at the exit of the nozzle as shown in Figure 21.13. In this figure, a variation of the stress is shown along the geometry of the 1/400th nozzle while the probe shows the value of σ_θ at the exit of the nozzle. It is worth noting that this value of 1.155×10^7 Pa obtained numerically is in close agreement with the analytical result as calculated above. An error of only 0.95% shows that our numerical results are correct.

21.3.3.3 Total Deformation

Figure 21.14 shows deformations for different loading steps focusing on the connecting region. As expected, after the first loading step, the deformation is negligible as only the bolts have been tightened at this point. At the same time, bolts are expected to show some deformation as seen in Figure 21.14(a). Moving on, as the hot gas is thrown out of the nozzle and propellant is made to flow through the regeneration channels leading to pressure loads and force (loading step 2), small gap is generated as shown in Figure 21.14(b). Also, the topmost portion of the setup remains fixed in space while the bottom moves away. Finally, with everything turned on at loading step 3, radial as well as axial growth is observed which is basically added up due to

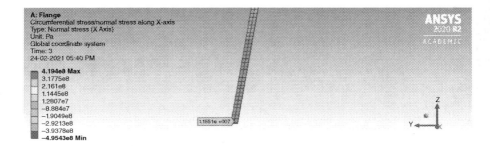

FIGURE 21.13 Circumferential stress.

A Self-sustained Machine Learning Model

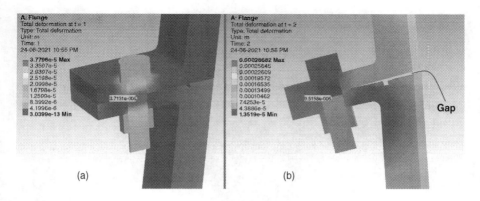

FIGURE 21.14 (a) Deformation at loading step 1 and (b) Deformation at loading step 2.

the thermal strain. Overall, a maximum deformation of 0.027918 m is found at the bottom of the nozzle.

21.3.3.4 Stress

The stresses generated in the rocket nozzle assembly comprising of 12 bolts are shown in Figure 21.15. The figure is a snapshot of the stresses after loading step 3 in which all loading conditions have been turned on.

At the first loading step, the stress generated in the bolt can be directly calculated analytically. The ratio of preload and the cross-sectional area of the bolt turns out to be 4.11×10^8 P which is very close to the stress marked by the probe in Figure 21.16(a). A little rise in the stresses at every point in the system is observed with loading step 2. Finally, toward the end, it is noticed that the stress does not change significantly but everything grows due to the increase in distance between the molecules. A dramatic version of the stress distribution is shown in Figure 21.16(b) which helps us noticing that the stress realized here is below the threshold of 1100 MPa and the rocket nozzle has the capability to withstand it.

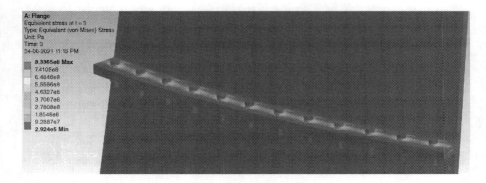

FIGURE 21.15 Stresses generated in rocket nozzle (showing 12 bolts and nuts only).

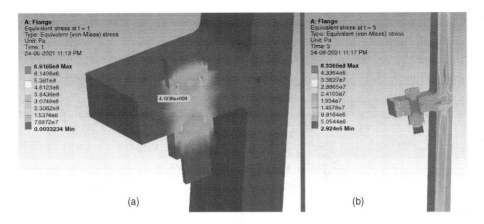

FIGURE 21.16 (a) Stress at loading step 1 and (b) Stress at loading step 3.

21.3.3.5 Gap Generated between the Lower and the Middle Parts of the Nozzle

Initially, just with the bolt preload, gap generation is not expected. We are most concerned with the inner edge at the contact region as this is the area that will initiate the leakage. As seen via the probe in Figure 21.17(a), the gap so produced is negligible (~10^{-6} m). But moving on to the second load step, the gap shows a 100-fold increase i.e., gap ~10^{-4} m (Figure 21.17(b)) which is at the verge of the allowable limit. A value higher than this will lead to the leakage of gas, and hence the unexpected event. Also, as we move away from the inner edge, toward the boundary of the system, the gap reduced to zero.

Finally, at loading step 3, it is interesting to note that the gap is pretty much the same as the previous step which means that the effect of thermal condition is minute (Figure 21.17(c)). This is ascribed to the small difference between the coefficient of thermal expansion of the bolt material (1.71×10^{-5}/°C) and the middle and the lower parts of the nozzle (1.8×10^{-5}/°C). If bolt had a significantly higher coefficient of thermal expansion than the coefficient of thermal expansion of the parts of the nozzle, then a higher gap would have been realized. This gap should be minimized as much

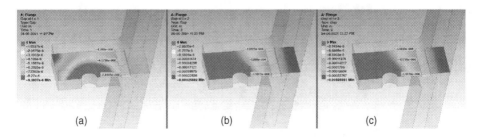

FIGURE 21.17 (a) Gap at loading step 1; (b) Gap at loading step 2; and (c) Gap at loading step 3.

A Self-sustained Machine Learning Model

as possible to avoid any fatal failures. Some methods to do so include increasing the bolt preload, adding a washer between the nut and the middle part of the nozzle, making the nozzle flange thicker (material contribution), and so on. But despite this, the failure might take place which leaves only one way to avoid it and that is to quickly predict the probability of the failure well in advance.

21.3.4 MOTIVATION FOR EMPLOYING MACHINE LEARNING

The core analysis carried out above, in reality, is actually done repeatedly for different sets of environmental conditions to test all the possible scenarios that might occur in-flight and what improvements can be done to avoid these. But even then, something unexpected might take place and lead to fatality. To circumvent such dangerous events, an ultra-quick and self-sustained model is needed that can be employed in-flight to predict the mechanical properties of the rocket nozzle. ML turns out to be a great candidate for solving this problem. To make this happen, we need to generate reasonable amount of data including the material properties, input environmental conditions, and the output mechanical properties that can be fed for training the ML model. Once trained, the ML model will be able to predict the mechanical properties of the rocket nozzle (total deformation, stresses, and gap formed) accurately based on the trend/behavior of the input conditions (bolt preload, pressure variation, forces from regeneration channels, and thermal condition) and also to warn us of any failure that might take place. Based on the warning(s), appropriate action such as reducing the propellant flow, periodically turning off one of the engines, etc. can be taken up to avoid the unexpected event.

21.3.4.1 Design of Experiments

To train the ML model, we create a series of data including all the input data and the corresponding outputs for which we employ design of experiments (DOE). We have five input variables namely the bolt preload (in %), pressure variation (in kPa) comprising of the maximum and the minimum pressure, forces from regeneration channels (in N), and the thermal condition (i.e., temperature in degree centigrade). The values for these parameters taken in the simulation above are 50%, 330 kPa to 85 kPa, 4450 N, and 400°C, respectively. Using a full factorial design, we take two values for the temperature and the bolt preload, and three values for each of the remaining parameters. These values are taken such that they are offset by +10% and −10% from the original values mentioned above. The levels of the input data generated using DOE are tabulated in Table 21.3.

Based on Table 21.3, it can be concluded that $2\times3\times3\times3\times2 = 108$ combinations of input data can be generated. The output mechanical properties for each of these combinations i.e., $108 \times 3 = 324$ output data points are calculated via simulation. The core process explained above was repeated 108 times inputting different environmental conditions so generated which took more than 20 days. Using this data, different ML models are employed to test which one predicts most accurately. Finally, a test data is also generated to check the variation of the predicted value from the actual (obtained via core analysis) value.

TABLE 21.3
Input Parameters Levels Generated Using Design of Experiments

	Levels		
Input Design Parameters	**Level 1**	**Level 2**	**Level 3**
Temperature (°C)	360	440	
Pressure Max (k*Pa*)	297	330	363
Pressure Min (k*Pa*)	76.5	85	93.5
Force due to regeneration channel (N)	4005	4450	4895
Bolt Preload (%)	45	55	

21.4 MACHINE LEARNING MODEL

ML makes a computer learn from a dataset using some defined algorithm. When the ML model is being deployed, since the machine (computer) has learnt from the dataset it gives a virtual feeling to the user that the machine is smart. In the crux, ML techniques are used wherever sufficient data (records, historical data, etc.) is present and that can be used to train an ML model.

The following are the most basic and crucial steps in solving any ML problem:

Acquiring the data – Mostly, the data in any ML model is acquired in two ways. First is by generating data points in a working mechanical, electrical, physical, or software system. The other idea is to use historical/recorded data to make future predictions.

Data pre-processing – In most cases, the data is just records put together. The type of problem and the model parameters are the deciding factors that demand cleaning and processing of the data. Removing null values, redundant values, removing outliers, etc. are the common steps that are carried out in this step.

Data visualization – Once the data has been processed, identifying trends in the dataset can be very much beneficial for selecting best features and creating insights from the data with respect to the problem that is to be solved.

Feature engineering – Feature engineering is a very important step in building an ML model. Selecting best features that will lead to a robust model is a big task in ML model building. Techniques like Principal Component Analysis (PCA), Linear Discriminant Analysis (LDA) are used for dimensionality reduction that helps building a robust ML model.

Identifying the problem – A given ML problem can be of multiple types. Broadly ML techniques are classified as follows:

a. *Supervised learning* – In this type of ML, the dataset used has both labeled data and unlabeled data. The model is made to learn to generate

A Self-sustained Machine Learning Model

what type of output given a set of input features. Supervised learning is further divided as follows:

Regression – A regression problem is when given some input features, one or more output parameters are needed to be predicted. For example, predicting house prices from historical data of house prices in an area is a standard regression problem. The common regression algorithms used in ML are linear regression, Support Vector Regression (SVR), Decision Tree Regression, Random Forest Regression, etc.

Classification – A classification problem is when a model is trained to predict a category out of some categories. For example, predicting gender of a person using his/her handwriting is a classification problem. The common classification algorithms used in ML are Logistic Regression, Support Vector Machine (SVM), Decision Tree Classification, Random Forest Classification, Gaussian Naïve Bayes, etc.

b. *Unsupervised learning* – In this type of ML technique, only labeled dataset is provided. The most common type of unsupervised learning is clustering. Clustering is a technique in which formation of separate clusters is done using some defined algorithmic paradigm. Clustering is often used to find hidden trends in the data. K-means clustering is the most popular clustering algorithm.

c. *Reinforcement learning* – This type of learning category is commonly used in games. Basically, it consists of labeled data only like unsupervised learning. But from a state, reward is provided to every sub-state that after some time makes the machine learn. In human vs computer chess, this learning is used at the backend to make a computer move. Reinforcement learning is still not that used unlike supervised and unsupervised learning, although it has specific use cases in AI (Artificial Intelligence) game development and robotics.

21.4.1 Tools Used

Python is an expressive, object-oriented, and very high-level programming language. Originated in 1991 by Guido Van Rossum (.py is its extension). It has various notable advantages:

- Supported by a multitude of platforms (Raspberry Pi, Windows, Mac, etc.)
- Syntax similar to that of English and helps in rapid prototyping
- Manipulation of data, analysis of data, and scripting is enhanced by it
- Indentation is a vital part of python and adds to the readability of code
- Interpreted language with vast libraries for usage in artificial intelligence

Windows, macOS, Linux, android, and iOS support various python GUI like Kivy, PyQT, Tkinter, etc. Environment of Python is free and open source,

448   Handbook of Sustainable Materials

TABLE 21.4
Libraries Used

S. No.	Library Name	Version	Function
1.	Numpy	1.19.4	Provides fast and easy computation of higher-order arrays
2.	Pandas	1.0.1	Manipulation of data in data frame
3.	Matplotlib	3.1.3	Plotting graphs
4.	Seaborn	0.10.0	Graphically attractive visualization library
5.	Scikit Learn	0.23.2	Contains pre-defined popular machine learning models
6.	Dtree	0.0.1	Used for visualizing decision-based ML algorithms

thus rendering ease of access and data manipulation. It has two modes namely Interactive and Script modes. Script mode is more advantageous when dealing with large chunks of code as compared to interactive mode. In the present work, Python version 3.9 was used.

21.4.1.1   IDE Used: Google Colab

Google Colab is a cloud IDE that can be used to run Python scripts and files. Anaconda Navigator is another famous Python IDE. Google Colab was preferred over Anaconda Navigator as it provides the following advantages over Anaconda Navigator:

*   Free computation power (free RAM and GPU usage)
*   Share notebooks via links through Google Drive
*   Convenient sharing of files and multi-user access service that helps in peer-programming

Table 21.4 contains the libraries that are provided in Python. These libraries help in faster computation of data, data visualization, data preprocessing, and some standard ML models that can be used for constructing the ML model.

21.4.2   MODEL TO PREDICT MECHANICAL PROPERTIES OF ROCKET NOZZLE IN-FLIGHT

The main motivation for building an ML model is that due to the applied loadings, the mechanical properties of the rocket nozzle arrangement vary significantly in flight that sometimes overshoot leading to disasters. Although the study of the distribution of stress and deformation is conventionally done using simulation software to optimize the dimensions and materials employed, but by using ML, it is possible to predict these mechanical properties along with a drastic reduction in computation time. More importantly, the ML systems can be deployed in real applications like an actual rocket nozzle during flight. The computationally generated data that has been

A Self-sustained Machine Learning Model

	Temp	Pressure max	Pressure min	Force	Bolt preload	Deformation	Stress	Gap
0	360	297	76.5	4005	45	0.027043	666150000	-0.000246
1	360	297	76.5	4005	55	0.027034	714040000	-0.000235
2	360	297	76.5	4450	45	0.027078	737210000	-0.000290
3	360	297	76.5	4450	55	0.027067	778080000	-0.000278
4	360	297	76.5	4895	45	0.027113	817460000	-0.000336
5	360	297	76.5	4895	55	0.027101	849050000	-0.000322
6	360	297	85.0	4005	45	0.027035	656270000	-0.000240
7	360	297	85.0	4005	55	0.027026	705860000	-0.000230
8	360	297	85.0	4450	45	0.027069	727090000	-0.000285
9	360	297	85.0	4450	55	0.027059	768130000	-0.000272
10	360	297	85.0	4895	45	0.027105	806320000	-0.000330

FIGURE 21.18   First ten rows of dataset.

generated following DOE principles are used to train an ML model that will be able to predict stress, deformation, and gap in-flight.

21.4.2.1   Dataset Acquisition

The data consists of 108 data points generated using DOE. The data has been acquired by performing multiple runs with different combinations of input parameters between each successive run giving a unique output for every data value sent as input. The data consists of unique input parameters along with their respective values, as tabulated in Table 21.3. Each unique combination of data values was simulated in the software to compute the corresponding values of total deformation (in meters), stress (in Pascal), and gap formed (in meters). A snapshot of the first ten rows of the dataset is shown in Figure 21.18.

21.4.2.2   Data Cleaning and Visualization

After acquiring the data, the next step is to normalize the data for model training purpose. The dataset consists of a range of following output values:

- Deformation is of the order of 10^{-2}
- Stress is of the order of 10^7
- Gap is negative due to its parity in the simulation and is of the order of 10^{-4}

These values can lead to poor model characteristics. Thus, these values were multiplied with suitable constant factor and all values turned to positive to have a normalized distribution of values. After the model is trained to its best accuracy, the inverse of these constants can be multiplied to get the exact prediction. The final data after cleaning and normalization is shown in Figure 21.19.

	Temp	Pressure max	Pressure min	Force	Bolt preload	Deformation	Stress	Gap
0	360	297	76.5	4005	45	270.43	66.615	24.552
1	360	297	76.5	4005	55	270.34	71.404	23.474
2	360	297	76.5	4450	45	270.78	73.721	28.972
3	360	297	76.5	4450	55	270.67	77.808	27.754
4	360	297	76.5	4895	45	271.13	81.746	33.554
5	360	297	76.5	4895	55	271.01	84.905	32.154
6	360	297	85.0	4005	45	270.35	65.627	24.043
7	360	297	85.0	4005	55	270.26	70.586	23.000
8	360	297	85.0	4450	45	270.69	72.709	28.456
9	360	297	85.0	4450	55	270.59	76.813	27.246
10	360	297	85.0	4895	45	271.05	80.632	33.021

FIGURE 21.19 First ten rows of cleaned and normalized dataset.

To ensure that the values are uniformly distributed over the selected limits as per DOE rules, the data was visualized. The visualization graphs are shown in Figure 21.20. The said visualizations show that the values depicted are uniformly distributed. Similar results were obtained for remaining input parameters i.e., maximum pressure, minimum pressure, bolt preload, and temperature.

21.4.3 Model Usage

When the data was completely free of redundancy, the next step was to train the model. For our dataset, the following two factors were taken into consideration for further steps:

1. Since the dataset has scarce number of rows due to higher computation time, the dataset was not split into training and testing data. Instead, the

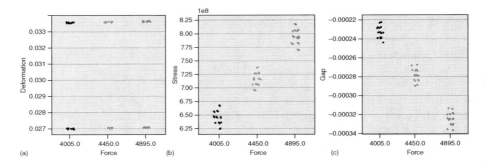

FIGURE 21.20 Visualization of output parameters: (a) Deformation, (b) Stress, and (c) Gap.

A Self-sustained Machine Learning Model

whole dataset has been used for training purposes and after a reliable model has been produced, the model was tested against randomly generated test values.

2. Due to DOE principles, the dataset does not portray a general trend. Thus, it alluded that decision-based ML algorithms, like Decision Tree, Random Forest, and XGBoost, are likely to fit the dataset and result in a reliable ML model. Following this, we test all the above-mentioned decision-based algorithms one by one.

21.4.3.1 Decision Tree Model

A decision tree is an algorithm that uses *Tree* data structure used in computer science to build a regression model. The tree is built recursively in a top-down fashion from root-leaf. The root node is the start, and the leaf nodes consist of the final prediction constituting the predictions of the output parameters. Whenever a decision is taken from a node in the tree, there is a cost related to that split known as the *cost of split*. Once, the tree has been built recursively, individual costs of every branch are combined to predict the value of the output variable. A *Gini Score (G)* is used to determine the feasibility of a split, which is given by $G = \sum p_k (1 - p_k)$. In this, p_k is the proportion of class inputs present in that group. The decision tree model is often prone to *overfitting* as it computes a single tree in the entire training process. Employing this model led to an average error of 2.6% on the deformation and gap predictions but on stress predictions, the average error was much more ~11.1%.

21.4.3.2 Random Forest Model

A Random Forest is also a decision-based algorithm like *Decision Tree*. A group of trees is called a forest. This algorithm builds multiple trees and combines their individual results as the final result. The advantage that Random Forest has over Decision Tree is that it uses subset of randomly generated trees each time when it combines the results. Thus, generally Random Forest is not prone to overfitting. *Random Forests* have the following advantages over *Decision Trees*:

- Random Forests use random samples for training each decision tree that produces accurate results over the entire data as a whole.
- Along with this, they have minimal variance error as each time a small subset of the data is used for training purpose resulting in a robust model.

The data was trained using the *Random Forest Regressor* package in Scikit learn. Figure 21.21b, & shows the training process of a single tree for each of the three output variables.

Overall, *Random Forest* yielded better results than the decision tree. So further hyperparameter tuning was done to tune the parameters to give the best possible accuracy. At the end, the average error for deformation and gap prediction was 2.6% while the average error for stress prediction turned out to be 5.7% which is much less than the one obtained using Decision Tree.

21.4.3.3 XGBoost Model

Gradient boosting is a technique in ML where results from the previous iteration are used to train the model faster from previously calculated trends in the data. Ingredient descent randomly selected subset of the data is used to build trees. *XGBoost* has the following advantages over *Random Forest*:

1. The learning rate is faster comparatively as it uses the past trends to learn moving ahead in the model training; hence saving on the computational costs.
2. Dataset with less noise trains well on *XGBoost* on an average, therefore, reducing the computation time even further if the data is free from noise.

Since the dataset generated using DOE for the rocket nozzle has minimal noise, *XGBoost* was used to train, and it resulted in an appreciable improvement in the model prediction as compared to the Random Forest.

Finally, the last two ML models were used to accurately predict the mechanical properties for a random combination of input parameters and they were compared with the values obtained via simulation. The exact predicted values and the corresponding errors have been discussed in Section 21.4.4.

21.4.4 SOME INTERESTING RESULTS

Finally, a test input data was arbitrarily created and the core analysis (*simulation*) as well as the ML models (*Random Forest* and *XGBoost*) were employed to calculate the output mechanical properties including the deformation, stress, and gaps formed. The input environmental conditions are shown in Table 21.5. While Figures 21.21 to 21.23 represent the corresponding outputs from simulation, Random Forest and XGBoost ML models for deformation, stress, and gap formed, respectively.

The average percentage error in computing deformation is found to be 4.27% and 4.26% for Random Forest and XGBoost ML models, respectively. The closeness in the error is evident from Figure 21.23, as the plots corresponding to Random Forest and XGBoost overlap each other. Moving on to the prediction of stress, error

---

TABLE 21.5
Input Conditions for Test Data

Dataset Number	Temperature (°C)	Pressure Variation (kPa)		Force from RC (N)	Bolt Preload (%)
		Pressure Maximum	Pressure Minimum		
1.	380	350	92	4100	50
2.	435	310	77	4800	47
3.	410	360	87	4400	53
4.	435	350	87	4100	50
5.	380	310	77	4100	53

A Self-sustained Machine Learning Model

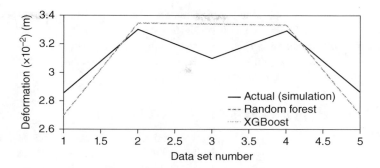

FIGURE 21.21 Comparison of deformation computed via simulation (black) and predicted from Random Forest (red) and XGBoost (blue) machine learning model.

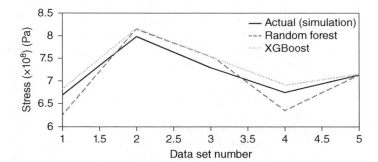

FIGURE 21.22 Comparison of stress computed via simulation (black) and predicted from Random Forest (red) and XGBoost (blue) machine learning model.

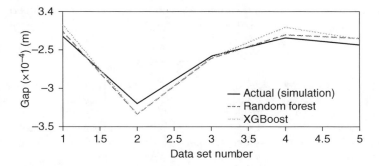

FIGURE 21.23 Comparison of gap computed via simulation (black) and predicted from Random Forest (red) and XGBoost (blue) machine learning model.

of 3.37% and 1.86% are obtained from Random Forest and XGBoost, respectively. As shown in Figure 21.23, the plot of XGBoost is closer to the one obtained via simulation; hence the error is relatively lower. Finally, comparing the gap computed from the three methods (Figure 21.23), the percentage error for Random Forest and XGBoost turns out to be 2.85% and 4.33%, respectively.

21.5 CONCLUSION, PROSPECTS, AND CHALLENGES

A rocket nozzle consisting of two parts clipped together with a series of 200 nuts and bolts was designed using design modeler. The numerous loading types and their methods of application along with the three loading steps were discussed before initiating the core analysis of the nozzle. Using ANSYS, simulations were done to carry out the FEA of the nozzle assembly for obtaining the total deformation, stresses generated, and the gap formed between the lower and the middle parts of the nozzle at each loading step. DOE was employed, and large dataset was produced by simulating the assembly for each of the combinations of loading types obtained from the levels generated via DOE. This at-home generated data was used to train various ML models including Decision Tree, Random Forest, and XGBoost. Following this, a self-sustained ML model was created that composed of Random Forest and XGBoost ascribed to their high accuracies. This model could predict the mechanical properties of the rocket nozzle within *one second* in-flight with an overall *error of less than* 3%. This system can be linked to the rocket system for the predictions and warnings of unexpected events that might take place.

The nozzle assembly is of utmost importance in rocket nozzle where several phenomena, such as deformation, generated stresses, and formed gap, need to be very precisely understood. Here sustainable materials can play a vital role both in designing and 3D printing. This will also help in designing computer experiments and for the production of sufficiently large data for simulation based on which ML models can be trained for in-flight prediction. However, it may turn out to be challenging to virtually design the rocket nozzle and carrying out the CFD and FEA for the whole body. This is worthwhile to note that transferring the CFD data to FEA platform is a rare problem and requires synchronization between the two. This would be quite interesting to revisit all these phenomena by using high-performance computing and to produce the data for ML for different materials for the sustained functioning of the nozzle. Doing it in-flight makes it quite challenging.

REFERENCES

1. Marchi, C.H., Laroca, F., Silva, A.F. and Hinckel, J.N., 2004 Apr 1. Numerical solutions of flows in rocket engines with regenerative cooling. *Numerical Heat Transfer, Part A: Applications*, 45(7), pp. 699–717.
2. Murugan, J.P., Kurian, T., Jayaprakash, J. and Sreedharapanickar, S., 2015 Oct. 3-D analysis of flanged joints through various preload methods using ANSYS. *Journal of the Institution of Engineers (India): Series C*, 96(4), pp. 407–417.
3. Malik, L. and Tevatia, A., 2021. Comparative analysis of aerodynamic characteristics of F16 and F22 combat aircraft using computational fluid dynamics. *Defence Science Journal*, 71(2), pp. 137–145.

A Self-sustained Machine Learning Model

4. Malik, L., Rawat, S., Kumar, M. and Tevatia, A., 2021. Simulation studies on aerodynamic features of Eurofighter Typhoon and Dassault Rafale combat aircraft. *Materials Today: Proceedings, 38*, pp. 191–197.

5. Malik, L., 2022. Tapered coils system for space propulsion with enhanced thrust: A concept of plasma detachment. *Propulsion and Power Research, 11*(2), pp.171–180.

6. Malik, L., Kumar, M. and Singh, I.V., 2021. A three-coil setup for controlled divergence in magnetic nozzle. *IEEE Transactions on Plasma Science, 49*(7), pp. 2227–2237.

7. Malik, L., Escarguel, A., Kumar, M., Tevatia, A. and Sirohi, R.S., 2021. Uncovering the remarkable contribution of lasers peak intensity region in holography. *Laser Physics Letters, 18*(8), p. 086003.

8. Lijo, V., Kim, H.D., Setoguchi, T. and Matsuo, S., 2010 Jun 1. Numerical simulation of transient flows in a rocket propulsion nozzle. *International Journal of Heat and Fluid Flow, 31*(3), pp. 409–417.

9. Peake, S., Ellis, R. and Broquere, B., 2006. Update: Sustainable Carbonized Rayon for Solid Rocket Motor Nozzles. In *42nd AIAA/ASME/SAE/ASEE Joint Propulsion Conference & Exhibit* (p. 4598).

10. Ali, M.M. and Nived, N., 2021. Composites materials for sustainable space industry: A review of recent developments. *World Review of Science, Technology and Sustainable Development, 17*(2–3), pp. 172–196.

11. Valero Sánchez, M., 2021. *Study of the use of lunar materials to produce rocket propellants* (Bachelor's thesis, Universitat Politècnica de Catalunya).

12. Olufsen, F. and Orbekk, E., 2017. Application of CMC materials in rocket propulsion. *Advances in High Temperature Ceramic Matrix Composites and Materials for Sustainable Development*. Wiley, Hoboken, pp. 367–374.

13. Baxi, P., Jain, R., Dhadke, Y., Chhabra, Y. and Khatawate, V.H., 2021 Jan. Design and Analysis of Bell-Parabolic De Laval Rocket Exhaust Nozzle. In *2021 4th Biennial International Conference on Nascent Technologies in Engineering (ICNTE)* (pp. 1–6). IEEE.

14. Thomas, D.J., 2022. Advanced active-gas 3D printing of 436 stainless steel for future rocket engine structure manufacture. *Journal of Manufacturing Processes, 74*, pp. 256–265.

15. Oztan, C. and Coverstone, V., 2021. Utilization of additive manufacturing in hybrid rocket technology: A review. *Acta Astronautica, 180*, pp. 130–140.

16. Fritz, T.J., Huebsch, W. and Wilhelm, J., 2017. Analysis of 3D Printed Titanium Rocket Nozzle. In *58th AIAA/ASCE/AHS/ASC Structures, Structural Dynamics, and Materials Conference* (p. 0510).

17. Oztan, C., Ginzburg, E., Akin, M., Zhou, Y., Leblanc, R.M. and Coverstone, V., 2021. 3D printed ABS/paraffin hybrid rocket fuels with carbon dots for superior combustion performance. *Combustion and Flame, 225*, pp. 428–434.

18. Ryakam, S. and Crispin, Y., 2021. Optimal Control of Hybrid Rocket for Sub-orbital Ascent Trajectory. In *AIAA Propulsion and Energy 2021 Forum* (p. 3508).

19. Kobayashi, T., Nishida, S.I., Nakamura, S. and Nakatani, S., 2021, Jan. Study of Image Processing Methods for Space Debris Capture. In *2021 IEEE/SICE International Symposium on System Integration (SII)* (pp. 304–309). IEEE.

20. Mehta, Y., Bhargav, V.N. and Kumar, R., 2021. Experimental characterization and control of an impinging jet issued from a rocket nozzle. *New Space, 9*(3), pp. 187–201.

21. Zhou, Z., Bao, Y., Sun, P. and Li, Y., 2022. Cooling of rocket plume using aqueous jets during launching. *Engineering Applications of Computational Fluid Mechanics, 16*(1), pp. 20–35.

22. Lu, C., Zhou, Z., Shi, Y., Bao, Y. and Le, G., 2021. Numerical simulations of water spray on launch pad during rocket launching. *Journal of Spacecraft and Rockets, 58*(2), pp. 566–574.

23. Almayas, A., Yaakob, M.S., Aziz, F.A., Yidris, N. and Ahmad, K.A., 2021. CFD application for solid propellant rocket simulation: A review. *CFD Letters*, *13*(1), pp. 84–95.
24. Leon-Cardona, D., Rodriguez-Ferreira, J., Rosso-Cerón, A. and del Jesús Matínez, M., 2021. Design of a Converging-Diverging nozzle for Kappa-DX experimental rocket engines by Computational Fluid Dynamics (CFD) method. *43rd COSPAR Scientific Assembly. Held 28 January-4 February*, *43*, (p. 2168).
25. Waxenegger-Wilfing, G., Sengupta, U., Martin, J., Armbruster, W., Hardi, J., Juniper, M. and Oschwald, M., 2020 Nov 25. Early Detection of Thermoacoustic Instabilities in a Cryogenic Rocket Thrust Chamber using Combustion Noise Features and Machine Learning. arXiv preprint arXiv:2011.14985.
26. Ma, H., Zhang, Y.X., Haidn, O.J., Thuerey, N. and Hu, X.Y., 2020 Oct 1. Supervised learning mixing characteristics of film cooling in a rocket combustor using convolutional neural networks. *Acta Astronautica*, *175*, pp. 11–18.
27. Min, A.T., Sagarna, R., Gupta, A., Ong, Y.S. and Goh, C.K. 2017 Oct 11. Knowledge transfer through machine learning in aircraft design. *IEEE Computational Intelligence Magazine*, *12*(4), pp. 48–60.
28. https://www.techsteel.net/alloy/stainless-steel/a286. Visited on 20/02/2022.
29. http://www.matweb.com/search/datasheettext.aspx?matguid=7a87941825a3463eaba79 79c4333721f. Visited on 20/02/2022.

22 Heat-Assisted Dieless Sheet Forming Techniques for Hard-to-Form Materials

Nikhil Kadian, Rakesh Rathee, and Ajay

CONTENTS

22.1 Introduction 457
22.2 Classifications of Incremental Sheet Forming 458
    22.2.1 Laser Assisted Incremental Sheet Forming (LA-ISF) 459
    22.2.2 Halogen Lamp Assisted Incremental Sheet Forming 459
    22.2.3 Hot Air Assisted Incremental Sheet Forming 460
    22.2.4 Oil Heated Incremental Sheet Forming 461
    22.2.5 Friction Stirred Incremental Sheet Forming 461
    22.2.6 Electrical Assisted Incremental Sheet Forming 461
    22.2.7 Induction Assisted Incremental Sheet Forming 463
    22.2.8 Ultrasonic Assisted Incremental Sheet Forming 463
22.3 Conclusions and Future Directions 466
References 467

22.1 INTRODUCTION

Incremental sheet forming (ISF) is one of the latest sheet metal forming processes without using process-specific dies and punches [1, 2]. ISF process eliminates the involvement of forming tools such as dies and punches that are necessarily required in conventional sheet metal forming processes; therefore, ISF provides flexibility in product design and reduces the product development cost, particularly for small batch size production [3–5]. Consequently, ISF has been given adequate attention by the researchers for enabling the low cost of tooling and flexibility in the product design. Researchers have also found the potential usage of ISF in aerospace industry to manufacture aircrafts' body panels, passenger sheet covers, instrument panels, and in automobile industry to produce engine cover, door inlet-outlet plates, hoods, and also for highly customized medical prosthetics products like human skull, denture plate, and cranial plate. It has also been revealed from all the available literature that ISF is a highly customizable process due to low setup cost and production cost, when batch size is very small or single product is to be produced.

DOI: 10.1201/9781003297772-22

ISF is based on the localized deformation of sheet metal at room temperature with the help of a special forming tool that is programmed to move along a CNC (computer numerical control) controlled trajectory and gradually advancing to deform the sheet layer by layer into the required shape which is normally designed by suitable CAD (computer-aided design) software [6–9]. It has been found that this process has been facing complexities in deformation of hard-to-deform materials like magnesium alloys and titanium alloys, which find several applications in biomedical, aerospace and automobile industries due to their unique properties of greater strength and light weight. In order to increase the formability of material that is hard-to-form, the material sheet is heated locally or globally and this heating of material is done using a special variant of ISF process known as heat-assisted incremental sheet forming (HA-ISF) process. ISF, executed at room temperature, possesses particular problems like spring back effect and greater amount of forming forces required to deform the sheet. HA-ISF overcomes such issues up to a great extent. With capability to reduce the problems encountered during conventional ISF, HA-ISF has attracted the attention of researchers in this decade [10]. Although some work related to HA-ISF has been performed but the literature is not available in well-organized manner.

Therefore, the proposed work aims to provide a review of different methodologies used by different researchers during HA-ISF to heat the work sheet to be deformed in desired shapes and to study the effect of these HA-ISF techniques on forming forces to accomplish this viable process. Authors have also provided future directions for researchers regarding the forming forces in relation to process variables.

22.2 CLASSIFICATIONS OF INCREMENTAL SHEET FORMING

ISF can be classified on the basis of the following parameters:

On the basis of forming method
- Single point incremental forming (SPIF)
- Two-point incremental forming (TPIF)
- Hybrid forming

On the basis of forming temperature
- Room temperature
- Elevated temperature

On the basis of forming tool
- Hemispherical-end tool
- Roller ball-end tool
- Flat-end tool

In single point incremental forming (SPIF), local stretching of sheet is executed with the help of a single point forming tool without direct or indirect involvement of forming dies. It is also called negative incremental forming. In two-point incremental forming (TPIF), the sheet is supported by partial die or full die for producing the

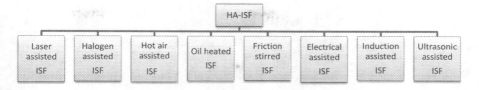

FIGURE 22.1 Classification of HA-ISF.

desired surface in sheet. Hybrid incremental forming is the combination of asymmetric incremental forming and stretch forming.

The ISF process can be performed by using two strategies i.e., first is single stage and second is multiple stages. In single stage forming, the sheet is clamped in ISF fixture and deformation is produced with the help of rigid tool that moves on the sheet along a predetermined tool path till the desired product is obtained. While in multistage strategy of ISF, the hemi-ellipsoidal parts are formed using two sequential stages. Some materials like magnesium alloys and titanium alloys are very often used in medical and aerospace industries due to their compatibility with biological conditions of humans and higher strength-to-weight ratio. These materials are hard-to-deform at room temperature due to their structure and nature. Furthermore, their formability can be increased by increasing the temperature of sheet globally or locally to deform the sheet in the desired shape easily and effectively. This kind of forming at higher temperature by using some schemes to heat the sheets locally or globally is known as HA-ISF. On the basis of these schemes, HA-ISF has been classified into six subcategories, as shown in Figure 22.1.

22.2.1 Laser Assisted Incremental Sheet Forming (LA-ISF)

In LA-ISF, local and pointed heating is accomplished with the help of lasers directed toward the point of deformation and thus increasing the formability of material to be deformed. Duflou et al. [11] used laser heated ISF process (shown in Figure 22.2) and observed reduction in forces, improved formability, and dimensional accuracy. Gottmann et al. [12] also performed laser assisted ISF on TiAl6V4 material and observed the maximum increase in formability at 4000 °C. The laser heat has been observed to provide a precise and more controlled heating zone for smooth deformation. LA-ISF has a drawback of involving higher hardware cost of setup.

22.2.2 Halogen Lamp Assisted Incremental Sheet Forming

In halogen lamp assisted ISF (shown in Figure 22.3), several halogen bulbs are moved with forming tool and thus supplying their heat energy to the material to be deformed. Kim et al. [13] used the experimental setup in which many halogen lamps were traversing with the tool. The influences of process parameters of AZ31 (an alloy of magnesium) sheet were investigated by heating the magnesium alloy

460   Handbook of Sustainable Materials

FIGURE 22.2   Experimental setup used by Duflou et al. [11].

sheet locally using halogen heat. There has been observed an increase in formability and thus increase in height of fracture while heating the sheet by halogen lamp. In halogen heated forming, much of the heat supplied is far from tool-sheet contact point and thus complete localized heating is not achieved in this process.

22.2.3   Hot Air Assisted Incremental Sheet Forming

In hot air assisted ISF, the air blowers are used to heat the material to be deformed. Ji et al. [14] used hot air for heating AZ31 (an alloy of magnesium) and observed

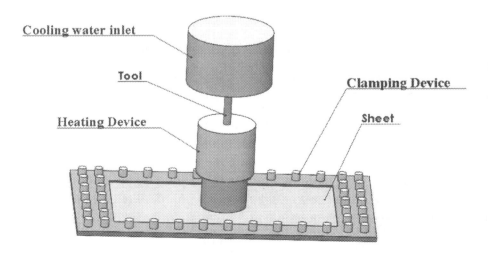

FIGURE 22.3   Schematic of halogen assisted ISF.

Heat-Assisted Dieless Sheet Forming Techniques for Hard-to-Form Materials **461**

a higher rate of formability above 1500 °C temperature. Leonhardt et al. [15] further investigated the mechanical properties and initial microstructure of AZ31 sheet. They employed hot air blowers for heating the sheet globally to employ the temperature uniformly over the sheet. They observed the maximum wall angles of 50° at 3000°C.

22.2.4 OIL HEATED INCREMENTAL SHEET FORMING

Galdos et al. [16] used the heated oil for global heating of AZ31B, a magnesium alloy sheet during ISF. They observed full recrystallization at 2500 °C and the maximum angle of deformation. Due to limited temperature of oil and other properties, the oil heating can only be used to heat aluminum and magnesium alloys. Zhang et al. [17] also used the method of oil heating the AZ31B magnesium alloy sheets. They observed that the maximum forming angle was decreasing linearly with the increase in step size.

22.2.5 FRICTION STIRRED INCREMENTAL SHEET FORMING

Rotation of a tool generates heat at the tool tip, and this heat further increases formability of material to be formed using ISF operation. Xu et al. [18] formed AZ31B magnesium alloy and applied two methods of heating, out of which one was frictional stir-up and other was electrical-assisted ISF. They further investigated and compared the effect on parameters like temperature and different feed rates. Uheida et al. [19] formed the titanium sheet by rotating the spindle in climb direction and in conventional direction. Forming forces F_X (in X-direction) and Fz (in Z-direction) were found to increase with the increase in step depth (Δz). Liu et al. [20] also used the friction stirred ISF to form AA7075O sheet of aluminum alloy. They observed rapid improvement in formability of AA7075O when the spindle speed was increased from 1000 rpm to 3000 rpm.

22.2.6 ELECTRICAL ASSISTED INCREMENTAL SHEET FORMING

When electric current flows through a conductor, it generates heat due to resistance faced by electrons to flow. This property of current persuaded many researchers to use current for heating the sheet to be formed during ISF. Li et al. [21] proposed a novel method of heating the TC4Ti alloy sheets (shown in Figure 22.4). Current was allowed to flow from one end of sheet to the other. MoS_2 was used as solid lubricant. They observed that the less dimensional accuracy of the formed component was due to excessive expansion of material or failure.

Liu et al. [22] developed a novel tool for heating purpose of Ti6Al4V sheets and didn't observe any major change in the microstructure of formed material. They were able to stabilize tool temperature at 1000 °C. By cooling tool, they observed the reduction in tool wear and minimization of adhesion of sheet material on tool tip.

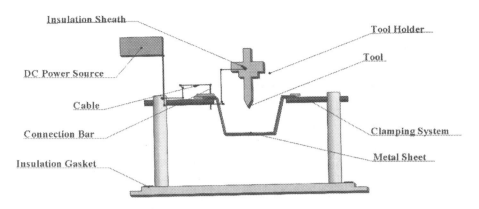

FIGURE 22.4 Schematic of electrical assisted ISF.

Xu et al. [23] also explored the cause of roughness in HA-ISF and improved the contact conditions of tool for refining the surface finish. They used the technique of passing current through the slave tool instead of master tool (shown in Figure 22.5). A maximum DC current of 800 A and 15 V was used to provide the required energy to heat the sheet materials. Surface roughness was observed to be decreased by 41%.

Magnus et al. [24] developed a robot-forming setup for HA-ISF. He used two six-axes industrial robots installed with force and torque sensor, forming tool and one clamping frame. They observed reduced spring back and greater dimensional accuracy using this setup of HA-ISF. Vahdani et al. [25] also investigated AA6061

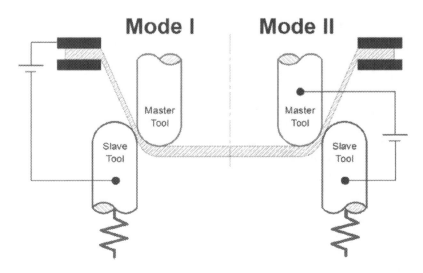

FIGURE 22.5 Scheme used by Xu et al. [23].

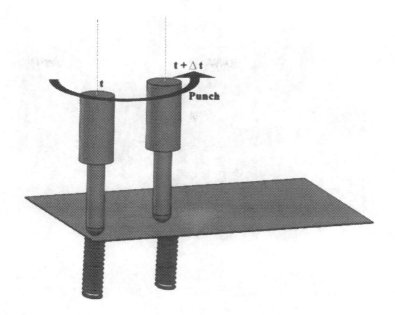

FIGURE 22.6 Schematic of induction assisted ISF.

aluminum, Ti-6Al-4V titanium, and DC01 steel sheets during HA-ISF, and studied the effect of lubricant on formability of these materials under different levels of vertical pitch, feed rate, and electric current. Electric current and lubricant had significant effects on formability.

22.2.7 INDUCTION ASSISTED INCREMENTAL SHEET FORMING

Ambrogio et al. [26] used induction heating and cryogenic cooling simultaneously on Ti-6Al4-V sheets and studied the effect of heating and cooling rate after deformation on microstructure evolution and microhardness. They coupled a coil with forming punch, passed magnetic field through Ti-6Al4-V sheets, and varied the magnetic field through sheet to produce eddy current for local heating (shown in Figure 22.6).

Al-Obaidi et al. [27] also used the method of induction heating during HA-ISF and formed the components with lower forming forces and reduced residual stresses in sheets. Smusz et al. [28] heated steel sheets using induction heaters (shown by Figure 22.7) and observed that the heat distribution was not uniform due to variation in cross-sectional area, while the temperature attainment was uniform throughout the process.

22.2.8 ULTRASONIC ASSISTED INCREMENTAL SHEET FORMING

In ultrasonic assisted ISF, ultrasonic vibrations along with tool movement are used to reduce the forming forces by repetitive force application and thus increasing

FIGURE 22.7 Scheme used by Smusz et al. [28].

temperature during forming in ISF. These vibrations increase the temperature and hence it can be considered as HA-ISF. Li et al. [29] performed HA-ISF using ultrasonic vibrations (shown in Figure 22.8) and observed the effect of various parameters. In 2018, Bai Lang et al. [30] also performed HA-ISF using ultrasonic vibrations. The simulation and experimentation results revealed that the ultrasonic vibrations can reduce the amount of forming forces required to deform the sheet material. The results obtained from experimentation were very close to those predicted by simulation.

The work of several researchers during HA-ISF is summarized in Table 22.1, which shows the method of heating the sheet material, types of material and lubricants used, and respective outcomes.

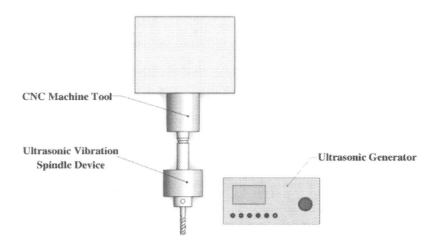

FIGURE 22.8 Schematic of ultrasonic assisted ISF.

Heat-Assisted Dieless Sheet Forming Techniques for Hard-to-Form Materials 465

TABLE 22.1
Summary of Hard-to-form Materials during HA-ISF

Method	Authors	Material Used	Lubrication Used	Achievements
Laser assisted ISF	Duflou et al. [11]	Al5182, TiAl6V4, 65Cr2	Graphite	Effective in pointed-heating with laser. Dimensional accuracy was increased due to lesser spring back.
	Gottmann et al. [12]	TiAl6V4	NA	Effective in precise local heating.
Halogen assisted ISF	Kim et al. [13]	AZ31	Synthetic oil	Low hardware cost and lesser spring back.
Hot air assisted ISF	Ji et al. [14]	AZ31	NA	Able to get trend of forming limit diagrams at different temperatures with low hardware cost.
	Leonhardt et al. [15]	AZ31	NA	Low hardware cost and better surface quality.
Oil heated ISF	Galdos et al. [16]	AZ31B	Dynalene 600	Globally heating was done with low hardware cost.
	Zhang et al. [17]	AZ31B magnesium alloy	Hydraulic oil	Able to predict a quadratic regression model for forming angle with low hardware cost.
Friction stirred ISF	Xu et al. [18]	AZ31B	ROCOL copper based anti-seize metal	Good correlation between friction stirred and electrical assisted ISF.
	Uheida et al. [19]	CP titanium sheets	98.5% pure MoS_2 powder	A great effort to identify the direction of tool rotation as important factor in forming of sheet.
	Liu [20]	AA7075O	Graphite mixed with mineral oil in ratio 1:4.	Able to study different effects on surface quality, mechanical properties, and formability at different speed of rotation with a very low hardware cost.
Electrical assisted ISF	Li et al.[21]	TC4Ti alloy	MoS_2	Applied an innovative and efficient approach of heating material to be deformed.
	Liu et al.[22]	Ti6Al4V	ROCOL copper based anti-seize compound	Able to develop a novel tool with reduced problem of sticking of material. Reached higher temperatures of about 500—600°C with instant local heating.
	Xu et al. [23]	AZ31B magnesium alloy	ROCOL copper based anti-seize compound	Compared process capabilities of electric assisted and friction stirred ISF with low hardware cost.
	Magnus et al. [24]	Ti6Al4V	MoS_2 coating was done tool side.	Designed a forming tool for doing local heating using joule heating.

(Continued)

TABLE 22.1 (*Continued*)
Summary of Hard-to-form Materials during HA-ISF

Method	Authors	Material Used	Lubrication Used	Achievements
Induction assisted incremental sheet forming	Ambrogio et al. [26]	Ti6Al4V	MoS_2	Induction heating coupled with cryogenic cooling improved dimensional accuracy and surface quality. Local heating was done with low hardware cost.
	Al-Obaidi et al. [27]	DC04, DP 980, 22MnB5	Graphite	Able to compare the formability of different materials with local heating.
Ultrasonic assisted ISF	Li et al. [29]	AA1050-H14, AA5052-H34	NA	Able to reduce forming force up to 56.58% for AA5052 and 37.08% for AA1050 with the help of ultrasonic vibration of tool.
	Bai et al. [30]	Al1060	NA	Developed a simulation model and compared the results with experimental data

22.3 CONCLUSIONS AND FUTURE DIRECTIONS

Observing the available literature, we can infer that HA-ISF can reduce the required deforming forces to deform the hard-to-deform materials and can increase the formability of the materials significantly to produce desirable shapes and sizes of components. However, enough literature is available on electrical assisted ISF due to easier availability of hardware, but dimensional inaccuracy, surface quality, and tool wear are major challenges in this process. Very limited literature is available on induction assisted ISF. It can be observed that induction assisted ISF is able to solve major challenges faced in electrical assisted ISF, but the synchronization of flux movement with tool is difficult to achieve in this process. Other HA-ISF processes like laser assisted ISF and halogen assisted ISF are uneconomical for this operation due to high hardware cost. Friction stirred incremental forming has given good results but it can increase temperature up to a limited range due to convective heat transfer through tool by its rotational speed. It was also observed that HA-ISF is very significant method of dieless sheet forming to process materials that are hard-to-form at room temperature, which further makes this process efficient, viable, and economical for batch production and job production applications of sheet metal forming. The better surface quality, reduced tool wear, and less spring back of the hard-to-form materials should be concern of future direction. Modeling and optimization of various algorithms of these hard materials are other thrust areas for future work.

REFERENCES

1. Kumar A, Gulati V (2019) Experimental investigation and optimization of surface roughness in negative incremental forming, Measurement 131:419–430.
2. Kumar A, Gulati V (2018) Experimental investigations and optimization of forming force in incremental sheet forming, Sādhanā 43(10):159.
3. Kumar A, Gulati V, Kumar P (2019) Experimental investigation of forming forces in single point incremental forming, Advances in Industrial and Production Engineering, pp. 423–430, ISBN 978-981-13-6411-2, https://doi.org/10.1007/978-981-13-6412-9_41
4. Jackson K, Allwood J (2009) The mechanics of incremental sheet forming, Journal of Materials Processing Technology 209(3):1158–1174.
5. Kumar A, Gulati V, Kumar P, Singh V, Kumar B, Singh H (2018) Parametric effects on formability of AA2024-O aluminum alloy sheets in single point incremental forming, Journal of Materials Research and Technology, ISSN: 2238-7854, https://doi.org/10.1016/j.jmrt.2018.11.001
6. Kumar A, Gulati V, Kumar P (2018) Investigation of surface roughness in incremental sheet forming, Procedia Computer Science, Elsevier, 133:1014–1020, ISSN: 1877-0509, https://doi.org/10.1016/j.procs.2018.07.074
7. Kumar A, Gulati V, Kumar P (2018) Investigation of process variables on forming forces in incremental sheet forming, International Journal of Engineering and Technology 10(3):680–684, e-ISSN: 0975-4024, p- ISSN: 2319-8613, https://doi.org/10.21817/ijet/2018/v10i3/181003021
8. Kumar A, Gulati V, Kumar P, Singh H, Singh V, Kumar S, Haleem A (2018) Parametric investigation of forming forces in single point incremental forming, Elsevier, Materials Today Proceedings.
9. Kumar A (2019) Incremental Sheet Forming Technologies: Principles, Merits, Limitations, and Applications, CRC Press, Taylor and Francis, ISBN: 978-0-367-27674-4.
10. Liu Zhaobing (2018) Heat-assisted incremental sheet forming: A state- of-the-art review, International Journal of Advanced Manufacturing Technology 98:2987–3003, https://doi.org/10.1007/s00170-018-2470-3
11. Duflou JR, Callebaut B, Verbert J, De Baerdemaeker H (2007) Laser assisted incremental forming: Formability and accuracy improvement, CIRP Annals 56(1):273–276.
12. Göttmann A, Diettrich J, Bergweiler G, Bambach M, Hirt G, Loosen P, Poprawe R (2011) Laser-assisted asymmetric incremental sheet forming of titanium sheet metal parts. Production Engineering 5(3):263–271.
13. Kim SW, Lee YS, Kang SH, Lee JH (2007) Incremental forming of Mg alloy sheet at elevated temperatures. Journal of Mechanical Science and Technology 21(10):1518–1522.
14. Ji YH, Park JJ (2008) Formability of magnesium AZ31 sheet in the incremental forming at warm temperature. Journal of Materials Processing Technology 201(1):354–358.
15. Leonhardt A, Kurz G, Hernandez J, Krausel V, Ladgrebe D, Letzig D (2018) Experimental study of incremental sheet forming of Magnesium alloy AZ31 with hot air heating. Procedia Manufacturing 15 (2018) 1192–1199.
16. Galdos L, Argandona ESD, Ulacia I, Arruebarrena G (2012) Warm incremental forming of magnesium alloys using hot fluid as heating media. Key Engineering Materials 504–506:815–820.
17. Zhang S, Tang G, Wang W, Jiang X (2020) Evaluation and optimization on the formability of AZ31B Mg alloy during warm incremental sheet forming assisted with oil bath heating, https://doi.org/10.1016/j.measurement.2020.107673
18. Xu D, Bin Lu, Tingting Cao, Jun Chen, Hui Long, Jian Cao (2014) A comparative study of process potential for frictional stir and electric hot assisted incremental sheet forming, Procedia Engineering 81 (2014) 2324–2329.

19. Uheida EH, Oosthuizen GA, Dimitrov DM, Bezuidenhout MB, Hugo PA (2018) Effects of the relative tool rotation direction on formability during the incremental forming of titanium sheets, International Journal of Advanced Manufacturing Technology 96:3311–3319.
20. Liu Z (2017), Friction stir incremental forming of AA7075-O sheets: Investigation on process feasibility, Procedia Engineering 207 (2017) 783–788.
21. Li Zhengfang, Lu Shihong, Zhang Tao, Zhang Chun, Mao Zhixiang (2018) Electric assistance hot incremental sheet forming: An integral heating design, International Journal of Advanced Manufacturing Technology 96:3209–3215.
22. Liu RZ, Lu B, Xu DK, Chen J, Chen F, Ou H, Long H (2015) Development of novel tools for electric heating assisted incremental sheet forming, International Journal of Advanced Manufacturing Technology, https://doi.org/10.1007/s00170-015-8011-4
23. Xu DK, Lu B, Cao TT, Zhang H, Chen J, Long H, Cao J (2015) Enhancement of process capabilities in electrically-assisted double sided incremental forming, Materials and Design 92 (2016) 268–280.
24. Magnus C (2016) Joule heating of the forming zone in incremental sheet metal forming: Part 2, International Journal of Advanced Manufacturing Technology 89:295–309.
25. Vahdani M, Mohammad J, Mohammad B, Gorji H (2019), Electric hot incremental sheet forming of Ti-6Al-4V titanium, AA6061 aluminum, and DC01 steel sheets, International Journal of Advanced Manufacturing Technology, https://doi.org/10.1007/s00170-019-03624-2
26. Ambrogio G, Gagliardi F, Chamanfar A, Misiolek WZ, Filice L (2017) Induction heating and cryogenic cooling in single point incremental forming of Ti-6Al-4V: Process setup and evolution of microstructure and mechanical properties, International Journal of Advanced Manufacturing Technology 91:803–812.
27. Al-Obaidi A, Kräusel V, Landgrebe D (2016) Hot single-point incremental forming assisted by induction heating. International Journal of Advanced Manufacturing Technology 82(5):1163–1171.
28. Smusz R, Trzepiecinski T, Malinowski T, Pieja T (2018) Experimental investigations of induction heating in warm forming of stainless steel sheets, Tehnički vjesnik 25(Supplement 2): 312–318.
29. Li P et al (2017) Evaluation of forming forces in ultrasonic incremental sheet metal forming, Aerospace Science and Technology, http://dx.doi.org/10.1016/j.ast.2016.12.028
30. Bai L, Li Y, Yan M, Yao Z, Yao Z (2018) Influences of process parameters and vibration parameters on forming force in ultrasonic assisted incremental forming process, Advances in Materials Science and Engineering, 2018, Article ID 5726845, 12 pages, https://doi.org/10.1155/2018/5726845

Index

Note: Locators in *italics* represent figures and **bold** indicate tables in the text

A

Abrasive wear, 198, 200
ABTS-free radical scavenging assay, 372
Acalypha indica, 179, 181
Acetaminophen, 185
Acetic anhydride, 74
Acetoacetyl-CoA, 81
Acetobacter, 75
Acetyl-CoA, 81
Acid attack on concrete, 113
Acid Black 1, 249
Acid Blue 92, 249
Acid Brown 214, 249
Acrylonitrile butadiene styrene (ABS), 221
Actinomycetes, 241
Additive manufacturing (AM), 315
Aerobic biodegradation, 44
Aeromonas, 22
Aerosols, 72
Agglomeration of particles, 204
Agrobacterium, 75
Agroindustrial wastes, 73
Ag-zirconia composite, 267
Air-foils, 401, *404*
Alcaligenes eutrophus, 23
Algae, 241–242
Algae-based bioplastic, 32
Alkali-silica reaction (ASR), 291
Allyl-terpene maleate polymer, 58–59, 61
Al7075 matrix, 200
Aloe vera extract, 265
Alternating current (AC) on, 142
Alumina, 344, 394
Aluminum-ceramic composites, 197
Aluminum-containing refractory, 247
Aluminum matrix self-lubricating composite, 199
Aluminum oxide nanoparticles, 344
Amalgamation of waste plastic, 221
Amino acids (AAs), 307
5-Aminosalicylic acid (Mesalamine, MES), 185
Anaerobic biodegradation, 44
Analytic hierarchy process (AHP), 4, 8
 multi-criteria decision-making methodology, 4, 8–9
Angustifolia haw agave fiber, 157
Animal-based biodegradable materials
 chitin and chitosan, 76–78
 collagen and elastin, 78–80

hydroxyapatite (HAp), 80–81
keratin and keratin derivatives, 78
Antibacterial activity by diffusion
 method, 373
Anticancer efficacy employing cell lines,
 373–374
 apoptosis evaluation by Annexin V-FITC/PI
 Staining, 374
 Caspase-3 expression study, 375
 cell cycle analysis, 374–375
 fluorescence microscopic analysis using
 AO/EB staining, 375
Anticancer evaluation, 376
 evaluation of survivability, 377
 organ histology, 378
 subacute oral toxicity, 376–377
 transformation in hematological profile
 and oxidative stress measurement,
 377–378
Anticancer remedies, 371
Anticancer threshold, 388–389
 accomplishment of tissue antioxidant activity,
 391–392
 hematological indices after treatment, 391
 organ histopathology, 392–394
 survivability and growth performance,
 389–391
Anti-Caspase-3 antibody, 375
Antiferromagnetism, 336
Antihypertensive elastin peptides, 80
Antimicrobial fabricant, 78
Antioxidants, 45, 370
 activities, in vitro evaluation, 372–373
 and antibacterial capability, 378–379
 antibacterial efficacy, 380–381
 decolorization activity of ABTS radical
 cation, 379
 DPPH radical scavenging, 380
 NO radical-scavenged activity, 380
 capacity, 378
 enzymes, 391
 migration, 45
Antiproliferation on cancer and non-cancer cell
 lines, 381–383
 apoptosis by FITC annexin V-PI, 383
 caspase-3 in MCF-7 cell line, 384–386
 dual stain analysis of MCF-7 cell line, 386
 propidium iodide-based cell cycle analysis,
 383–384

469

470   Index

Antitumor activity, 381
Apoptosis evaluation by annexin V-FITC/PI
    staining, 374
Arar, 250
Artificial fertilizers, 94
Artificial neural network, 263
Ascorbic acid, 185
Aspergillus fumigatus fresenius fungi, 242
Aspergillus niger (A. niger), 181
Aspergillus oryzae NRRL447 proteins, 264
Association of Plastics Manufacturers
    in Europe (APME), 103
Astaxanthin, 76
Atomic adsorption spectroscopy (AAS), 268–269
Atomic bombardment electrospray
    ionization, 268
ATSM C 642–97, 117
Aulosira fertilissima, 28
2,2'-Azino-bisethylbenzo thiazoline-6-sulphonic
    acid (ABTS), 372
Azotobacter, 22

B

Bacillus indicus, 241
Bacillus megaterium, 23
Bacillus megaterium D01, 241
Back-biting reaction, 84
Bacteria, 241
Bagasse, 3, 244
Ball milling, 346
Banana bunches, 3
Band gap energy, 343
Bark powder of Thespesia populnea, 262
Benzimidazole, 262
Benzothiazole, 262
Benzoxazole, 262
Bhopal gas tragedy, 99
Bicyclic anhydride monomers, 63
Bio-based biopolymers, 22, 76
Bio-based plastics, 20–21, 40, *41*, 42, 47, 49
    and biodegradation, 43–44
Biodegradability, 33
Biodegradable and environmental friendly
    lubrication, 194
Biodegradable materials, 32
    advantage, 72
    animal-based
        chitin and chitosan, 76–78
        collagen and elastin, 78–80
        hydroxyapatite (HAp), 80–81
        keratin and keratin derivatives, 78
    challenges in commercialization, 84–85
    microbe-based
        levan, 82–83
        polyhydroxyalkanoates (PHAs), 81–82
        polylactides, 83–84

needs, 72–73
    plant-based
        cellulose and cellulose derivatives, 73–75
        starch and starch derivatives, 75–76
    plastic production, 73
Biodegradable plastics, 21–22, 40, 42
Biodegradable polymers, 44
Biodegradation, 40, 44
Bio-derived feedstocks, 57
Biological oxygen demand (BOD), 142
Biomimetics, 194
Biomineralization, 242
Bioplastics, 21, 28–29, 31, 40
    from crab shells and tree discards, 30
    from food/vegetative waste, 27–28
    from nature, 26–27
    and petroleum-based counterparts, *20*
    seaweed polysaccharide
        bioplastics, 29–30
    waste management, 31
Biopolymers
    bio-based/biodegradable polymers
        cellulose, 25
        polyesters, 22–24
        polylactic acid (PLA), 24
        starch, 24–25
    classification, 22
    global consumption and production, 31–32
    life cycle assessment (LCA)
        analysis, 30–31
    limitations and challenges, 32
    sustainability, 25–26
        bioplastics, 28–29
        bioplastics from crab shells and tree
            discards, 30
        from food/vegetative waste, 27–28
        from nature, 26–27
        seaweed polysaccharide bioplastics,
            29–30
Biosensors, 178
Blending, 28
Blue-green algae, 28
Bottom-up process, 345
Bowman's capsule, 393
Box-Behnken design (BBD), 263
BP oil well's offshore failure, 99
Breast tumor progression, 80
Brown seaweeds, 29
Brunauer, Emmett, and Teller (BET) analysis,
    267, 275
Brundtland Commission, 99–100
Building's insulation, 2
Butylated hydroxyanisole, 378
Butylated hydroxytoluene, 378
Butylene adipate, 24
By-catch marine organisms, 80
By-product valorization, 73

Index

471

C

Calcium aluminate, 113
Calcium silicate hydrate (CSH), 113
Carbon-based nanoparticles, 343
Carbon dioxide (CO_2)
  emissions, 30, 104–105, 111, 217
  as monomer, 237–239
Carbon emissions, 105, 214
Carbon nanotubes (CNTs), 199, 203, 342
Carbon paste electrode (CPE), 184
Carbon quantum dots, 262
Carbon reduction waste segregation, 214
Carboxylic acid synthesis, 28
Carboxymethyl cellulose (CMC), 74
3-Carene, 56
β-Carotene, 76
Carrot-based bioplastics, 27
Caspase-3 expression, 384–386
Caspase-3 protein, 375
Catalytic reforming, 57
Catechin, 370
Catechol (CC) carbon, 185
C-C coupling reactions, 236
Cefazolin (CEZ), 142
Cefpodoxime, 380
Celastrus paniculatus, 348, 354, 373, 380, 394
Cellulose, 25, 40, 58, 73
  bioplastics, 25
  fiber, 5
  polymer, 73
Cellulose acetate (CA), 21, 74
Cellulose and cellulose derivatives, 73–74
  carboxymethyl cellulose (CMC), 74
  cellulose acetate (CA), 74
  cellulose nitrate (CN), 74
  cellulose sulfate (CS), 74
  ethyl cellulose (EC), 74
  microbial cellulose (MC), 75
  nanocellulose (NC), 75
Cellulose-based insulation, 7
Cellulose nitrate (CN), 74
Cellulose sulfate (CS), 74
Cement, 280
Cementitious materials in concrete, 112
Central composites design (CCD), 263–264
CeO2/graphene oxide (GO) – polylactic (PLA)
    composite, 186
Ceramics, 260
Ceramics, green synthesis, 244
  agricultural wastes as raw material, 244–246
  industrial wastes as raw material, 246–249
  plant extracts as raw material, 249–250
Cerium
  nanoparticles, synthesis, 178, **178**
  transition of oxidation states of, *183*
Cerium (III) nitrate hexahydrate, 181

Cerium NPs composites, 185
Cerium oxide (CeO_2), 185
  green synthesis of, 177–179, **180**
    electrochemical sensing applications, 183
    electrochemical sensor, 183–187
    fungus-mediated, 181–182
    nutrient-mediated, 179, 182–183
    plant-mediated, 179–181
    polymer-mediated, 182
  nanoparticles (CNPs), 178
    in electrochemical sensing, **186**–**187**
Chain-growth polymerization, terpene, 62–63
Chemical-based synthesis, 234–235
Chemical oxygen demand (COD), 141–142
Chemical vapor deposition (CVD), 343
Chitin, 73, 76–78
  extraction, **77**, 77
  properties, 76
Chitosan, 73, 76–78
  biopolymer, 267
  extraction, 76, **77**
  food preservative, 78
  preparation, 77
  properties, 76, 78
Chloride
  attack, 113
  ion permeability, 289
Chlorophyta, 29
Cholesterin, 65
Cholesterol oxidase (ChOx), 185
Class F fly ash, 112, 114
Clausius-Clapeyron relation, 331
Clean fabrication technic, 2
Clostridium, 23
CNPs (CeO2-NPs) fabrication, 182
CNPs with graphene oxide (CeNP/GO), 184
Coal, 247
Coal fly ash, 248
Coconut coir, 3
Coconut husk, 3
Coefficient of friction (COF), 196
Coffee grounds waste, 246
Cold pressing, 249
Collagen, 73, 78–80
  applications, 79
  fibers, 78
  supplemented food, 79
Colletotrichum sp. fungi, 242
Combustion chamber, *433*
Compressive strength (CS), 224–225, 279–281
Computational fluid dynamics (CFD),
    404–406, 434
  meshing, 406–407
  numerical solution, 407
  outcome, 407–410
Computer-aided design (CAD), 315, 434, 458
Computer-aided manufacturing (CAM), 315

472 Index

Concrete shrinkage, 288
Concrete with fly ash, 111–112
  acid attack studies
    percentage compressive strength
      reduction, 129–131
    percentage weight reduction, 126–129
  acid resistance of fly ash concrete, 113
    acid attack, 113
  compressive strength studies, 121–123
  experimental methodology
    materials, 114
    mix design and test specimen
      preparation, 114
    test method, 114–120
  permeable voids of fly ash concrete, 112
    effect, 112–113
    mechanism of, 112
  permeable voids studies, 123–126
Condensation-polymerization method, 237
Conductivity, 290
Congestions, 393
Consistency limits, 155
Construction
  market, 216
  scope-growth-impacts in India, *217*
Conventional manufacturing (CM) processes, 315
Conventional materials, 260
Copper-oxide nanoparticles, 262, 266–267
Corn starch, 262
Corporate social responsibility (CSR), 100–101
Cost of split, 451
Co-terephthalate, 42
CO_2 utilization, 105
COVID-19 pandemic, 40, 217, 274
Cradle-to-grave analysis, 101
Criticality index, 326
Crop evapotranspiration (ETC), 307
Crustacean shells, 76
Crystallization, 182
Crytallinity, 265
Cubical and cylindrical specimens, 172–173
Cupriavidus, 23
Cupriavidus necator, 22
Cupric oxide nanoparticles, 345
Curcumin, 370
Curie temperature, 331, 336
Current density (CD), 139–140
Cyanobacteria, 28
Cyclic anhydride, 61
Cyclic voltammetry (CV), 184
β-Cyclodextrin, 264
Cyclopentyl methyl ether (CPME), 236
Cyrene (dihydrolevoglucosenone), 262

D

Dairy wastewater, 142

Darrieus turbines, 402
2D-dichalcogenides, 197, 200
Deacetylation, 76
Decentralized wastewater treatment systems
  (DWTS), 137–138
Decision-making problem, 8
Decision tree, 422–423, 451
Decomposition-based strategy, 264
Degermination, 75
Degradation, 20
  kinetics, 46
Dehydrogenative coupling, 262
Deinococcus radiodurans, 241
Demagnetization, 324
Demineralization, 76
Density of concrete, 285–286
Density of sewage sludge, 286
Depolymerization, 44
Deproteinization, 76
Design Expert® software, 263
Detoxification, 234
1D-graphene, 194, 197
2D-graphite, 194, 197
2D-hexagonal boron nitride (h-BN), 194, 197,
  205
1,4-Diaminobutane (BDA), 268
1,2-Diaminoethane (EDA), 268
1,6-Diaminohexane (HDA), 268
Diels–Alder reaction, 64
Dietary flavanones, 376
Differential pulse voltammetry (DPV), 184
Differential scanning calorimetry (DSC), 332
2,5-Dihydroxy benzoic acid (DHB), 268
Dimethyl sulfoxide (DMSO), 373
Dioxins, 72
Dipentene, 58
2,2-Diphenyl-1-pikryl-hydrazyl (DPPH), 372
Direct arylation polymerization (DArP), 236
Direct current (DC), 142
Direct metal additive manufacturing (DMAM),
  315
Distillery wastewater, 142
2D materials, 199
2D-molybdenum disulfide (MoS_2), 194
DNA plastics, 26
Dopamine (DA), 185
Dose-dependent anticancer, 382
Drug carriers, *369*
Durability of constructions, 288
Durian peel fibers, 3
$Dy_2Ce_2O_7$ ceramic nanostructure, 249

E

Earth houses, 152
Earth Summit, 100
EC, *see* Electrocoagulation

Index

Eco-aesthetics, 306
Eco-efficiency, **95**, 96, 102, *102*, 103
Economy sustainability, 94
Ecosystem
  resilience, 97
  services, 94, 97
Efficient Consumer Response (ECR), 105
Egg white, 182
Egyptians blue, 342
Elastic modulus, 7, 199
Elastin, 78–80
Elastin-derived peptides (EDPs), 80
Electrical and electronic equipment
  (EEE) waste, 218
Electrical assisted incremental sheet forming,
  461–463
Electrical conductivity, 141
Electrical resistivity and corrosion
  resistance, 290
Electrochemical impedance spectroscopy
  (EIS), 185
Electrochemical redox procedures, 262
Electrochemical sensors, 178, 184–185
Electrochemistry, 262
Electrocoagulation (EC), 138–139
  current density (CD), 140
  dairy wastewater, 142
  distillery wastewater, 142
  efficiency, 139
  electrical conductivity, 141
  electrochemical and chemical reactions, 139
  electrode material, 141–142, **143–145**
  electrolysis time, 140
  energy and electrode consumption, 141
  Fe, Al, Mg, SS, Ti, and Cu electrodes,
    application of, **143–145**
  hospital wastewater, 142
  paper mill wastewater, 141
  petroleum wastewater, 141–142
  pH of solution, 139
  pollutant removal mechanism, *138*
  real graywater, 142
  removal of pharmaceutical compounds, 142
  tannery wastewater, 141
  textile wastewater, 141
Electrode material, 141–142, **143–145**
Electrode systems, 184
Electrolysis, 138
  time, 140
Electronic Data Interchange (EDI), 104
Electronic gadgets (E-plastic), 221
Ellman's reagent, 378
Emerging contaminants (ECs), 137
Emission pollution, 1
Emulsion polymerization, 58
Energy conservation, 2, 194
Energy consumption, 1

Energy dispersive X-ray analysis (EDX),
  266–267, 332, 357, 360
Energy-effective natural fibers, 1, 5
  alternative sustainable insulation
    materials, 2–5
  analytic hierarchy process
    methodology, 8–10
  selection, 5–8
Enterprise Resource Planning (ERP), 104
Entner-Doudoroff pathway, 28
Environmental degradation, 94, 234
Environmental pollution, 106, 194
Environmental Product Declaration
  (EPD), 102
Environmental sustainability,
  94–95, 193
Environment health key indicators, 94
Enzyme-based biodegradable plastic, 21
Enzyme β-ketoacyl-CoA thiolase, 81
E-plastic concrete; *see also* E-waste
  feasible possibilities, 220–221
  incorporation, 226
  potential, 227
  properties, **222–223**
  srength characteristics, 221–224
    compressive strength, 224–225
    flexural strength test, 226–227
    splitting tensile strength, 225–226
  waste, 228
Epochs, 425
Epoxy resins, 64
Erythrosine pollutants, 249
Escherichia coli, 373
  bacteria, 22
Ethanol, 178
Ethyl cellulose (EC), 74
Ethylene carbonate, 268
Ethylhexyl palmitate, 264
Ethylhexyl stearate, 264
Ettringite, 113
Eukaryotic algae, 242
European Union Product Environmental
  Footprint (EU PEF), 42
Evapotranspiration (ETC), 307
E-waste, 215
  components, *219*
  electronic appliances, 218
  fibers, 224
  generation, 218–220
  generators rank, *219*
  India, 218
Expanded polystyrene (EPS), 3
Expert Choice software, 4
Extracellular biosynthesis of nanoparticles, 241
Extrudates, 46
Extruded polystyrene (XPS), 3
Extrusion, 45–46

F

FA, *see* Fly ash concrete
Factorial analysis, 263
Faraday's law, 140
FEA, *see* Finite element analysis
Fenton reaction, 184
Fe-Rh alloys, 325
Fiberboards, 3
Fibers, 156–157
Fibers of Agave Americana and Eulaliopsis
   Binata, 153
   physical attributes of, **154**
   testing, 157–158
Fibrous proteins, 79
Finite element analysis (FEA), 404, 435
   building algebraic equations, 411–412
   meshing, 412–413
   numerical solution strategy, 410
   plate theory, 410–411
   pressure load from CFD to FEA, 413–415
   shell theory, 411
Fish biomass, 80
Fisher esterification, 64
Five capitals model, **95**, 96
Flame AAS (FAAS), 269
Flexural strength, 226
Flexure strength, 284–285
5-Fluorouracil, 377
Fluxing wastes, 244
Fly ash (FA) concrete, 111–112, 203
   acid attack studies
      percentage compressive strength
         reduction, 129–131
      percentage weight reduction, 126–129
   acid resistance, 113
      acid attack, 113, 117
   compressive strength reduction in HCl and
      H_2SO_4 Solution, **120**
   compressive strength studies, 121–123,
      121–123
   compressive strength test, 114
   durability, 125
   experimental methodology
      materials, 114
      mix design and test specimen preparation,
         114
      test method, 114–120
   Hopper No. and percentage weight reduction,
      128
   hydrochloric acid on, *119*
   immersed apparent mass, 116–117
   permeable voids of fly ash concrete, 112
      effect, 112–113, **117–118**
      mechanism of, 112
   permeable voids studies, 123–126
   permeable void test, 114–115

physical properties, **114**
saturated mass after boiling, 116
saturated mass after immersion, 115
strength test results, **115**, 131
sulfuric acid on concrete cubes, *120*
Food security, 309
Forest, 451
Fossil-based glass fiber, 4
Fossil-based plastics, 42
Fossil-fuel-derived succinic acid, 27
Foundation for Rehabilitation of Local Health
   Tradition (FRLHT), 348
Fourier transform infrared (FTIR) spectroscopy,
   25, 265
Fourier transform ion cyclotron resonance, 268
Frattini test, 277
Free radicals, 391
Frictionless, 199
Friction stirred incremental sheet forming, 461
Frost resistance, concrete, 290–291
Fructose, 82
Fucus species, 29
Fuel wastes, 244
Full-Heusler alloy (X_2YZ), 326
Fungal mycelium, 242
Fungi, 242
2-Furanylquinazolines, 262
2-Furfurylidene derivatives, 262

G

Gain in body weight (BW), 377
Gel pores, 112
German Technical Inspection Association, 103
GHG, *see* Greenhouse gases
Glass
   ceramics, 249
   cullet, 246
   insolation, 3
   wool, 3
Global Commerce Initiative (GCI), 105
Global warming, 193, 324
Gloriosa superba, 179
Gloriosa superba L. leaf, 179
Glucose, 82, 184
Glucose syrup, 76
Glutamate, 178
Glutamate oxidase, 178
Glutathione nanoparticles, 370
Glycerol, 25, 27
Glycol, 27
Grafting of terpenes, 65
Grain size analysis, 154–155
Gram-negative (G-) bacteria, 182
Gram-positive bacteria, 182, 371
Graphene oxide, 178
Graphenes (GNPs), 197, 199

Index

Graphene-silver nanocomposites, 265
Graphite (Gr), 197
Graphite reinforced self-lubricating hybrid composite (Al-SiC-Gr), 198
Graphite solid lubricant, 206
Green building, 2
  energy performance, 2
  insulation, 2
  materials, 7, 13
Green chemistry, 95, **95**, 101, *261*
Green engineering, 95, **95**
Greenhouse gases (GHG), 1, 56, 274, 324
  emissions, 7, 103, 215, 274
Green insulation material, 2
Green materials, 259–260
  characterization techniques, 265
    atomic adsorption spectroscopy (AAS), 268–269
    Brunauer, Emmett, and Teller (BET) analysis, 267
    energy dispersive X-ray analysis, 266–267
    Fourier transform infrared (FTIR) spectroscopy, 265
    inductive coupled plasma mass spectroscopy (ICP-MS), 269
    mass spectroscopy, 268
    nuclear magnetic resonance, 267–268
    scanning electron microscopy, 266
    UV-visible spectroscopy, 267
    X-ray diffraction, 265–266
  chemical waste, reducing, 263
    response surface methodology, 263–264
  synthesis, 260–263
Green revolution in agriculture, 94
Green seaweeds, 29
Green synthesis, nanoparticles (NP), 346–347
Green synthesis of cerium oxide, 177–179, *179*
  electrochemical sensing applications, 183
    electrochemical sensor, 183–187
  fungus-mediated, 181–182
  nanostructures, 178
  nutrient-mediated, 182–183
  plant-mediated, 179–181
  polymer-mediated, 182
Green synthesis of sustainable materials, 233–234, *235*
  ceramics, 244
    agricultural wastes as raw material, 244–246
    industrial wastes as raw material, 246–249
    plant extracts as raw material, 249–250
  metal and metal oxide nanoparticles, 239–241, *240*
    algae, 241–242
    bacteria, 241
    fungi, 242

microbes, 241–243
plants, 242–243
yeast, 242
polymers, 234–235
  carbon dioxide (CO_2) as monomer, 238–239
  industrial wastes, 239
  renewable raw materials, 235–236
  sustainable solvents/methods, 236–237
principle, 234
Green synthesized spinel lithium titanate (GSLTO), 265–266
Green tribology, 194
Grenelle Environment meeting, France, 104
Griess reagent, 380
Gun cotton, *see* Cellulose nitrate
Gypsum, 113

H

Half-Heusler alloy (XYZ), 326
Halogen lamp assisted incremental sheet forming, 459–460
Hazardous substances, 95
HDPE, 45
Heat-assisted incremental sheet forming (HA-ISF), 458, *459*
Heat conductivity, 3
Heat exchangers, 106
Heat-stable biopolymer, 82
Hematoxylin, 378
Hemicellulose, 58
Hemolysis, 376
  activity of synthesized nanoparticles, 375–376
    green synthesized nanoparticles on RBCs morphology, 376
  assay, 387
  prevention of, 387
    nanoparticles on RBC morphology, 387–388
HepG2 cells, 83
Hesperetin, 376
Heterocycles compounds, 262
Heuristic search technique, 263
Heusler alloys, 326–327
  applications, 327–328, *328*
  magnetic refrigeration, 330–331
    challenges, 334–337
    preparation and properties, 331–334
  properties, *327*
  with reversible martensitic transition, 328–330
Hexagonal boron nitride, 200
Hibiscus sabdariffa, 181
Hierarchy process methodology, 4
High-density polyethylene (HDPE), 221

Index

Higher density fibers, 7
High impact PS (HIPS) plastic, 224
High-sulfur-content possessing polyenes, 239
Histopathology, 392
Homopolymer, 82
Homopolysaccharide, 82
Hospital wastewater, 142
Hot air assisted incremental sheet forming, 460–461
    hard-to-form materials, 465–466
Human environment's accelerating deterioration, 99
Human-made capital, 98
Humicola sp., 181
HVAC (Heating, Ventilating, and Air-Conditioning), 304
Hyaline casts, 393
Hybrid metal matrix, 205
Hybrid photovoltaic-thermal (PVT) solar systems, 106
Hydrochloric acid (HCl) attack, 113
Hydrodesulfurization process, 239
Hydrogels, 83
Hydrogen bonding stabilizer, 45
Hydrophila, 22
Hydrophobic anhydrides, 64
Hydrophobicity, 344
Hydrophobic protein, 80
Hydroquinone (HQ), 185
Hydroxyalkanoates (HAs), 81
Hydroxyapatite (HAp), 80–81
Hydroxyl radicals, 178
    detection, 184
Hygiene, sanitation, sewage treatment (HSST) systems, 136
Hygromorphic wood composites, 307
Hypercellularity, 393
Hysteresis, 330

I

Immersed apparent, 116
Increase in life span (ILS), 377
Incremental sheet forming (ISF), 457–458
    classifications, 458–465
    electrical assisted incremental sheet forming, 461–463
    friction stirred incremental sheet forming, 461
    halogen lamp assisted incremental sheet forming, 459–460, *460*
    hot air assisted incremental sheet forming, 460–461
    induction assisted incremental sheet forming, 463
    laser assisted incremental sheet forming (LA-IS F), 459

oil heated incremental sheet forming, 461
ultrasonic assisted incremental sheet forming, 463–465, *464*
India
    construction scope-growth-impacts in, *217*
    e-waste, 218
    production profile for electronics in, *219*
    waste management, 229
Indian Codal provisions, 158
Induction assisted incremental sheet forming, 463
Inductive coupled plasma mass spectroscopy (ICP-MS), 269
Industrialism, 107
Industrialization, 259
Industrial wastes, 111, 239
    polymers, 239
Injection molding process, 47
Inorganic nanoparticles, 343
Insect cuticles, 76
Insulation, 2
Inverse magnetocaloric effect, **325**, 327, 334
Irganox 1010, 45
ISF, *see* Incremental sheet forming
Iterative mathematical search technique, 263

J

Jania rubens, 241–242
Joint Committee on Powder Diffraction Standards (JCPDS), 355
Juniperus phoenicea, 250

K

Kaolin clay, 249
Kappa-carrageenan, 264
Kappa-DX experimental rocket engines, 434
Kappaphycus alvarezii, 29
Kenaf, 4
Keratin, 73
    derivatives, 78, **79**
    natural source of, 78, **79**
    protein, 78
Keratinocytes, 78
Kernel powder crosslinker, 264
Ketones, 45
Kevlar fiber, 4
Kigelia Africana, 262
Kigelia Africana stem extract, 265
Klebsiella pneumonia, 241
K-nearest Neighbors (KNN), 421–422

L

Lactate, 178, 184
Lactate dehydrogenase (LDH), 184

Index

477

Lactate oxidase, 178
Lactic acid (LA), 24, 83
    production by microbial fermentation, 84
    purification, 84
Lambert-Beer's law, 269
Laminaria pallida species, 29
L-ascorbic acid, 250
Laser ablation, 346
Laser assisted incremental sheet forming
    (LA-IS F), 459
Leachability, 291–294
Lead phthalate, 44
Leica Rotary Microtome, 378
Lemon juice, 250
Levan, 73, 82–83
    biosynthetic pathway, *83*
Levansucrase, 82
Life-cycle assessment (LCA), 101
Life-cycle chain, 103
Life-cycle impact assessment (LCIA), 102
Life-cycle interpretation, 102
Life-cycle inventory (LCI), 102–103
Life-cycle management tool, 102
Life-support systems, 95
Lignin, 8, 58
Lignin-polymer blends and oligomers,
    64–65
Limit of detection (LOD), 183–185
Limonene, 56
Lipid-based nanoparticles, 342
Lipids, 42
Lithium silicate-based ceramics, 244
Load Frame testing, 157
Logistic regression, 420–421
Loss in ignition (LOI) value, 275
Low-density polyethylene (LDPE), 221
Lubricants, 193

M

Machine learning (ML), 404, 443–447
    data cleaning and visualization, 449–451
    dataset acquisition, 449
    decision tree, 449–451
    Google Colab, 448
    mechanical properties, 448–449
    random forest model, 451–452
    tools used, 447–448
    XGBoost model, 452
Machine learning model and WTB, 415–417
    dataset acquisition, 417–420
    fault detection, 417
    model usage, 420
        decision tree, 422–423
        K-nearest Neighbors (KNN), 421–422
        logistic regression, 420–421
        neural network, 424–428

random forest, 423–424
    support vector machine (SVM), 421
    tools used, 417
Macromolecules, 260
Macro-porous alumina ceramics, 246
Magnetic entropy, 336
Magnetic Moringa oleifera, 264–265
Magnetic refrigerant, 324
Magnetic refrigeration, 324, 330–331
Magnetization curves, 332
Magnetocaloric effect, **325**, 326
Malondialdehyde (MDA), 377
Maltol, 76
Maltose syrups, 76
Mannitol, 76
Marine algae, 242
Marine cyanobacterium, 242
Marine product wastes, 73
Marinobacter pelagius, 241
Mass spectroscopy, 268
Materials, classification of, *261*
Maximum dry density (MDD), 153
Maxwell relation, 332
MCF-7 cell line, 374
Mean survival time (MST), 377
Mechanical energy, 433
Melanin nanostructures, 370
Meshing, 406–407, 412–413, 437–438
Mesoporous ceramics, 342
Metal and metal oxide nanoparticles, green
    synthesis, 239–241, *240*
    affected by, 240
    algae, 241–242
    bacteria, 241
    fungi, 242
    microbes, 241–243
    plants, 242–243
    whole plants/plant parts, **243**
    yeast, 242
Metal-based nanoparticles
    as anticancer entities, 371–372
    as antimicrobial agents, 371
    as antioxidant agents, 370–371
Metal hydroxide coagulant, 141
Metallic materials, 260
Methylene blue, 249
Methylobacterium, 23
Methylobacterium rhodesianum, 23
Methyl Orange, 249
Methylorubrum extorquens, 23
Mice model-bearing tumors, 389
Michael addition, 239
Microbe-based biodegradable materials
    levan, 82–83
    polyhydroxyalkanoates (PHAs),
        81–82
    polylactides, 83

478   Index

lactic acid production by microbial
fermentation, 84
lactic acid purification, 84
ring-opening polymerization (ROP), 84
Microbes, 241–243
Microbial biofilm, 82
Microbial cellulose (MC), 75
Microfibrillar angle, 8
Microplastics in water bodies, 72
Micropollutants, 137
ML, *see* Machine learning
Modulus of elasticity, 285
Modulus of rupture (MR), 226
Molybdenum, 199
Monochromatic X-rays, 265
Mono-functionalized fatty acids, 235
Monomers, 62
β-myrcene, 59
Monostroma species, 29
Monoterpenes, 56, 63, 66
types of, **67**
Montreal Protocol, 302
Mud houses, 152
Mullite ceramics, 248
Multi-criteria decision-making (MCDM)
methods, 303–304
Multi-layer graphene (MLG), 199
Municipal wastewater, 138
Myrcene, 56, 59

N

Nacelle, 400
Nano-based drug delivery, 368
Nanocellulose (NC), 75
Nanofabrication, 178
Nanofibers (NFs), 312
Nanolithography, 346
Nanoparticle-related drug delivery, 371
Nanoparticles (NPs), 177, 312, 342
BET, 358–359
dimensions of, 342
EDX, 356–357
forms of, 343
aluminum oxide nanoparticles, 344
cupric oxide nanoparticles, 345
silver nanoparticles, 344–345
tin dioxide nanoparticles, 343
SEM, 356–357
synthesis of, 345–346
chemical synthesis, 346
green synthesis, 346–347
plant extract mediated green synthesis of
Ag and CuO NPs, 349–350
plant materials, 347–349
SnO_2, 347
synthesized nanoparticles, 350–352

TEM, 357–358
XRD analysis, 355–356
Nanoribbons (NRBs), 312
Nanorods (NRs), 312
Nanosheets (NSs), 312
Nanostructures in biotechnology, *369*
Nanostructures (NStr), 312
Nanotechnology, 342, 346
Nanotechnology-related drugs, 369
Nanotubes (NTs), 312
Nanowires (NWs), 312
National economic models, 1, 103
Natural capital, 98
Natural capitalism, **95**, 96
Natural fibers, 2–4, 10
decision matrix, **11**
density, 7
factor priorities., *13*
factor weights, *10*
in insulation materials, 3, 5–8, **6**, 13
natural resources, 5
priority values, **12**
Natural indigenous fibers, 153
Natural insulating materials, 3
Natural resources, 106, 259
Natural-sourced building materials, 3
Negative incremental forming, 458
Neural network, 424–428, *428*
Nickel hydroxide (Ni(OH)2), 184
Nicotinamide adenine dinucleotide
(NADH), 184
Ni_2-Mn-In Heusler alloys, 325
Niobium disulfides, 199
Nonbiodegradable biopolymers, 22
Nonbiodegradable materials
plastics, 40–41
problems, 72
Nonbiodegradable wastes, 215
amalgamation, 229
construction industry and waste generation,
215–217
e-plastic waste generation, 218–220
population growth, 215
Noncytotoxic ability, 382
Non-isocyanate polyurethanes, 268
Nonrenewable plastics, 40
Nonrenewable resources, 97
Nontoxic biopolymer, 182
Normalized difference vegetation index
(NDVI), 308
Nostoc muscorum, 28
Notogenia striata, 29
Novozym enzymes, 264
NP, *see* Nanoparticles
Nuclear magnetic resonance (NMR),
27, 267–268
Nucleic acid of nucleotides, 22

Index

479

O

Off-stoichiometric Heusler alloys, 331
O-H elongation, 353
Oil-based plastics, 32
Oil heated incremental sheet forming, 461
One pot synthesis of honeycomb biomass
 adsorbent, 262
Optimization, 263–265
Optimum moisture content (OMC), 153
Ordinary portland cement (OPC), 226
Organizational impact assessment (OEF), 42
Organization for Economic Co-operation and
 Development (OECD), 376
Origami-based architecture materials, 307
Oxidase enzymes tyrosinase, 178
Oxidative damage, 370
Oxidative disruption, 375
Oxidative stress, 370–371
Ozone depletion, 324

P

Paper mill wastewater, 141
Paris Agreement COP21, 324
Particle size distribution, 154–156
Pectin, 8, 182
PerkinElmer Lambda-750, 351
Peroxides, 45
Pesticides, 94, 99
PET, *see* Polyethylene terephthalate
Petrochemical industry, 57
Petroleum-based chemicals, 3
Petroleum-based plastics, 30, 32
Petroleum-based polypropylene (PP), 23
Petroleum refineries, 239
Petroleum wastewater, 141–142
Petroplastics, 20, *20*
Petroselinum crispum, 179
Phaeophyta, 29
Phellandrene, 63
PH of solution, 139
Phosphate buffer saline (PBS), 372
Photodynamic cancer therapy, 372
Photosynthesis, 3
Phytonutrients of diverse selection, 370
Pineapple leaves and fibers, 3
Pinene
 cationic polymerization of, *62*, 62–63
 step-growth polymerization, *63*
α-Pinene, 56, 58
β-Pinene, 56, 58
Plant-based biodegradable materials
 cellulose and cellulose derivatives, 73–74
  carboxymethyl cellulose (CMC), 74
  cellulose acetate (CA), 74
  cellulose nitrate (CN), 74

cellulose sulfate (CS), 74
  ethyl cellulose (EC), 74
  microbial cellulose (MC), 75
  nanocellulose (NC), 75
 starch and starch derivatives, 75–76
Plant-based cellulose, 29
Plant-based fiber, 4
Plant extract mediated green synthesis
 of Ag and CuO NPs, 349–350
Plant materials, 347–349
Plants, 242–243
Plasmastone, 249
Plasticize polymer, 45
Plasticizer, 27
Plastics, 19, 72, 215
 advantages of, 39
 biodegradable plastics, 40, 42
 classification, 40, *41*
 consumption, 20
 in electronic equipment, 220
 environmental sustainability of, 309
 fossil-based plastics, 42
 in health care, 40
 life cycle, 40
 nonbiodegradable waste, 215
 nonrenewable plastics, 40
 polymers, 221
 thermoplastic, 220
 thermosetting, 220
 types, 220–221, **221**
 WEEE, 220
Plate theory, 410–411
Plectonema boryanum UTEX 485, 241–242
Pollutant concentration, 139
Pollutant removal efficiency (RE), *140*
Pollution control authority, 137
Polyacrylic acid that is coated with nanoceria
 (PAA–CNPs), 186
Poly-alloocimene, 58, 60–61
 synthesis, *59*
Polyamides, 22
Polybutylene adipate (PBA), 42
Poly(butylene-co-decylene terephthalate)
 copolyesters, 237
Polybutylene succinate, 42, 48
Polybutylene succinate adipate, 48
Poly(butylene terephthalate), 237
Polycaprolactone (PCL), 26–27, 42
Polycarbonates, 22, 58, 238
Polycondensation-coupling ring-opening
 polymerization, 237
Polycrystalline ingot, 331
Polycyclic aromatic hydrocarbons, 72
Polyepoxides-sheathed CNTs, 342
Polyesters, 5, 22–24, 58
 poly (butylene adipate-co-terephthalate)
  (PBAT), 24

480 Index

polyhydroxyalkanoates (PHA), 22–23, 42, 81–82
polyhydroxybutyrate (PHB), 23
Polyethylene glycol, 247
Polyethylenes (PE), 39, 41, 239
Polyethylene terephthalate (PET), 39, 42, 221
Poly-hdroxyurethanes (PHUs), 268
Polyhydroxyalkanoates (PHAs), 22–23, 42, 81–82
    fermentation, 81
    synthase enzyme, 81
Poly-3-hydroxybutyrate (PHB), 42
Polyhydroxybutyric acid, 73
1,4-Polyisoprene, 62
Polylactic acid (PLA), 24, 83–84
Polylactides, 73, 83
    lactic acid production by microbial fermentation, 84
    lactic acid purification, 84
    production, 83–84
    ring-opening polymerization (ROP), 84
Polymer biodegradation, 44
Polymeric substances, 72
Polymerization, 56, 236
Polymer-layered materials, 342
Polymer (PAllo), 58
Polymer production, 237
Polymers, green synthesis, 234–235
    carbon dioxide (CO_2) as monomer, 238–239
    industrial wastes, 239
    renewable raw materials, 235–236
    sustainable solvents/methods, 236–237
Polymyrcene, 59, 61
Polyol synthesis from epoxy resins, 65
Polyphenols, 179
Polypropylene (PP), 5, 39, 41, 43, 221
    carbonates, 238
    fibers, 239
Polypyrrole (PPy), 185
Polysaccharides, 22, 42
    based hydrogel, 264
    Ulvan, 29
Polysiloxanes, 65
Poly(stearyl methacrylate), 59–61
Polystyrene, 45
Poly (styrene-*co*-2-ethylhexyl acrylate), 45
Polystyrene (PS), 39, 221
Polyterpene hydrocarbons, 62
Polyurethanes (PE), 221, 268
Polyvinyl alcohol (PVA), 27, 246
Polyvinyl chloride (PVC), 39, 42, 221
Population-driven economic growth, 215
Population growth and nonbiodegradable wastes, 215–217, *216*
Porosity, 112
    in concrete, 112
    structure, 112
Porous glass ceramics, 246

Porous silicon carbide (SiC) ceramics, 247
Porphyra capensis, 29
Portieria *hornemannii,* 242
Portland cement, 113
Potential energy, 411
Powder metallurgy, *205*
Power generation networks, 264
Power utilization, 141
Pozzolan hydration products, 112
Pozzolanic materials in cement concrete, 111
Pozzolanic strength, 277
Prebiotic colonization of pig gut, 82
Proctor compaction tests, 153, 159
Production profile for electronics in India, *219*
Prokaryotic bacteria, 241
Propellant, high-pressure helium system, *433*
Propelling nozzle, *see* Rocket engine nozzle
Property-affecting wastes, 244
Propidium iodide (PI), 374
Prosopis juliflora, 181
Proteins, 22, 40, 42
Pseudomonas, 22–23
Pseudomonas aeruginosa, 373
Pseudomonas antarctica, 241
Pseudomonas fluorescens, 241
Pseudomonas meridian, 241
Pseudomonas putida, 23
Punica granatum, 249
Purine, 262
Pyrolysis of biomass, 262

Q

Quantum dots (QD), 342
Quercetin, 370
Quinoxalines, 262

R

Radioactive waste, 215
Ralstonia, 23
Ralstonia eutropha, 81
Random forest, 423–424, 451
Raw materials, 111
Real graywater, 142
Recycled and bio-based plastics, 39–40
    applications, 47–48
    biodegradation, 43–44
    challenges and limitations, 48–49
    classification, 40–42
    cradle-to-gate methodology, 42
    life cycle assessment, 42
    polymer properties, impact of recycling on, 43
    processing
        additives, 44–45
        compression molding, 46–47

Index

extrusion, 45–46
injection molding, 47
Recycling of polymers, 43, 45
Red blood corpuscles (RBCs), 375
Redox reaction, 177
Red seaweeds, 29
Refrigeration capacity (RC), 325, 331
Regression equation, 263
Renal lesions, 393
Renewable energy resources, 305
Renewable feedstock, 235
Renewable natural resources, 56
Renewable raw materials, 235–236
Renewable resources, 97
Renewable/sustainable energy resources, 194
Response surface methodology (RSM), 263–264
Resveratrol, 370
Rhodophyta, 29
Rhodopseudomonas palustris, 22
(R)-3-hydroxybutyrate monomer, 81
Rice husk, 246
Rifampicin, 371
Ring-opening polymerization (ROP), 64, 84
of lactide, 24
Rio Declaration on Environment and
Development, 100
Rio Earth Summit, 302
Rocketdyne F-1 engine, 432, *435*
Rocket engine nozzle, 432
functioning, 432–433
machine learning, 443–447
data cleaning and visualization, 449–451
dataset acquisition, 449
decision tree, 449–451
Google Colab, 448
mechanical properties, 448–449
random forest model, 451–452
tools used, 447–448
XGBoost model, 452
Rocketdyne F-1 engine, 435
circumferential stress, 442
core analysis, 441
design, 436–437
FEA, 437
force reaction in the axial (Z)
direction, 441
gap, 443–445
loading types and boundary
conditions, 437
meshing, 437–438
model setup, 438–440
stress, 443
total deformation, 442–443
Space Shuttle Challenger Disaster, 433–434
Space Shuttle Columbia Disaster, 434
work on, 434–435
Rock wool, 3

Rural mud houses, 152
R values, 2, 7

S

Saccharomyces cerevisiae yeast, 242
Salwa, 4
Sand-clay soil for soils, *155*
consistency limits, 155
Sansevieria fiber, 3
Sarcina ventriculi, 75
Sargassum wightii algae, 241
Saturated lime test, 277
Saturated mass after immersion, 115
Saturn V rocket, 432
Sauropus androgynus extract, 262
SCADA dataset, 417
Scanning electron microscopy (SEM), 241, 266,
332, 356–357
Screen-printed carbon electrode (SPCE), 186
Seaweed polysaccharide bioplastics, 29–30
Second-order polynomial model, 263
Self-lubricating composite (SLC), 194, 205
Self-lubricating Cu matrix hybrid composites, 198
Self-lubricating hybrid metal matrix composite
(SLHMMCs), 193–197
characteristics, 195
defined, 194
fabrication route, 204–205
sustainability, 205–206
tribological aspects, **202–203**
dichalcogenides, 199–200
graphite and graphene, 198–199
hexagonal boron nitride, 200–204
Self-lubricating mechanism, *195*
Sewage sludge
ash, *277*
physical characteristics, *276*
pozzolanic materials, 277
Shell theory, 411
Sieve analysis, 154
Silicon kerf waste, 247
Silver (Ag) nanowires, 262
Silver-decorated graphene-base
nanocomposite, 262
Silver nanoparticles, 344–345
Silver nitrate ($AgNO_3$), 349
Silver-zirconia composite, 262, 267
Simarouba glauca, 348, 354, 373, 380, 394
Single point incremental forming (SPIF), 458
Single-use materials, 259
Sinter crystallization, 249
Sisal fibers, 153
Sludge-based concrete, strength property of,
282–283
Smart net-zero energy communities, 304–305
Social sustainability, 94

Index

Societies, collapse of, 93
Society's politico-economic arrangements, 107
Soil composition check, 154
Soil-fiber compositions, 156
Soil–preliminary tests, 153–154
Soils
    Jandot (Bilaspur), 153–173
    Kashmir (Hamirpur), 159–173
    physical attributes of, **154**
Solar collectors, 106
Sol-gel combustion method, 250
Sol-gel method, 182
Solid lubricants, 194–195, 197
Solid state fabrication, 204
Solid-waste fabrication, 2
SolidWorks, 404, *428*
Solvent-free/green solvent, 262
Solvent-free method, 264
Sorbitol, 76
Sorptivity, 289–290
Soybeans, 40
Space Shuttle Challenger Disaster, 433–434
Space Shuttle Columbia Disaster, 434
Speciespolysaccharide-alginate, 29
Sphaerotilus natans, 23
Spinel lithium titanate (GSLTO) nanoparticles,
    265–266
Spirulina platensis, polymer from, 28
Split tensile strength, 225
Sputtering, 346
Stainless steel (SS), 146
Standard Proctor compaction tests, 153
Starch, 24–25, 40, 73, 268
    derivatives, 75–76
    durable bioplastics, 27
    extraction, *75*
    milk, 75
Starch-based sweeteners, 76
Step-growth polymerization, terpene, 63–64
Stockholm Declaration, 99
Strength activity index, 277
Strength tests, 158–164
Streptococcus pyogenes, 373
Stress-carrying capacity, 170–171
Subtractive manufacturing (SM) processes, 315
Sulfate attack, 113
Sulfuric acid (H_2SO_4) attack, 113
Super-antifriction, 193
Superhydrophobicity, 194
Super magnetic iron oxide nanoparticles
    (SPION), 342
Superplasticizer, 284
Support vector machine (SVM), 421
Surface plasmon resonance, 241
Sustainability of materials, 93–96, *94*, 309
    applications, 105–106
    challenges, **107**

concepts, 96–99
defined, 97
design space (SDS), 311
history, 99–101
interpretation, 98
metrics, 101–103
prospects and challenges, 107–108
sustainability compliance index (SCI), 311
sustainable frameworks, **95**
sustainable logistics, 103–105
Sustainable biodegradable plastic, 21
Sustainable chemistry, 95
Sustainable development, 96, 108, 112, 301
Sustainable development goals (SDGs), 323–324
Sustainable insulation materials, 1–2
Sustainable material and technology, 323–326
Sustainable materials, 40, 301–303
    in aerospace, 310–312
    in agriculture, 307–310
    in architecture, 305–307
    in biomedical and health-care, 312–314
    in engineering applications, 303–305
    future applications aspects, 316–318
    in green technology, 314–315
Sustainable polymers, 56
Sustainable product development (SPD), 311
Sustainable production, 73
Sustainable social net product (SSNP), 302
Sustainable solvents/methods, 236–237
Sustainable synthesis, 260
Synergetic effect, 184
Synthesis of materials, 260
Synthetic antioxidants, 378
Synthetic biopolymers, 22
Synthetic Minority Over-sampling Technique
    (SMOTE), 418
Synthetic polymers, 28, 56
Syntrophomonas, 23

T

Tabu search, 263
Taguchi, 263
Tanneries, 80
Tannery wastewater, 141
T-dependent lymphocytes, 394
Tensile cracking, 225
Tensile strength, 7, 284
Terpene-based monomers, 67
Terpene-based polymers, 55–56, 66
    applications, 65–66
    characterization
        allyl-terpene maleate polymer, 61
        poly-alloocimene, 60–61
        polymyrcene, 61
        poly(stearyl methacrylate), 61
        tetrahydrogeraniol acrylate (THGA), 62

Index

green chemistry and sustainable approach, 57–58

modification, 64

grafting of terpenes, 65

lignin-polymer blends and oligomers, 64–65

polymer formation

chain-growth polymerization, 62–63

step-growth polymerization, 63–64

synthesis, 58

allyl-terpene maleate polymer, 58–59

poly-alloocimene, 58

polymyrcene, 59

poly(stearyl methacrylate), 59–60

tetrahydrogeraniol acrylate (THGA), 60

terpenes and terpenoids, 56

Terpene-diallyl maleate, *60*, 61

Terpenes and terpenoids, 56, 62

cationic polymerization, 64

grafting, 65, *66*

types, *57*

Terpenoids, 179

α-Terpinene, 56

Terpinolene, 58

Tetraethylenepentamine, 249

Tetrahydrogeraniol acrylate (THGA), 60, 62

1,1,3,3-Tetramethylguanidine (TMG), 235

Tetrapods (TPs), 312

Textile wastewater, 141

Therapeutic index, 379

Thermal conductivity, 199, 290

Thermal decomposition, 346

Thermal insulation, 3

materials, 3–4

Thermal stability, 200

Thermocouples, 327

Thermoelectric devices, 327

Thermo-oxidative degradation, 43

Thermoplastic, 5

Thermoplastic plastics, 220

Thermoplastic polymers, 235

Thermoplastic starch, 21

Thermo-pressing technique, 239

Thermoset, 5

Thermosetting plastics, 220

Thermostability tests, 65

Thespesia populnea, 262

Thespesia populnea tree extract, 265

Three-bladed wind turbines, 401

Three-component system possessing, 239

Tin dioxide (SnO_2), 347

nanoparticles, 343

Tin mercaptide, 44

TiO_2 nanoparticles, 242

Top-down process, 345

TOPSIS, 4

Transfructosylation reaction, 82

Transition metal dichalcogenides (TMDs), 199

Transmission electron microscopy (TEM), 241, 357–358

Triazabicyclodecene, 63

Tributyrin, 185

Trimethylolpropane triacrylate (TMPTA), 43, 45

Triple bottom line (TBL), **95**, 96

Tryptophan, 185

Tubers, 75

Tubular necrosis, 393

Tumor, 376

Tungsten, 199

Turbinaria conoides, 242

Turbo pump in engine, *432*

Two-point incremental forming (TPIF), 458

U

Ultrasonic assisted incremental sheet forming, 463–465

Ulva fasciata algae, 241

Unconfined compressive strength (UCS) tests, 158, 166–170

United Nations World Summit on Sustainable Development, 100

Universal testing equipment (UTM), 226

Unzipping mechanism, 84

Urbanization, 1

Uric acid, 185

USPV, 287–288

UV-visible spectroscopy, 267

V

Value-driven design (VDD), 310

Van der Waals bonds, 197

Verification tests, 158

Vernacular architectural style, 152

Vernacular fibers of Agave Americana and Eulaliopsis Binata, 152–153

experimental investigation, 158

proctor compaction tests, 159

strength tests, 158–164

material attributes

consistency limits, 155

fibers, 156–157

particle size distribution, 154–156

proctor compaction test, 155

soil–preliminary tests, 153–154

materials, 153

stress-carrying capacity, 170–171

cubical and cylindrical specimens, 172–173

test samples, 157

unconfined compressive strength (UCS) Tests, 166–170

V-fluorescein isothiocyanate (FITC), 374

484 Index

Vitamin C, 250
Voluntary Inter-industry Commerce Solutions (VICS), 105
Von-Mises stresses, 428
Vulcanization, 239

W

Waste
 classification, *214*
 generating reagents, 262
 valorization, 73
Waste electrical and electronic equipment (WEEE), 218
Waste recycling, 244
Waste sewage sludge, 274
 characterization
  mineralogical and morphological properties, 275–277
  physiochemical properties, 275
  pozzolanic properties, 277–278
Waste sewage sludge ash, 277, 284
Waste sewage sludge concrete
 durability properties
  alkali-silica reaction (ASR), 291
  chloride ion permeability, 289
  density, 285–286
  electrical resistivity and corrosion resistance, 290
  frost resistance, 290–291
  shrinkage, 288
  sorptivity, 289–290
  thermal conductivity, 290
  USPV, 287–288
  water absorption, 286–287
 ecological and economic analysis, 294–296
 fresh properties, workability, 278–279
 leachability, 291–294
 mechanical properties
  compressive strength, 279–281
  flexure strength, 284–285
  modulus of elasticity, 285
  tensile strength, 284
Wastewater generation, 136
Wastewater treatment, 136
 decentralized wastewater treatment systems (DWTS), 137–138
 electrocoagulation (EC) process, 138–139
  current density (CD), 140
  dairy wastewater, 142
  distillery wastewater, 142
  efficiency, 139
  electrical conductivity, 141
  electrochemical and chemical reactions, 139
  electrode material, 141–142, **143–145**

electrolysis time, 140
energy and electrode consumption, 141
hospital wastewater, 142
petroleum wastewater, 141–142
pH of solution, 139
real graywater, 142
removal of pharmaceutical compounds, 142
tannery wastewater, 141
textile wastewater, 141
 goal, 137
 levels of, *137*
 technologies, 137
Water; *see also* Wastewater treatment
 absorption, concrete, 286–287
 demand, 136
 quality, 137
 treatment, 264
Water-binder ratio, 277
Wax, 8
Wear performance of Al-SiC-hBN hybrid composite, *201*
WEEE, *see* Waste electrical and electronic equipment
Wet-chemical deposition, 183
Wind energy, 400
Wind turbine blades (WTB), 400
 air-foils, *404*
 blade velocity vectors, 407–410
 computational fluid dynamics (CFD) of, 404
 core analysis, 404
  building algebraic equations, 411–412
  computational fluid dynamics (CFD) of, 405–410
  design of, 404–405
  finite element analysis (FEA) of, 410–411
  meshing, 412–413
  pressure load from CFD to FEA, 413–415
 deformation of blade, *414*
 developments in, 402, *403*
 failure of, 402–403
  acoustic emission, 403
  machine vision detection methods, 404
  strain measurement detection, 403
  ultrasound detection methods, 403
 finite element analysis (FEA), 404
 historical evolution, 401–402
 interior components, *401*
 machine learning model, 415–417
  dataset acquistion, 417–420
  fault detection, 417
  model usage, 420–428
  tools, 417
 material properties, **405**
 pressure contours, *409*
 pressure load imported from CFD to FEA, *413*

Index

thickness variation, *413*
time dependence of stress distribution, *415*
types of, 401–402
velocity streamlines, *408*
Von-Mises stress, *414*
Wind turbines with vertical axis (VAWT), 402
Wood pulp, 74
World Business Council for Sustainable
       Development (WBCSD), 100
WTB, *see* Wind turbine blades

X

XGBoost, 452, 454
X-ray diffraction (XRD), 265–266, 332,
       355–356
X-ray Photoemission Spectroscopy (XPS), 181
X-ray powder diffraction (XRD),
       241, 266, 275

Y

Yeast, 242
Young's modulus, 7, 43

Z

Z-conduction electrons, 327
Zero-dimensional nanomaterials, 342
Zero waste, 73
Zero waste emission, 264
Zero-wear phenomenon, 193
Zinc oxide (ZnO), 312
    nanocrystals (NCs), 313
    nanoparticles, 142, 313
    NStr-based novel biomaterials, 312
    polymeric nanocomposites, 313
Zonaria species, 29
Zone of inhibition (ZOI), 373, 378–380, 394

Printed in the United States
by Baker & Taylor Publisher Services